Classics in Mathematics

O. Timothy O'Meara   Introduction to Quadratic Forms

Springer
Berlin
Heidelberg
New York
Barcelona
Hong Kong
London
Milan
Paris
Singapore
Tokyo

O. Timothy O'Meara

Introduction to Quadratic Forms

Reprint of the 1973 Edition

Springer

O. Timothy O'Meara
University of Notre Dame
Department of Mathematics
Notre Dame, ID 46556
USA

Originally published as Vol. 117 of the
Grundlehren der mathematischen Wissenschaften

Mathematics Subject Classification (1991): Primary 11D04, 11D09, 11D79, 11D88, 11Exx, 11H55, 11Rxx, 11Sxx, 12D15, 12Exx, 12Fxx, 12Jxx, 13A15, 13A18, 13A20, 15A03, 15A04, 15A06, 15A09, 15A15, 15A18, 15A21, 15A36, 15A63, 15A66, 15A69, 15A72, 16K20, 17A45
Secondary 13Fxx, 20Gxx, 20Hxx

Cataloging-in-Publication Data applied for

Die Deutsche Bibliothek - CIP-Einheitsaufnahme
O'Meara, Onorato Timothy:
Introduction to quadratic forms / O. Timothy O'Meara. - Reprint of the 1973 ed.- Berlin; Heidelberg; New York; Barcelona; Hong Kong; London; Milan; Paris; Singapore; Tokyo: Springer, 2000
(Classics in mathematics)
ISBN 3-540-66564-1

ISSN 1431-0821
ISBN 3-540-66564-1 Springer-Verlag Berlin Heidelberg New York

SPIN 10734083 41/3143CK-5 4 3 2 1 0 – Printed on acid-free paper

O. T. O'Meara

Introduction to Quadratic Forms

Third Corrected Printing

With 10 Figures

Springer-Verlag
Berlin Heidelberg New York 1973

O. T. O'Meara
University of Notre Dame, Department of Mathematics,
Notre Dame, ID 46556/USA

AMS Subject Classifications (1970)

Primary 1002, 10 B 40, 10 C 05, 10 C 20, 10 C 30, 10 E 45, 1202, 12 A 10, 12 A 40, 12 A 45, 12 A 50, 12 A 90, 13 C 10, 13 F 05, 13 F 10, 1502, 15 A 33, 15 A 36, 15 A 57, 15 A 63, 15 A 66, 20 G 15, 20 G 25, 20 G 30, 20 G 40, 20 H 20, 20 H 25, 20 H 30,
Secondary 12 A 65, 12 Jxx

ISBN 3-540-02984-2 Springer-Verlag Berlin Heidelberg New York

ISBN 0-387-02984-2 Springer-Verlag New York Heidelberg Berlin

---

Printed in Germany. Offsetprinting and bookbinding: Brühlsche Universitätsdruckerei, Gießen

In Memory of my Parents

Preface

The main purpose of this book is to give an account of the fractional and integral classification problem in the theory of quadratic forms over the local and global fields of algebraic number theory. The first book to investigate this subject in this generality and in the modern setting of geometric algebra is the highly original work *Quadratische Formen und orthogonale Gruppen* (Berlin, 1952) by M. EICHLER. The subject has made rapid strides since the appearance of this work ten years ago and during this time new concepts have been introduced, new techniques have been developed, new theorems have been proved, and new and simpler proofs have been found. There is therefore a need for a systematic account of the theory that incorporates the developments of the last decade.

The classification of quadratic forms depends very strongly on the nature of the underlying domain of coefficients. The domains that are really of interest are the domains of number theory: algebraic number fields, algebraic function fields in one variable over finite constant fields, all completions thereof, and rings of integers contained therein. Part One introduces these domains via valuation theory. The number theoretic and function theoretic cases are handled in a unified way using the Product Formula, and the theory is developed up to the Dirichlet Unit Theorem and the finiteness of class number. It is hoped that this will be of service, not only to the reader who is interested in quadratic forms, but also to the reader who wishes to go deeper into algebraic number theory and class field theory. In Part Two there is a discussion of topics from abstract algebra and geometric algebra which will be used later in the arithmetic theory. Part Three treats the theory of quadratic forms over local and global fields. The direct use of local class field theory has been circumvented by introducing the concept of the quadratic defect (which is needed later for the integral theory) right at the start. The quadratic defect gives, in effect, a systematic way of refining certain types of quadratic approximations. However, the global theory of quadratic forms does present a dilemma. Global class field theory is still so inaccessible that it is not possible merely to quote results from the literature. On the other hand a thorough development of global class field theory cannot be included in a book of this size and scope. We have therefore decided to compromise by specializing the methods of global class field theory to the case of quadratic extensions, thereby obtaining all that is needed for the global theory of quadratic forms. Part Four starts with a systematic development of the formal aspects of integral quadratic forms over Dedekind domains. These techniques are then applied, first to solve the local integral classification problem, then to investigate the global integral theory, in particular to establish the relation between the class, the genus, and the spinor genus of a quadratic form.

It must be emphasized that only a small part of the theory of quadratic forms is covered in this book. For the sake of simplicity we confine ourselves entirely to quadratic forms and the orthogonal group, and then to a particular part of this theory, namely to the classification problem over arithmetic fields and rings. Thus we do not even touch upon the theory of hermitian forms, reduction theory and the theory of minima, composition theory, analytic theory, etc. For a discussion of these matters the reader is referred to the books and articles listed in the bibliography.

O. T. O'MEARA

February, 1962.

I wish to acknowledge the help of many friends and mathematicians in the preparation of this book. Special thanks go to my former teacher EMIL ARTIN and to GEORGE WHAPLES for their influence over the years and for urging me to undertake this project; to RONALD JACOBOWITZ, BARTH POLLAK, CARL RIEHM and HAN SAH for countless discussions and for checking the manuscript; and to Professor F. K. SCHMIDT and the Springer-Verlag for their encouragement and cooperation and for publishing this book in the celebrated Yellow Series. I also wish to thank Princeton University, the University of Notre Dame and the Sloan Foundation[1] for their generous support.

O. T. O'MEARA

December, 1962.

[1] ALFRED P. SLOAN FELLOW, 1960—1963.

Contents

Part Three

Arithmetic Theory of Quadratic Forms over Fields

Part Four

Arithmetic Theory of Quadratic Forms over Rings

Prerequisites and Notation

If X and Y are any two sets, then $X \subset Y$ will denote strict inclusion, $X - Y$ will denote the difference set, $X \twoheadrightarrow Y$ will denote a surjection of X onto Y, $X \rightarrowtail Y$ an injection, $X \rightarrowtail\!\!\!\rightarrow Y$ a bijection, and $X \to Y$ an arbitrary mapping. By "almost all elements of X" we shall mean "all but a finite number of elements of X".

N denotes the set of natural numbers, Z the set of rational integers, Q the set of rational numbers, R the set of real numbers, P the set of positive numbers, and C the set of complex numbers.

We assume a knowledge of the elementary definitions and facts of general topology, such as the concepts of continuity, compactness, completeness and the product topology.

From algebra we assume a knowledge of 1) the elements of group theory and also the fundamental theorem of abelian groups, 2) galois theory up to the fundamental theorem and including the description of finite fields, 3) the rudiments of linear algebra, 4) basic definitions about modules.

If X is any additive group, in particular if X is either a field or a vector space, then $\dot{X}$ will denote the set of non-zero elements of X. If H is a subgroup of a group G, then $(G : H)$ is the index of H in G. If E/F is an extension of fields, then $[E : F]$ is the degree of the extension. The characteristic of F will be written $\chi(F)$. If α is an element of E that is algebraic over F, then $\operatorname{irr}(x, \alpha, F)$ is the irreducible monic polynomial in the variable x that is satisfied by α over the field F. If E_1 and E_2 are subfields of E, then E_1E_2 denotes the compositum of E_1 and E_2 in E. If E/F is finite, then $N_{E/F}$ will denote the norm mapping from E to F; and $S_{E/F}$ will be the trace.

Part One

Arithmetic Theory of Fields

Chapter I

Valuated Fields

The descriptive language of general topology is known to all mathematicians. The concept of a valuation allows one to introduce this language into the theory of algebraic numbers in a natural and fruitful way. We therefore propose to study some of the connections between valuation theory, algebraic number theory, and topology. Strictly speaking the topological considerations are just of a conceptual nature and in fact only the most elementary results on metric spaces and topological groups will be used; nevertheless these considerations are essential to the point of view taken throughout this chapter and indeed throughout the entire book.

§ 11. Valuations

§ 11 A. The definitions

Let F be a field. A valuation on F is a mapping $|\ |$ of F into the real numbers $\mathbf{R}$ which satisfies

$$(V_1)\quad |\alpha| > 0 \quad \text{if} \quad \alpha \neq 0, \quad |0| = 0$$

$$(V_2)\quad |\alpha\beta| = |\alpha| \cdot |\beta|$$

$$(V_3)\quad |\alpha + \beta| \leqq |\alpha| + |\beta|$$

for all α, β in F. A mapping which satisfies (V_1), (V_2) and

$$(V_{3'})\quad |\alpha + \beta| \leqq \max(|\alpha|, |\beta|)$$

will satisfy (V_3) and will therefore be a valuation. Axiom (V_3) is called the triangle law, axiom $(V_{3'})$ is called the strong triangle law. A valuation which satisfies the strong triangle law is called non-archimedean, a valuation which does not satisfy the strong triangle law is called archimedean. Non-archimedean valuations will be used to describe certain properties of divisibility in algebraic number theory.

The mapping $\alpha \to |\alpha|$ is a multiplicative homomorphism of $\dot{F}$ into the positive real numbers, and so the set of images of $\dot{F}$ forms a multiplicative subgroup of R. We call the set

$$|F| = \{|\alpha| \in \mathsf{R} \,|\, \alpha \in \dot{F}\}$$

the value group of F under the given valuation. We have the equations

$$|1| = 1, \quad |-\alpha| = |\alpha|, \quad |\alpha^{-1}| = |\alpha|^{-1},$$

and also

$$|\,|\alpha| - |\beta|\,| \leqq |\alpha - \beta|\,.$$

Every field F has at least one valuation, the trivial valuation obtained by putting $|\alpha| = 1$ for all α in $\dot{F}$. Such a valuation satisfies the strong triangle law and is therefore non-archimedean. A finite field can possess only the trivial valuation since, if we let q stand for the number of elements in F, we have

$$|\alpha|^{q-1} = |\alpha^{q-1}| = |1| = 1 \qquad \forall \alpha \in \dot{F}\,.$$

Any subfield F of the complex numbers C can be regarded as a valuated field by restricting the ordinary absolute value from C to F. Conversely, it will follow from the results of § 12 that every field with an archimedean valuation is obtained essentially in this way. A valuated field which contains the rational numbers Q and which induces the ordinary absolute value on Q must be archimedean since $|1 + 1| = 2 > 1$.

Now a few words about the topological properties of the valuated field F. First we notice that F can be regarded as a metric topological space in a natural way: define the distance between two points α and β of F to be $|\alpha - \beta|$. If we take this topology on F and the product topology on $F \times F$, then it is easily seen by elementary methods that the mappings

$$(\alpha, \beta) \to \alpha + \beta \quad \text{and} \quad (\alpha, \beta) \to \alpha\beta$$

of $F \times F$ into F are continuous. So are the mappings

$$\alpha \to -\alpha \quad \text{and} \quad \alpha \to \alpha^{-1}$$

of F into F and of $\dot{F}$ into $\dot{F}$, respectively. These four facts simply mean, in the language of topological groups, that F is a topological field. Hence the mappings

$$(\alpha_1, \alpha_2, \ldots, \alpha_n) \to \alpha_1 + \alpha_2 + \cdots + \alpha_n$$

and

$$(\alpha_1, \alpha_2, \ldots, \alpha_n) \to \alpha_1 \alpha_2 \ldots \alpha_n$$

of $F \times \cdots \times F$ into F are continuous. Hence a polynomial with coefficients in F determines a continuous function of $F \times \cdots \times F$ into F; and a rational function is continuous at any point of $F \times \cdots \times F$ at which its denominator is not zero. The inequality $||\alpha| - |\alpha_0|| \leqq |\alpha - \alpha_0|$ shows that the mapping

$$\alpha \to |\alpha|$$

of F into R is continuous.

The limit of a sequence and the sum of a series can be defined as it is usually defined in a first course on the calculus. We find that if $\alpha_n \to \alpha$ and $\beta_n \to \beta$ as $n \to \infty$, then

$$\alpha_n \pm \beta_n \to \alpha \pm \beta, \quad \alpha_n \beta_n \to \alpha\beta$$

$$\alpha_n^{-1} \to \alpha^{-1} \quad \text{if} \quad \alpha \neq 0$$

$$|\alpha_n| \to |\alpha| .$$

Similarly if $\sum\limits_1^\infty a_\lambda$ and $\sum\limits_1^\infty b_\lambda$ converge, then so do

$$\sum_1^\infty (a_\lambda \pm b_\lambda) = \sum_1^\infty a_\lambda \pm \sum_1^\infty b_\lambda .$$

The terms of any convergent series must tend to 0.

The closure $\hat{G}$ of a subfield G of F is again a subfield of F. For we can find $\alpha_n \to \alpha$ and $\beta_n \to \beta$ with α_n and β_n in G whenever α and β are given in $\hat{G}$; then

$$\alpha + \beta = \lim (\alpha_n + \beta_n) \in \hat{G} .$$

Hence $\hat{G}$ is closed under addition. Similarly with multiplication and inversion. Hence $\hat{G}$ is a field.

Closely related to the concept of a valuation is the concept of an analytic map. An analytic map is an isomorphism φ of the valuated field F onto a valuated field F' such that $|\varphi\alpha| = |\alpha|$ holds for all α in F. In other words, an analytic map preserves the valuation as well as the algebraic structure. An analytic map φ is therefore a topological isomorphism between the topological fields F and F'. Suppose now that F' is just any abstract field, but that F is still the valuated field under discussion. Also suppose that we have an isomorphism φ of F onto F'. We can then define a valuation on F' by putting $|\beta| = |\varphi^{-1}\beta|$ for all β in F'. When we perform this construction we shall say that φ has carried the valuation from F to F'. Clearly the valuation just defined makes φ analytic.

We conclude this subparagraph with an important example. Consider the rational numbers Q and a fixed prime number p. A typical $\alpha \in \dot{\mathsf{Q}}$ can be written in the form

$$\alpha = p^i\left(\frac{m}{n}\right)$$

with m and n prime to p. Do this with each α and put

$$|\alpha|_p = \left(\frac{1}{p}\right)^i .$$

It is easy to show that this defines a non-archimedean valuation on $\mathbf{Q}$. To say that α is small under this valuation means that it is highly divisible by p. (Here is our first glimpse of the connection between valuations and number theory; we shall return to this example in a more general setting in Chapter III.)

§ 11B. Non-archimedean valuations

11:1. *A valuation on a field F is non-archimedean if and only if it is bounded on the natural integers of F.*

Proof. We recall that the natural integers in an arbitrary field F are the finite sums of the form $1 + \cdots + 1$. We need only do the sufficiency. Thus we are given a fixed positive bound M such that $|m| \leqq M$ holds for any natural integer m in F. Then

$$\begin{aligned} |\alpha+\beta|^n &= |(\alpha+\beta)^n| \\ &= |\alpha^n + \tbinom{n}{1}\alpha^{n-1}\beta + \cdots + \beta^n| \\ &\leqq |1|\,|\alpha|^n + |\tbinom{n}{1}|\,|\alpha|^{n-1}\,|\beta| + \cdots + |1|\,|\beta|^n \\ &\leqq M\,\{|\alpha|^n + |\alpha|^{n-1}\,|\beta| + \cdots + |\beta|^n\} \\ &\leqq M\,(n+1)\,\{\max(|\alpha|, |\beta|)\}^n. \end{aligned}$$

Hence

$$|\alpha+\beta| \leqq M^{1/n}(n+1)^{1/n}\max(|\alpha|, |\beta|) .$$

If we let $n \to \infty$ we obtain the result. **q. e. d.**

This result has two immediate consequences. First, a field of characteristic $p > 0$ can have no archimedean valuations. Second, a valuated extension field E of F is non-archimedean if and only if F is non-archimedean under the induced valuation.

11:2. **Principle of Domination.** *In a non-archimedean field we have*

$$|\alpha_1 + \cdots + \alpha_n| = |\alpha_1|$$

if $|\alpha_\lambda| < |\alpha_1|$ for $\lambda > 1$.

Proof. It suffices to prove $|\alpha+\beta| = |\alpha|$ when $|\alpha| > |\beta|$. We have

$$|\alpha| = |-\beta + \alpha + \beta| \leqq \max(|\beta|, |\alpha+\beta|) ,$$

and so $|\alpha| \leqq |\alpha+\beta|$. But

$$|\alpha+\beta| \leqq \max(|\alpha|, |\beta|) = |\alpha| .$$

Hence $|\alpha+\beta| = |\alpha|$. **q. e. d.**

11:2a. *Suppose that* $\sum_1^\infty \alpha_\lambda$ *is convergent. If* $|\alpha_\lambda| < |\alpha_1|$ *for* $\lambda > 1$, *then*

$$|\sum_1^\infty \alpha_\lambda| = |\alpha_1| .$$

11:3. *Let* E/F *be an algebraic extension of fields. Suppose a valuation on* E *induces the trivial valuation on* F. *Then the valuation is trivial on* E.

Proof. For suppose that the given valuation is non-trivial on E. Then we can find $\alpha \in E$ with $|\alpha| > 1$. Let us write

$$\alpha^n + a_1\alpha^{n-1} + \cdots + a_{n-1}\alpha + a_n = 0$$

with all a_i in F. Now all $|a_i|$ are either 0 or 1 since the valuation is trivial on F. And $|\alpha^n| > |\alpha^i|$ whenever $n > i$. Hence

$$|\alpha^n| = |\alpha^n + a_1\alpha^{n-1} + \cdots + a_n|$$

by the Principle of Domination. Hence $|\alpha^n| = 0$, and this is absurd.

q. e. d.

We shall see later in Chapter III that the above result does not hold if the extension E/F is transcendental.

§ 11C. Equivalent valuations

Consider two valuations $|\ |_1$ and $|\ |_2$ on the same field F. We say that $|\ |_1$ and $|\ |_2$ are equivalent valuations if they define the same topology on F. It is clear that equivalence of valuations is an equivalence relation on the set of all valuations on F.

11:4. *Let* $|\ |_1$ *and* $|\ |_2$ *be two valuations on the same field* F. *Then the following assertions are equivalent:*

(1) *The two valuations are equivalent,*

(2) $|\alpha|_1 < 1 \Leftrightarrow |\alpha|_2 < 1$,

(3) *There is a positive number* ϱ *such that* $|\alpha|_1^\varrho = |\alpha|_2$ *for all* α *in* F.

Proof. (1) $\Rightarrow$ (2). On grounds of symmetry it is enough to consider an α in F with $|\alpha|_1 < 1$ and to prove that $|\alpha|_2 < 1$. Let

$$N = \{x \in F \mid |x|_2 < 1\} .$$

This set N is a neighborhood of 0 under the topology induced by either valuation. Now $|\alpha^n|_1 = |\alpha|_1^n$ can be made arbitrarily small by choosing n large enough, in particular we can choose an n such that $\alpha^n \in N$. But then $|\alpha|_2^n < 1$. And so $|\alpha|_2 < 1$.

(2) $\Rightarrow$ (3). By taking inverses we deduce from (2) that $|\alpha|_1 > 1$ if and only if $|\alpha|_2 > 1$. Hence $|\alpha|_1 = 1$ if and only if $|\alpha|_2 = 1$. In particular, if one of the valuations is trivial then so is the other. We may therefore assume that neither valuation is trivial.

Take α_0 in F with $0 < |\alpha_0|_2 < 1$. Then $0 < |\alpha_0|_1 < 1$ by hypothesis. Hence we have $|\alpha_0|_2 = |\alpha_0|_1^\varrho$ where

$$\varrho = \log |\alpha_0|_2 / \log |\alpha_0|_1 > 0 .$$

We claim that $|\alpha|_2 = |\alpha|_1^\varrho$ for all α in F. For suppose if possible that there is an α for which $|\alpha|_2$ and $|\alpha|_1^\varrho$ are not equal. Replacing α by its inverse if necessary allows us to assume that $|\alpha|_2 < |\alpha|_1^\varrho$. Now choose a rational number m/n with $n > 0$ such that

$$|\alpha|_2 < |\alpha_0|_2^{m/n} = |\alpha_0|_1^{\varrho m/n} < |\alpha|_1^\varrho .$$

This gives

$$|\alpha^n/\alpha_0^m|_2 < 1 \quad \text{and} \quad |\alpha^n/\alpha_0^m|_1 > 1$$

which denies our hypothesis. Hence our supposition about α is false. Hence (3) follows.

(3) $\Rightarrow$ (1). This part is clear. **q. e. d.**

11:4a. *Suppose* $|\ |_1$ *is non-trivial. Then* $|\ |_1$ *is equivalent to* $|\ |_2$ *if*

$$|\alpha|_1 < 1 \Rightarrow |\alpha|_2 < 1 .$$

Proof. If $|\alpha|_1 > 1$, then $|\alpha|_2 > 1$ by taking inverses. It is therefore enough to prove

$$|\alpha|_1 = 1 \Rightarrow |\alpha|_2 = 1 .$$

Choose $\beta \in F$ with $0 < |\beta|_1 < 1$. Then

$$|\alpha^n \beta|_1 < 1 \Rightarrow |\alpha^n \beta|_2 < 1 \Rightarrow |\alpha|_2^n\, |\beta|_2 < 1 .$$

It follows from the last inequality, by letting $n \to \infty$, that $|\alpha|_2 \leqq 1$. Replace α by α^{-1}. This gives $|\alpha|_2 \geqq 1$. Hence $|\alpha|_2 = 1$. **q. e. d.**

11:4b. *The trivial valuation is equivalent to itself and itself alone.*

11:5. *Let* $|\ |_1$ *and* $|\ |_2$ *be two equivalent valuations on a field* E *and let* F *be an arbitrary subfield of* E. *Suppose the two valuations induce the same non-trivial valuation on* F. *Then* $|\ |_1$ *and* $|\ |_2$ *are equal on* E.

Proof. We have a positive number ϱ such that $|\alpha|_1^\varrho = |\alpha|_2$ for all α in E. Choose $\alpha_0 \in F$ with

$$0 < |\alpha_0|_1 = |\alpha_0|_2 < 1 .$$

Then $|\alpha_0|_1^\varrho = |\alpha_0|_1$. Hence $\varrho = 1$. **q. e. d.**

Consider the valuation $|\ |$ on our field F and let ϱ be any positive number. We know that $|\ |^\varrho$, if it is a valuation, will be equivalent to $|\ |$. Of course $|\ |^\varrho$ need not be a valuation at all; for instance the ordinary absolute value on $\mathbf{Q}$ with $\varrho > 1$ gives

$$|1 + 1|^\varrho = 2^\varrho > 2 = |1|^\varrho + |1|^\varrho .$$

However $|\ |^\varrho$ is a valuation whenever $0 < \varrho \leqq 1$. To see this we observe that $|\alpha + \beta|^\varrho \leqq (|\alpha| + |\beta|)^\varrho$; it therefore suffices to prove that

$$(|\alpha| + |\beta|)^\varrho \leqq |\alpha|^\varrho + |\beta|^\varrho .$$

But

$$1 = \frac{|\alpha|}{|\alpha| + |\beta|} + \frac{|\beta|}{|\alpha| + |\beta|} \leqq \left(\frac{|\alpha|}{|\alpha| + |\beta|}\right)^{\varrho} + \left(\frac{|\beta|}{|\alpha| + |\beta|}\right)^{\varrho},$$

since $0 < \varrho \leqq 1$. So it is true.

In the non-archimedean case things are simpler. The strong triangle law must obviously hold for $|\ |^{\varrho}$ if it holds for $|\ |$, even if ϱ is greater than 1. Hence $|\ |^{\varrho}$ is a valuation if $|\ |$ is non-archimedean and $\varrho > 0$. It is clear that $|\ |^{\varrho}$ is non-archimedean if and only if $|\ |$ is.

§ 11D. Prime spots

Consider a field F. By a prime spot, or simply a spot, on F we mean a single class of equivalent valuations on F; thus a spot is a certain set of maps of F into R. Consider a prime spot $\mathfrak{p}$ on F. Each valuation $|\ |_{\mathfrak{p}} \in \mathfrak{p}$ defines the same topology on F by the definition of a prime spot. We call this the $\mathfrak{p}$-adic topology on F. If $\mathfrak{p}$ contains the trivial valuation (in which case it can contain no other) we call $\mathfrak{p}$ the trivial spot on F. In the same way we can define archimedean and non-archimedean spots. If $\mathfrak{p}$ is non-trivial it will contain an infinite number of valuations. Two spots on F are equal if and only if their topologies are the same.

Suppose $\sigma\colon F \rightarrowtail F'$ is an isomorphism of a field F with a spot $\mathfrak{p}$ onto an abstract field F'. It is easily seen that *there is a unique spot* $\mathfrak{q}$ *on* F' *which makes* σ *topological:* the existence of $\mathfrak{q}$ is obtained by letting σ carry some valuation in $\mathfrak{p}$ over to F', and the uniqueness of $\mathfrak{q}$ follows from the fact that both σ and σ^{-1} will be topological. In this construction we say that σ carries the spot $\mathfrak{p}$ to F'. The unique spot on F' that makes σ topological will be written

$$\mathfrak{p}^{\sigma}.$$

To each $|\ |_{\mathfrak{p}} \in \mathfrak{p}$ there corresponds a valuation $|\ |_{\mathfrak{p}^{\sigma}} \in \mathfrak{p}^{\sigma}$ such that

$$|\beta|_{\mathfrak{p}^{\sigma}} = |\sigma^{-1}\beta|_{\mathfrak{p}} \quad \forall\, \beta \in F',$$

namely the valuation obtained by carrying $|\ |_{\mathfrak{p}}$ to F'.

11:6. *Let F and G be two fields provided with $\mathfrak{p}$-adic and $\mathfrak{q}$-adic topologies respectively. Let σ be a topological isomorphism of F onto G. Then $\mathfrak{q} = \mathfrak{p}^{\sigma}$. And for each $|\ |_{\mathfrak{p}} \in \mathfrak{p}$ there is a $|\ |_{\mathfrak{q}} \in \mathfrak{q}$ which makes σ analytic.*

Proof. Clearly $\mathfrak{q} = \mathfrak{p}^{\sigma}$ by definition of $\mathfrak{p}^{\sigma}$. Then $|\ |_{\mathfrak{q}}$ is simply the valuation $|\ |_{\mathfrak{p}^{\sigma}}$ defined above. **q. e. d.**

Let $\mathfrak{P}$ be a prime spot on an extension E of F. Each valuation in $\mathfrak{P}$ induces a valuation on F, and all valuations of F that are obtained from $\mathfrak{P}$ in this way are equivalent. Hence $\mathfrak{P}$ determines a unique spot $\mathfrak{p}$ on F. We say that $\mathfrak{P}$ induces $\mathfrak{p}$, or that $\mathfrak{P}$ divides $\mathfrak{p}$, and we write

$$\mathfrak{P}|\mathfrak{p}.$$

Whenever we refer to the spot $\mathfrak{P}$ on F we shall really mean that spot $\mathfrak{p}$ on F which is divisible by $\mathfrak{P}$. Here the $\mathfrak{P}$-adic topology on E induces the $\mathfrak{p}$-adic topology on F. We refer to this induced topology as the $\mathfrak{P}$-adic topology on F.

Now consider a set of prime spots S on F and another set T on E. We say that T divides S and write $T|S$ if the spot induced on F by each spot $\mathfrak{P}$ in T is in S. It is clear that there is an absolutely largest set of spots T on E which divides a given set S on F; we then say that T fully divides S and we write $T||S$. One often uses the same letter S to denote the set of spots on E which fully divides the given set S on F.

§ 11E. The Weak Approximation Theorem

11:7. *Let* $|\ |_\lambda$ $(1 \leqq \lambda \leqq n)$ *be a finite number of inequivalent non-trivial valuations on a field* F. *Then there is an* $\alpha \in F$ *such that* $|\alpha|_1 > 1$ *and* $|\alpha|_\lambda < 1$ *for* $2 \leqq \lambda \leqq n$.

Proof. If $n = 1$ it is simply the fact that $|\ |_1$ is non-trivial. Next let $n = 2$. Since $|\ |_1$ and $|\ |_2$ are inequivalent we can find b, c in F such that

$$|b|_1 < 1\,,\quad |b|_2 \geqq 1\,,\quad |c|_1 \geqq 1\,,\quad |c|_2 < 1\,.$$

Then $\alpha = c/b$ does the job.

We continue by induction to n. First choose b with $|b|_1 > 1$ and $|b|_\lambda < 1$ $(2 \leqq \lambda \leqq n-1)$, then c with $|c|_1 > 1$ and $|c|_n < 1$. If $|b|_n < 1$ we are through. If $|b|_n = 1$, form cb^r and observe that for sufficiently large values of r we have

$$|cb^r|_1 > 1\,,\ |cb^r|_\lambda < 1 \quad (2 \leqq \lambda \leqq n)\,;$$

take $\alpha = cb^r$. Finally consider $|b|_n > 1$. Using the fact that $1 + b^r \to 1$ if $|b| < 1$ we easily see that

$$\left|\frac{cb^r}{1+b^r}\right|_\lambda \to \begin{cases} |c|_\lambda & \text{if } \lambda = 1 \text{ or } \lambda = n \\ 0 & \text{if } 2 \leqq \lambda \leqq n-1\,. \end{cases}$$

This time take $\alpha = cb^r/(1+b^r)$ with a sufficiently large r. **q. e. d.**

11:8. **Theorem.** *Let* $|\ |_\lambda$ $(1 \leqq \lambda \leqq n)$ *be a finite number of inequivalent non-trivial valuations on a field* F. *Consider* n *field elements* α_λ $(1 \leqq \lambda \leqq n)$. *Then for each* $\varepsilon > 0$ *there is an* $\alpha \in F$ *such that* $|\alpha - \alpha_\lambda|_\lambda < \varepsilon$ *for* $1 \leqq \lambda \leqq n$.

Proof. For each i $(1 \leqq i \leqq n)$ we can find $b_i \in F$ such that $|b_i|_i > 1$ and $|b_i|_\lambda < 1$ when $\lambda \neq i$. If we let $r \to \infty$ we see that

$$\frac{b_i^r}{1+b_i^r} \to \begin{cases} 1 & \text{under } |\ |_i \\ 0 & \text{under } |\ |_\lambda \text{ if } \lambda \neq i\,. \end{cases}$$

Hence

$$c_r = \sum_{i=1}^{n} \frac{\alpha_i b_i^r}{1+b_i^r} \to \alpha_i$$

under the topology defined by $|\ |_i$. Then $\alpha = c_r$ with a sufficiently large r is the α we require. **q. e. d.**

§ 11F. Complete valuations and complete spots

Consider the distance function $d(\alpha, \beta) = |\alpha - \beta|$ associated with the valuation $|\ |$ on F. We can follow the language of metric topology and introduce the concept of a Cauchy sequence and completeness with respect to $d(\alpha, \beta)$. Completeness of $|\ |$ then means, by definition, that every Cauchy sequence converges to a limit in F.

11:9. **Example.** We have already mentioned that the terms of any convergent series over a valuated field must tend to 0. If F is a field with a complete non-archimedean valuation there is the following remarkable converse: every infinite series whose terms tend to 0 is convergent. For if we form the partial sums $A_1, \ldots, A_n, \ldots$ of $\sum_1^\infty \alpha_\lambda$ we see from the strong triangle law that

$$|A_m - A_n| \leqq \max(|a_{n+1}|, \ldots, |a_m|),$$

hence the partial sums form a Cauchy sequence, hence $\sum_1^\infty \alpha_\lambda$ has a limit in F.

Let $\mathfrak{p}$ be a spot on the field F. We say that F is complete at $\mathfrak{p}$, or simply that F is complete, if there is at least one complete valuation in $\mathfrak{p}$. Because of the formula $|\ |_1^\varrho = |\ |_2$ relating equivalent valuations we see that if F is complete at $\mathfrak{p}$, then every valuation in $\mathfrak{p}$ is complete.

By a completion of a field F at one of its spots $\mathfrak{p}$ we mean a composite object consisting of a field E and a prime spot $\mathfrak{P}$ on E with the following properties:

1. E is complete at $\mathfrak{P}$,
2. F is a subfield of E and $\mathfrak{P}|\mathfrak{p}$,
3. F is dense in E.

We shall often shorten the terminology and just refer to a completion E of a given field F; this will of course mean that we have a certain prime spot $\mathfrak{p}$ on F in mind and that E is really a composite object consisting of the field E and a prime spot $\mathfrak{P}$ on E.

11:10. **Example.** A complete field is its own completion. It has no other completion.

11:11. **Example.** Consider the trivial spot $\mathfrak{p}$ on F. Here every Cauchy sequence has the form

$$\alpha_1, \ldots, \alpha_n, \alpha, \ldots, \alpha, \ldots$$

and this converges to α. Hence F is complete.

11:12. **Example.** Let F and G be two fields provided with $\mathfrak{p}$-adic and $\mathfrak{q}$-adic topologies respectively. Let σ be a topological isomorphism of F onto G. Then F is complete at $\mathfrak{p}$ if and only if G is complete at $\mathfrak{q}$.

11:13. **Theorem.** *A field F has a completion at each of its spots.*

Proof. Consider a spot $\mathfrak{p}$ on F and a valuation $|\ | \in \mathfrak{p}$. It is enough to construct a field $E \supseteq F$ and to provide it with a complete valuation $|\ |$ which induces the original valuation on F and is such that F is dense in E. The required spot $\mathfrak{P}$ on E is then the one to which $|\ |$ belongs.

Let $d(\alpha, \beta) = |\alpha - \beta|$ be the metric associated with the given valuation. We know from topology that the metric space F has a completion, i. e. that there is a metric space E which is complete and contains F as a dense subset, and such that the metric d on E induces the original metric on F. We have to make this metric space into a valuated field.

In order to define addition and multiplication let us consider two typical elements α and β of E. Since E is the closure of F we can find sequences $\{a_n\}$ and $\{b_n\}$ of elements of F such that $a_n \rightarrow \alpha$ and $b_n \rightarrow \beta$ under the metric d. Now these sequences are Cauchy sequences in F; this implies that $\{a_n + b_n\}$ and $\{a_n b_n\}$ are Cauchy sequences too; hence they converge to limits in E. Define

$$\alpha + \beta = \lim (a_n + b_n)\,, \quad \alpha\,\beta = \lim (a_n b_n)\,.$$

Take the original 0 and 1 of F as the 0 and 1 of E. One may check that these definitions are independent of the original choice of $\{a_n\}$ and $\{b_n\}$. Clearly the new laws of composition agree with the original ones on F. The field axioms should now be checked for E. For instance,

$$\alpha + \beta = \lim (a_n + b_n) = \lim (b_n + a_n) = \beta + \alpha$$

proves commutativity of addition; and the limit of the Cauchy sequence $\{-a_n\}$ gives the negative of α.

Finally define $|\alpha| = d(\alpha, 0)$ for all α in E. This gives the original valuation on F. Note that if $a_n \rightarrow \alpha$ then

$$|\,|a_n| - |\alpha|\,| = |d(a_n, 0) - d(\alpha, 0)| \leqq d(a_n, \alpha)$$

so that $|\alpha|$ is then $\lim |a_n|$. Hence

$$|\alpha\beta| = \lim |a_n b_n| = \lim |a_n|\,|b_n| = |\alpha|\,|\beta|\,.$$

Similarly $|\alpha + \beta| \leqq |\alpha| + |\beta|$. Hence $|\ |$ is a valuation on E. And the metric associated with this valuation is d since

$$|\alpha - \beta| = \lim |a_n - b_n| = \lim d(a_n, b_n) = d(\alpha, \beta)\,.$$

q. e. d.

11:14. *Let E be a completion of F and let φ be a topological isomorphism of F into some complete field G. Then there is a unique prolongation of φ to a topological isomorphism of E into G.*

Proof. Let $\mathfrak{P}|\mathfrak{p}$ and $\mathfrak{q}$ be the given spots on E, F and G. It follows from Proposition 11:6 that there exist valuations $|\ |_{\mathfrak{P}} \in \mathfrak{P}$ and $|\ |_{\mathfrak{q}} \in \mathfrak{q}$ which make φ analytic on F.

We now define $\varphi\alpha$ for a typical $\alpha \in E$. Approximate to α by elements of F, say $a_n \rightarrow \alpha$ with all $a_n \in F$. Then $\{a_n\}$ is a Cauchy sequence, hence $\{\varphi a_n\}$ is too. The latter Cauchy sequence has a limit in G; define

$$\varphi\alpha = \lim_n \varphi a_n .$$

The definition of $\varphi\alpha$ is independent of the choice of the a_n and it agrees with φ on F. If we now consider a typical $\beta \in E$ and an approximation $b_n \rightarrow \beta$ by elements of F, we can easily check that

$$\varphi(\alpha + \beta) = \varphi\alpha + \varphi\beta\,, \quad \varphi(\alpha\beta) = (\varphi\alpha)\,(\varphi\beta)\,, \quad |\varphi\alpha|_{\mathfrak{q}} = |\alpha|_{\mathfrak{P}}\,.$$

Hence φ is analytic. This proves the existence part of the proposition.

Let ψ be another prolongation of φ. Then

$$\varphi\alpha = \lim_n \varphi a_n = \lim_n \psi a_n = \psi\alpha\,.$$

So φ is unique. **q. e. d.**

11:15. *Let E_1 and E_2 be completions of F at the same spot $\mathfrak{p}$. Then there is a unique topological isomorphism of E_1 onto E_2 which is the identity on F.*

Proof. This is a special case of the last proposition, obtained by taking φ as the identity map on F. **q. e. d.**

The important results of this subparagraph have now been established: a field has a completion at each of its prime spots, and this completion is unique up to a natural topological isomorphism. There are two instances where we can be more specific. First suppose that E/F is an extension with spots $\mathfrak{P}|\mathfrak{p}$, and let E be complete at $\mathfrak{P}$. In this event the closure $\hat{F}$ of F in E (at the prime spot induced by $\mathfrak{P}$ on $\hat{F}$, and hence with the topology induced from E) is a completion of F at $\mathfrak{p}$. This is true since a closed subset of a complete metric space is complete. In the second instance consider an extension E/F with spots $\mathfrak{P}|\mathfrak{p}$, but do not necessarily assume that E is complete at $\mathfrak{P}$. Suppose there is a subfield E_1 with a spot $\mathfrak{P}_1$ induced by $\mathfrak{P}$ which is a completion of F at $\mathfrak{p}$. Then E_1 is absolutely unique (not just up to an isomorphism); indeed E_1 is closed in E since it is complete; but E_1 is part of the closure of F in E since F is dense in E_1; hence E_1 is the closure of F in E.

11:16. **Notation.** The same letter $\mathfrak{p}$ will be used for the prime spot of any completion of F at $\mathfrak{p}$. We usually use $F_{\mathfrak{p}}$ to denote a completion of F at $\mathfrak{p}$. Thus $\mathfrak{p}$ will also refer to the spot on $F_{\mathfrak{p}}$. If E/F is an extension with spots $\mathfrak{P}|\mathfrak{p}$ we can form a completion $E_{\mathfrak{P}}$ with its spot $\mathfrak{P}$. We let $F_{\mathfrak{P}}$ denote the closure of F in $E_{\mathfrak{P}}$. We know that $F_{\mathfrak{P}}$ provided with the spot induced by $\mathfrak{P}$ is a completion of F at $\mathfrak{p}$; we call it the completion in $E_{\mathfrak{P}}$ of F at $\mathfrak{p}$. According to our convention the spot induced by $\mathfrak{P}$ on $F_{\mathfrak{P}}$ is also written $\mathfrak{p}$. We have a natural topological isomorphism $F_{\mathfrak{P}} \rightarrowtail F_{\mathfrak{p}}$. If H/F is a subextension of E/F, and if $\mathfrak{P}$ induces $\mathfrak{P}_0$ on H, then the closure $H_{\mathfrak{P}}$ with its spot $\mathfrak{P}_0$ is the completion in $E_{\mathfrak{P}}$ of H at $\mathfrak{P}_0$. The $\mathfrak{P}$-adic topology on $E_{\mathfrak{P}}$ induces the $\mathfrak{P}_0$-adic topology on $H_{\mathfrak{P}}$ since $\mathfrak{P}|\mathfrak{P}_0$, hence $F_{\mathfrak{P}}$ is the closure of F in $H_{\mathfrak{P}}$ as well as in $E_{\mathfrak{P}}$. Furthermore, the spot $\mathfrak{p}$ on $F_{\mathfrak{P}}$ is induced by $\mathfrak{P}_0$ on $H_{\mathfrak{P}}$ as well as by $\mathfrak{P}$ on $E_{\mathfrak{P}}$. Hence $F_{\mathfrak{P}}$ with its spot $\mathfrak{p}$ is the completion of F in $H_{\mathfrak{P}}$ as well as in $E_{\mathfrak{P}}$.

§ 11G. Normed spaces over complete fields

Let V be a vector space over a valuated field F. A norm on V is a real valued function $\| \ \|$ with the following properties:

(1) $\|x\| > 0$ if $x \in \dot{V}$, and $\|0\| = 0$,
(2) $\|\alpha x\| = |\alpha| \, \|x\| \quad \forall \alpha \in F, \quad x \in V$,
(3) $\|x + y\| \leq \|x\| + \|y\| \quad \forall x, y \in V$.

We can introduce a distance function on V by defining $\|x - y\|$ to be the distance between two typical points x and y of V; this makes V into a metric space and the various topological concepts are thereby introduced into V.

11:17. *Let V be a finite dimensional vector space over a field with a complete valuation. Then all norms with respect to the given valuation induce the same topology on V. And V is complete under any one of them.*

Proof. 1) Let F be the given field, $| \ |$ the given valuation, and n the dimension of V over F. We consider a typical norm $\| \ \|$ on V which we shall call the given norm; the topology associated with this norm will be called the given topology. Fix a base $y_1, \ldots, y_n$ for V. Introduce a new norm $\| \ \|_0$ by defining

$$\|\alpha_1 y_1 + \cdots + \alpha_n y_n\|_0 = \max_i |\alpha_i|$$

for a typical vector in V. We shall refer to this as the constructed norm, and its topology will be called the constructed topology. If $n = 1$ we can easily find a constant K such that

$$\|x\| = K \|x\|_0 \quad \forall x \in V .$$

The entire result then follows. We proceed by induction to any $n > 1$.

2) We claim that we can find constants A and B such that

$$A \, \|x\|_0 \leq \|x\| \leq B \, \|x\|_0 \quad \forall x \in V .$$

For B it is easy: just take $B = \sum_1^n \|y_i\|$. Now let us find A. Consider the subspace $U = F y_2 + \cdots + F y_n$. By the inductive hypothesis U is complete under the given norm, hence it is closed in V. Now additive translation in V is clearly continuous, hence $U + y_1$ is a closed subset of V. There is therefore a neighborhood of 0 which contains no vector whose first coordinate in the base $y_1, \ldots, y_n$ is 1. Hence there is a neighborhood N of 0 (in the given topology) such that every vector with at least one coordinate equal to 1 falls outside N. In fact we can suppose that N is an open circular neighborhood, of radius A say, in the metric derived from $\| \ \|$. Now consider a typical x in $\dot{V}$, say

$$x = \alpha_1 y_1 + \cdots + \alpha_n y_n \quad \text{with} \quad \|x\|_0 = |\alpha_1| \, .$$

Then the first coordinate of $y = x/\alpha_1$ is 1, hence y is not in N, i. e. $\|y\| \geqq A$, hence $\|x\| \geqq A\,|\alpha_1| = A\,\|x\|_0$. So we have our A.

3) It is clear from step 2) that every neighborhood of one topology contains a neighborhood of the other. So the given topology is the same as the constructed topology on V.

4) Finally the question of completeness. Put $W = F y_1$ so that V is the direct sum $V = W \oplus U$. Consider a Cauchy sequence $x_1, \ldots, x_\nu, \ldots$ of vectors under the given norm. Then step 2) says that this is a Cauchy sequence under the constructed norm. Write

$$x_\nu = w_\nu + u_\nu \, , \quad w_\nu \in W \, , \quad u_\nu \in U \, .$$

Then $\|w_\nu - w_\mu\|_0 \leqq \|x_\nu - x_\mu\|_0$, so that the w_ν form a Cauchy sequence of vectors of W under the constructed norm. Similarly with the u_ν. Hence by the inductive hypothesis,

$$\exists \lim_\nu w_\nu = w \in W \, , \quad \exists \lim_\nu u_\nu = u \in U \, .$$

Hence $x_\nu \rightarrow w + u$. So V is complete. **q. e. d.**

11:18. **Theorem.** *Let E/F be a finite extension of fields with spots $\mathfrak{P}|\mathfrak{p}$. Suppose that F is complete at $\mathfrak{p}$. Then*

(1) *E is complete at $\mathfrak{P}$,*

(2) *$\mathfrak{P}$ is the only spot on E which divides $\mathfrak{p}$.*

Proof. Consider two spots $\mathfrak{P}|\mathfrak{p}$ and $\mathfrak{P}'|\mathfrak{p}$ on E. Pick $|\ |_\mathfrak{P} \in \mathfrak{P}$ and $|\ |_{\mathfrak{P}'} \in \mathfrak{P}'$ in such a way that they both induce the same $|\ |_\mathfrak{p} \in \mathfrak{p}$ on F. Regard E as a finite dimensional vector space over F in the natural way; then $|\ |_\mathfrak{P}$ and $|\ |_{\mathfrak{P}'}$ become norms with respect to the complete valuation $|\ |_\mathfrak{p}$ on F. Hence $|\ |_\mathfrak{P}$ is complete by the last proposition. And also $|\ |_\mathfrak{P}$ and $|\ |_{\mathfrak{P}'}$ induce the same topology on E, again by the last proposition. So E is complete at $\mathfrak{P}$ and $\mathfrak{P} = \mathfrak{P}'$. **q. e. d.**

11:18a. **Corollary.** *Let φ be an isomorphism of E into a field G with a spot $\mathfrak{q}$. Suppose that φ is topological on F. Then it is topological on E.*

Proof. We can assume that $\varphi E = G$. Let φ^{-1} carry $\mathfrak{q}$ to a spot $*$ on E. Then $\varphi\colon E \rightarrowtail G$ is topological under the $*$-adic topology on E and the $\mathfrak{q}$-adic one on G. It is therefore enough to prove that $* = \mathfrak{P}$. Now the restriction $\varphi\colon F \rightarrowtail \varphi F$ is topological with the $*$-adic topology on F; it is also topological with the $\mathfrak{p}$-adic topology on F by hypothesis. Hence the $*$-adic and $\mathfrak{p}$-adic topologies are equal on F. Hence $*|\mathfrak{p}$ by § 11D. But $\mathfrak{P}|\mathfrak{p}$. Hence $\mathfrak{P} = *$ by the theorem. **q. e. d.**

11:19. *Let E/F be a finite extension of fields and let $\mathfrak{P}$ be a spot on E. Then*

(1) $E_{\mathfrak{P}} = EF_{\mathfrak{P}}$,

(2) $[E_{\mathfrak{P}}: F_{\mathfrak{P}}] \leqq [E:F] < \infty$.

Proof. Recall the notation: $E_{\mathfrak{P}}$ is a completion of E at $\mathfrak{P}$, $F_{\mathfrak{P}}$ is the completion of F at $\mathfrak{P}$ obtained by taking the closure of F in $E_{\mathfrak{P}}$, and $EF_{\mathfrak{P}}$ is the compositum of E and $F_{\mathfrak{P}}$ in $E_{\mathfrak{P}}$.

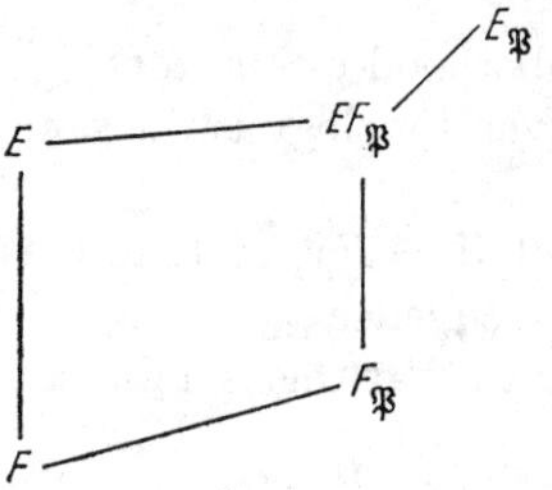

First we prove that $[EF_{\mathfrak{P}}: F_{\mathfrak{P}}] \leqq [E:F]$ by the following argument of field theory: take a tower of simple extensions from F to E, translate the tower by $F_{\mathfrak{P}}$, thereby obtain a tower from $F_{\mathfrak{P}}$ to $EF_{\mathfrak{P}}$ in which the successive layers have smaller degrees than before their translation; then $[EF_{\mathfrak{P}}: F_{\mathfrak{P}}] \leqq [E:F]$.

We now prove the first part. Since $EF_{\mathfrak{P}}/F_{\mathfrak{P}}$ is finite, and since $F_{\mathfrak{P}}$ is complete at $\mathfrak{P}$, we deduce from Theorem 11:18 that $EF_{\mathfrak{P}}$ is complete at $\mathfrak{P}$, hence it is closed in $E_{\mathfrak{P}}$ under the $\mathfrak{P}$-adic topology. But $E_{\mathfrak{P}}$ is the closure of E and $E \subseteq EF_{\mathfrak{P}} \subseteq E_{\mathfrak{P}}$. Hence $EF_{\mathfrak{P}} = E_{\mathfrak{P}}$.

The second part is now immediate. **q. e. d.**

§ 12. Archimedean valuations

The purpose of § 12 is to show that there is exactly one archimedean spot on $\mathbf{Q}$, namely the one determined by the ordinary absolute value, and that there are essentially two complete archimedean fields, namely the real and complex numbers $\mathbf{R}$ and $\mathbf{C}$. We shall use $|\ |_\infty$ for the ordinary absolute value on $\mathbf{Q}$ (see § 31 for further discussion of the spot ∞).

12:1. *There is exactly one archimedean spot $\mathfrak{p}$ on $\mathbf{Q}$. Every valuation $|\ |$ in $\mathfrak{p}$ is of the form $|\ | = |\ |_\infty^\varrho$ where $|\ |_\infty$ is the ordinary absolute value on $\mathbf{Q}$ and $0 < \varrho \leqq 1$.*

Proof. 1) Consider any two rational integers $m > 1$ and $n > 1$. Suppose we express m to the base n as follows:

$$m = a_0 + a_1 n + \cdots + a_r n^r$$

with

$$0 \leqq a_i < n , \quad a_r > 0 , \quad r \geqq 0 .$$

Then

$$|a_i| = |1 + \cdots + 1| < n .$$

Hence

$$|m| < n \, (1 + |n| + \cdots + |n|^r) .$$

2) We can draw two conclusions from the above inequality. The first is that $|n| \geqq 1$ holds for any rational integer $n > 1$. For suppose not. Then we have an $n > 1$ with $|n| < 1$. Consider any rational integer $m > 1$. Then by step 1) we have

$$|m| < n\,(1 + |n| + \cdots + |n|^r + \cdots) = \frac{n}{1 - |n|}\,.$$

Hence $|\ |$ is bounded on the natural integers of $\mathbf{Q}$ and is therefore non-archimedean, contrary to hypothesis. Hence we do indeed have $|n| \geqq 1$ whenever $n > 1$.

The second conclusion is this: if $m > 1$ and $n > 1$ are two rational integers, then

$$|m| \leqq n\left(\frac{\log m}{\log n} + 1\right)|n|^{\log m/\log n}.$$

To see this express m to the base n and use the inequality in step 1). We obtain

$$|m| < n\,(r + 1)\,|n|^r$$

since $|n| \geqq 1$. But $r \leqq \log m/\log n$ since $n^r \leqq m$. The second conclusion then follows on replacing r by $\log m/\log n$.

3) Now use the inequality proved in step 2), substitute m^k for m, and let $k \to \infty$. This gives

$$|m| \leqq |n|^{\log m/\log n}\,.$$

Take logarithms and then interchange m and n: we get

$$\frac{\log|m|}{\log m} = \frac{\log|n|}{\log n}\,.$$

Hence $\log|n|/\log n$ is a constant, ϱ say, for all rational integers $n > 1$. Hence $|n| = n^{\varrho}$ for these n. Hence

$$|x| = |x|_\infty^{\varrho} \quad \forall\, x \in \mathbf{Q}\,.$$

But there is at least one rational integer $n > 1$ such that $|n| > 1$ since $\mathfrak{p}$ is archimedean. For this n we have $1 < |n| \leqq n$. Hence $0 < \varrho \leqq 1$.

4) We have therefore proved that a valuation $|\ |$ in an arbitrary archimedean spot $\mathfrak{p}$ is equivalent to $|\ |_\infty$. Hence there is just one archimedean spot on $\mathbf{Q}$. **q. e. d.**

12:2. **Lemma.** *Let F be a field with a complete archimedean valuation $|\ |$. Then $1 + a \in F^2$ whenever $|a| < \frac{1}{4}|4|$.*

Proof. Since $\chi(F)$ must be 0 we can define $r = 4\,|a|/|4| < 1$. It is enough to prove that the polynomial $x^2 + 2x - a$ has a root in F; for if it does, then its discriminant will be a square in F, and its discriminant is actually $4\,(1 + a)$, hence $1 + a \in F^2$.

Define a sequence $x_0, x_1, \ldots, x_\nu, \ldots \in F$ by means of the formulas

$$x_0 = 1\,, \quad x_{\nu+1} = \frac{a}{x_\nu} - 2\,.$$

Then we must have $|x_\nu| \geqq \frac{1}{2}\,|2|$: for this already holds when $\nu = 0$ and by an inductive argument we obtain

$$|x_{\nu+1}| = \left|2 - \frac{a}{x_\nu}\right| \geqq |2| - \frac{|a|}{|x_\nu|} \geqq \frac{1}{2}\,|2|\,.$$

From this follows

$$\frac{|x_{\nu+2} - x_{\nu+1}|}{|x_{\nu+1} - x_\nu|} \leqq \frac{4\,|a|}{|4|} = r < 1\,.$$

Hence the Ratio Test says that

$$\sum_0^\infty |x_{\nu+1} - x_\nu|$$

is convergent. Hence for each $\varepsilon > 0$ there is a ν_0 such that

$$|x_\lambda - x_\mu| \leqq \sum_\mu^{\lambda-1} |x_{\nu+1} - x_\nu| < \varepsilon$$

if $\lambda \geqq \mu \geqq \nu_0$. But this means that $\{x_\nu\}$ is a Cauchy sequence, so it must have a limit $x \neq 0$ in F. But

$$x_{\nu+1} x_\nu + 2x_\nu - a = 0\,.$$

Letting $\nu \to \infty$ in this equation shows that $x \in F$ is a solution to the equation $x^2 + 2x - a = 0$. **q. e. d.**

12:3. **Lemma.** *Let F be a field with a complete archimedean valuation. Then there is a prolongation of the valuation from F to $F(i)$ where $i^2 = -1$.*

Proof. We can assume that i is not in F. Let $|\ |$ denote the given valuation on F. Let N denote the norm $N_{F(i)/F}$. Define

$$|\alpha| = |N\alpha|^{1/2} \quad \forall\, \alpha \in F(i)\,.$$

This is possible since $N\alpha \in F$ for all $\alpha \in F$. This new function is a prolongation of the original valuation since

$$|\alpha| = |N\alpha|^{1/2} = |\alpha^2|^{1/2} = |\alpha| \quad \forall\, \alpha \in F\,.$$

How do we know that the prolongation is a valuation? Only the triangle law needs checking, and here it is enough to verify that

$$|1 + \alpha| \leqq 1 + |\alpha| \quad \forall\, \alpha \in F(i)\,.$$

This holds for all $\alpha \in F$. So consider $\alpha \in F(i) - F$. Let

$$x^2 + bx + c = \operatorname{irr}(x, \alpha, F)\,.$$

Then $c = N\alpha$ and so $|c|^{1/2} = |\alpha|$. If we had $|b|^2 > 4\,|c|$, the quantity

$$b^2 - 4c = b^2\left(1 - \frac{4c}{b^2}\right)$$

would be in F^2 by Lemma 12:2, and so $x^2 + bx + c$ would be reducible. Hence $|b|^2 \leqq 4\,|c|$. Now

$$x^2 + (b-2)\,x + (1 + c - b) = \operatorname{irr}(x, 1+\alpha, F)\,.$$

Hence

$$\begin{aligned} |1+\alpha|^2 &= |1 + c - b| \\ &\leqq 1 + |c| + |b| \\ &\leqq 1 + |c| + 2\,|c|^{1/2} \\ &= (1 + |c|^{1/2})^2 \\ &= (1 + |\alpha|)^2. \end{aligned}$$

Hence $|1+\alpha| \leqq 1 + |\alpha|$. **q. e. d.**

12:4. **Theorem.** *Let F be a complete archimedean field. Then there is a topological isomorphism of F onto either the real or complex numbers.*

Proof. 1) Let $\mathfrak{p}$ be the given complete archimedean spot on F. Consider the complex numbers C at the spot $\mathfrak{q}$ determined by the ordinary absolute value. We let Q be the prime field of F and R the closure of Q in F under the $\mathfrak{p}$-adic topology; thus R is the completion of Q in F.

2) First a reduction of the problem. Suppose we have proved our theorem in the case where $\sqrt{-1} \in F$. Let us show how to derive the general case from this. We can consider $\sqrt{-1} \notin F$. Put $E = F(\sqrt{-1})$. Then there is a prime spot $\mathfrak{P}$ on E which divides $\mathfrak{p}$ by Lemma 12:3. And E is complete by Theorem 11:18. But we are assuming that our result holds in this situation. Hence there is a topological isomorphism φ of E onto C. Now φR is complete, hence closed in C. But $\varphi R \neq \mathsf{C}$. Hence $\varphi R = \mathsf{R}$. Hence $\varphi F = \mathsf{R}$.

3) Therefore assume that $\sqrt{-1} \in F$. Put $C = R(\sqrt{-1})$. Let φ be the natural isomorphism $\varphi\colon Q \rightarrowtail \mathsf{Q}$. This is topological since Q has just one archimedean spot. By Proposition 11:14 there is a prolongation of φ to a topological isomorphism $\varphi\colon R \rightarrowtail \mathsf{R}$. By field theory there is a prolongation of φ to an algebraic isomorphism $\varphi\colon C \rightarrowtail \mathsf{C}$, and this prolongation must be topological by Corollary 11:18a since R is complete. If $F = C$ we are through. We therefore assume that $C \subset F$ and use this to produce a contradiction.

$$\begin{array}{ccc} F & & \\ | & & \\ C & \overset{\varphi}{\rightarrowtail} & \mathsf{C} \\ | & & | \\ R & \rightarrowtail & \mathsf{R} \\ | & & | \\ Q & \rightarrowtail & \mathsf{Q} \end{array}$$

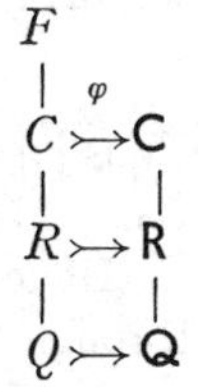

4) Fix $|\;| \in \mathfrak{p}$. Then by Proposition 11:6 there is a valuation in $\mathfrak{q}$ which makes $\varphi\colon C \rightarrowtail \mathsf{C}$ analytic. We therefore have the following additional information about the field C: the value group $|C|$ is equal to P, and every closed bounded subset of C is compact.

Consider a point $x \in F - C$, i. e. a point x in F that is not in C. We shall minimize the distance from x to points of C. To do this consider a closed sphere M which meets C and has center at x. Then the function

$$\alpha \rightarrow |\alpha - x|$$

of $M \cap C$ into R is continuous; and $M \cap C$ is a closed bounded subset of C and is therefore compact; so our function attains its minimum; we therefore have $\alpha_0 \in M \cap C$ such that

$$|\alpha_0 - x| \leqq |\alpha - x| \quad \forall\, \alpha \in M \cap C\,.$$

So in fact we really have

$$0 < |\alpha_0 - x| \leqq |\alpha - x| \quad \forall\, \alpha \in C\,.$$

Replace x by $\alpha_0 - x$ and then scale by a suitable element of C (recalling that $|C| = \mathsf{P}$); in this way we obtain $z \in F - C$ such that

$$2 = |z| \leqq |\alpha - z| \quad \forall\, \alpha \in C\,. \tag{1}$$

5) Consider any $z \in F - C$ which satisfies equation (1), and any natural number n. We find

$$\begin{aligned} 2^n \left(1 + \frac{1}{2^n}\right) &= |z^n| + 1 \\ &\geqq |z^n - 1| \\ &= \prod_{k=1}^{n} |z - \zeta_k| \\ &\geqq |z - 1|\, 2^{n-1} \end{aligned}$$

where $1 = \zeta_1, \zeta_2, \ldots, \zeta_n$ are the n-th roots of unity in C. Divide by 2^{n-1} and let $n \rightarrow \infty$; this gives $|z - 1| \leqq 2$, hence $|z - 1| = 2$. Therefore

$$2 = |z - 1| \leqq |\alpha - (z - 1)| \quad \forall\, \alpha \in C\,.$$

So $z - 1 \in F - C$ has the property of equation (1) whenever z does. Hence so does $z - n$ for any $n \in \mathsf{N}$. So for a fixed z and for all $n \in \mathsf{N}$ we have

$$2 = |z - n| \geqq |n| - |z|\,.$$

Thus $|\ |$ is bounded on the rational integers in F; this is impossible since the valuation is archimedean.

Hence our assumption that $F \neq C$ is false. So $F = C$ and we are through. **q. e. d.**

Consider a field F and an archimedean spot $\mathfrak{p}$ on F. Let $F_\mathfrak{p}$ be a completion of F at $\mathfrak{p}$. We now know that there is a topological isomorphism φ of $F_\mathfrak{p}$ onto either R or C. Let φ^{-1} carry the ordinary absolute value back to $F_\mathfrak{p}$. This gives us a valuation $|\ | \in \mathfrak{p}$ on F such that $|m| = m$ holds for all natural numbers m in F. If $|\ |' \in \mathfrak{p}$ is another valuation with the same property, then $|\ |$ and $|\ |'$ are equal on the prime field and equivalent on F, hence they are equal on F by Proposition 11:5. So we can make the following definition.

12:5. **Definition.** Let $\mathfrak{p}$ be an archimedean spot on a field F. By the ordinary absolute value on F at $\mathfrak{p}$ we mean that unique valuation $|\ | \in \mathfrak{p}$ with the property that $|m| = m$ holds for all natural numbers m in F.

12:6. *Let $\mathfrak{p}$ be an archimedean spot on a field F. Then every valuation in $\mathfrak{p}$ is of the form $|\ |^\varrho$ with $0 < \varrho \leqq 1$, where $|\ |$ is the ordinary absolute value on F at $\mathfrak{p}$.*

Proof. Consider typical $|\ |_* \in \mathfrak{p}$. Then there is a ϱ $(0 < \varrho \leqq 1)$ such that

$$|\alpha|_* = |\alpha|^\varrho \quad \forall\, \alpha \in Q$$

where Q is the prime field of F. Now $|\ |_*$ and $|\ |^\varrho$ are equivalent on F. Hence $|\ |_* = |\ |^\varrho$ by Proposition 11:5. **q. e. d.**

12:7. **Remark.** Let E/F be an extension of fields with spots $\mathfrak{P}\,|\,\mathfrak{p}$ which are not necessarily archimedean nor necessarily complete. However, suppose that $\mathfrak{p}$ is non-trivial. Then restricting a valuation in $\mathfrak{P}$ to F gives a valuation in $\mathfrak{p}$; by Proposition 11:5 this sets up an injection of the valuations in $\mathfrak{P}$ into those in $\mathfrak{p}$. In fact this is a bijection: in the non-archimedean case surjectivity follows easily from the strong triangle law and the formula $|\ |_* = |\ |^\varrho$ relating equivalent valuations, in the archimedean case it follows from the last proposition.

12:8. **Definition.** Let $\mathfrak{p}$ be an archimedean spot on a field F. We call $\mathfrak{p}$ real or complex according as $F_\mathfrak{p}$ is isomorphic to $\mathbf{R}$ or $\mathbf{C}$.

12:9. **Definition.** Let $\mathfrak{p}$ be an archimedean spot on a field F. By the normalized valuation on F (or on $F_\mathfrak{p}$) at $\mathfrak{p}$ we mean the function

$$|\alpha|_\mathfrak{p} = \begin{cases} |\alpha| & \text{if } \mathfrak{p} \text{ real,} \\ |\alpha|^2 & \text{if } \mathfrak{p} \text{ complex,} \end{cases}$$

where $|\ |$ is the ordinary absolute value on F (or on $F_\mathfrak{p}$) at $\mathfrak{p}$.

12:10. **Remark.** The normalized valuation is a true valuation at a real spot. It is not a true valuation at a complex spot; there the triangle law must be replaced by

$$|\alpha + \beta|_\mathfrak{p} \leqq 2\,(|\alpha|_\mathfrak{p} + |\beta|_\mathfrak{p})$$

or more generally,

$$\left|\sum_1^r \alpha_i\right|_\mathfrak{p} \leqq 2^{r-1}\left(\sum_1^r |\alpha_i|_\mathfrak{p}\right).$$

Normalized valuations will also be introduced over the local fields of Chapter III. They provide one way of regularizing the behavior of the product formula of Chapter III.

12:11. **Definition.** Let $\mathfrak{p}$ be a real spot on a field F, and let α be any element of F (or of $F_\mathfrak{p}$). We say that α is positive at $\mathfrak{p}$ if $\alpha \in \dot{F}^2_\mathfrak{p}$. We say that α is negative at $\mathfrak{p}$ if $\alpha \in -\dot{F}^2_\mathfrak{p}$.

12:12. **Example.** Let $\mathfrak{p}$ be a real spot on the field F. So there is a topological isomorphism of $F_\mathfrak{p}$ onto the field of real numbers R. This isomorphism carries the positive elements of $F_\mathfrak{p}$ onto the positive real numbers, and the negative elements of $F_\mathfrak{p}$ onto the negative real numbers. The positive elements of $F_\mathfrak{p}$ are an open subgroup of index 2 in $\dot{F}_\mathfrak{p}$. The negative elements of $F_\mathfrak{p}$ are an open subset of $F_\mathfrak{p}$. We have the disjoint union

$$F_\mathfrak{p} = -\dot{F}^2_\mathfrak{p} \cup 0 \cup \dot{F}^2_\mathfrak{p}$$

of $F_\mathfrak{p}$ into negative elements, zero, and positive elements.

12:13. **Example.** Let F be a prime field of characteristic 0. Then there is exactly one archimedean spot on F. This spot is real. The positive elements of F are the elements of the form m/n with m and n natural numbers in F.

§ 13. Non-archimedean valuations

§ 13A. The residue class field

We let F be an arbitrary field, $\mathfrak{p}$ any non-archimedean prime spot on F. We define

$$\mathfrak{o}(\mathfrak{p}) = \{\alpha \in F \mid |\alpha|_\mathfrak{p} \leqq 1\}$$

$$\mathfrak{u}(\mathfrak{p}) = \{\alpha \in F \mid |\alpha|_\mathfrak{p} = 1\}$$

$$\mathfrak{m}(\mathfrak{p}) = \{\alpha \in F \mid |\alpha|_\mathfrak{p} < 1\},$$

where $|\ |_\mathfrak{p}$ denotes a valuation in $\mathfrak{p}$; these definitions are clearly independent of the choice of $|\ |_\mathfrak{p}$ in $\mathfrak{p}$. The elements of $\mathfrak{o}(\mathfrak{p})$ are called the integers of F at $\mathfrak{p}$, or simply the integers of F when there is no risk of confusion; every rational integer in F is an integer of F at $\mathfrak{p}$ by the strong triangle law. It is easily verified that $\mathfrak{o}(\mathfrak{p})$ is a subring containing the identity of F, and that F is the quotient field of $\mathfrak{o}(\mathfrak{p})$. We call $\mathfrak{o}(\mathfrak{p})$ the ring of integers of F (at $\mathfrak{p}$) or the valuation ring of $\mathfrak{p}$. If $\mathfrak{q}$ is some other spot on F, then it follows from Proposition 11:7 that

$$\mathfrak{p} = \mathfrak{q} \Leftrightarrow \mathfrak{o}(\mathfrak{p}) = \mathfrak{o}(\mathfrak{q}).$$

Note that $\mathfrak{o}(\mathfrak{p}) = F$ if and only if $\mathfrak{p}$ is the trivial spot on F. Thus $\mathfrak{o}(\mathfrak{p})$ is only exceptionally a field – if and only if $\mathfrak{p}$ is trivial. The set $\mathfrak{u}(\mathfrak{p})$ is a multiplicative subgroup of $\mathfrak{o}(\mathfrak{p})$; it consists precisely of all invertible elements of the ring $\mathfrak{o}(\mathfrak{p})$; accordingly we shall call $\mathfrak{u}(\mathfrak{p})$ the group of units of F at $\mathfrak{p}$, or simply the group of units of F when there is no risk of confusion. Now let us comment on $\mathfrak{m}(\mathfrak{p})$. Again we see that

$$\mathfrak{p} = \mathfrak{q} \Leftrightarrow \mathfrak{m}(\mathfrak{p}) = \mathfrak{m}(\mathfrak{q});$$

and $\mathfrak{m}(\mathfrak{p}) = 0$ if and only if $\mathfrak{p}$ is the trivial spot. It is easily verified using the strong triangle law that $\mathfrak{m}(\mathfrak{p})$ is an ideal in $\mathfrak{o}(\mathfrak{p})$. We call $\mathfrak{m}(\mathfrak{p})$ the maximal ideal of F at $\mathfrak{p}$. This name is justified by the fact that $\mathfrak{m}(\mathfrak{p})$ is the absolutely largest proper ideal in $\mathfrak{o}(\mathfrak{p})$. Why is this true? Consider an ideal $\mathfrak{a}$ of $\mathfrak{o}(\mathfrak{p})$ that is not contained in $\mathfrak{m}(\mathfrak{p})$; pick $\alpha \in \mathfrak{a} - \mathfrak{m}(\mathfrak{p}) \subseteq \mathfrak{u}(\mathfrak{p})$. Then $1 = \alpha\alpha^{-1} \in \mathfrak{a}$. Hence $\mathfrak{a} = \mathfrak{o}(\mathfrak{p})$ and $\mathfrak{a}$ is not proper.

We are ready to define the residue class field of F at $\mathfrak{p}$. Essentially it is the field $\mathfrak{o}(\mathfrak{p})/\mathfrak{m}(\mathfrak{p})$. But we would like a little more flexibility than the usual definition of the quotient ring $\mathfrak{o}(\mathfrak{p})/\mathfrak{m}(\mathfrak{p})$ will permit. So we frame our definition as follows: by a residue class field of F at $\mathfrak{p}$ we mean any composite object (φ, H) consisting of a field H and a ring homomorphism φ of $\mathfrak{o}(\mathfrak{p})$ onto H which has kernel $\mathfrak{m}(\mathfrak{p})$. The field F will always have at least one residue class field at $\mathfrak{p}$, namely the field $\mathfrak{o}(\mathfrak{p})/\mathfrak{m}(\mathfrak{p})$ together with the naturally associated homomorphism of $\mathfrak{o}(\mathfrak{p})$ onto it. Also the residue class field is essentially unique. More precisely: let (φ, H) and (φ', H') be two residue class fields of F at $\mathfrak{p}$; then there is a unique ring isomorphism ψ of H onto H' such that $\psi \circ \varphi = \varphi'$. (The existence of ψ is one of the elementary isomorphism theorems of ring theory.)

A residue class field is usually referred to without explicitly mentioning the other half of the composite object, namely the homomorphism φ. The bar symbol is often used for the mapping φ and when this is done one lets $\bar{F}$ denote the corresponding field. (Of course $\bar{F}$ is the image of $\mathfrak{o}(\mathfrak{p})$, not of F, under the bar mapping.) When several spots are under discussion at the same time we can use $F(\mathfrak{p})$ for a residue class field of F at $\mathfrak{p}$.

13:1. **Example.** As an example of the notation let us see what is meant by $F_\mathfrak{p}(\mathfrak{p})$. Here $F_\mathfrak{p}$ is a completion of F at $\mathfrak{p}$; and we have agreed to use the same letter $\mathfrak{p}$ for the spot on the completion $F_\mathfrak{p}$; hence $F_\mathfrak{p}(\mathfrak{p})$ means a residue class field of the completion $F_\mathfrak{p}$ at $\mathfrak{p}$.

13:2. **Example.** $\chi(F(\mathfrak{p})) = \chi(F)$ whenever $\chi(F) \neq 0$.

13:3. **Notation.** Let $\mathfrak{a}$ be an additive subgroup of F. For any two elements $\alpha, \beta \in F$ the congruence

$$\alpha \equiv \beta \bmod \mathfrak{a}$$

takes on its usual meaning, namely $\alpha - \beta \in \mathfrak{a}$. This defines an equivalence relation in the usual way. Furthermore, if

$$\alpha \equiv \beta \bmod \mathfrak{a}, \quad \text{and} \quad \alpha' \equiv \beta' \bmod \mathfrak{a},$$

then

$$\alpha + \alpha' \equiv \beta + \beta' \bmod \mathfrak{a} \quad \text{and} \quad \lambda\alpha \equiv \lambda\beta \bmod \lambda\mathfrak{a}$$

for any $\lambda \in F$, where $\lambda\mathfrak{a}$ is the additive subgroup

$$\lambda\mathfrak{a} = \{\lambda x \mid x \in \mathfrak{a}\}.$$

For any $\gamma \in F$ we shall write

$$\alpha \equiv \beta \bmod \gamma$$

instead of $\alpha \equiv \beta \bmod \gamma\,\mathfrak{o}(\mathfrak{p})$.

§ 13B. The fundamental invariants *e* and *f*

Consider an arbitrary extension of fields E/F provided with non-archimedean spots $\mathfrak{P}|\mathfrak{p}$. Fix completions $F_\mathfrak{p}$ and $E_\mathfrak{P}$, and let $F_\mathfrak{P}$ be the completion of F at $\mathfrak{p}$ that is obtained by taking the closure of F in $E_\mathfrak{P}$ in the usual way. According to our conventions the spot $\mathfrak{p}$ can refer to any one of the fields $F, F_\mathfrak{p}$ or $F_\mathfrak{P}$; similarly with $\mathfrak{P}$. Now this presents a problem: does $\mathfrak{o}(\mathfrak{p})$ refer to F or does it refer to $F_\mathfrak{p}$? This point should be cleared up right away even though it will not be needed until much later. We therefore make the following convention: $\mathfrak{o}(\mathfrak{p})$ is to be the valuation ring of F at $\mathfrak{p}$, while the valuation ring of $F_\mathfrak{p}$ at $\mathfrak{p}$ is to be denoted by the new symbol $\mathfrak{o}_\mathfrak{p}$. Similarly we introduce new symbols $\mathfrak{u}_\mathfrak{p}$ and $\mathfrak{m}_\mathfrak{p}$ for the group of units and the maximal ideal of $F_\mathfrak{p}$ at $\mathfrak{p}$. Similarly with E and $E_\mathfrak{P}$ at $\mathfrak{P}$. Note that

$$\mathfrak{o}(\mathfrak{P}) \cap F = \mathfrak{o}(\mathfrak{p}) \subseteq \mathfrak{o}(\mathfrak{P})$$
$$\mathfrak{u}(\mathfrak{P}) \cap F = \mathfrak{u}(\mathfrak{p}) \subseteq \mathfrak{u}(\mathfrak{P})$$
$$\mathfrak{m}(\mathfrak{P}) \cap F = \mathfrak{m}(\mathfrak{p}) \subseteq \mathfrak{m}(\mathfrak{P})\,.$$

And similar formulas hold between F and $F_\mathfrak{p}$.

13:4. *In the above situation consider valuations* $|\ |_\mathfrak{p} \in \mathfrak{p}$, $|\ |_\mathfrak{P} \in \mathfrak{P}$ *and* $|\ |'_\mathfrak{P} \in \mathfrak{P}$. *Then*

(1) $|F|_\mathfrak{P} \subseteq |E|_\mathfrak{P}, \quad |F|_\mathfrak{p} = |F_\mathfrak{p}|_\mathfrak{p}$

(2) $(|E_\mathfrak{P}|_\mathfrak{P} : |F_\mathfrak{P}|_\mathfrak{P}) = (|E|_\mathfrak{P} : |F|_\mathfrak{P})$

(3) $(|E|'_\mathfrak{P} : |F|'_\mathfrak{P}) = (|E|_\mathfrak{P} : |F|_\mathfrak{P})$.

Proof. (1) Clearly every element in the value group $|F|_\mathfrak{P}$ is in the value group $|E|_\mathfrak{P}$. In particular, $|F|_\mathfrak{p} \subseteq |F_\mathfrak{p}|_\mathfrak{p}$. Conversely, take $\alpha \in F_\mathfrak{p}$ and write it $\alpha = \lim a_n$ with $a_n \in F$. For large enough n we have $|a_n - \alpha|_\mathfrak{p} < |\alpha|_\mathfrak{p}$, hence $|a_n|_\mathfrak{p} = |\alpha|_\mathfrak{p}$ by the Principle of Domination. Thus $|\alpha|_\mathfrak{p} \in |F|_\mathfrak{p}$. So $|F|_\mathfrak{p} = |F_\mathfrak{p}|_\mathfrak{p}$.

(2) is clear from (1).

(3) It is enough to prove that $(G : H) = (G^\varrho : H^\varrho)$ holds for multiplicative subgroups $H \subseteq G$ of the positive real numbers P, where ϱ is any positive number. This follows from the fact that the multiplicative isomorphism $x \rightarrowtail x^\varrho$ carries G to G^ϱ and H to H^ϱ. **q. e. d.**

We now define the ramification index of the extension E/F at $\mathfrak{P}|\mathfrak{p}$ to be the group index

$$e(\mathfrak{P}|\mathfrak{p}) = (|E|_\mathfrak{P} : |F|_\mathfrak{P})\,;$$

this is either a natural number or ∞. The last proposition shows that the ramification index is independent of the choice of valuation in $\mathfrak{P}$, and also that it has the same value for the extension $E_\mathfrak{P}/F_\mathfrak{P}$ at $\mathfrak{P}|\mathfrak{p}$.

We go through a somewhat similar procedure to get the degree of inertia $f(\mathfrak{P}|\mathfrak{p})$. Consider the composite object $(^-, \bar{E})$ comprising the residue class field of E at $\mathfrak{P}$. Here $\bar{E}$ is the image of $\mathfrak{o}(\mathfrak{P})$ under the bar mapping. Let $\bar{F}$ be the image of $\mathfrak{o}(\mathfrak{p}) = \mathfrak{o}(\mathfrak{P}) \cap F$ in $\bar{E}$. The kernel of the bar mapping on $\mathfrak{o}(\mathfrak{p})$ is $\mathfrak{m}(\mathfrak{p})$. Hence $(^-, \bar{F})$ is a residue class field of F at $\mathfrak{p}$. We shall call the residue class field obtained in this way the residue class field of F at $\mathfrak{p}$ that is obtained by natural restriction from the residue class field of E at $\mathfrak{P}$. We introduce the following convention along the lines of the $E_{\mathfrak{P}}/F_{\mathfrak{P}}$ convention for completions: if $E(\mathfrak{P})$ denotes a residue class field of E at $\mathfrak{P}$, then $F(\mathfrak{P})$ will denote the residue class field of F at $\mathfrak{p}$ that is obtained from $E(\mathfrak{P})$ by natural restriction. We know that we have a natural ring isomorphism

$$F(\mathfrak{p}) \rightarrowtail F(\mathfrak{P}) .$$

In particular, a residue class field of F at $\mathfrak{p}$ can be naturally imbedded in any residue class field of E at $\mathfrak{P}$ when $\mathfrak{P}|\mathfrak{p}$; in other words there is a natural isomorphism $F(\mathfrak{p}) \rightarrowtail E(\mathfrak{P})$.

13:5. (1) $F(\mathfrak{p}) = F_{\mathfrak{p}}(\mathfrak{p})$, (2) $[E_{\mathfrak{P}}(\mathfrak{P}) : F_{\mathfrak{P}}(\mathfrak{P})] = [E(\mathfrak{P}) : F(\mathfrak{P})]$.

Proof. (1) Here it is understood that $F(\mathfrak{p})$ is obtained from $F_{\mathfrak{p}}(\mathfrak{p})$ by natural restriction. Use bar for the natural homomorphism of $\mathfrak{o}_{\mathfrak{p}}$ onto $F_{\mathfrak{p}}(\mathfrak{p})$. Then

$$\overline{\mathfrak{o}(\mathfrak{p})} = F(\mathfrak{p}), \quad \overline{\mathfrak{o}_{\mathfrak{p}}} = F_{\mathfrak{p}}(\mathfrak{p}) .$$

We have to prove that $\overline{\mathfrak{o}_{\mathfrak{p}}} \subseteq \overline{\mathfrak{o}(\mathfrak{p})}$. To this end consider typical $\alpha \in \mathfrak{o}_{\mathfrak{p}}$. Then $\alpha = \lim a_n$ with $a_n \in F$. So there is an $a \in F$ with $|a - \alpha| < 1$. Then $a - \alpha \in \mathfrak{m}_{\mathfrak{p}}$. So $a \in \mathfrak{o}(\mathfrak{p})$, and $\bar{\alpha} = \bar{a}$. Hence $\overline{\mathfrak{o}_{\mathfrak{p}}} \subseteq \overline{\mathfrak{o}(\mathfrak{p})}$.

(2) It is understood that $F_{\mathfrak{P}}(\mathfrak{P})$ is obtained from $E_{\mathfrak{P}}(\mathfrak{P})$ and that $F(\mathfrak{P})$ is obtained from $E(\mathfrak{P})$ by natural restriction. Now it is easily seen that $E(\mathfrak{P})$ can be replaced by any other residue class field of E at $\mathfrak{P}$ without changing the degree $[E(\mathfrak{P}) : F(\mathfrak{P})]$, hence we can assume that $E(\mathfrak{P})$ is obtained from $E_{\mathfrak{P}}(\mathfrak{P})$ by natural restriction. So all residue class fields under discussion are now obtained from $E_{\mathfrak{P}}(\mathfrak{P})$ by natural restric-

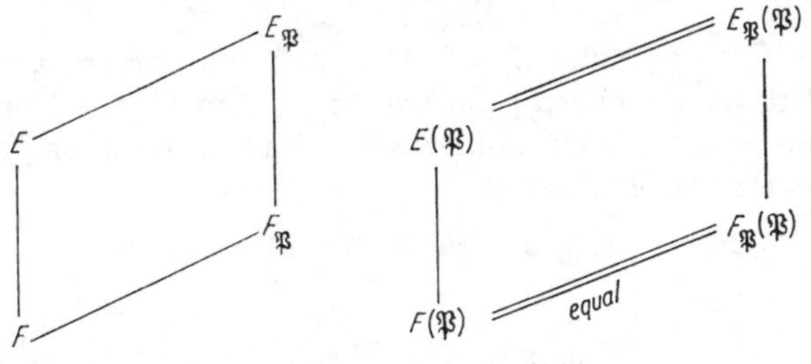

tion. The result follows from the above diagrams since, by the first part of this proposition, we have $F(\mathfrak{P}) = F_{\mathfrak{P}}(\mathfrak{P})$ and $E(\mathfrak{P}) = E_{\mathfrak{P}}(\mathfrak{P})$. **q. e. d.**

We define the degree of inertia of the extension E/F at $\mathfrak{P}|\mathfrak{p}$ to be the field degree

$$f(\mathfrak{P}|\mathfrak{p}) = [E(\mathfrak{P}) : F(\mathfrak{P})];$$

this is either a natural number or ∞. The last proposition shows that it has the same value for the extension $E_{\mathfrak{P}}/F_{\mathfrak{P}}$ at $\mathfrak{P}|\mathfrak{p}$.

13:6. *Let E/F be a finite extension of degree n with non-archimedean spots $\mathfrak{P}|\mathfrak{p}$. Then*

$$e(\mathfrak{P}|\mathfrak{p})\, f(\mathfrak{P}|\mathfrak{p}) \leq n\,.$$

Proof. Fix $|\ | \in \mathfrak{P}$. Let $(^-, \bar{E})$ be a residue class field of E at $\mathfrak{P}$ and let $(^-, \bar{F})$ be obtained from it by natural restriction. Put

$$e = (|E| : |F|)\,, \quad f = [\bar{E} : \bar{F}]\,.$$

We must prove $ef \leq n$.

Consider r elements $\bar{\omega}_1, \ldots, \bar{\omega}_r$ of $\bar{E}$ which are independent over $\bar{F}$. Choose representatives $\omega_1, \ldots, \omega_r$ of these elements in $\mathfrak{o}(\mathfrak{P})$. Then all these representatives must be in $\mathfrak{u}(\mathfrak{P})$. Consider a non-zero element $A = \sum a_\lambda \omega_\lambda$ with the a_λ in F and suppose that $|a_1| = \max |a_\lambda|$. Then

$$|A| = |a_1| \cdot \left|\omega_1 + \frac{a_2}{a_1}\,\omega_2 + \cdots + \frac{a_r}{a_1}\,\omega_r\right|.$$

Now we cannot have

$$\left|\omega_1 + \frac{a_2}{a_1}\,\omega_2 + \cdots + \frac{a_r}{a_1}\,\omega_r\right| < 1\,,$$

for this would lead to a dependence relation among the $\bar{\omega}$'s over $\bar{F}$. Hence $|A| = |a_1|$. In other words, we have

$$|a_1\omega_1 + \cdots + a_r\omega_r| = \max |a_\lambda| \in |F|\,.$$

Now consider representatives $|\pi_1|, \ldots, |\pi_s| \in |E|$ taken from s different cosets of $|E|$ modulo $|F|$. Here the π_i are chosen in E. We claim that the rs elements $\{\omega_\lambda \pi_\mu\}$ are independent over F. Suppose we have a relation

$$\sum \sum a_{\lambda\mu}\omega_\lambda\pi_\mu = 0\,, \quad a_{\lambda\mu} \in F\,.$$

Write this in the form $\sum A_\mu \pi_\mu = 0$ with $A_\mu = \sum_\lambda a_{\lambda\mu}\omega_\lambda$. We just saw that $|A_\mu| \in |F|$ for any such $A_\mu \neq 0$. Therefore, since the $|\pi_\mu|$ are in different cosets of $|E|$ modulo $|F|$, no two $|A_\mu \pi_\mu|$ can be equal unless they are 0. Applying the Principle of Domination to the equation $\sum A_\mu \pi_\mu = 0$ then shows that all A_μ must be 0. Hence

$$\sum_\lambda a_{\lambda\mu}\omega_\lambda = 0 \quad \text{for all } \mu.$$

But

$$\max_\lambda |a_{\lambda\mu}| = |\sum_\lambda a_{\lambda\mu}\omega_\lambda|\,.$$

Hence all $a_{\lambda\mu} = 0$. Thus the $\{\omega_\lambda\pi_\mu\}$ are independent, as asserted.

This proves that both e and f must be finite. Hence we can take $s = e$ and $r = f$. This proves the proposition. **q. e. d.**

§ 13C. Transcendental prolongation of a valuation

There are two points worth remembering when constructing a valuation on an abstract field. First, if a function $|\ |$ satisfies the first two valuation axioms (V_1) and (V_2) of § 11A, and if

$$|\alpha| \leqq 1 \Rightarrow |1+\alpha| \leqq 1\,,$$

then $|\ |$ must satisfy $(V_{3'})$ and it is therefore a non-archimedean valuation. Secondly, consider a field F which is the quotient field of one of its subrings Z. Let a real valued function $|\ |: Z \rightarrow \mathsf{R}$ be given and suppose that it satisfies the valuation axioms (V_1), (V_2), $(V_{3'})$ of § 11A for elements of Z. Then there is a unique prolongation of this function to a non-archimedean valuation on F. This prolongation is defined as follows: take a typical $\alpha \in F$, write it $\alpha = a/b$ with $a, b \in Z$, and define $|\alpha| = |a|/|b|$. We leave it to the reader to verify that the new function is well-defined, that it is a prolongation of the original function, and that it satisfies the valuation axioms (V_1), (V_2), $(V_{3'})$.

It is always possible to prolongate a valuation from a non-archimedean field F to the field of rational functions $F(x)$, i. e. to the simple transcendental extension $F(x)$ of F. Here is one important way of doing it. For a typical polynomial

$$f(x) = a_0 x^n + a_1 x^{n-1} + \cdots + a_n \in F[x]\,,$$

define

$$|f(x)| = |f| = \max |a_\lambda|\,.$$

We start by verifying the valuation axioms for the ring $F[x]$. To do this we must consider some other typical polynomial

$$g(x) = b_0 x^m + b_1 x^{m-1} + \cdots + b_m \in F[x]\,.$$

Axioms (V_1) and $(V_{3'})$ are immediate. So is the inequality $|f \cdot g| \leqq |f| \cdot |g|$. In order to establish equality here we consider the last coefficient a_i for which $|f| = |a_i|$, and the last b_j for which $|g| = |b_j|$. It follows from the Principle of Domination that the coefficient of x^{i+j} in $f(x)\,g(x)$ has value $|a_i b_j| = |f| \cdot |g|$. Hence $|f \cdot g| = |f| \cdot |g|$ as asserted. Hence the valuation axioms hold in $F[x]$. But $F(x)$ is the quotient field of $F[x]$; so there is a unique prolongation of $|\ |$ from $F[x]$ to $F(x)$ given by $|f/g| = |f|/|g|$. This is the valuation we are after.

We shall call a polynomial with coefficients in the non-archimedean field F an integral polynomial if all its coefficients are integers in F (under the given valuation). Thus

$$f \text{ integral} \Leftrightarrow |f| \leqq 1$$

where $|\ |$ is the preceding prolongation of the valuation on F to $F(x)$.

13:7. *The monic irreducible factors of a monic polynomial with integral coefficients in a non-archimedean field also have integral coefficients.*

Proof. Let $|\ |$ be a valuation determining the given non-archimedean prime spot on F, and prolongate it to the valuation $|\ |$ on $F(x)$. Consider the integral polynomial $f = f(x) \in F[x]$ and let

$$f = p_1 p_2 \ldots p_r, \quad p_i \in F[x],$$

be its factorization into irreducible monic factors. Then $|p_i| \geqq 1$ for all i. And

$$|p_1| \cdot |p_2| \ldots |p_r| = |f| = 1 .$$

Hence all $|p_i| = 1$. Hence all p_i are integral. **q. e. d.**

§ 13D. Complete refinement of an approximation.

In certain problems it is important to have systematic methods for refining approximate equations to precise equality. The first time one encounters this is usually with Newton's method in an elementary calculus course. In fact we have already used this procedure to show that $1 + a$ is a square for sufficiently small a in § 12. Newton's method for roots of polynomials can be carried over to complete non-archimedean fields, but we shall not go into that here. Instead we give an analogous procedure that concerns itself with the factors, not the roots, of polynomials. Refinements of approximations will turn up again in Chapter VI in connection with the arithmetic theory of quadratic forms.

13:8. **Hensel's lemma.** *Let F be a complete non-archimedean field and consider the polynomial $f(x)$ with integral coefficients. Suppose that $f(x)$, when read in the residue class field, factors into two relatively prime polynomials one of which has degree $i > 0$. Then $f(x)$ has a factor of degree i in $F[x]$.*

Proof. 1) Let $\mathfrak{p}$ be the underlying prime spot. Fix a valuation $|\ |$ in $\mathfrak{p}$. Let $(\bar{\ }, \bar{F})$ denote the residue class field of F at $\mathfrak{p}$. For any integral polynomial $g(x)$ let $\bar{g}(x)$, or simply $\bar{g}$, denote the polynomial obtained by reading $g(x)$ in $\bar{F}$, i. e. by barring all coefficients of $g(x)$.

The given factors of $\bar{f}(x)$ can be obtained from integral polynomials over F; in fact we can write

$$\bar{f}(x) = \bar{\varphi}_1(x)\, \bar{\psi}_1(x)$$

where $\varphi_1(x)$ and $\psi_1(x)$ are integral polynomials over F with

$$\deg \varphi_1 = \deg \bar{\varphi}_1 = i, \quad \deg \psi_1 = \deg \bar{\psi}_1,$$

with the leading term of φ_1 a unit, and with

$$\deg \varphi_1 + \deg \psi_1 \leqq \deg f .$$

We then have $\bar{f} = \bar{\varphi}_1 \bar{\psi}_1 = \overline{\varphi_1 \psi_1}$ and, since $\bar{\varphi}_1$ and $\bar{\psi}_1$ are relatively prime, there are integral polynomials A, B such that $\bar{A}\bar{\varphi}_1 + \bar{B}\bar{\psi}_1 = 1$. If we now read this last sentence back in F we get

$$|f-\varphi_1\psi_1|<1\,,\quad |A\varphi_1+B\psi_1-1|<1\,,$$

where the valuation in question is the valuation on $F(x)$ that was introduced in § 11C. We therefore have a field element π such that

$$\left.\begin{array}{l}|f-\varphi_1\psi_1|\leqq|\pi|<1\\ |A\varphi_1+B\psi_1-1|\leqq|\pi|<1\,.\end{array}\right\}\tag{1}$$

2) So far we have just interpreted the given conditions as an approximation $\varphi_1\psi_1$ to f. The idea of the proof is to keep finding polynomials φ and ψ which give better and better approximations to f, and then to obtain a precise factorization by taking the limits of the approximating polynomials. The completeness of the field ensures that the limits in question do indeed exist. With this in mind let us consider integral polynomials φ and ψ such that

$$\left.\begin{array}{l}|f-\varphi\psi|\leqq|\pi|^n\quad(n\in\mathsf{N})\\ |\varphi-\varphi_1|\leqq|\pi|\,,\quad|\psi-\psi_1|\leqq|\pi|\\ \deg\varphi=i\,,\quad\deg\varphi+\deg\psi\leqq\deg f\end{array}\right\}\tag{2}$$

We show how to sharpen φ and ψ, i. e. we find integral polynomials φ' and ψ' such that

$$\left.\begin{array}{l}|f-\varphi'\psi'|\leqq|\pi|^{n+1}\\ |\varphi'-\varphi_1|\leqq|\pi|\,,\quad|\psi'-\psi_1|\leqq|\pi|\\ \deg\varphi'=i\,,\quad\deg\varphi'+\deg\psi'\leqq\deg f\,.\end{array}\right\}\tag{3}$$

And the sharpening will be done in such a way that

$$|\varphi'-\varphi|\leqq|\pi|^n,\quad|\psi'-\psi|\leqq|\pi|^n\,.\tag{4}$$

By equations (1) and (2) we can find integral polynomials C and D with

$$A\varphi+B\psi=1+\pi C\,,\quad f=\varphi\psi+\pi^nD\,.$$

Multiplying the first of these by π^nD and substituting in the second gives

$$f=\varphi\psi+\pi^nDA\varphi+\pi^nDB\psi-\pi^{n+1}DC\,.\tag{5}$$

By equation (2) the leading coefficient of φ must be a unit. Therefore if we divide φ into any integral polynomial we obtain an integral quotient and remainder. Hence we can find integral polynomials β,σ,γ,τ such that

$$DB=\beta\varphi+\sigma\,,\quad DC=\gamma\varphi+\tau\,,$$

where $\deg\sigma<\deg\varphi$ and $\deg\tau<\deg\varphi$. Then a computation gives

$$\left.\begin{array}{l}f=\varphi\psi+\pi^nE\varphi+\pi^n\sigma\psi-\pi^{n+1}\tau\\ \quad=(\varphi+\pi^n\sigma)(\psi+\pi^nE)-\pi^{n+1}(\tau+\pi^{n-1}\sigma E)\end{array}\right\}\tag{6}$$

where $E = DA + \beta\,\psi - \pi\gamma$. If we now put $\varphi' = \varphi + \pi^n\sigma$ and $\psi' = \psi + \pi^n E$ we have integral polynomials φ', ψ' which satisfy equation (4) and the first part of equation (3). The second part of equation (3) is immediate. Now $\deg \varphi' = \deg \varphi$ since $\deg \sigma < \deg \varphi$. By inspecting every term in the sum in equation (6) except $\pi^n E\,\varphi$ we see that this exceptional term must itself have degree $\leqq \deg f$. Hence the last part of equation (3) is proved.

3) In step 2) we have given the refinement procedure to be used at each stage of the approximation. We must now apply it all the way. Starting with φ_1, ψ_1 we successively improve our estimates and we obtain two sequences of integral polynomials $\{\varphi_n\}$ and $\{\psi_n\}$ by the preceding process. These polynomials will satisfy the following inequalities:

$$|f - \varphi_n\,\psi_n| \leqq |\pi|^n$$

$$|\varphi_n - \varphi_1| \leqq |\pi|\,, \quad |\psi_n - \psi_1| \leqq |\pi|\,,$$

$$\deg \varphi_n = i\,, \quad \deg \varphi_n + \deg \psi_n \leqq \deg f\,,$$

$$|\varphi_{n+1} - \varphi_n| \leqq |\pi|^n, \quad |\psi_{n+1} - \psi_n| \leqq |\pi|^n.$$

Hence the two sequences are Cauchy sequences of polynomials; and their degrees are bounded; from which it readily follows, in virtue of the completeness of F, that both sequences converge to polynomials in $F[x]$. We define

$$\Phi = \lim \varphi_n\,, \quad \Psi = \lim \psi_n\,.$$

We have $\deg \Phi \leqq i$ since $\deg \varphi_n = i$ for all n. On the other hand $|\varphi_1 - \varphi_n| < 1$ and, for large n, $|\varphi_n - \Phi| < 1$. Hence $|\varphi_1 - \Phi| < 1$. But the leading coefficient of φ_1 is a unit. So we get $\deg \Phi = \deg \varphi_1 = i$. Finally

$$|f - \Phi\,\Psi| = |\lim\,(f - \varphi_n\,\psi_n)| = \lim |f - \varphi_n\,\psi_n| = 0\,.$$

So $f = \Phi\,\Psi$. We have the desired factorization. **q. e. d.**

13:9. **Reducibility Criterion.** *The polynomial* $f(x) = a_0 x^n + a_1 x^{n-1} + \cdots + a_n$ *over a field with a complete non-archimedean valuation is reducible if* $0 < |a_0| < |f| = |a_i|$ *holds for some* $i < n$.

Proof. By scaling f we can assume that $|f| = 1$. We choose the smallest i for which $|a_i| = |f| = 1$. So $0 < i < n$. Put $\varphi(x) = a_i x^{n-i} + \cdots + a_n$ and let $\psi(x)$ be the polynomial identically 1. In the residue class field we have $\bar{f} = \bar{\varphi}\,\bar{\psi}$ with $\deg \bar{\varphi} = n - i$. But $\bar{\varphi}$ and $\bar{\psi}$ are relatively prime. Hence f has a factor of degree $n - i$. So f is reducible. **q. e. d.**

§ 14. Prolongation of a complete valuation to a finite extension

14:1. **Theorem.** *Let E/F be a finite extension of degree n of a field F with a complete valuation $|\ |$. Then there is a unique prolongation of the valua-*

ation from F to E and it is defined by the formula

$$|\alpha| = |N_{E/F}\,\alpha|^{1/n} \text{ for all } \alpha \in E\,.$$

The prolongation is complete.

Proof. Let us use N for the norm $N_{E/F}$. We know from Theorem 11:18 that the prolongation, if it exists, is unique and complete. Furthermore the function $|\alpha| = |N\alpha|^{1/n}$, if it is a valuation at all, will be a prolongation of the original valuation since $N\alpha = \alpha^n$ for all $\alpha \in F$. It remains for us to prove that this new function is indeed a valuation.

First the archimedean case. We can assume that $[E:F] > 1$. Since the given valuation is complete and archimedean there is a topological isomorphism of F onto either $\mathbf{R}$ or $\mathbf{C}$ by Theorem 12:4, hence onto $\mathbf{R}$; so $E = F(\sqrt{-1})$. But in the proof of Lemma 12:3 we showed that $|N\alpha|^{1/2}$ prolonged the valuation from F to E. Hence the archimedean case is established.

Therefore assume that the given valuation is non-archimedean[1]. Only the strong triangle law really needs proof, and in fact we just have to show that

$$|N\alpha| \leqq 1 \Rightarrow |N(1+\alpha)| \leqq 1\,.$$

Let

$$f(x) = x^m + a_1 x^{m-1} + \cdots + a_m = \operatorname{irr}(x, \alpha, F)\,.$$

Now $N\alpha$ is a certain power of $N_{F(\alpha)/F}\,\alpha$, hence

$$|a_m| = |N_{F(\alpha)/F}\,\alpha| \leqq 1\,.$$

If we now apply the Reducibility Criterion of § 13D to the irreducible polynomial $f(x)$ over F we find that

$$|a_\lambda| \leqq 1 \quad \text{for} \quad 1 \leqq \lambda \leqq m\,.$$

But $f(x-1) = \operatorname{irr}(x, 1+\alpha, F)$, and this polynomial in x has constant term

$$\pm (1 + \textstyle\sum (\pm a_\lambda))\,.$$

Since the strong triangle law holds in F we have $|1 + \sum (\pm a_\lambda)| \leqq 1$. But $N(1+\alpha)$ is a power of the above term. Hence $|N(1+\alpha)| \leqq 1$, as required. **q. e. d.**

---

[1] Here we use the classical method of deriving the prolongation theorem from Hensel's lemma. It is also possible to obtain this result, in fact a more general result, in an entirely different way. One introduces three new equivalent concepts, general valuations, general valuation rings, and places, then one uses Zorn's lemma to prove a prolongation theorem for places from which one obtains a prolongation theorem for general valuations, and finally one proves that the prolongation of an ordinary valuation is an ordinary valuation and that it satisfies the formula of the theorem. Hensel's lemma can then be derived from the prolongation theorem. For a detailed account of this method see G. WHAPLES, *Class field theory*, (University of Indiana lectures, 1959).

§ 15. Prolongation of any valuation to a finite separable extension

In this paragraph we consider a finite separable extension E/F of degree n, and a prime spot $\mathfrak{p}$ on F. The spot in question can be quite arbitrary, but the extension must be separable. For inseparable extensions the results involve more technicalities, and are fortunately unnecessary for our subsequent use. We prove that $\mathfrak{p}$ is divisible by at least one and at most n spots on E, and we investigate the behavior of the degree and the norm at all the $\mathfrak{P}$ dividing $\mathfrak{p}$.

§ 15 A. Construction and notation

As we have just said, E/F is a separable extension of degree n and $\mathfrak{p}$ is a prime spot on F. This is fixed for the entire paragraph. We let T be the set of prime spots on E which divide $\mathfrak{p}$. Since E/F is separable there will be a primitive element δ such that $E = F(\delta)$; we put

$$f(x) = \operatorname{irr}(x, \delta, F)\ .$$

We fix a completion $F_\mathfrak{p}$ of F at $\mathfrak{p}$; as usual, $\mathfrak{p}$ will be used to denote the prime spot on $F_\mathfrak{p}$. We define $E_\mathfrak{p}$ to be some fixed splitting field of $f(x)$ over $F_\mathfrak{p}$. Since $F_\mathfrak{p}$ is complete we know from Theorems 14:1 and 11:18 that there is exactly one spot $\mathfrak{P}_0$ on $E_\mathfrak{p}$ which divides $\mathfrak{p}$ on $F_\mathfrak{p}$.

Since E/F is separable we know from field theory that there are exactly n distinct isomorphisms of E into $E_\mathfrak{p}$ which are the identity on F. We let Σ denote the set of these isomorphisms. For each $\sigma \in \Sigma$ the isomorphism $\sigma^{-1}\colon \sigma E \rightarrowtail E$ will carry the spot $\mathfrak{P}_0$ on σE back to E; in § 11D we agreed to write this spot on E in the form $\mathfrak{P}_0^{\sigma^{-1}}$. Clearly

$$\mathfrak{P}_0^{\sigma^{-1}} \in T \qquad\qquad \forall\, \sigma \in \Sigma,$$

since $\mathfrak{P}_0^{\sigma^{-1}} \mid \mathfrak{p}^{\sigma^{-1}}$ on E/F and $\mathfrak{p}^{\sigma^{-1}} = \mathfrak{p}$. In particular T is not empty: *every spot on F is divisible by at least one spot on E.*

For each $\mathfrak{P} \in T$ we fix a completion $E_\mathfrak{P}$ of E at $\mathfrak{P}$; as usual, we let $\mathfrak{P}$ denote the corresponding spot on $E_\mathfrak{P}$. We take the completion $F_\mathfrak{P}$ of F at $\mathfrak{p}$ that is contained in $E_\mathfrak{P}$. Then there is a unique topological isomorphism of $F_\mathfrak{P}$ onto $F_\mathfrak{p}$ which is the identity on F. We shall need this map so we agree to denote it by $I_\mathfrak{P}$. Proposition 11:19 informs us that $E_\mathfrak{P} = EF_\mathfrak{P} = F_\mathfrak{P}(\delta)$.

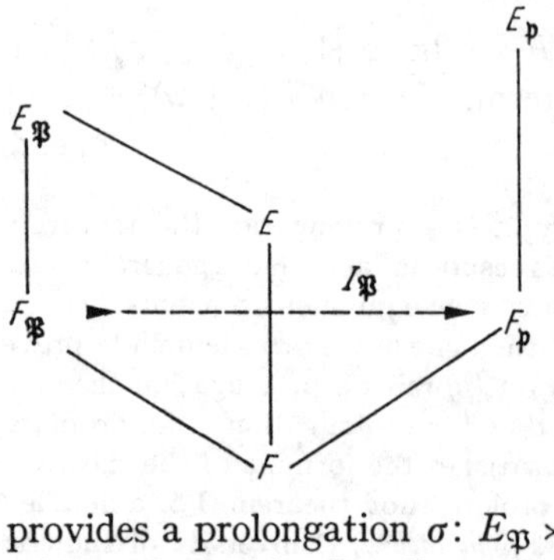

The map $I_\mathfrak{P}$ can be used to prove that *every $\mathfrak{P} \in T$ has the form $\mathfrak{P}_0^{\sigma^{-1}}$ for some $\sigma \in \Sigma$.* Namely, $I_\mathfrak{P}\colon F_\mathfrak{P} \rightarrowtail F_\mathfrak{p}$ is an isomorphism; so field theory provides a prolongation $\sigma\colon E_\mathfrak{P} \rightarrowtail E_\mathfrak{p}$ of $I_\mathfrak{P}$. But $I_\mathfrak{P}$ is topological, hence σ is topological by Corollary 11:18a. Restrict σ to E. Then $\sigma^{-1}\colon \sigma E \rightarrowtail E$

is a topological isomorphism. Hence $\mathfrak{P} = \mathfrak{P}_0^{\sigma^{-1}}$ by Proposition 11:6. This proves the assertion. Incidentally the rule $\sigma \rightarrowtail \mathfrak{P}_0^{\sigma^{-1}}$ provides a surjection

$$\Sigma \twoheadrightarrow T .$$

In particular *there are at most n spots on E which divide* $\mathfrak{p}$.

Suppose $\sigma \in \Sigma$ determines the spot $\mathfrak{P} = \mathfrak{P}_0^{\sigma^{-1}}$. Then $\sigma: E \rightarrowtail \sigma E$ is a topological isomorphism under the $\mathfrak{P}$-adic topology on E. Hence by Proposition 11:14 there is a unique prolongation of σ to a topological isomorphism of $E_{\mathfrak{P}}$ into $E_{\mathfrak{p}}$. We shall denote this prolongation by $\sigma_{\mathfrak{P}}$. Now σ is the identity on F; hence by Proposition 11:14 again we have

$$\sigma_{\mathfrak{P}} = I_{\mathfrak{P}} \quad \text{on} \quad F_{\mathfrak{P}} .$$

15:1. **Example.** If $(\sigma E)_{\mathfrak{P}_0}$ denotes the closure of σE in $E_{\mathfrak{p}}$, then

$$(\sigma E)_{\mathfrak{P}_0} = \sigma_{\mathfrak{P}} E_{\mathfrak{P}} = (F_{\mathfrak{p}})(\sigma E) = F_{\mathfrak{p}}(\sigma \delta) .$$

§ 15B. Local degrees and local norms

15:2. $\mathfrak{P}_0^{\sigma^{-1}} = \mathfrak{P}_0^{\tau^{-1}}$ *if and only if* $\sigma\delta$ *and* $\tau\delta$ *are conjugate over* $F_{\mathfrak{p}}$.

Proof. 1) First the necessity. Put $\mathfrak{P} = \mathfrak{P}_0^{\sigma^{-1}} = \mathfrak{P}_0^{\tau^{-1}}$. The topological isomorphisms

$$\sigma_{\mathfrak{P}}: E_{\mathfrak{P}} \rightarrowtail \sigma_{\mathfrak{P}} E_{\mathfrak{P}} , \quad \tau_{\mathfrak{P}}: E_{\mathfrak{P}} \rightarrowtail \tau_{\mathfrak{P}} E_{\mathfrak{P}}$$

are both equal to $I_{\mathfrak{P}}$ on $F_{\mathfrak{P}}$, hence $\tau_{\mathfrak{P}} \sigma_{\mathfrak{P}}^{-1}$ is the identity on $F_{\mathfrak{p}}$. But

$$\tau_{\mathfrak{P}} \sigma_{\mathfrak{P}}^{-1}(\sigma\delta) = \tau_{\mathfrak{P}}(\delta) = \tau(\delta) .$$

Hence $\sigma\delta$ and $\tau\delta$ are conjugate over $F_{\mathfrak{p}}$.

2) Now the sufficiency. Let us put $\mathfrak{P} = \mathfrak{P}_0^{\sigma^{-1}}$ and $\mathfrak{P}' = \mathfrak{P}_0^{\tau^{-1}}$. Then

$$\sigma_{\mathfrak{P}} E_{\mathfrak{P}} = F_{\mathfrak{p}}(\sigma\delta) , \quad \tau_{\mathfrak{P}'} E_{\mathfrak{P}'} = F_{\mathfrak{p}}(\tau\delta) .$$

Since $\sigma\delta$ and $\tau\delta$ are conjugate over $F_{\mathfrak{p}}$ we can find an abstract isomorphism

$$\varphi: \sigma_{\mathfrak{P}} E_{\mathfrak{P}} \rightarrowtail \tau_{\mathfrak{P}'} E_{\mathfrak{P}'},$$

which is the identity on $F_{\mathfrak{p}}$ and carries $\sigma\delta$ to $\tau\delta$. Since $F_{\mathfrak{p}}$ is complete, φ will be topological by Corollary 11:18a. Now a typical $\alpha \in E$ has the form

$$\alpha = a_0 + a_1\delta + \cdots + a_{n-1}\delta^{n-1}, \quad a_i \in F ,$$

from which it follows that

$$\varphi(\sigma_{\mathfrak{P}}\alpha) = \tau_{\mathfrak{P}'}(\alpha) \quad \forall \alpha \in E .$$

Hence $\tau_{\mathfrak{P}'}^{-1} \varphi \sigma_{\mathfrak{P}}$ is a topological isomorphism of $E_{\mathfrak{P}}$ onto $E_{\mathfrak{P}'}$ which is the identity on E; but this means that the $\mathfrak{P}$-adic and $\mathfrak{P}'$-adic topologies are equal on E. Hence $\mathfrak{P} = \mathfrak{P}'$. **q. e. d.**

Consider our extension E/F with spots $\mathfrak{P}|\mathfrak{p}$. By the local degree of the extension at $\mathfrak{P}$ (or at $\mathfrak{P}|\mathfrak{p}$) we mean the field degree

$$n(\mathfrak{P}|\mathfrak{p}) = [E_{\mathfrak{P}}: F_{\mathfrak{P}}] .$$

This quantity clearly depends on $\mathfrak{P}$, but it does not depend on the particular completion $E_{\mathfrak{P}}$ that is chosen at the given $\mathfrak{P}$.

By the local norm at $\mathfrak{P}|\mathfrak{p}$ we mean the multiplicative homomorphism

$$N_{\mathfrak{P}|\mathfrak{p}}: E_{\mathfrak{P}} \rightarrowtail F_{\mathfrak{p}}$$

that is defined by the equation

$$N_{\mathfrak{P}|\mathfrak{p}}\,\alpha = I_{\mathfrak{P}}(N_{E_{\mathfrak{P}}/F_{\mathfrak{P}}}\alpha) \quad \forall\, \alpha \in E_{\mathfrak{P}}.$$

Similarly we define the local trace at $\mathfrak{P}|\mathfrak{p}$ to be the additive homomorphism

$$S_{\mathfrak{P}|\mathfrak{p}}: E_{\mathfrak{P}} \rightarrowtail F_{\mathfrak{p}}$$

that is defined by the equation

$$S_{\mathfrak{P}|\mathfrak{p}}\,\alpha = I_{\mathfrak{P}}\,(S_{E_{\mathfrak{P}}/F_{\mathfrak{P}}}\alpha) \quad \forall\, \alpha \in E_{\mathfrak{P}}\,.$$

Clearly the local norm and the local trace depend on $E_{\mathfrak{P}}$ and $F_{\mathfrak{p}}$. The local trace is just mentioned in passing and will not be used in the sequel.

15:3. **Theorem.** *Let E/F be a finite separable extension of degree n, and let $\mathfrak{p}$ be a spot on F. Then for all α in E we have the formulas*

(1) $\sum_{\mathfrak{P}|\mathfrak{p}} n(\mathfrak{P}|\mathfrak{p}) = n$

(2) $\prod_{\mathfrak{P}|\mathfrak{p}} N_{\mathfrak{P}|\mathfrak{p}}\,\alpha = N_{E/F}\,\alpha$

(3) $\sum_{\mathfrak{P}|\mathfrak{p}} S_{\mathfrak{P}|\mathfrak{p}}\,\alpha = S_{E/F}\,\alpha\,.$

Proof. Define an equivalence relation on Σ by saying $\sigma \sim \tau$ if $\mathfrak{P}_0^{\sigma^{-1}} = \mathfrak{P}_0^{\tau^{-1}}$. Thus $\sigma \sim \tau$ if and only if $\sigma\delta$ and $\tau\delta$ are conjugate over $F_{\mathfrak{p}}$. Let $\bar{\sigma}$ denote the coset of $\sigma \in \Sigma$ under this equivalence relation, and let $\#(\sigma)$ denote the number of elements in the coset $\bar{\sigma}$. Let P be the subset of Σ that is obtained by picking exactly one representative from each coset in Σ.

By definition $\#(\sigma)$ is the number of $\tau \in \Sigma$ for which $\tau\delta$ and $\sigma\delta$ are conjugate over $F_{\mathfrak{p}}$; and this number is equal to the number of conjugates of $\sigma\delta$ over $F_{\mathfrak{p}}$; hence

$$\#(\sigma) = [F_{\mathfrak{p}}(\sigma\delta): F_{\mathfrak{p}}] = n\,(\mathfrak{P}_0^{\sigma^{-1}}|\mathfrak{p})\,.$$

These are the preliminaries to the proof. The three formulas now follow quickly.

(1) First the formula for the local degrees:

$$\sum_{\mathfrak{P}\in T} n(\mathfrak{P}|\mathfrak{p}) = \sum_{\sigma\in P} n\,(\mathfrak{P}_0^{\sigma^{-1}}|\mathfrak{p}) = \sum_{\sigma\in P} \#(\sigma) = n\,.$$

(2) Next the local norms. Consider $\sigma \in P$, $\tau \in \bar{\sigma}$. Put $\mathfrak{P} = \mathfrak{P}_0^{\sigma^{-1}} = \mathfrak{P}_0^{\tau^{-1}}$. Then as τ runs through $\bar{\sigma}$ we obtain $\#(\sigma)$ distinct isomorphisms $\tau_{\mathfrak{P}}\,\sigma_{\mathfrak{P}}^{-1}$

of $F_{\mathfrak{p}}(\sigma\delta)$ into $E_{\mathfrak{p}}$ which are identity on $F_{\mathfrak{p}}$; since $\#(\sigma) = [F_{\mathfrak{p}}(\sigma\delta) : F_{\mathfrak{p}}]$ we must therefore have all such isomorphisms. So for each $\alpha \in E$,

$$\begin{aligned} N_{\mathfrak{P}|\mathfrak{p}}\,\alpha &= I_{\mathfrak{P}}(N_{E_{\mathfrak{P}}/F_{\mathfrak{P}}}\,\alpha) \\ &= \sigma_{\mathfrak{P}}(N_{E_{\mathfrak{P}}/F_{\mathfrak{P}}}\,\alpha) \\ &= N_{F_{\mathfrak{p}}(\sigma\delta)/F_{\mathfrak{p}}}\,(\sigma_{\mathfrak{P}}\,\alpha) \\ &= \prod_{\tau\in\bar\sigma} (\tau_{\mathfrak{P}}\sigma_{\mathfrak{P}}^{-1})\,(\sigma_{\mathfrak{P}}\,\alpha) \\ &= \prod_{\tau\in\bar\sigma} \tau\alpha\,. \end{aligned}$$

Hence

$$\prod_{\mathfrak{P}|\mathfrak{p}} N_{\mathfrak{P}|\mathfrak{p}}\,\alpha = \prod_{\sigma\in P}\Bigl(\prod_{\tau\in\bar\sigma}\tau\alpha\Bigr) = \prod_{\tau\in\Sigma}\tau\alpha = N_{E/F}\,\alpha\,.$$

(3) The result for the local trace follows just as it did for the local norm. In fact, one need only replace $\prod$ by $\sum$ and N by S throughout step (2). **q. e. d.**

15:4. *Let E/F be a finite separable extension and let $\mathfrak{p}$ be a spot on F. Consider the $\mathfrak{p}$-adic topology on F, and a topology on E which is finer than all $\mathfrak{P}$-adic topologies for all $\mathfrak{P}|\mathfrak{p}$. Then*

$$N_{E/F} : E \rightarrow F$$

is continuous.

Proof. For each $\sigma \in \Sigma$ the mapping $\sigma : E \rightarrow E_{\mathfrak{p}}$ is continuous under the $\mathfrak{P}_0^{\sigma^{-1}}$-adic topology on E, hence under the given topology on E. Hence the mapping $\alpha \rightarrow (\sigma\alpha)_{\sigma\in\Sigma}$ of E into n-space $E_{\mathfrak{p}} \times \cdots \times E_{\mathfrak{p}}$ is continuous; but the multiplication map of $E_{\mathfrak{p}} \times \cdots \times E_{\mathfrak{p}}$ into $E_{\mathfrak{p}}$ is continuous since $E_{\mathfrak{p}}$ is a topological field; hence

$$\alpha \rightarrow \prod_{\sigma\in\Sigma} \sigma\alpha = N_{E/F}\,\alpha$$

is continuous. **q. e. d.**

15:4a. *Let α be an element of E. Given $\varepsilon > 0$ there is a $\delta > 0$ such that $|N_{E/F}x - N_{E/F}\alpha|_{\mathfrak{p}} < \varepsilon$ holds whenever $x \in E$ satisfies $|x-\alpha|_{\mathfrak{P}} < \delta$ for all $\mathfrak{P}|\mathfrak{p}$. (Here $|\ |_{\mathfrak{p}}$ and $|\ |_{\mathfrak{P}}$ are valuations in $\mathfrak{p}$ and $\mathfrak{P}$ respectively.)*

Proof. This follows from the proposition by considering the topology on E that is defined by the new distance function

$$d(x,y) = \max_{\mathfrak{P}|\mathfrak{p}} |x-y|_{\mathfrak{P}}\,.$$

q. e. d.

15:5. *Let E/F be a finite extension and let $\mathfrak{p}$ be a real spot on F. Suppose there are exactly r ($r \geqq 0$) real spots $\mathfrak{P}$ on E which divide $\mathfrak{p}$ and at which a given $\alpha \in \dot{E}$ is negative. Then $(-1)^r N_{E/F}\,\alpha$ is positive at $\mathfrak{p}$.*

Proof. Let P be the real spots on E which divide $\mathfrak{p}$ and at which α is positive, let N be those at which α is negative, and let C be the complex spots dividing $\mathfrak{p}$. Then

$$N_{\mathfrak{P}|\mathfrak{p}}\,\alpha \in N_{\mathfrak{P}|\mathfrak{p}}\,E_{\mathfrak{P}}^2 \subseteq F_{\mathfrak{p}}^2 \quad \forall\, \mathfrak{P} \in P \cup C\,.$$

If $\mathfrak{P} \in N$, then $\mathfrak{P}$ is real so that $E_{\mathfrak{P}} = F_{\mathfrak{P}}$, hence

$$N_{\mathfrak{P}|\mathfrak{p}}\,\alpha \in N_{\mathfrak{P}|\mathfrak{p}}(-E_{\mathfrak{P}}^2) \subseteq -F_{\mathfrak{p}}^2 \quad \forall\, \mathfrak{P} \in N\,.$$

Hence

$$N_{E/F}\,\alpha = \prod_{\mathfrak{P}|\mathfrak{p}} N_{\mathfrak{P}|\mathfrak{p}}\,\alpha \in (-1)^r F_{\mathfrak{p}}^2\,.$$

q. e. d.

15:6. **Example.** Take a valuation $|\ |_{\mathfrak{p}}$ in $\mathfrak{p}$ and a valuation $|\ |_{\mathfrak{P}}$ at each $\mathfrak{P}$ dividing $\mathfrak{p}$. Consider $\alpha \in E$ such that

$$|\alpha|_{\mathfrak{P}} \leqq 1 \quad \forall\, \mathfrak{P}\,|\,\mathfrak{p}\,.$$

Then it follows easily from the norm formula of Theorem **15**:3 that

$$|N_{E/F}\,\alpha|_{\mathfrak{p}} \leqq 1$$

and that

$$|N_{E/F}\,\alpha|_{\mathfrak{p}} = 1 \quad \text{if and only if} \quad |\alpha|_{\mathfrak{P}} = 1 \quad \forall\, \mathfrak{P}\,|\,\mathfrak{p}\,.$$

15:7. **Example.** Recall that $f(x) = \operatorname{irr}(x, \delta, F)$. Let

$$f(x) = f_1(x) \cdot f_2(x) \ldots f_r(x)$$

be a factorization of $f(x)$ into irreducible factors over $F_{\mathfrak{p}}$; all these factors are distinct because of the general assumption of separability. Take exactly one root δ_i of $f_i(x)$ in $E_{\mathfrak{p}}$ for each i $(1 \leqq i \leqq r)$. Let σ_i be the isomorphism of $E = F(\delta)$ onto $F(\delta_i)$ which is identity on F and carries δ to δ_i. Then $\sigma_i\delta$ and $\sigma_j\delta$ are conjugate over $F_{\mathfrak{p}}$ if and only if $i = j$. Moreover $\sigma\delta$ is conjugate to some $\sigma_i\delta$ over $F_{\mathfrak{p}}$ for every $\sigma \in \Sigma$. Hence

$$\mathfrak{P}_0^{\sigma_i^{-1}} \quad (1 \leqq i \leqq r)$$

are all the distinct prime spots on E which divide the given $\mathfrak{p}$. And the local degree at the spot corresponding to σ_i is equal to

$$[F_{\mathfrak{p}}(\delta_i) : F_{\mathfrak{p}}] = \deg f_i\,.$$

We have therefore established a correspondence

$$f_i \leftrightarrow \delta_i \leftrightarrow \sigma_i \leftrightarrow \mathfrak{P}_0^{\sigma_i^{-1}}$$

between the irreducible factors of $f(x)$ over $F_{\mathfrak{p}}$ and the spots on E dividing a given $\mathfrak{p}$, in which the degree of an irreducible factor is equal to the local degree at the corresponding spot.

15:8. **Example.** Suppose F is the field of rational numbers $\mathbf{Q}$ and $E = \mathbf{Q}(\sqrt{a})$ with $a \in \mathbf{Q}$. Consider the archimedean spot $\mathfrak{p}$ on $\mathbf{Q}$. If $a > 0$ with $a \notin \mathbf{Q}^2$ there are two spots $\mathfrak{P} | \mathfrak{p}$ on E, both of local degree 1. If $a < 0$ there is just one spot, of local degree 2.

15:9. **Example.** Suppose $F = \mathbf{Q}$ and $E = \mathbf{Q}(\sqrt[3]{2})$, and again consider the archimedean spot $\mathfrak{p}$ on $\mathbf{Q}$. Then $x^3 - 2$ has the irreducible decomposition

$$x^3 - 2 = \left(x - \sqrt[3]{2}\right)\left(x^2 + \sqrt[3]{2}\, x + \sqrt[3]{4}\right)$$

over $\mathsf{R} = \mathbf{Q}_\mathfrak{p}$. Hence there are two spots on E which divide $\mathfrak{p}$, the one of local degree 1, the other of local degree 2. Incidentally the first of these spots must be real and the second complex.

§ 15 C. The decomposition field

The situation that we are working with throughout § 15 is that of a finite separable extension E/F of degree n. We shall now introduce the concept of a decomposition field under the additional assumption that the extension in question is abelian. (A finite extension is called abelian if it is a galois extension with abelian galois group; a galois extension is one that is normal and separable.) We therefore assume that E/F is abelian and we let $\mathfrak{G} = \mathfrak{G}(E/F)$ denote its galois group.

We still have the same fixed prime spot $\mathfrak{p}$ on F. Let $\mathfrak{P}$ be any spot on E which divides $\mathfrak{p}$. Then $\mathfrak{P}^\sigma$ has been defined for all $\sigma \in \mathfrak{G}$: it is that unique spot on E which makes the automorphism σ of E (under $\mathfrak{P}$) onto E (under $\mathfrak{P}^\sigma$) topological. We easily see that

$$\mathfrak{P}^\sigma | \mathfrak{p} \quad \text{for all} \quad \sigma \in \mathfrak{G}$$

and that

$$\mathfrak{P}^{\sigma\tau} = (\mathfrak{P}^\tau)^\sigma \quad \text{for all} \quad \sigma, \tau \in \mathfrak{G}\,.$$

Consider the completions $F_\mathfrak{P} \subseteq E_\mathfrak{P}$. Then $E_\mathfrak{P} = F_\mathfrak{P} E = F_\mathfrak{P}(\delta)$ is a splitting field of $f(x)$ over $F_\mathfrak{P}$ since E/F is normal. And $F_\mathfrak{P}$ is a completion of F at $\mathfrak{p}$. Hence we can regard $E_\mathfrak{P}$ as the $E_\mathfrak{p}$ and $\mathfrak{P}$ as the $\mathfrak{P}_0$ of our discussion to date. In particular we have the following facts which we shall put to immediate use: if $\mathfrak{P}$ is any spot dividing $\mathfrak{p}$, then every spot dividing $\mathfrak{p}$ is of the form $\mathfrak{P}^\sigma$ for some $\sigma \in \mathfrak{G}$; and $\mathfrak{P}^\sigma = \mathfrak{P}^\tau$ if and only $\sigma^{-1}\delta$ and $\tau^{-1}\delta$ are conjugate over $F_\mathfrak{P}$.

We define the decomposition group of E/F at $\mathfrak{p}$ to be the subgroup

$$\mathfrak{Z} = \{\sigma \in \mathfrak{G} \,|\, \mathfrak{P}^\sigma = \mathfrak{P}\}$$

of $\mathfrak{G}$. The decomposition group at $\mathfrak{p}$ depends only on $\mathfrak{p}$, not on $\mathfrak{P}$. For if we start with some other spot on E which divides $\mathfrak{p}$, say with the spot $\mathfrak{P}^\tau$ where τ is in $\mathfrak{G}$, we obtain

$$\mathfrak{Z}' = \{\sigma \in \mathfrak{G} \,|\, (\mathfrak{P}^\tau)^\sigma = \mathfrak{P}^\tau\}$$

and it is easily seen, since $\mathfrak{G}$ is abelian, that this group is equal to $\mathfrak{Z}$. We define the decomposition field of E/F at $\mathfrak{p}$ to be the fixed field of the group $\mathfrak{Z}$.

15:10. *Let E/F be an abelian extension with spots $\mathfrak{P}|\mathfrak{p}$ and $\mathfrak{P}'|\mathfrak{p}$, and decomposition field Z at $\mathfrak{p}$. Then*

(1) $n(\mathfrak{P}|\mathfrak{p}) = [E:Z] = n(\mathfrak{P}'|\mathfrak{p})$,

(2) $Z_{\mathfrak{P}} = F_{\mathfrak{P}}$, $Z = F_{\mathfrak{P}} \cap E$,

(3) $e(\mathfrak{P}|\mathfrak{p}) = e(\mathfrak{P}'|\mathfrak{p})$, $f(\mathfrak{P}|\mathfrak{p}) = f(\mathfrak{P}'|\mathfrak{p})$ *if non-archimedean.*

Proof. (1) The local degree $n(\mathfrak{P}|\mathfrak{p})$ is $[F_{\mathfrak{P}}(\delta):F_{\mathfrak{P}}]$ and this is the number of conjugates of δ over $F_{\mathfrak{P}}$; this is clearly equal to the number of $\sigma \in \mathfrak{G}$ for which $\sigma\delta$ and δ are conjugate over $F_{\mathfrak{P}}$. But σ is in $\mathfrak{Z}$ if and only if σ^{-1} is in $\mathfrak{Z}$, hence if and only if $\sigma\delta$ and δ are conjugate over $F_{\mathfrak{P}}$. Hence the order of $\mathfrak{Z}$ is equal to $n(\mathfrak{P}|\mathfrak{p})$, hence so is $[E:Z]$. Hence $n(\mathfrak{P}|\mathfrak{p}) = n(\mathfrak{P}'|\mathfrak{p})$.

(2) Consider the extension E/Z with its galois group $\mathfrak{Z}$. The decomposition group of this at the spot induced by $\mathfrak{P}$ on Z is simply $\mathfrak{Z}$. Hence by the first part of this proposition, $[E_{\mathfrak{P}}:Z_{\mathfrak{P}}] = [E:Z]$. But $F_{\mathfrak{P}} \subseteq Z_{\mathfrak{P}}$ has $[E_{\mathfrak{P}}:F_{\mathfrak{P}}] = [E:Z]$. Hence $F_{\mathfrak{P}} = Z_{\mathfrak{P}}$.

Since $Z_{\mathfrak{P}} = F_{\mathfrak{P}}$ we must have $Z \subseteq F_{\mathfrak{P}} \cap E$. But $[E:F_{\mathfrak{P}} \cap E] \geqq [E_{\mathfrak{P}}:(F_{\mathfrak{P}} \cap E)_{\mathfrak{P}}] = [E_{\mathfrak{P}}:F_{\mathfrak{P}}] = [E:Z]$. Hence $F_{\mathfrak{P}} \cap E = Z$.

(3) Now $\mathfrak{p}$ becomes non-archimedean. Write $\mathfrak{P}' = \mathfrak{P}^{\tau}$ with a suitable $\tau \in \mathfrak{G}$. First let us do the ramification index. Fix $|\ |_{\mathfrak{P}} \in \mathfrak{P}$. Then using the definition of $\mathfrak{P}^{\tau}$ along with Proposition 11:6 we can find $|\ |_{\mathfrak{P}^{\tau}} \in \mathfrak{P}^{\tau}$ with the following property: the mapping τ of E (under $|\ |_{\mathfrak{P}}$) onto E (under $|\ |_{\mathfrak{P}^{\tau}}$) is analytic. Then

$$|E|_{\mathfrak{P}} = |\tau E|_{\mathfrak{P}^{\tau}} = |E|_{\mathfrak{P}^{\tau}}$$

and

$$|F|_{\mathfrak{P}} = |\tau F|_{\mathfrak{P}^{\tau}} = |F|_{\mathfrak{P}^{\tau}}.$$

Hence $e(\mathfrak{P}|\mathfrak{p}) = e(\mathfrak{P}^{\tau}|\mathfrak{p})$.

Now for the degree of inertia. Clearly

$$\tau(\mathfrak{o}(\mathfrak{P})) = \mathfrak{o}(\mathfrak{P}^{\tau}) \quad \text{and} \quad \tau(\mathfrak{m}(\mathfrak{P})) = \mathfrak{m}(\mathfrak{P}^{\tau}).$$

Take a residue class field $E(\mathfrak{P}^{\tau})$ and let $F(\mathfrak{P}^{\tau})$ be the residue class field of F at $\mathfrak{p}$ that is thereby obtained by natural restriction. Consider the composite homomorphism

$$\mathfrak{o}(\mathfrak{P}) \rightarrowtail \mathfrak{o}(\mathfrak{P}^{\tau}) \to E(\mathfrak{P}^{\tau}).$$

This has kernel $\mathfrak{m}(\mathfrak{P})$. Hence the composite homomorphism in conjunction with $E(\mathfrak{P}^{\tau})$ is a residue class field of E at $\mathfrak{P}$; write $E(\mathfrak{P}) = E(\mathfrak{P}^{\tau})$. But $\mathfrak{o}(\mathfrak{p})$ is carried onto $F(\mathfrak{P}^{\tau})$ by the composite homomorphism. Hence $F(\mathfrak{P}) = F(\mathfrak{P}^{\tau})$. Hence

$$f(\mathfrak{P}|\mathfrak{p}) = [E(\mathfrak{P}):F(\mathfrak{P})] = [E(\mathfrak{P}^{\tau}):F(\mathfrak{P}^{\tau})] = f(\mathfrak{P}^{\tau}|\mathfrak{p}).$$

q. e. d.

The preceding discussion shows that the local degree $n(\mathfrak{P}|\mathfrak{p})$ of an abelian extension depends only on $\mathfrak{p}$ and not on $\mathfrak{P}$. (This is not true for general separable extensions; for instance, see Example 15:9.) Accordingly one refers to the common value of the $n(\mathfrak{P}|\mathfrak{p})$ as the local degree of the abelian extension E/F at $\mathfrak{p}$, and one denotes this common value by $n_\mathfrak{p}$. The same simplification applies to the ramification index and the degree of inertia in the non-archimedean case, and they are denoted by $e_\mathfrak{p}$ and $f_\mathfrak{p}$ respectively.

15:10a. *Suppose d is a factor of $n_\mathfrak{p}$. Then there is a field H with $Z \subseteq H \subseteq E$ such that the local degree of H/F at $\mathfrak{p}$ is equal to d.*

Proof. The galois group $\mathfrak{G}(E_\mathfrak{P}/F_\mathfrak{P})$ of the galois extension $E_\mathfrak{P}/F_\mathfrak{P}$ is naturally isomorphic (by restriction) to a subgroup of $\mathfrak{G}(E/F)$. Hence $\mathfrak{G}(E_\mathfrak{P}/F_\mathfrak{P})$ is abelian; so it contains a subgroup $\mathfrak{H}$ of order $n_\mathfrak{p}/d$. The natural image of $\mathfrak{H}$ in $\mathfrak{G}(E/F)$ is of order $n_\mathfrak{p}/d$; so the fixed field H of this group has $[E:H] = n_\mathfrak{p}/d$. Now every element of Z is left fixed by $\mathfrak{H}$ since $Z_\mathfrak{P} = F_\mathfrak{P}$, hence $Z \subseteq H \subseteq E$. But

$$[E_\mathfrak{P}: H_\mathfrak{P}] \leqq [E: H] = n_\mathfrak{p}/d\,, \quad [H_\mathfrak{P}: Z_\mathfrak{P}] \leqq [H: Z] = d;$$

and we know that $[E_\mathfrak{P}: Z_\mathfrak{P}] = n_\mathfrak{p}$; hence $[H_\mathfrak{P}: Z_\mathfrak{P}] = d$. Thus the field H has all the desired properties. **q. e. d.**

§ 16. Discrete valuations

The value group $|F|$ of a non-trivial valuation on a field F is clearly infinite. In fact it is either a discrete subset or an everywhere dense subset of the set of positive numbers P; in the first instance it is infinite cyclic while in the second it is not. In order to verify these assertions consider the topological mapping $\log: \mathsf{P} \rightarrowtail \mathsf{R}$ which sends the multiplicative structure on P to the additive structure on R. The value group $|F|$ then becomes the additive subgroup $\log|F|$ of R. But every non-trivial additive subgroup of R is either a discrete infinite cyclic group, or else an everywhere dense non-cyclic subgroup of R. Hence $|F|$ has the property stated.

We shall call a valuation discrete if it is non-trivial and if its value group is a discrete infinite cyclic subgroup of P.[1] Thus an archimedean valuation cannot be discrete. The formula relating equivalent valuations shows that if a valuation is discrete, then so is every valuation that is equivalent to it. Accordingly we say that a prime spot $\mathfrak{p}$ on a field F is discrete if it contains at least one discrete valuation. So if $\mathfrak{p}$ is discrete it is non-trivial and non-archimedean, and every valuation in it is discrete.

[1] A discrete valuation does not yield the discrete topology. In fact it is easy to see that *the topology on an arbitrary valuated field is the discrete topology if and only if the valuation is trivial.*

16:1. *Let E/F be a finite extension of fields with spots $\mathfrak{P}|\mathfrak{p}$. Then $\mathfrak{P}$ is discrete if and only if $\mathfrak{p}$ is.*

Proof. Fix a valuation $|\ | \in \mathfrak{P}$. If $\mathfrak{P}$ is discrete, then $\mathfrak{p}$ is non-trivial by Proposition 11:3, and $|F|$ is infinite cyclic being a subgroup of $|E|$, hence $\mathfrak{p}$ is discrete. Conversely, suppose $\mathfrak{p}$ is discrete. Then

$$|E|^e \subseteq |F| \subseteq |E|$$

where $e = e(\mathfrak{P}|\mathfrak{p})$ is the ramification index of the extension. Hence $|E|$ must be a discrete subgroup of P. Hence $\mathfrak{P}$ is discrete. **q. e. d.**

Consider a discrete spot $\mathfrak{p}$ on F. Recall from ring theory that an element π in the integral domain $\mathfrak{o}(\mathfrak{p})$ is called a prime element of $\mathfrak{o}(\mathfrak{p})$ if it is a non-unit such that in every factorization $\pi = \alpha\beta$ with $\alpha, \beta \in \mathfrak{o}(\mathfrak{p})$ either α or β is a unit. By a prime element of F at $\mathfrak{p}$ we shall mean a prime element of the integral domain $\mathfrak{o}(\mathfrak{p})$ in the above sense. Suppose π is any element of $\mathfrak{o}(\mathfrak{p})$; pick a valuation $|\ | \in \mathfrak{p}$; then it is easily seen that π is a prime element of F at $\mathfrak{p}$ if and only if $|\pi|$ is that element of $|F|$ with largest value less than 1; and this is equivalent to saying that $|\pi|$ generates $|F|$. In particular, this shows that there is always at least one prime element $\dot{\pi}$ of F at $\mathfrak{p}$. Now suppose that π actually is a prime element of F at $\mathfrak{p}$. Then the following three facts are true: first, $\pi' \in F$ is a prime element at $\mathfrak{p}$ if and only if π/π' is a unit; second, $\mathfrak{m}(\mathfrak{p})$ is a principal ideal, in fact

$$\mathfrak{m}(\mathfrak{p}) = \pi\,\mathfrak{o}(\mathfrak{p});$$

and third, π is also a prime element of $F_\mathfrak{p}$ at $\mathfrak{p}$ since $|F| = |F_\mathfrak{p}|$.

It follows from the description of a prime element π in terms of valuations that every $\alpha \in \dot{F}$ can be expressed in the form

$$\alpha = \varepsilon\pi^\nu$$

with ε a unit at $\mathfrak{p}$ and $\nu \in \mathsf{Z}$. If π is fixed, then the representation is unique; if the prime element is allowed to vary, then ε will vary but ν will not. We can therefore define the order of α at $\mathfrak{p}$ to be

$$\operatorname{ord}_\mathfrak{p} \alpha = \nu\,.$$

We formally put $\operatorname{ord}_\mathfrak{p} 0 = \infty$.

The following rules for operating with the order function are evident:

$$|\alpha|_\mathfrak{p} < |\beta|_\mathfrak{p} \Leftrightarrow \operatorname{ord}_\mathfrak{p}\alpha > \operatorname{ord}_\mathfrak{p}\beta\,,$$

$$\operatorname{ord}_\mathfrak{p}\alpha\beta = \operatorname{ord}_\mathfrak{p}\alpha + \operatorname{ord}_\mathfrak{p}\beta\,,$$

$$\operatorname{ord}_\mathfrak{p}\alpha > 0 \Leftrightarrow \alpha \in \mathfrak{m}(\mathfrak{p})\,,$$

$$\operatorname{ord}_\mathfrak{p}\alpha \geqq 0 \Leftrightarrow \alpha \in \mathfrak{o}(\mathfrak{p})\,.$$

16:2. *Let E/F be a finite extension of fields with discrete spots $\mathfrak{P}|\mathfrak{p}$. Then*

$$\operatorname{ord}_\mathfrak{P} \alpha = e(\mathfrak{P}|\mathfrak{p}) \operatorname{ord}_\mathfrak{p} \alpha \textit{ for all } \alpha \in F\,.$$

Proof. Fix $|\ | \in \mathfrak{P}$. Then

$$(|E|:|F|) = e(\mathfrak{P}|\mathfrak{p}) \quad (= e)\,.$$

Hence if Π, $\dot{\pi}$ are prime elements of E, F we must have

$$|\Pi|^e = |\pi|\,.$$

Therefore $\pi = \Delta\, \Pi^e$ with $\Delta \in \mathfrak{u}(\mathfrak{P})$. So

$$\alpha = \varepsilon \pi^{\mathrm{ord}_\mathfrak{p}\alpha} = (\varepsilon\, \Delta^{\mathrm{ord}_\mathfrak{p}\alpha})\, \Pi^{e\,\mathrm{ord}_\mathfrak{p}\alpha},$$

for some $\varepsilon \in \mathfrak{u}(\mathfrak{p})$. This gives the result. **q. e. d.**

Consider $\mathfrak{o}(\mathfrak{p})$. In each coset of $\mathfrak{o}(\mathfrak{p})$ modulo $\mathfrak{m}(\mathfrak{p})$ pick exactly one representative c; always agree to pick $c = 0$ as the representative of $\mathfrak{m}(\mathfrak{p})$. Call any set C which is so obtained a representative set in $\mathfrak{o}(\mathfrak{p})$ of the residue class field of F at $\mathfrak{p}$.

Suppose a representative set C has been fixed. For each ν in $\mathbf{Z}$ pick π_ν in F with $\mathrm{ord}_\mathfrak{p}\, \pi_\nu = \nu$ (for instance the π_ν could be the powers π^ν of a fixed prime element π at $\mathfrak{p}$; we choose π_ν instead of π^ν in order to provide greater flexibility for applications). The immediate significance of the sets C and $\{\pi_\nu\}$ is that they give unique power series expansions for the elements of F.

16:3. *Let C be a fixed representative set of the residue class field of F at a discrete spot $\mathfrak{p}$. Suppose $\pi_\nu \in F$ is chosen at each $\nu \in \mathbf{Z}$ with $\mathrm{ord}_\mathfrak{p}\, \pi_\nu = \nu$. Then every element α of $\dot{F}$ can be expressed uniquely in the form*

$$\alpha = \sum_n^\infty c_\nu \pi_\nu$$

with $c_\nu \in C$, $n = \mathrm{ord}_\mathfrak{p}\, \alpha$, and $c_n \neq 0$.

Proof. 1) We can put $\alpha = \varepsilon \pi_n$ for some $\varepsilon \in \mathfrak{u}(\mathfrak{p})$ since $\mathrm{ord}_\mathfrak{p}\, \alpha = n$. Choose $c_n \in C$ with $\varepsilon \equiv c_n \bmod \mathfrak{m}(\mathfrak{p})$. Then

$$\alpha = c_n \pi_n + \alpha' \quad \text{with} \quad \mathrm{ord}_\mathfrak{p}\, \alpha' > n\,.$$

Next apply this procedure to α' to obtain α'', then to α'', and so on. After $m + 1$ steps we obtain an expression

$$\alpha = c_n \pi_n + \cdots + c_{n+m} \pi_{n+m} + \alpha^{(m+1)}$$

with $\mathrm{ord}_\mathfrak{p}\, \alpha^{(m+1)} > n + m$. In this way we can define

$$c_n, c_{n+1}, \ldots, c_{n+m}, \ldots \in C\,.$$

The partial sums

$$c_n \pi_n + \cdots + c_{n+m} \pi_{n+m}$$

clearly converge to α. Hence

$$\alpha = \sum_n^\infty c_\nu \pi_\nu\,.$$

2) In order to prove uniqueness we consider two expressions

$$\sum_{n}^{\infty} c_\nu \pi_\nu = \sum_{n}^{\infty} d_\nu \pi_\nu$$

with the c_ν and d_ν in C. Let i be the first integer for which $c_i \neq d_i$; then $|c_i - d_i|_\mathfrak{p} = 1$ since c_i and d_i will fall in different residue classes of $\mathfrak{o}(\mathfrak{p})$ modulo $\mathfrak{m}(\mathfrak{p})$ when they are not equal; hence

$$\left| \sum_{i}^{\infty} (c_\nu - d_\nu)\, \pi_\nu \right|_\mathfrak{p} = |\pi_i|_\mathfrak{p}$$

by Corollary 11:2a. But

$$\sum_{i}^{\infty} (c_\nu - d_\nu)\, \pi_\nu = 0$$

by hypothesis. This gives a contradiction. So we do indeed have $c_i = d_i$ for all i. q. e. d.

16:4. **Theorem.** *Let E/F be a finite extension of degree n with discrete spots $\mathfrak{P} \mid \mathfrak{p}$. Suppose further that the spots are complete. Then*

$$e(\mathfrak{P}\mid\mathfrak{p})\, f(\mathfrak{P}\mid\mathfrak{p}) = n\,.$$

Proof. 1) Write $e = e(\mathfrak{P}\mid\mathfrak{p})$ and $f = f(\mathfrak{P}\mid\mathfrak{p})$. Let $(^-, \bar{E})$ be a residue class field of E at $\mathfrak{P}$, and let $(^-, \bar{F})$ be obtained from it by natural restriction. Choose $\omega_1, \ldots, \omega_f$ in $\mathfrak{o}(\mathfrak{P})$ in such a way that $\bar{\omega}_1, \ldots, \bar{\omega}_f$ form a base for $\bar{E}$ over $\bar{F}$. If we consider all elements of the form $b_1\omega_1 + \cdots + b_f\omega_f$ with the b's in $\mathfrak{o}(\mathfrak{p})$ we fall in each residue class of $\mathfrak{o}(\mathfrak{P})$ modulo $\mathfrak{m}(\mathfrak{P})$ at least once. Hence we can select a representative set C of the residue class field of E at $\mathfrak{P}$ in which every element is of the form

$$b_1\omega_1 + \cdots + b_f\omega_f\,, \quad \text{with all} \quad b_i \in \mathfrak{o}(\mathfrak{p})\,.$$

2) Let Π, π be fixed prime elements of E, F at $\mathfrak{P}, \mathfrak{p}$ respectively. For each $\varkappa \in Z$ write $\varkappa = \mu e + \nu$ with $\mu, \nu \in Z$ and $0 \leqq \nu \leqq e - 1$. Define $\Pi_\varkappa = \pi^\mu \Pi^\nu$. Then

$$\operatorname{ord}_\mathfrak{P} \Pi_\varkappa = \mu \operatorname{ord}_\mathfrak{P} \pi + \nu = \mu e + \nu = \varkappa\,.$$

3) We have already proved in Proposition 13:6 that $ef \leqq n$. It therefore suffices to prove that the ef elements

$$\omega_\lambda \Pi^\nu \quad (1 \leqq \lambda \leqq f\,, \quad 0 \leqq \nu \leqq e-1)$$

span E over F. Consider a typical $\alpha \in E$. By Proposition 16:3 we can express it in the series expansion

$$\alpha = \sum_{\varkappa = se}^{\infty} C_\varkappa \Pi_\varkappa$$

with $\operatorname{ord}_{\mathfrak{P}}\alpha \geqq se$ and the $C_\varkappa$ in C. If we group the terms we obtain

$$\alpha = \sum_{\mu=s}^{\infty} (C_{\mu e}\Pi_{\mu e} + C_{\mu e+1}\Pi_{\mu e+1} + \cdots + C_{\mu e+(e-1)}\Pi_{\mu e+(e-1)})$$

$$= \left(\sum_{\mu=s}^{\infty} C_{\mu e}\Pi_{\mu e}\right) + \cdots + \left(\sum_{\mu=s}^{\infty} C_{\mu e+(e-1)}\Pi_{\mu e+(e-1)}\right).$$

It therefore suffices for us to prove that each

$$\sum_{\mu=s}^{\infty} C_{\mu e+\nu}\Pi_{\mu e+\nu} = \left(\sum_{\mu=s}^{\infty} C_{\mu e+\nu}\pi^{\mu}\right)\Pi^{\nu}$$

is in the space spanned by the above $\omega_\lambda \Pi^\nu$ over F. But by the choice of the set C we can write

$$\sum_{\mu=s}^{\infty} C_{\mu e+\nu}\pi^{\mu} = \sum_{\mu=s}^{\infty}\left(\sum_{\lambda=1}^{f} b_{\lambda\mu}\omega_\lambda\right)\pi^{\mu}$$

$$= \sum_{\lambda=1}^{f}\left(\sum_{\mu=s}^{\infty} b_{\lambda\mu}\pi^{\mu}\right)\omega_\lambda,$$

with all $b_{\lambda\mu}$ in $\mathfrak{o}(\mathfrak{p})$. Hence

$$\sum_{\mu=s}^{\infty} C_{\mu e+\nu}\Pi_{\mu e+\nu} \in \sum_{\lambda=1}^{f} F\omega_\lambda \Pi^{\nu}.$$

q. e. d.

Chapter II

Dedekind Theory of Ideals

In Chapter I we studied the ring of integers $\mathfrak{o}(\mathfrak{p})$ of a single non-archimedean spot $\mathfrak{p}$. We shall see in § 33J that the set of algebraic integers of a number field F can be expressed in the form

$$\mathfrak{o}(S) = \bigcap_{\mathfrak{p}\in S} \mathfrak{o}(\mathfrak{p})$$

where S consists of all non-archimedean spots on F. This exhibits a strong connection between the algebraic integers and the prime spots of a number field, and we shall start to exploit it here. Specifically, we shall use the theory of prime spots to set up an ideal theory in $\mathfrak{o}(S)$. For the present we can be quite general and we consider an arbitrary field F that is provided with a set of spots satisfying certain axioms. We shall call these axioms the Dedekind axioms for S since they lead to Dedekind's ideal theory in $\mathfrak{o}(S)$.

The general assumption throughout this chapter is that we have a field F and a non-empty set of spots S on F which satisfies the axioms (D_1), (D_2) and (D_3) given immediately below. We fix $|\ |_{\mathfrak{p}} \in \mathfrak{p}$ at each $\mathfrak{p}$ in the given set S.

§ 21. Dedekind axioms for S.

We call the non-empty set of spots S on the field F a Dedekind set of spots if it satisfies the following three axioms:

(D_1) every spot in S is discrete

(D_2) for each $\alpha \in F$ we have $|\alpha|_\mathfrak{p} \leqq 1$ for almost all $\mathfrak{p} \in S$

(D_3) whenever $\mathfrak{q}$ and $\mathfrak{q}'$ are distinct spots in S, there corresponds to each $\varepsilon > 0$ an $\alpha \in F$ such that

$$|\alpha - 1|_\mathfrak{q} < \varepsilon\,, \quad |\alpha|_{\mathfrak{q}'} < \varepsilon\,, \quad |\alpha|_\mathfrak{p} \leqq 1$$

for all $\mathfrak{p} \in S - (\mathfrak{q} \cup \mathfrak{q}')$.

By the ring of integers of F at S we shall mean the subring

$$\mathfrak{o}(S) = \bigcap_{\mathfrak{p} \in S} \mathfrak{o}(\mathfrak{p})$$

of F. The multiplicative subgroup

$$\mathfrak{u}(S) = \bigcap_{\mathfrak{p} \in S} \mathfrak{u}(\mathfrak{p})$$

of $\mathfrak{o}(S)$ consists precisely of the invertible elements of $\mathfrak{o}(S)$ and so we shall refer to it as the group of units of F at S. At times it is convenient to relax the notation and to write $\mathfrak{o}$ and $\mathfrak{u}$ instead of $\mathfrak{o}(S)$ and $\mathfrak{u}(S)$.

21:1. **Example.** The set of all non-archimedean spots on the field of rational numbers $\mathbf{Q}$ is Dedekind. This will be proved in § 31.

The axioms are clearly independent of the choice of $|\ |_\mathfrak{p} \in \mathfrak{p}$. By taking inverses we note that (D_2) actually implies the stronger assertion of equality: if $\alpha \in \dot{F}$, then $|\alpha|_\mathfrak{p} = 1$ for almost all $\mathfrak{p} \in S$. If S consists of just a single discrete spot, then (D_2) is automatically satisfied while (D_3) is vacuously true; hence a single discrete spot is always Dedekind. Similarly if S satisfies (D_1) and is at the same time a finite set, then it is Dedekind; this follows at once from the Weak Approximation Theorem of § 11E. Approximation is of course the key to the third axiom; we can use this axiom to derive an approximation theorem which, in certain important situations, is stronger than the one given in § 11E; this we now do.

21:2. **Strong Approximation Theorem.** *Let T be a finite subset of the Dedekind set of spots S on the field F. Suppose that $\alpha_\mathfrak{p} \in F$ is given, one for each $\mathfrak{p} \in T$. Then for each $\varepsilon > 0$ there is an $A \in F$ such that*

$$\begin{cases} |A - \alpha_\mathfrak{p}|_\mathfrak{p} < \varepsilon & \forall\, \mathfrak{p} \in T \\ |A|_\mathfrak{p} \leqq 1 & \forall\, \mathfrak{p} \in S - T\,. \end{cases}$$

Proof. 1) By the Weak Approximation Theorem of § 11E we can assume that S is an infinite set of spots. We can also assume that

$$|\alpha_\mathfrak{p}|_\mathfrak{q} \leqq 1 \quad \forall\, \mathfrak{p} \in T, \quad \forall\, \mathfrak{q} \in S - T:$$

for if necessary we can adjoin to T all those $\mathfrak{q} \in S - T$ at which $|\alpha_\mathfrak{p}|_\mathfrak{q} > 1$ for at least one $\alpha_\mathfrak{p}$; and then define $\alpha_\mathfrak{q} = 0$ at the new $\mathfrak{q}$; the new set T which is obtained by this adjunction is still finite because of the axiom (D_2); if we can prove our theorem for the new T we will have it for the original one. Hence we can indeed make the above assumption. In the same way we can assume, by adjoining a single spot to T if necessary, that T consists of at least two spots.

2) Consider a spot $\mathfrak{p} \in T$ that is fixed for the moment. For each $\mathfrak{q} \in T - \mathfrak{p}$ we can find an element of $\mathfrak{o}(S)$ that is arbitrarily close to 1 at $\mathfrak{p}$ and to 0 at $\mathfrak{q}$. Do this for each $\mathfrak{q} \in T - \mathfrak{p}$ and multiply all these answers together. Using the fact that multiplication is continuous in the $\mathfrak{p}$-adic topology we can obtain in this way an element of $\mathfrak{o}(S)$ that is arbitrarily close to 1 at $\mathfrak{p}$ and to 0 at all $\mathfrak{q} \in T - \mathfrak{p}$. Let us denote such an element by $A_\mathfrak{p}$. Then obtain an $A_\mathfrak{p}$ for each $\mathfrak{p} \in T$.

3) The rest is easy. Just form $\sum_{\mathfrak{p} \in T} \alpha_\mathfrak{p} A_\mathfrak{p}$. This element satisfies

$$\Big|\sum_{\mathfrak{p} \in T} \alpha_\mathfrak{p} A_\mathfrak{p}\Big|_\mathfrak{q} \leqq 1 \quad \forall\, \mathfrak{q} \in S - T$$

by choice of T and the $A_\mathfrak{p}$. And by continuity of addition and multiplication it can be made arbitrarily close to $\alpha_\mathfrak{p}$ simultaneously at all $\mathfrak{p} \in T$ just by making the approximations $A_\mathfrak{p}$ in step 2) sharp enough.

q. e. d.

21:2a. **Corollary.** *Let $\xi_\mathfrak{p}$ be given in the value group $|F|_\mathfrak{p}$ at each $\mathfrak{p} \in S$, with almost all $\xi_\mathfrak{p} = 1$. Then there is an $A \in F$ such that*

$$|A|_\mathfrak{p} = \xi_\mathfrak{p} \quad \forall\, \mathfrak{p} \in T\,, \quad |A|_\mathfrak{p} \leqq \xi_\mathfrak{p} \quad \forall\, \mathfrak{p} \in S\,.$$

Proof. We can assume that $\xi_\mathfrak{p} = 1$ for all $\mathfrak{p} \in S - T$ by enlarging T if necessary. Pick $\alpha_\mathfrak{p} \in \dot{F}$ for each $\mathfrak{p} \in T$ in such a way that $|\alpha_\mathfrak{p}|_\mathfrak{p} = \xi_\mathfrak{p}$. Choose $A \in F$ with

$$\begin{cases} |A - \alpha_\mathfrak{p}|_\mathfrak{p} < |\alpha_\mathfrak{p}|_\mathfrak{p} & \forall\, \mathfrak{p} \in T \\ \quad |A|_\mathfrak{p} \leqq 1 & \forall\, \mathfrak{p} \in S - T\,. \end{cases}$$

This A is the required element. **q. e. d.**

21:3. *Let S be a Dedekind set of spots on F. Then $\mathfrak{o}(S) \subset F$. And F is the quotient field of $\mathfrak{o}(S)$.*

Proof. We have $\mathfrak{o}(\mathfrak{p}) \subset F$ for any $\mathfrak{p} \in S$ since all spots in S are non-trivial. Hence $\mathfrak{o}(S) \subset F$.

Consider a typical $\alpha \in \dot{F}$ which we wish to express as a quotient of elements of $\mathfrak{o}(S)$. Put

$$T = \{\mathfrak{p} \in S \mid |\alpha|_\mathfrak{p} > 1\}\,.$$

We can clearly assume that T is not empty. So T will be a finite set by Axiom (D_2). Now we have $\alpha^{-1} \in \mathfrak{o}(\mathfrak{p})$ for all $\mathfrak{p} \in T$, hence we can use the

Strong Approximation Theorem to find $b \in \mathfrak{o}(S)$ such that

$$|b - \alpha^{-1}|_{\mathfrak{p}} < |\alpha^{-1}|_{\mathfrak{p}} \quad \forall\, \mathfrak{p} \in T\,.$$

But then $|b|_{\mathfrak{p}} = |\alpha^{-1}|_{\mathfrak{p}}$ for all $\mathfrak{p} \in T$. Hence $\alpha b \in \mathfrak{o}(S)$. Then $\alpha = (\alpha b)/b$ expresses α as a quotient of two elements in $\mathfrak{o}(S)$, so $\mathfrak{o}(S)$ has F as a quotient field. **q. e. d.**

21:4. *Let S be a Dedekind set of spots on F. Then $\mathfrak{o}(\mathfrak{p})$ is the closure of $\mathfrak{o}(S)$ in F at each $\mathfrak{p} \in S$. In particular $\mathfrak{o}(S)$ contains a representative set of the residue class field at $\mathfrak{p}$.*

Proof. $\mathfrak{o}(\mathfrak{p})$ is closed since the mapping $\alpha \to |\alpha|_{\mathfrak{p}}$ is continuous. Consider an ε-neighborhood $(0 < \varepsilon < 1)$ of a typical element α of $\mathfrak{o}(\mathfrak{p})$. Choose $A \in F$ such that

$$\begin{cases} |A - \alpha|_{\mathfrak{p}} < \varepsilon \\ \qquad |A|_{\mathfrak{q}} \leqq 1 \quad \forall\, \mathfrak{q} \in S - \mathfrak{p}\,. \end{cases}$$

Then $A \in \mathfrak{o}(S)$. Hence every neighborhood of α meets $\mathfrak{o}(S)$ and so $\mathfrak{o}(\mathfrak{p})$ is its closure in F at $\mathfrak{p}$.

The second part of the proposition is now clear. **q. e. d.**

§ 22. Ideal theory

S is a Dedekind set of spots on the field F. We use $\mathfrak{o}$ and $\mathfrak{u}$ instead of $\mathfrak{o}(S)$ and $\mathfrak{u}(S)$.

§ 22A. Operations with fractional ideals

The field F is of course a vector space over itself in the obvious natural way. And it is an $\mathfrak{o}$-module under the induced laws. We define a fractional ideal $\mathfrak{a}$ of F at S to be a non-zero $\mathfrak{o}$-module $\mathfrak{a} \subseteq F$ which has the following property: there is a non-zero $\lambda \in \mathfrak{o}$ such that $\lambda\mathfrak{a} \subseteq \mathfrak{o}$. This property simply asserts that the elements of the $\mathfrak{o}$-module $\mathfrak{a}$ are to have bounded denominators. We use $I(S)$, or simply I, to denote the set of all fractional ideals of F at S.

We note that every finitely generated non-zero $\mathfrak{o}$-module in F is a fractional ideal at S (the converse is also true and it will be established in Corollary 22:5b). In particular, the set $\alpha\mathfrak{o}$ is a fractional $\mathfrak{o}$-ideal for any $\alpha \in \dot{F}$. We call any such ideal a principal ideal and we let $P(S)$, or just P, denote the set of principal ideals in $I(S)$.

We shall call the fractional ideal $\mathfrak{a}$ integral if $\mathfrak{a} \subseteq \mathfrak{o}$. Thus the integral ideals are the ideals of $\mathfrak{o}$ in the usual sense, with the exception of 0. We say that $\mathfrak{a}$ divides $\mathfrak{b}$, or that $\mathfrak{b}$ is a multiple of $\mathfrak{a}$, if $\mathfrak{b} \subseteq \mathfrak{a}$; we denote this fact by writing $\mathfrak{a} | \mathfrak{b}$. We define the greatest common divisor, the least common multiple, and the product of any two fractional ideals $\mathfrak{a}$, $\mathfrak{b}$ as follows:

g.c.d.: $\mathfrak{a}+\mathfrak{b}=\{\alpha+\beta \mid \alpha\in\mathfrak{a},\ \beta\in\mathfrak{b}\}$,

l.c.m.: $\mathfrak{a}\cap\mathfrak{b}$,

product: $\mathfrak{a}\mathfrak{b}=\left\{\sum_{\text{finite}}\alpha\beta \mid \alpha\in\mathfrak{a},\ \beta\in\mathfrak{b}\right\}$.

These new objects are clearly $\mathfrak{o}$-modules; and one easily verifies that their elements are of bounded denominator; in order to verify that all are in I we must still show that they are non-zero; for $\mathfrak{a}+\mathfrak{b}$ and $\mathfrak{a}\mathfrak{b}$ this is clear; for $\mathfrak{a}\cap\mathfrak{b}$ we first choose λ, μ in $\mathfrak{o}$ such that $\lambda\mathfrak{a}$ and $\mu\mathfrak{b}$ are integral; then

$$\mathfrak{a}\cap\mathfrak{b}\supseteq\lambda\mathfrak{a}\cap\mu\mathfrak{b}\supseteq(\lambda\mathfrak{a})(\mu\mathfrak{b})\supset 0\,.$$

Hence we have proved that

$$\mathfrak{a}+\mathfrak{b}\,,\quad \mathfrak{a}\cap\mathfrak{b}\,,\quad \mathfrak{a}\mathfrak{b}\ \in I\,.$$

Clearly $\alpha\mathfrak{a}\in I$ for any $\alpha\in\dot{F}$.

We say that $\mathfrak{a}$ and $\mathfrak{b}$ are relatively prime if $\mathfrak{a}+\mathfrak{b}=\mathfrak{o}$.

Product formation provides I with an associative and commutative multiplicative law in which $\mathfrak{o}$ acts as an identity. We shall see that every $\mathfrak{a}\in I$ has an inverse under this law so that I is in fact a group.

§ 22B. Valuations on I

For any $\mathfrak{a}\in I$ and any $\mathfrak{p}\in S$ we define

$$|\mathfrak{a}|_\mathfrak{p}=\max_{\alpha\in\mathfrak{a}}|\alpha|_\mathfrak{p}\,.$$

Clearly $|\mathfrak{a}|_\mathfrak{p}>0$. If we take $\lambda\in\dot{F}$ such that $\lambda\mathfrak{a}\subseteq\mathfrak{o}$, then $\mathfrak{a}\subseteq\lambda^{-1}\mathfrak{o}$ and so $|\mathfrak{a}|_\mathfrak{p}\leqq|\lambda^{-1}|_\mathfrak{p}$. Hence the maximum in the definition of $|\mathfrak{a}|_\mathfrak{p}$ is attained, so $|\mathfrak{a}|_\mathfrak{p}\in\mathbf{P}$. This also shows that $|\mathfrak{a}|_\mathfrak{p}\leqq 1$ for almost all $\mathfrak{p}\in S$ since it is true for λ^{-1}; hence

$$|\mathfrak{a}|_\mathfrak{p}=1\quad\text{for almost all}\quad\mathfrak{p}\in S\,.$$

Clearly $|\mathfrak{o}|_\mathfrak{p}=1$. And for any $\alpha\in\dot{F}$ we have

$$|\alpha\mathfrak{a}|_\mathfrak{p}=|\alpha|_\mathfrak{p}\,|\mathfrak{a}|_\mathfrak{p}\,.$$

One easily proves that

$$|\mathfrak{a}\mathfrak{b}|_\mathfrak{p}=|\mathfrak{a}|_\mathfrak{p}\,|\mathfrak{b}|_\mathfrak{p}\,,\quad |\mathfrak{a}+\mathfrak{b}|_\mathfrak{p}=\max\,(|\mathfrak{a}|_\mathfrak{p},|\mathfrak{b}|_\mathfrak{p})$$

hold for any $\mathfrak{a}$, $\mathfrak{b}\in I$. It is also true that

$$|\mathfrak{a}\cap\mathfrak{b}|_\mathfrak{p}=\min\,(|\mathfrak{a}|_\mathfrak{p},|\mathfrak{b}|_\mathfrak{p})\,,$$

but this is a little more difficult, so we offer a proof. Clearly $|\mathfrak{a}\cap\mathfrak{b}|_\mathfrak{p}\leqq\min\,(|\mathfrak{a}|_\mathfrak{p},|\mathfrak{b}|_\mathfrak{p})$. We must reverse this inequality. Take $|\mathfrak{a}|_\mathfrak{p}\leqq|\mathfrak{b}|_\mathfrak{p}$. Pick $a\in\mathfrak{a}$ and $b\in\mathfrak{b}$ in such a way that

$$|a|_\mathfrak{p}=|\mathfrak{a}|_\mathfrak{p}\leqq|\mathfrak{b}|_\mathfrak{p}=|b|_\mathfrak{p}\,.$$

By Corollary 21:2a there is a $\mu \in F$ such that

$$\begin{cases} |\mu|_{\mathfrak{p}} = \left|\dfrac{a}{b}\right|_{\mathfrak{p}} \leqq 1 \\ |\mu|_{\mathfrak{q}} \leqq \min\left(1, \left|\dfrac{a}{b}\right|_{\mathfrak{q}}\right) \qquad \forall\, \mathfrak{q} \in S - \mathfrak{p}\,. \end{cases}$$

Hence $\mu b \in a\mathfrak{o}$. And $\mu \in \mathfrak{o}$ so that also $\mu b \in \mathfrak{b}$. Hence $\mu b \in \mathfrak{a} \cap \mathfrak{b}$. Hence

$$|\mathfrak{a} \cap \mathfrak{b}|_{\mathfrak{p}} \geqq |\mu b|_{\mathfrak{p}} = |a|_{\mathfrak{p}} = |\mathfrak{a}|_{\mathfrak{p}} = \min\left(|\mathfrak{a}|_{\mathfrak{p}}, |\mathfrak{b}|_{\mathfrak{p}}\right).$$

This proves the assertion.

22:1. *Let S be a Dedekind set of spots on a field F and let $\mathfrak{a}, \mathfrak{b}$ be fractional ideals at S. Then*

(1) $\mathfrak{a} = \{\alpha \in F \mid |\alpha|_{\mathfrak{p}} \leqq |\mathfrak{a}|_{\mathfrak{p}} \quad \forall\, \mathfrak{p} \in S\}$

(2) $\mathfrak{a} \subseteq \mathfrak{b} \;\Leftrightarrow\; |\mathfrak{a}|_{\mathfrak{p}} \leqq |\mathfrak{b}|_{\mathfrak{p}} \quad \forall\, \mathfrak{p} \in S\,.$

Proof. 1) The second part follows at once from the first. In order to prove the first we have to show that an α in $\dot{F}$ which satisfies $|\alpha|_{\mathfrak{p}} \leqq |\mathfrak{a}|_{\mathfrak{p}}$ for all $\mathfrak{p} \in S$ is in $\mathfrak{a}$. This reduces (on replacing $\mathfrak{a}$ by $\alpha^{-1}\mathfrak{a}$) to proving that $1 \in \mathfrak{a}$ if $1 \leqq |\mathfrak{a}|_{\mathfrak{p}}$ for all $\mathfrak{p} \in S$. Now it is clearly enough to prove that $1 \in \mathfrak{a} \cap \mathfrak{o}$; and furthermore $|\mathfrak{a} \cap \mathfrak{o}|_{\mathfrak{p}} = 1$. So we have reduced the question to the following: given $\mathfrak{a} \subseteq \mathfrak{o}$ with $|\mathfrak{a}|_{\mathfrak{p}} = 1$ for all $\mathfrak{p} \in S$, prove that $1 \in \mathfrak{a}$. This is the form we need to establish the proof.

2) Pick an arbitrary non-zero γ in $\mathfrak{a}$ and fix it. Consider the set

$$T = \{\mathfrak{p} \in S \mid |\gamma|_{\mathfrak{p}} < 1\}\,.$$

If T is empty, then $\gamma^{-1} \in \mathfrak{o}$ and so $1 = \gamma^{-1}\gamma \in \mathfrak{a}$ as asserted. Hence assume that T is not empty. So T is a finite subset of S.

Consider a $\mathfrak{p} \in T$, fixed for the moment. Choose $\alpha_{\mathfrak{p}} \in \mathfrak{a}$ such that $|\alpha_{\mathfrak{p}}|_{\mathfrak{p}} = |\mathfrak{a}|_{\mathfrak{p}} = 1$. Then the Strong Approximation Theorem gives us an element of F which will be written $\beta_{\mathfrak{p}}$ to denote its dependence on the temporarily fixed $\mathfrak{p}$, such that

$$\begin{cases} \left|\dfrac{1}{\alpha_{\mathfrak{p}}} - \beta_{\mathfrak{p}}\right|_{\mathfrak{p}} \leqq |\gamma|_{\mathfrak{p}} \\ \qquad |\beta_{\mathfrak{p}}|_{\mathfrak{q}} \leqq |\gamma|_{\mathfrak{q}} \quad \forall\, \mathfrak{q} \in S - \mathfrak{p}\,. \end{cases}$$

Since $\gamma \in \mathfrak{o}(S)$ and $|\alpha_{\mathfrak{p}}|_{\mathfrak{p}} = 1$ it is clear that $\beta_{\mathfrak{p}}$ is in $\mathfrak{o}(S)$.

Now do this once at each $\mathfrak{p} \in T$ to obtain an $\alpha_{\mathfrak{p}}$ and a $\beta_{\mathfrak{p}}$ as above. For each $\mathfrak{q} \in T$ we have

$$\left|1 - \sum_{\mathfrak{p} \in T} \alpha_{\mathfrak{p}}\beta_{\mathfrak{p}}\right|_{\mathfrak{q}} = \left|(1 - \alpha_{\mathfrak{q}}\beta_{\mathfrak{q}}) - \sum_{\mathfrak{p} \in T - \mathfrak{q}} \alpha_{\mathfrak{p}}\beta_{\mathfrak{p}}\right|_{\mathfrak{q}} \leqq |\gamma|_{\mathfrak{q}}\,.$$

And for each $\mathfrak{q} \in S - T$ we have

$$\left|1 - \sum_{\mathfrak{p} \in T} \alpha_{\mathfrak{p}}\beta_{\mathfrak{p}}\right|_{\mathfrak{q}} \leqq 1 = |\gamma|_{\mathfrak{q}}\,.$$

Hence

$$\left(1 - \sum_{\mathfrak{p} \in T} \alpha_\mathfrak{p} \beta_\mathfrak{p}\right) \gamma^{-1} \in \mathfrak{o}(S),$$

by definition of $\mathfrak{o}(S)$. So

$$1 - \sum_{\mathfrak{p} \in T} \alpha_\mathfrak{p} \beta_\mathfrak{p} \in \gamma\mathfrak{o} \subseteq \mathfrak{a}.$$

But $\alpha_\mathfrak{p} \in \mathfrak{a}$ and $\beta_\mathfrak{p} \in \mathfrak{o}$. Hence $1 \in \mathfrak{a}$. **q. e. d.**

22:2. *Let S be a Dedekind set of spots on F. At each $\mathfrak{p} \in S$ let there be given an $\alpha_\mathfrak{p} \in \dot{F}$ with almost all $|\alpha_\mathfrak{p}|_\mathfrak{p}$ equal to 1. Then*

$$\mathfrak{a} = \{x \in F \mid |x|_\mathfrak{p} \leqq |\alpha_\mathfrak{p}|_\mathfrak{p} \quad \forall\, \mathfrak{p} \in S\}$$

is a fractional ideal at S. And

$$|\mathfrak{a}|_\mathfrak{p} = |\alpha_\mathfrak{p}|_\mathfrak{p} \quad \forall\, \mathfrak{p} \in S.$$

Proof. Clearly $\mathfrak{a}$ is an $\mathfrak{o}$-module. We deduce from Corollary 21:2a that it is not 0. Again by Corollary 21:2a there is a $\lambda \in \dot{F}$ such that

$$|\lambda|_\mathfrak{p} \leqq \min(1, |\alpha_\mathfrak{p}^{-1}|_\mathfrak{p}) \quad \forall\, \mathfrak{p} \in S.$$

So there is a non-zero $\lambda \in \mathfrak{o}$ with

$$|\lambda x|_\mathfrak{p} \leqq |\lambda \alpha_\mathfrak{p}|_\mathfrak{p} \leqq 1 \quad \forall \mathfrak{p} \in S, \quad \forall\, x \in \mathfrak{a}.$$

This means that $\lambda x \in \mathfrak{o}$ for all $x \in \mathfrak{a}$; hence $\mathfrak{a}$ is a fractional ideal.

Clearly $|\mathfrak{a}|_\mathfrak{p} \leqq |\alpha_\mathfrak{p}|_\mathfrak{p}$ holds for all $\mathfrak{p} \in S$. But Corollary 21:2a gives us an $x \in \mathfrak{a}$ such that $|x|_\mathfrak{p} = |\alpha_\mathfrak{p}|_\mathfrak{p}$ at a fixed $\mathfrak{p} \in S$. Hence $|\mathfrak{a}|_\mathfrak{p} = |\alpha_\mathfrak{p}|_\mathfrak{p}$. **q. e. d.**

22:3. **Example.** The following identities are true for all $\mathfrak{a}, \mathfrak{b}, \mathfrak{c} \in I$ and all $n \in \mathbf{N}$:

$$\mathfrak{a}(\mathfrak{b} + \mathfrak{c}) = \mathfrak{a}\mathfrak{b} + \mathfrak{a}\mathfrak{c}, \quad \mathfrak{a}(\mathfrak{b} \cap \mathfrak{c}) = \mathfrak{a}\mathfrak{b} \cap \mathfrak{a}\mathfrak{c}$$

$$(\mathfrak{a} + \mathfrak{b})^n = \mathfrak{a}^n + \mathfrak{b}^n, \quad (\mathfrak{a} \cap \mathfrak{b})^n = \mathfrak{a}^n \cap \mathfrak{b}^n$$

$$\mathfrak{a} \cap (\mathfrak{b} + \mathfrak{c}) = (\mathfrak{a} \cap \mathfrak{b}) + (\mathfrak{a} \cap \mathfrak{c}), \quad \mathfrak{a} + (\mathfrak{b} \cap \mathfrak{c}) = (\mathfrak{a} + \mathfrak{b}) \cap (\mathfrak{a} + \mathfrak{c})$$

$$\mathfrak{a}\mathfrak{b} = (\mathfrak{a} + \mathfrak{b})(\mathfrak{a} \cap \mathfrak{b}).$$

Also

$$\mathfrak{a}^n = \mathfrak{b}^n \quad \text{for some} \quad n \in \mathbf{N} \Rightarrow \mathfrak{a} = \mathfrak{b}.$$

All proofs by inspection using the properties of $|\ |_\mathfrak{p}$ on I.

§ 22C. Properties of $\mathfrak{o}$ and I

The factorization theory of ideals follows quickly from the results of § 22B. For instance I is a group: namely, for any $\mathfrak{a} \in I$ take that $\mathfrak{b} \in I$ for which $|\mathfrak{b}|_\mathfrak{p} = |\mathfrak{a}|_\mathfrak{p}^{-1}$ for all $\mathfrak{p} \in S$; then $|\mathfrak{a}\mathfrak{b}|_\mathfrak{p} = 1 = |\mathfrak{o}|_\mathfrak{p}$ for all $\mathfrak{p} \in S$ and so $\mathfrak{a}\mathfrak{b} = \mathfrak{o}$. Hence every $\mathfrak{a} \in I$ has an inverse and I is a group. The inverse of $\mathfrak{a}$ will be written $\mathfrak{a}^{-1}$.

22:4. **Example.** $(\mathfrak{a}+\mathfrak{b})^{-1}=\mathfrak{a}^{-1}\cap\mathfrak{b}^{-1}$, $(\mathfrak{a}\cap\mathfrak{b})^{-1}=\mathfrak{a}^{-1}+\mathfrak{b}^{-1}$.

The principal ideals P are a subgroup of I. Hence we can form the quotient group I/P; this is called the ideal class group of F at S. The class number $h_F(S)$ is the order of the ideal class group. Clearly $h_F(S)=1$ means that every fractional ideal is principal, and this is equivalent to saying that $\mathfrak{o}$ is a principal ideal domain. The class number is 1 when S is a finite set: given $\mathfrak{a}\in I$ use Corollary 21:2a to pick $\alpha\in\dot{F}$ such that $|\alpha|_\mathfrak{p}=|\mathfrak{a}|_\mathfrak{p}$ for all $\mathfrak{p}\in S$; then $\mathfrak{a}=\alpha\mathfrak{o}$ by Proposition 22:1. It will be shown in § 33H that the class number is finite over algebraic number fields.

22:5. *Let S be a Dedekind set of spots on F and let $\mathfrak{a}$, $\mathfrak{b}$, $\mathfrak{c}$ be fractional ideals at S. Then there are non-zero field elements α and β such that*

$$\mathfrak{c}=\alpha\mathfrak{a}+\beta\mathfrak{b}\,.$$

And in fact β can be anything for which $\beta\mathfrak{b}\subseteq\mathfrak{c}$.

Proof. Take $\beta\in\dot{F}$ such that $\beta\mathfrak{b}\subseteq\mathfrak{c}$ (there is always at least one such β; for instance any $\beta\in\mathfrak{b}^{-1}\mathfrak{c}$ will do). Let T be any finite subset of S with the property that

$$|\mathfrak{a}|_\mathfrak{p}=|\mathfrak{b}|_\mathfrak{p}=|\mathfrak{c}|_\mathfrak{p}=|\beta|_\mathfrak{p}=1\quad\forall\,\mathfrak{p}\in S-T\,.$$

By Corollary 21:2a there is an $\alpha\in\dot{F}$ such that

$$\begin{cases}|\alpha|_\mathfrak{p}=|\mathfrak{c}|_\mathfrak{p}|\mathfrak{a}|_\mathfrak{p}^{-1} & \forall\,\mathfrak{p}\in T\\ |\alpha|_\mathfrak{p}\le 1 & \forall\,\mathfrak{p}\in S-T\,.\end{cases}$$

Then

$$|\alpha\mathfrak{a}+\beta\mathfrak{b}|_\mathfrak{p}=\max\left(|\alpha\mathfrak{a}|_\mathfrak{p}\,,\,|\beta\mathfrak{b}|_\mathfrak{p}\right)=|\mathfrak{c}|_\mathfrak{p}\quad\forall\,\mathfrak{p}\in S\,.$$

Hence $\alpha\mathfrak{a}+\beta\mathfrak{b}=\mathfrak{c}$ by Proposition 22:1. **q. e. d.**

22:5a. *Every fractional ideal is of the form $\alpha\mathfrak{o}+\beta\mathfrak{o}$ with α, β in $\dot{F}$.*

22:5b. *Every fractional ideal is a finitely generated $\mathfrak{o}$-module.*

22:5c. *Suppose $\mathfrak{b}$ is actually integral. Then there is an integral ideal $\mathfrak{a}'$ in the ideal class of $\mathfrak{a}$ such that $\mathfrak{a}'$ and $\mathfrak{b}$ are relatively prime.*

Consider an ideal $\mathfrak{a}$ with $0\subset\mathfrak{a}\subset\mathfrak{o}$. We call $\mathfrak{a}$ decomposable if there are integral ideals $\mathfrak{b}$ and $\mathfrak{c}$ which are distinct from $\mathfrak{o}$, and hence from $\mathfrak{a}$, such that $\mathfrak{a}=\mathfrak{b}\mathfrak{c}$; if there are no such ideals we call $\mathfrak{a}$ indecomposable. We recall from elementary algebra that an ideal $\mathfrak{a}$ of $\mathfrak{o}$ is called a prime ideal if the residue class ring $\mathfrak{o}/\mathfrak{a}$ is an integral domain. We shall prove shortly that the indecomposable ideals, the prime ideals, and the maximal ideals of $\mathfrak{o}$ are the same (provided of course that we omit the ideals 0 and $\mathfrak{o}$). It is convenient to introduce

$$\mathfrak{o}(S)\cap\mathfrak{m}(\mathfrak{p})\,;$$

this is an integral ideal and Corollary 21:2a shows that

$$|\mathfrak{o}(S)\cap\mathfrak{m}(\mathfrak{p})|_\mathfrak{q}=\begin{cases}|\pi_\mathfrak{p}|_\mathfrak{p} & \text{if}\quad \mathfrak{q}=\mathfrak{p}\\ 1 & \text{if}\quad \mathfrak{q}\in S-\mathfrak{p}\end{cases}$$

for any $\mathfrak{p}$ in S, where $\pi_\mathfrak{p}$ is a prime element of F at $\mathfrak{p}$.

22:6. *Let* $\mathfrak{a}$ *be a fractional ideal with respect to a Dedekind set of spots* S *on the field* F. *Suppose* $0 \subset \mathfrak{a} \subset \mathfrak{o}$. *Then the following assertions about* $\mathfrak{a}$ *are equivalent:*

(1) $\mathfrak{a} = \mathfrak{o} \cap \mathfrak{m}(\mathfrak{p})$ *for some* $\mathfrak{p} \in S$

(2) $\mathfrak{a}$ *is a prime ideal in* $\mathfrak{o}$

(3) $\mathfrak{a}$ *is indecomposable*

(4) $\mathfrak{a}$ *is a maximal ideal in* $\mathfrak{o}$.

Proof. (1) $\Rightarrow$ (2). Consider typical α, β in $\mathfrak{o} - \mathfrak{a}$. Then

$$|\alpha\beta|_{\mathfrak{p}} = |\alpha|_{\mathfrak{p}} |\beta|_{\mathfrak{p}} = 1 ,$$

hence $\alpha\beta \in \mathfrak{o} - \mathfrak{a}$, hence $\mathfrak{o}/\mathfrak{a}$ is an integral domain, hence $\mathfrak{a}$ is a prime ideal.

(2) $\Rightarrow$ (3). Suppose we had a factorization $\mathfrak{a} = \mathfrak{b}\mathfrak{c}$ with $0 \subset \mathfrak{b} \subset \mathfrak{o}$ and $0 \subset \mathfrak{c} \subset \mathfrak{o}$. Then $\mathfrak{a} \subset \mathfrak{b}$ and $\mathfrak{a} \subset \mathfrak{c}$. Pick $\beta \in \mathfrak{b} - \mathfrak{a}$ and $\gamma \in \mathfrak{c} - \mathfrak{a}$. Then $\beta\gamma \in \mathfrak{b}\mathfrak{c} = \mathfrak{a}$. And this denies the primeness of $\mathfrak{a}$.

(3) $\Rightarrow$ (4). Suppose we had $\mathfrak{b} \in I$ with $\mathfrak{a} \subset \mathfrak{b} \subset \mathfrak{o}$. Then $\mathfrak{a} = \mathfrak{b}(\mathfrak{b}^{-1}\mathfrak{a})$ would be a factorization in which $\mathfrak{b} \subset \mathfrak{o}$ and $\mathfrak{b}^{-1}\mathfrak{a} \subset \mathfrak{o}$. Thus $\mathfrak{a}$ would be decomposable.

(4) $\Rightarrow$ (1). By Proposition 22:1 there is a $\mathfrak{p} \in S$ such that $|\mathfrak{a}|_{\mathfrak{p}} < 1$. So

$$|\mathfrak{a}|_{\mathfrak{p}} \leqq |\pi_{\mathfrak{p}}|_{\mathfrak{p}} = |\mathfrak{o}(S) \cap \mathfrak{m}(\mathfrak{p})|_{\mathfrak{p}}$$

with $\pi_{\mathfrak{p}}$ a prime element of F at $\mathfrak{p}$. Hence $|\mathfrak{a}|_{\mathfrak{q}} \leqq |\mathfrak{o}(S) \cap \mathfrak{m}(\mathfrak{p})|_{\mathfrak{q}}$ holds for all $\mathfrak{q} \in S$. Hence $\mathfrak{a} \subseteqq \mathfrak{o}(S) \cap \mathfrak{m}(\mathfrak{p})$, again by Proposition 22:1. Hence $\mathfrak{a} = \mathfrak{o}(S) \cap \mathfrak{m}(\mathfrak{p})$ by the hypothesis that $\mathfrak{a}$ is a maximal ideal in $\mathfrak{o}$.

q. e. d.

The preceding discussion provides a natural bijection of the set of prime spots S onto the set of proper prime ideals of $\mathfrak{o}(S)$ that is determined by the rule

$$\mathfrak{p} \rightarrowtail \mathfrak{o}(S) \cap \mathfrak{m}(\mathfrak{p}) .$$

In order to cut down on notation we shall refer to $\mathfrak{o}(S) \cap \mathfrak{m}(\mathfrak{p})$ as the prime ideal $\mathfrak{p}$ of F at S whenever it is convenient to do so, and provided there is no risk of confusing the two $\mathfrak{p}$'s. When we use this convention we can refer to S as either the given set of prime spots, or the set of prime ideals of $\mathfrak{o}$ at the given set of prime spots. As an example of the convention let us see what is meant by the equation

$$|\mathfrak{p}|_{\mathfrak{q}} = \begin{cases} |\pi_{\mathfrak{p}}|_{\mathfrak{p}} & \text{if} \quad \mathfrak{q} = \mathfrak{p} \\ 1 & \text{if} \quad \mathfrak{q} \in S - \mathfrak{p} . \end{cases}$$

Here $\pi_{\mathfrak{p}}$ is a prime element of F at the prime spot $\mathfrak{p}$; the first $\mathfrak{p}$ is a prime ideal; all other $\mathfrak{p}$'s and $\mathfrak{q}$'s are prime spots.

22:7. **Unique Factorization Theorem.** *Let* S *be a Dedekind set of spots on a field* F. *Then the fractional ideals under product formation form a free abelian group. The set of prime ideals* S *is a base for this group.*

Proof. Consider typical $\mathfrak{a} \in I$. At each $\mathfrak{p} \in S$ choose $\nu_\mathfrak{p} \in \mathbf{Z}$ in such a way that $|\mathfrak{a}|_\mathfrak{p} = |\mathfrak{p}|_\mathfrak{p}^{\nu_\mathfrak{p}}$. Then

$$|\mathfrak{a}|_\mathfrak{q} = |\prod_S \mathfrak{p}^{\nu_\mathfrak{p}}|_\mathfrak{q} \quad \forall\, \mathfrak{q} \in S\,.$$

Hence

$$\mathfrak{a} = \prod_S \mathfrak{p}^{\nu_\mathfrak{p}}$$

by Proposition 22:1. Thus the multiplicative group I is generated by its prime ideals. The uniqueness is also clear from the equation

$$\left|\prod_S \mathfrak{p}^{\nu_\mathfrak{p}}\right|_\mathfrak{q} = |\mathfrak{q}|_\mathfrak{q}^{\nu_\mathfrak{q}}\,.$$

q. e. d.

For example consider $\mathfrak{a}, \mathfrak{b} \in I$ in their prime factorizations

$$\mathfrak{a} = \prod_S \mathfrak{p}^{\nu_\mathfrak{p}}\,, \quad \mathfrak{b} = \prod_S \mathfrak{p}^{\mu_\mathfrak{p}}\,.$$

Then

$$\mathfrak{a} \subseteq \mathfrak{b} \iff \nu_\mathfrak{p} \geqq \mu_\mathfrak{p} \quad \forall\, \mathfrak{p} \in S\,.$$

Also

$$\mathfrak{a} + \mathfrak{b} = \prod_S \mathfrak{p}^{\min(\nu_\mathfrak{p},\, \mu_\mathfrak{p})}, \quad \mathfrak{a} \cap \mathfrak{b} = \prod_S \mathfrak{p}^{\max(\nu_\mathfrak{p},\, \mu_\mathfrak{p})}\,.$$

And $\mathfrak{a}$ and $\mathfrak{b}$ are relatively prime if and only if $\min(\nu_\mathfrak{p}, \mu_\mathfrak{p}) = 0$ for all $\mathfrak{p} \in S$.

22:8. **Definition.** The order $\operatorname{ord}_\mathfrak{p} \mathfrak{a}$ of the fractional ideal $\mathfrak{a}$ at $\mathfrak{p}$ is the power of $\mathfrak{p}$ that appears in the prime factorization

$$\mathfrak{a} = \prod_S \mathfrak{p}^{\nu_\mathfrak{p}}\,.$$

Thus $\operatorname{ord}_\mathfrak{p} \mathfrak{a}\mathfrak{b} = \operatorname{ord}_\mathfrak{p} \mathfrak{a} + \operatorname{ord}_\mathfrak{p} \mathfrak{b}$, and $\operatorname{ord}_\mathfrak{p} \alpha\mathfrak{o} = \operatorname{ord}_\mathfrak{p} \alpha$, and

$$|\mathfrak{a}|_\mathfrak{p} = |\pi_\mathfrak{p}|_\mathfrak{p}^{\operatorname{ord}_\mathfrak{p} \mathfrak{a}}\,.$$

22:9. **Example.** Given fractional ideals $\mathfrak{a}, \mathfrak{b}, \mathfrak{r}_1, \mathfrak{r}_2$ we claim that there are non-zero scalars α, β such that

$$\begin{cases} \alpha\mathfrak{a} + \beta\mathfrak{b} = \mathfrak{o} \\ \alpha\mathfrak{a}\,\mathfrak{r}_1 + \beta\mathfrak{b}\,\mathfrak{r}_2 = \mathfrak{r}_1 + \mathfrak{r}_2\,. \end{cases}$$

It is enough to prove this with $\mathfrak{r}_1 + \mathfrak{r}_2 = \mathfrak{o}$. By Corollary 22:5c we can pick $\alpha \in \dot{F}$ such that $\alpha\mathfrak{a} + \mathfrak{r}_2 = \mathfrak{o}$, then we can pick $\beta \in \dot{F}$ with $\beta\mathfrak{b} + \alpha\mathfrak{a}\,\mathfrak{r}_1 = \mathfrak{o}$. So we clearly must have $\beta\mathfrak{b} + \alpha\mathfrak{a} = \mathfrak{o}$. And it follows, for instance by the Unique Factorization Theorem, that $\alpha\mathfrak{a}\,\mathfrak{r}_1 + \beta\mathfrak{b}\,\mathfrak{r}_2 = \mathfrak{o}$.

§ 22D. Three residue class isomorphisms

Consider two fractional ideals $\mathfrak{a}$ and $\mathfrak{b}$ with $\mathfrak{b} \subseteq \mathfrak{a}$. Being ideals they are also additive groups and as such we can form their quotient group

$\mathfrak{a}/\mathfrak{b}$. Our purpose in this subparagraph is to demonstrate three additive group isomorphisms. These are

1. $\mathfrak{o}/\mathfrak{p} \rightarrowtail F(\mathfrak{p})$ for each $\mathfrak{p} \in S$.
2. $\mathfrak{a}/\mathfrak{b} \rightarrowtail \mathfrak{o}/\mathfrak{a}^{-1}\mathfrak{b}$ if $\mathfrak{b} \subseteq \mathfrak{a}$.
3. $\mathfrak{o}/\mathfrak{a}\mathfrak{b} \rightarrowtail (\mathfrak{o}/\mathfrak{a}) \oplus (\mathfrak{o}/\mathfrak{b})$ if $\mathfrak{a} + \mathfrak{b} = \mathfrak{o}$.

For the first we consider the residue class field $(^-, \bar{F})$ at $\mathfrak{p}$ and restrict the bar mapping from $\mathfrak{o}(\mathfrak{p})$ to $\mathfrak{o}$. This restriction has kernel $\mathfrak{o}(S) \cap \mathfrak{m}(\mathfrak{p}) = \mathfrak{p}$; and its image is $\bar{F}$ since there is a representative of every residue class of F at $\mathfrak{p}$ in $\mathfrak{o}$ by Proposition 21:4. One of the elementary isomorphism theorems of group theory now provides the desired isomorphism $\mathfrak{o}/\mathfrak{p} \rightarrowtail \bar{F}$.

Next consider $\mathfrak{a}/\mathfrak{b}$. Here let bar denote the natural homomorphism of $\mathfrak{a}$ onto $\mathfrak{a}/\mathfrak{b}$. Pick $\alpha \in \mathfrak{a}$ such that $\mathfrak{a} = \alpha\mathfrak{o} + \mathfrak{b}$; this is possible by Proposition 22:5. Define the composite homomorphism

$$x \to \alpha x \to \overline{\alpha x}$$

of $\mathfrak{o}$ into $\mathfrak{a}/\mathfrak{b}$. This composite map is clearly surjective; and its kernel is $\mathfrak{a}^{-1}\mathfrak{b}$. Hence there is a natural isomorphism $\mathfrak{o}/\mathfrak{a}^{-1}\mathfrak{b} \rightarrowtail \mathfrak{a}/\mathfrak{b}$.

In the third instance we let bar denote the natural homomorphism of $\mathfrak{o}$ onto $\mathfrak{o}/\mathfrak{a}\mathfrak{b}$. The restriction of bar to $\mathfrak{a}$ has kernel $\mathfrak{a} \cap \mathfrak{a}\mathfrak{b} = \mathfrak{a}\mathfrak{b}$. Hence

$$\bar{\mathfrak{a}} \rightarrowtail \mathfrak{a}/\mathfrak{a}\mathfrak{b} \rightarrowtail \mathfrak{o}/\mathfrak{b}\,.$$

Similarly $\bar{\mathfrak{b}} \rightarrowtail \mathfrak{o}/\mathfrak{a}$. Now $\bar{\mathfrak{o}} = \bar{\mathfrak{a}} + \bar{\mathfrak{b}}$ since $\mathfrak{a}$ and $\mathfrak{b}$ are relatively prime. Let us show that $\bar{\mathfrak{a}} \cap \bar{\mathfrak{b}} = 0$. For any $\bar{x} \in \bar{\mathfrak{a}} \cap \bar{\mathfrak{b}}$ we have $a \in \mathfrak{a}$ and $b \in \mathfrak{b}$, such that $\bar{x} = \bar{a} = \bar{b}$. Thus

$$x - a \in \mathfrak{a}\mathfrak{b} \subseteq \mathfrak{a}\,, \quad x - b \in \mathfrak{a}\mathfrak{b} \subseteq \mathfrak{a}, \quad b - a \in \mathfrak{a}\,.$$

So $b = (b - a) + a \in \mathfrak{a}$. Hence $b \in \mathfrak{a} \cap \mathfrak{b}$. But

$$\mathfrak{a} \cap \mathfrak{b} = (\mathfrak{a} \cap \mathfrak{b})\,(\mathfrak{a} + \mathfrak{b}) = \mathfrak{a}\mathfrak{b}\,.$$

Hence $b \in \mathfrak{a}\mathfrak{b}$. Hence $\bar{b} = 0$. Hence $\bar{x} = 0$. So $\bar{\mathfrak{a}} \cap \bar{\mathfrak{b}} = 0$. So $\bar{\mathfrak{o}} = \bar{\mathfrak{a}} \oplus \bar{\mathfrak{b}}$. Therefore

$$\mathfrak{o}/\mathfrak{a}\mathfrak{b} \rightarrowtail (\mathfrak{o}/\mathfrak{a}) \oplus (\mathfrak{o}/\mathfrak{b})$$

as asserted.

§ 22E. Discrete valuation rings

Suppose S is a Dedekind set consisting of exactly one discrete spot. In this event $\mathfrak{o}(S) = \mathfrak{o}(\mathfrak{p})$ and we obtain an ideal theory corresponding to the discrete valuation ring $\mathfrak{o}(\mathfrak{p})$. Our conventions now read as follows:

$$\mathfrak{o} = \mathfrak{o}(S) = \mathfrak{o}(\mathfrak{p})\,, \quad \mathfrak{p} = \mathfrak{m}(\mathfrak{p}) = \pi\mathfrak{o}$$

where π is a prime element for F at $\mathfrak{p}$. Every ideal is principal since S is a finite set of spots. But here we can say more. For the Unique Factorization Theorem tells us that every fractional ideal has the form $\mathfrak{p}^\nu$ for exactly one $\nu \in \mathbf{Z}$. Hence the ideals $\pi^\nu\mathfrak{o}$ with ν in $\mathbf{Z}$ are all the distinct fractional ideals of F at $\mathfrak{p}$.

§ 22F. Definition of a Dedekind domain

We shall call a subring R of a field F a Dedekind domain[1] if there is a Dedekind set of spots S on F such that $R = \mathfrak{o}(S)$. We shall prove in the next proposition that S, if it exists, is unique. We then call S the underlying set of spots of the Dedekind domain R.

22:10. *Let S and T be Dedekind sets of spots on the field F. If $\mathfrak{o}(S) = \mathfrak{o}(T)$, then $S = T$.*

Proof. It is enough to prove that $* \in S$ whenever $*$ is a discrete spot on F for which $\mathfrak{o}(S) \subseteqq \mathfrak{o}(*)$. Let $\mathfrak{o}$ stand for $\mathfrak{o}(S)$. Pick $|\ |_*$ in $*$ and define

$$\mathfrak{a} = \{\alpha \in \mathfrak{o} \mid |\alpha|_* < 1\}.$$

Then $\mathfrak{a}$ is clearly an ideal in $\mathfrak{o}$ since $\mathfrak{o} \subseteqq \mathfrak{o}(*)$. And $0 \subset \mathfrak{a}$ since $|\ |_*$ is non-trivial on F and hence on $\mathfrak{o}$. And $\mathfrak{a} \subset \mathfrak{o}$ since 1 is not in $\mathfrak{a}$. Hence $0 \subset \mathfrak{a} \subset \mathfrak{o}$. It is easily seen that $\mathfrak{a}$ is actually a prime ideal in $\mathfrak{o}$. So there is a prime spot $\mathfrak{p}$ in S such that $\mathfrak{a}$ is equal to the prime ideal $\mathfrak{p}$ of F at S.

Hence for any α in $\mathfrak{o}$ we have $|\alpha|_\mathfrak{p} < 1$ if and only if $|\alpha|_* < 1$. If we can prove that every β in F which satisfies $|\beta|_\mathfrak{p} < 1$ also satisfies $|\beta|_* < 1$ we shall have $\mathfrak{p} = *$ by Corollary 11:4a, and we shall be through. So consider such a β. By Corollary 21:2a there is a γ such that $|\gamma|_\mathfrak{p} = 1$ and $|\gamma|_\mathfrak{q} \leqq \min(1, |\beta^{-1}|_\mathfrak{q})$ for all $\mathfrak{q} \in S - \mathfrak{p}$. Here γ is in $\mathfrak{o}$ and $|\gamma|_* = 1$. Now $\beta\gamma$ is also in $\mathfrak{o}$ with $|\beta\gamma|_\mathfrak{p} < 1$. Hence $|\beta\gamma|_* < 1$. Hence $|\beta|_* < 1$. **q. e. d.**

§ 23. Extension fields

Consider a finite extension E/F, a set of spots S on F, and a set T on E. Recall from § 11D that $T|S$ means that the spot induced on F by each spot in T is in S, and $T||S$ means that T consists of all spots on E which induce a spot in S on F. Our main purpose here is to show that if S is Dedekind and if $T|S$, then T is also Dedekind. This will be done under the assumption of separability of E/F since this is how we developed the theory of prolongations in § 15, but as a matter of fact the result is true in general. The reader will naturally ask for the connection between the ideal theory at S and that at T. This is a classical question with classical answers, but unfortunately we cannot go into it here[2].

23:1. **Theorem.** *Let $T|S$ be sets of spots on the finite separable extension E/F. If S is Dedekind then so is T.*

Proof. We can assume that $T||S$. Fix $|\ |_\mathfrak{P} \in \mathfrak{P}$ at each $\mathfrak{P} \in T$, and $|\ |_\mathfrak{p} \in \mathfrak{p}$ at each $\mathfrak{p} \in S$; let this be done in such a way that $|\ |_\mathfrak{p}$ is the restriction of $|\ |_\mathfrak{P}$ whenever $\mathfrak{P}|\mathfrak{p}$. Now every spot in S is discrete, hence

[1] There are several equivalent definitions of a Dedekind domain. See O. Zariski and P. Samuel, *Commutative algebra* (Princeton, 1958) and E. Artin, *Theory of algebraic numbers* (Göttingen lectures, 1956).

[2] For a discussion of classical ideal theory see E. Hecke, *Vorlesungen über die Theorie der algebraischen Zahlen* (New York, 1948).

the same is true in T by Proposition 16:1. Hence T satisfies Axiom (D_1) of § 21. Consider a typical $\alpha \in E$ and write

$$\alpha^m + a_1\alpha^{m-1} + \cdots + a_m = 0$$

with all $a_i \in F$. Let S_0 be a subset consisting of almost all spots of S such that

$$|a_i|_{\mathfrak{p}} \leqq 1 \quad \forall\, a_i, \quad \forall\, \mathfrak{p} \in S_0 .$$

Let $T_0 || S_0$. So T_0 is a subset of T which consists of almost all spots in T by § 15A. By the Principle of Domination it is impossible to have $|\alpha|_{\mathfrak{P}} > 1$ at any $\mathfrak{P}$ in T_0. Hence $|\alpha|_{\mathfrak{P}} \leqq 1$ for all $\mathfrak{P} \in T_0$. So Axiom (D_2) holds in T.

Now we do Axiom (D_3). Here we have two distinct spots $\mathfrak{P}_1$ and $\mathfrak{P}_2$ of T under consideration, and also a real number ε with $0 < \varepsilon < 1$. We ask for an $A \in E$ such that

$$|1 - A|_{\mathfrak{P}_1} < \varepsilon , \quad |A|_{\mathfrak{P}_2} < \varepsilon$$

and $|A|_{\mathfrak{P}} \leqq 1$ for all $\mathfrak{P} \in T$. Let $\mathfrak{p}_1$ and $\mathfrak{p}_2$ be the spots in S which are divisible by $\mathfrak{P}_1$ and $\mathfrak{P}_2$ respectively. First suppose $\mathfrak{p}_1$ and $\mathfrak{p}_2$ are distinct. Pick $\alpha \in \mathfrak{o}(S)$ with

$$|1 - \alpha|_{\mathfrak{p}_1} < \varepsilon , \quad |\alpha|_{\mathfrak{p}_2} < \varepsilon .$$

Then $A = \alpha$ is the element we are looking for. Hence we can assume that $\mathfrak{p}_1 = \mathfrak{p}_2 = \mathfrak{p}$, say. Define $T_{\mathfrak{p}} \subseteq T$ such that $T_{\mathfrak{p}} || \mathfrak{p}$. Here $T_{\mathfrak{p}}$ is finite by § 15A. By the Weak Approximation Theorem of § 11E we have a $B \in E$ such that $|B|_{\mathfrak{P}} \leqq 1$ for all $\mathfrak{P} \in T_{\mathfrak{p}}$ with $|1 - B|_{\mathfrak{P}_1}$ small and $|B|_{\mathfrak{P}_2} < \varepsilon$. Choose $\alpha \in \mathfrak{o}(S)$ in such a way that $|1 - \alpha|_{\mathfrak{P}_1}$ is small and $|B\alpha|_{\mathfrak{P}} \leqq 1$ for all $\mathfrak{P} \in T - T_{\mathfrak{p}}$. If all approximations are good enough we will have $|1 - \alpha B|_{\mathfrak{P}_1} < \varepsilon$ by continuity of multiplication. Here $A = \alpha B$ is the element we are looking for. **q. e. d.**

23:2. *Let E/F be a finite separable extension with Dedekind sets of spots $T || S$. Then the following assertions for a typical $\alpha \in E$ are equivalent:*

(1) $\alpha \in \mathfrak{o}(T)$

(2) $\operatorname{irr}(x, \alpha, F)$ *has all coefficients in* $\mathfrak{o}(S)$

(3) α *satisfies a monic polynomial with coefficients in* $\mathfrak{o}(S)$.

Proof. (1) $\Rightarrow$ (2). There is no loss of generality in assuming that $E = F(\alpha)$. Let

$$f(x) = \operatorname{irr}(x, \alpha, F) = x^m + a_1 x^{m-1} + \cdots + a_m .$$

We have to prove that

$$a_i \in \mathfrak{o}(\mathfrak{p}) \quad \forall\, a_i, \quad \forall\, \mathfrak{p} \in S .$$

So consider a $\mathfrak{p} \in S$ and perform the following construction of § 15: take the completion $F_{\mathfrak{p}}$ of F at $\mathfrak{p}$ and the splitting field $E_{\mathfrak{p}}$ of $f(x)$ over $F_{\mathfrak{p}}$, and let $\mathfrak{P}_0$ be the unique spot on $E_{\mathfrak{p}}$ which divides $\mathfrak{p}$ on $F_{\mathfrak{p}}$. Let $\alpha_1, \ldots, \alpha_m$

be all the roots of $f(x)$ in $E_{\mathfrak{p}}$. Let σ_i be the isomorphism of E into $E_{\mathfrak{p}}$ which is the identity on F and which carries α to α_i. Then $\mathfrak{P}_0^{\sigma_i^{-1}}$ is a spot on E which divides $\mathfrak{p}$, hence it is in T, hence α is an integer of E at $\mathfrak{P}_0^{\sigma_i^{-1}}$. Now σ_i is a topological isomorphism of E (under $\mathfrak{P}_0^{\sigma_i^{-1}}$) into $E_{\mathfrak{p}}$, hence $\sigma_i\alpha$ is an integer of $\sigma_i E$ at $\mathfrak{P}_0$. So we have proved that $\alpha_1, \ldots, \alpha_m$ are all in $\mathfrak{o}(\mathfrak{P}_0)$. But

$$f(x) = (x - \alpha_1) \ldots (x - \alpha_m) .$$

Hence all a_i are in $\mathfrak{o}(\mathfrak{P}_0) \cap F = \mathfrak{o}(\mathfrak{p})$. This proves the first implication.

The implication (2)⇒(3) is obvious. An easy application of the Principle of Domination will show that (3) ⇒ (1). **q. e. d.**

Chapter III

Fields of Number Theory

The first two chapters have been done in great generality. Before we can move on to the deeper results of number theory we shall have to make additional assumptions about the underlying field F. We shall do this by explicitly stating our fields of interest. They are the field of rational numbers or any field of rational functions in one variable over a finite field of coefficients, all finite extensions of these fields, and all completions thereof. By restricting ourselves to these fields we obtain two additional properties. Roughly speaking, the first of these properties is one of finiteness of the residue class field and the second is one of dependence among the valuations. These are actually the decisive properties that distinguish the rest of the arithmetic theory from the first two chapters. In fact it is possible to axiomatize these properties[1] and to show that they lead directly to the fields of number theory, but we shall not go into that here.

§ 31. Rational global fields

By a rational number field Q we shall mean any prime field of characteristic 0. Thus the field of rational numbers $\mathbf{Q}$ is essentially the only rational number field. By the ring of rational integers Z of Q we mean the subset

$$0, \pm 1, \pm 2, \ldots$$

of Q. Similarly the prime numbers of Q are the elements

$$2, 3, 5, 7, 11, \ldots .$$

[1] This is done by E. Artin and G. Whaples, *Bull. Am. Math. Soc.* (1945), pp. 469—492.

We call a field Q a rational function field if there is a subfield k containing a finite number of elements, and an element x of Q which is transcendental over k, such that $Q = k(x)$. Strictly speaking we should call such a field a rational function field in one variable over a finite constant field, but all rational function fields used in this book will satisfy the additional assumptions so we settle with the shorter terminology. By the integers of Q we mean the polynomial ring $k[x]$ contained in Q; by a prime function or prime polynomial we mean a monic irreducible polynomial in $k[x]$. Note that these concepts depend on the choice of x so that we should really refer to them "with respect to x". However k is independent of the choice of x: in fact we shall characterize k as the intersection of all valuations rings in Q. We call k the constant field of the rational function field Q.

A field Q will be called a rational global field if it is either a rational number field or a rational function field. By the ring of integers Z of Q we mean Z in the first instance and $k[x]$ in the second. Similarly a prime is either a prime number or a prime polynomial as the case may be.

Each prime p of a rational global field Q determines a spot called the p-adic spot at Q. It is defined as follows: take a typical $\alpha \in \dot{Q}$ and write it in the form

$$\alpha = p^i u/v$$

with both u and v in Z and with neither divisible by p. Define

$$|\alpha|_p = \lambda^i$$

where λ is some real constant with $0 < \lambda < 1$. Then $|\ |_p$ is a non-archimedean valuation on Q and hence determines a prime spot which we write as p. A different choice of λ gives the same spot p; and every valuation in p can be obtained by varying λ. Later in § 33A we shall make a canonical choice of the valuation $|\ |_p$ in p by making a specific choice of λ.

Let us define the spot ∞ on the rational global field Q. If Q is a rational number field there is a unique archimedean spot on Q by Example 12:13; this is the spot ∞ in this case. If Q is a rational function field we let ∞ be the spot determined by the valuation $|\ |_\infty$ which is defined in the following way: for each $\alpha \in \dot{Q}$ write $\alpha = f/g$ with $f, g \in Z = k[x]$ and put

$$|\alpha|_\infty = \lambda^{-(\deg f - \deg g)}$$

where λ is a real constant with $0 < \lambda < 1$ and deg stands for the degree of a polynomial. This definition of the spot ∞ is again independent of the choice of λ. Note that ∞ is archimedean in the rational number case, and non-archimedean in the rational function case.

We shall prove in a moment that every non-trivial spot on a rational global field is either a p-adic spot for some prime p, or else the spot ∞; the first type is called a finite spot, the second infinite. But note that in the function field case a spot may be regarded as infinite under one choice of x, but finite under another; for instance the infinite spot under x is a finite spot under $y = 1/x$. Whenever we refer to a spot p without further qualification it can be either finite or infinite.

31:1. *The finite p-adic spots and the spot ∞ are all the non-trivial spots on the rational global field Q. All these spots are different.*

Proof. 1) If p and q are distinct primes in Q then $p \in \mathfrak{m}(p)$ and $p \in \mathfrak{u}(q)$ so that the spots p and q are distinct. Can the spot p be equal to the spot ∞? This is clearly impossible in the number theoretic case since ∞ is archimedean and p is not. It is also impossible in the function theoretic case since $x \notin \mathfrak{o}(\infty)$ and $x \in \mathfrak{o}(p)$. This proves the second part of the proposition.

2) Now let a non-trivial spot $\mathfrak{p}$ be given on Q. We want to prove that $\mathfrak{p}$ is one of the spots under consideration. If $\mathfrak{p}$ is archimedean we are in the number theoretic case and so $\mathfrak{p} = \infty$ by Example 12:13. Hence assume that $\mathfrak{p}$ is non-archimedean. Pick a valuation $|\ |$ in $\mathfrak{p}$.

If Q is a rational function field with $x \notin \mathfrak{o}(\mathfrak{p})$, then for any $f(x)$ in $k[x]$ we have

$$|f(x)| = |x|^{\deg f}$$

by the Principle of Domination since the valuation is trivial on the finite field k, hence

$$\left|\frac{f(x)}{g(x)}\right| = \left|\frac{1}{x}\right|^{\deg g - \deg f}$$

with $|1/x| < 1$, so $\mathfrak{p} = \infty$.

We are therefore left with the following possibility: $\mathfrak{p}$ is non-archimedean and Q is either a rational number field or a rational function field in which $|x| \leq 1$. In either event we have $Z \subseteq \mathfrak{o}(\mathfrak{p})$. This, together with the fact that $\mathfrak{p}$ is non-trivial, implies that $0 \subset \mathfrak{m}(\mathfrak{p}) \cap Z \subset Z$. Hence

$$\mathfrak{m}(\mathfrak{p}) \cap Z = pZ$$

for some $p \in Z$ which is neither 0 nor a unit of the ring Z. We claim that p can be chosen a prime: for if we have a factorization $p = rs$ with $r, s \in Z$ then $|rs| < 1$ implies that $|r| < 1$, say; hence $r \in \mathfrak{m}(\mathfrak{p}) \cap Z = pZ$; hence there is an $r_1 \in Z$ such that $1 = r_1 s$; in other words s is a unit of the ring Z, so p is a prime times a unit. We can therefore assume that p is a prime. A typical $\alpha \in \dot{Q}$ can be written $\alpha = p^i u/v$ with u, v in $Z - pZ$. Then $|u| = |v| = 1$, hence

$$|\alpha| = |p|^i \quad \text{with} \quad |p| < 1\,,$$

hence $|\ |$ is in the spot p, hence $\mathfrak{p} = p$. **q. e. d.**

31:2. *The set of all finite spots S on a rational global field Q is Dedekind. And $Z = \mathfrak{o}(S)$.*

Proof. The Dedekind axioms (D_1) and (D_2) of § 21 are immediate. Let us verify (D_3). Consider the spots determined by two distinct primes p and q of Z. For each $r \in \mathsf{N}$ write

$$1 = \xi p^r + \eta q^r \quad \text{with} \quad \xi, \eta \in Z,$$

and put

$$\alpha_r = \xi p^r, \quad 1 - \alpha_r = \eta q^r.$$

Then $|\alpha_r|_l \leqq 1$ for all primes l since $\alpha_r \in Z$. And

$$|\alpha_r|_p \leqq |p|_p^r, \quad |1 - \alpha_r|_q \leqq |q|_q^r.$$

If r is large, $|p|_p^r$ and $|q|_q^r$ are small. Hence (D_3) holds.

Clearly $Z \subseteq \mathfrak{o}(S)$. Suppose $\alpha \in Q - Z$. Then there is a prime p such that $\alpha = p^{-i} u/v$ with $i \in \mathsf{N}$ and both u and v prime to p. But then $|\alpha|_p = |p|_p^{-i} > 1$. So $\alpha \notin \mathfrak{o}(S)$. Hence $Z = \mathfrak{o}(S)$. **q. e. d.**

31:3. **Example.** Let Q be a rational function field. Then the only polynomials that are integral at ∞ are constants. Hence

$$k = \bigcap_{\mathfrak{p}} \mathfrak{o}(\mathfrak{p}),$$

the intersection being taken over all spots on Q. Thus the field of constants k is independent of x, as we asserted earlier. Incidentally this shows that the set of all spots on Q, even though it satisfies (D_1) and (D_2), is not Dedekind.

31:4. **Remark.** Consider the finite spot p determined by the prime p of the rational global field Q. A representative set of the residue class field at p can be found in Z by Proposition 21:4; this representative set can then be refined to the following:

1. the elements $0, 1, \ldots, p-1$ of Q when Q is a rational number field,
2. the set of all polynomials over k (include 0) of degree $< \deg p$ when Q is a rational function field.

The number of elements in the residue class field of Q at p is therefore equal to

1. p in the first case,
2. $q^{\deg p}$ in the second, where q is the order of k.

In the function theoretic case one easily shows that Q has k as a representative set of the residue class field at ∞.

31:5. **Example.** Consider the field of rational numbers $\mathbf{Q}$ and a prime number p in $\mathbf{Q}$. The prime number p determines the p-adic spot p on $\mathbf{Q}$. Take a completion $\mathbf{Q}_p$ of $\mathbf{Q}$ at p and fix it. As usual let p stand for the prime spot associated with $\mathbf{Q}_p$. The field $\mathbf{Q}_p$ provided with its spot p is called the field of p-adic numbers. The ring of integers of $\mathbf{Q}_p$ is called the ring of p-adic integers and is written $\mathbf{Z}_p$. The spot p on $\mathbf{Q}$ or $\mathbf{Q}_p$ is discrete with prime element p. And $0, 1, \ldots, p-1$ is a representative set of the residue class field of $\mathbb{Q}$ or $\mathbb{Q}_p$ at p. Hence by Proposition 16:3 every p-adic number has the form

$$\sum_u^\infty c_\nu p^\nu \quad (0 \leqq c_\nu < p,\ c_u \neq 0)$$

where the sum is taken in the p-adic topology.

31:6. **Definition.** An algebraic number field, or just a number field, is a finite extension of a rational number field.

31:7. **Definition.** An algebraic function field, or just a function field, is a finite extension of a rational function field.

31:8. **Definition.** A global field is a finite extension of a rational global field. In other words, the global fields are the algebraic number fields and the algebraic function fields.

31:9. *Every global field is a finite separable extension of a rational global field.*

Proof. 1) Let F be the given global field. By definition F contains a rational global field Q such that F/Q is finite. In the number theoretic case all extensions are separable. Hence we can assume that Q is a rational function field $Q = k(x)$. We shall show that for a suitable choice of x the extension F/Q is separable[1]. The method of finding the new x will be the following: keep taking p-th roots of x in F until x no longer has a p-th root in F. Here p denotes the characteristic of the field F.

2) The next step is in fact a lemma: prove that every function field F satisfies $[F:F^{p^\nu}] = p^\nu$ for any natural number ν, where F^{p^ν} is the field consisting of the p^ν-th powers of the elements of F. First consider the rational function field $Q = k(x)$. Then Q^p is the rational function field $Q^p = k(x^p)$ since $k^p = k$. And clearly $[Q:Q^p] \leqq p$. Now any relation of the form

$$f_0(x^p) + x f_1(x^p) + \cdots + x^{p-1} f_{p-1}(x^p) = 0$$

with all $f_i(x^p) \in k[x^p]$ would make x algebraic over k, hence $1, x, \ldots, x^{p-1}$ are independent over Q^p, hence $[Q:Q^p] = p$.

From this we can deduce that $[F:F^p] = p$: let φ be the isomorphism of F onto F^p that is defined by $\varphi\xi = \xi^p$; then $\varphi(Q) = Q^p$; hence $[F:Q] = [F^p:Q^p]$; but

$$[F:F^p]\,[F^p:Q^p] = [F:Q]\,[Q:Q^p];$$

hence

$$[F:F^p] = [Q:Q^p] = p$$

as asserted.

---

[1] F is then said to be separably generated over k. See O. Zariski and P. Samuel, *Commutative algebra* (Princeton, 1958), p. 105, for the corresponding result in an algebraic function field in several variables over a perfect ground field.

Finally we observe that F^p is a finite extension of the rational function field Q^p, hence it is an algebraic function field, hence $[F^p:F^{p^2}] = p$; etc.; hence $[F:F^{p^\nu}] = p^\nu$. This proves the lemma.

3) x is clearly not a p-th power in $k(x)$. If x is in F^p, replace x by its p-th root, thereby obtaining a larger Q. Repeat. Ultimately we can assume that $x \notin F^p$. We claim that $F/k(x)$ is separable for this choice of x. Suppose not: let F_0 be the separable part of the extension F/Q; thus $Q \subseteq F_0 \subset F$; now $[F:F_0] = p^e$ for some $e \geqq 1$; but we know from galois theory that the p^e-th power of every element of F is in F_0, i. e. $F^{p^e} \subseteq F_0$. hence $F^{p^e} = F_0$ by step 2); hence $x \in F^{p^e} \subseteq F^p$, and this is a contradiction. So F/Q is indeed separable. **q. e. d.**

§ 32. Local fields

§ 32A. The normalized valuation

Let $\mathfrak{p}$ be any non-archimedean spot on a field F. We define $N\mathfrak{p}$ to be the number of elements in the residue class field of F at $\mathfrak{p}$. Suppose that

$$\mathfrak{p} \text{ is discrete and } N\mathfrak{p} < \infty .$$

Under these assumptions it is possible to canonically select a valuation $|\ |_\mathfrak{p}$ at the spot $\mathfrak{p}$. (The fact that such a canonical choice is possible is an important part of the development of algebraic number theory; in particular it makes possible the Product Formula of § 33B.) How is the canonical choice to be made? Each valuation in $\mathfrak{p}$ is completely determined by the generator of its value group; we choose that valuation $|\ |_\mathfrak{p}$ in $\mathfrak{p}$ under which $N\mathfrak{p}$ becomes a generator of the value group $|F|_\mathfrak{p}$; in other words, $|\ |_\mathfrak{p}$ is defined by the equation

$$|\alpha|_\mathfrak{p} = \left(\frac{1}{N\mathfrak{p}}\right)^{\mathrm{ord}_\mathfrak{p}\,\alpha} \qquad \forall\, \alpha \in F .$$

The valuation defined in this way is called the normalized valuation at $\mathfrak{p}$.

From now on, whenever the spot $\mathfrak{p}$ in question is discrete with finite residue class field, we let $|\ |_\mathfrak{p}$ stand for the normalized valuation at $\mathfrak{p}$.

If $\mathfrak{p}$ is a spot having the above properties, then so is the spot $\mathfrak{p}$ on the completion $F_\mathfrak{p}$. And $N\mathfrak{p}$ has the same value on the completion as on F. In particular, the normalized valuation on $F_\mathfrak{p}$ induces the normalized valuation on F at $\mathfrak{p}$. We let $|\ |_\mathfrak{p}$ stand for both.

32:1. **Definition.** A local field is a composite object consisting of a spot $\mathfrak{p}$ on a field F such that

(1) $\mathfrak{p}$ is complete and discrete,

(2) the residue class field at $\mathfrak{p}$ is finite.

32:2. **Example.** The field of p-adic numbers $\mathbb{Q}_p$ is a local field.

32:3. *Let E/F be a finite extension of fields with spots $\mathfrak{P}|\mathfrak{p}$. Suppose F is a local field at $\mathfrak{p}$. Then*

(1) *E is a local field at $\mathfrak{P}$*

(2) $N\mathfrak{P} = (N\mathfrak{p})^{f(\mathfrak{P}|\mathfrak{p})}$

(3) $|\alpha|_{\mathfrak{P}} = |N_{E/F}\alpha|_{\mathfrak{p}} \quad \forall\ \alpha \in E$ *(both valuations normalized).*

Proof. $E(\mathfrak{P})/F(\mathfrak{P})$ is a finite extension of degree $f(\mathfrak{P}|\mathfrak{p})$, and $F(\mathfrak{P})$ contains $N\mathfrak{p}$ elements, hence $E(\mathfrak{P})$ contains $N\mathfrak{P} = (N\mathfrak{p})^{f(\mathfrak{P}|\mathfrak{p})}$ elements. In particular $E(\mathfrak{P})$ is finite. But E is complete by Theorem 11:18, and it is discrete by Proposition 16:1, hence it is a local field at $\mathfrak{P}$. We are left with the third part of the proposition. Before proving this let us first consider a typical α in F; then the equation $e(\mathfrak{P}|\mathfrak{p})\, f(\mathfrak{P}|\mathfrak{p}) = n = [E:F]$ of Theorem 16:4 gives us

$$|\alpha|_{\mathfrak{P}} = \left(\frac{1}{N\mathfrak{P}}\right)^{\operatorname{ord}_{\mathfrak{P}}\alpha} = \left(\frac{1}{N\mathfrak{p}}\right)^{f(\mathfrak{P}|\mathfrak{p})\, e(\mathfrak{P}|\mathfrak{p})\, \operatorname{ord}_{\mathfrak{p}}\alpha} = |\alpha|_{\mathfrak{p}}^{n}\,.$$

Then by Theorem 14:1 we have for any $\alpha \in E$

$$|\alpha|_{\mathfrak{P}} = |N_{E/F}\alpha|_{\mathfrak{P}}^{1/n} = |N_{E/F}\alpha|_{\mathfrak{p}}$$

q. e. d.

32:4. *The completion of a global field at any one of its non-trivial non-archimedean spots is a local field.*

Proof. Let F be the field, $\mathfrak{p}$ the spot. Take a rational global field $Q \subseteq F$ over which F is finite. Form the completions $F_{\mathfrak{p}}$ and $Q_{\mathfrak{p}} \subseteq F_{\mathfrak{p}}$. Then Q is discrete with finite residue class field at the spot induced by $\mathfrak{p}$ by Remark 31:4, hence $Q_{\mathfrak{p}}$ is a local field. But $F_{\mathfrak{p}}/Q_{\mathfrak{p}}$ is finite since F/Q is. Hence $F_{\mathfrak{p}}$ is a local field by Proposition 32:3. **q. e. d.**

32:5. **Example.** Let F be a local field with maximal ideal $\mathfrak{p}$. Then by § 22E every integral $\mathfrak{o}$-ideal is of the form $\mathfrak{p}^{\nu}$ for some $\nu \in \mathsf{N}$. Use the results of § 22D to show that the group index $(\mathfrak{o}:\mathfrak{p}^{\nu})$ is $(N\mathfrak{p})^{\nu}$.

§ 32B. Ramified and unramified extensions of local fields[1]

We consider a finite extension E/F of local fields at the spots $\mathfrak{P}|\mathfrak{p}$. We shorten the notation and put

$$e = e(\mathfrak{P}|\mathfrak{p})\,, \quad f = f(\mathfrak{P}|\mathfrak{p})\,, \quad n = [E:F]\,.$$

We know from Theorem 16:4 that

$$ef = n\,.$$

We shall say that the extension E/F (of local fields at $\mathfrak{P}|\mathfrak{p}$) is unramified if

$$e = 1\,, \quad f = n\,.$$

If E/F is not unramified we call it ramified. We call E/F fully ramified if

$$e = n\,, \quad f = 1\,.$$

---

[1] For a general theory of ramification see, for instance, O. F. G. SCHILLING, *The theory of valuations* (New York, 1950).

Thus a fully ramified extension is ramified when $n > 1$. Let us make the following observations: given finite extensions $F \subseteq E \subseteq H$ of local fields, then

$$H/F \text{ is unramified} \Leftrightarrow H/E \text{ and } E/F \text{ are,}$$

$$H/F \text{ is fully ramified} \Leftrightarrow H/E \text{ and } E/F \text{ are;}$$

these facts follow easily from the multiplicative behavior of e, f, n in a tower of fields.

32:6. *Let E/F be a finite extension of local fields. Suppose that $E = F(\alpha)$ for some α which satisfies a monic integral polynomial $f(x)$ over F with $f'(\alpha)$ a unit. Then α is integral. And E/F is unramified. And the residue class field $\bar{E}$ is equal to $\bar{F}(\bar{\alpha})$.*

Proof. 1) The Principle of Domination says that α must be an integer of the local field E. Let $g(x) = \operatorname{irr}(x, \alpha, F)$. Write $f(x) = g(x)\,h(x)$ with $g(x)$ and $h(x)$ monic; then $g(x)$ and $h(x)$ have integral coefficients in F by Proposition 13:7; and the formula for differentiation gives $f'(\alpha) = g'(\alpha)\,h(\alpha)$; so $g'(\alpha)$ is a unit. We have therefore shown that the given polynomial $f(x)$ can be taken irreducible over F.

2) Next we claim that $\bar{f}(x)$ is irreducible over $\bar{F}$. Suppose not. Let $\varphi(x) = \operatorname{irr}(x, \bar{\alpha}, \bar{F})$. Since $\bar{f}(\bar{\alpha}) = 0$ there is a polynomial $\psi(x)$ over $\bar{F}$ such that $\bar{f}(x) = \varphi(x)\,\psi(x)$ and

$$1 \leq \deg\varphi < \deg\bar{f} = \deg f\,.$$

Now $\bar{f}'(\bar{\alpha}) \neq 0$ since $f'(\alpha)$ is a unit of E; hence $\bar{\alpha}$ is not a multiple root of $\bar{f}(x)$; hence $\varphi(x)$ and $\psi(x)$ are relatively prime. Then Hensel's lemma implies that $f(x)$ has a factor of degree equal to $\deg\varphi$. But this is impossible since $f(x)$ is irreducible. Hence $\bar{f}(x)$ is indeed irreducible over $\bar{F}$.

3) We have $\bar{\alpha} \in \bar{E}$ since α is an integer of E. But $\bar{f}(\bar{\alpha}) = 0$. Hence

$$[\bar{E}:\bar{F}] \geq [\bar{F}(\bar{\alpha}):\bar{F}] = \deg\bar{f} = \deg f = n\,.$$

But $[\bar{E}:\bar{F}] = n/e \leq n$. Hence

$$[\bar{E}:\bar{F}] = n\,, \quad e = 1\,.$$

So E/F is unramified. And also by comparing degrees in the above inequalities we obtain $\bar{E} = \bar{F}(\bar{\alpha})$. **q. e. d.**

32:6a. *Suppose m is a unit in F. Then the local field $F(\zeta)$ obtained by adjoining an m-th root of unity ζ to F is unramified over F.*

32:7. **Lemma.** *Let C be a cyclic subgroup of order m of $\dot{F}$ with m relatively prime to $N\mathfrak{p}$. Then distinct elements of C fall in distinct cosets of $\mathfrak{o}(\mathfrak{p})$ modulo $\mathfrak{m}(\mathfrak{p})$.*

Proof. Let ζ generate C. Then C consists precisely of all m-th roots of unity $1, \zeta, \ldots, \zeta^{m-1}$ in F. Hence

$$x^m - 1 = \prod_{i=0}^{m-1} (x - \zeta^i) .$$

Computing derivatives at $x = \zeta^j$ gives

$$m\zeta^{j(m-1)} = \prod_{i \neq j} (\zeta^j - \zeta^i) .$$

But the left hand side is a unit since m is prime to $N\mathfrak{p}$, i. e. since m is not 0 in the residue class field; and every term on the right is an integer; hence every term $\zeta^i - \zeta^j$ is a unit. Thus ζ^i and ζ^j fall in different cosets whenever $i \neq j$. **q. e. d.**

32:8. *Let F be a local field at $\mathfrak{p}$. Then the group C of $(N\mathfrak{p} - 1)^{th}$ roots of unity in F is cyclic of order $N\mathfrak{p} - 1$. So F contains all $(N\mathfrak{p} - 1)^{th}$ roots of unity. It contains no other roots of unity of period prime to $N\mathfrak{p}$.*

Proof. Consider any extension $F(\zeta)/F$ where ζ is a root of the polynomial $x^{N\mathfrak{p}-1} - 1$. By Proposition 32:6 we obtain $[F(\zeta):F] = [\bar{F}(\bar{\zeta}):\bar{F}]$. But $\bar{\zeta}^{N\mathfrak{p}} = \bar{\zeta}$; and $\bar{F}$, being a finite field of $N\mathfrak{p}$ elements, consists of all solutions of $x^{N\mathfrak{p}} = x$; hence $\bar{\zeta} \in \bar{F}$; hence $[F(\zeta):F] = 1$ and ζ is in F. So F contains all roots of the polynomial $x^{N\mathfrak{p}-1} - 1$, and these roots are $N\mathfrak{p} - 1$ in number since the polynomial in question has distinct roots over F. Hence C has order $N\mathfrak{p} - 1$. Since C is a finite subgroup of $\dot{F}$ it is cyclic[1].

The second part of the proposition is now obvious. Let us do the third. Consider a root of unity $\eta \in F$ of period m where m is prime to $N\mathfrak{p}$. Then Lemma 32:7 tells us that the elements $1, \eta, \ldots, \eta^{m-1}$ of the cyclic group generated by η fall in distinct cosets of $\mathfrak{o}(\mathfrak{p})$ modulo $\mathfrak{m}(\mathfrak{p})$. Hence $\bar{\eta}$ has period m in the residue class field $\bar{F}$ of F at $\mathfrak{p}$. But there are $N\mathfrak{p} - 1$ elements in the multiplicative group of $\bar{F}$. So m divides $N\mathfrak{p} - 1$. Hence η is an $(N\mathfrak{p} - 1)^{th}$ root of unity. **q. e. d.**

32:8a. *The $(N\mathfrak{p} - 1)^{th}$ roots of unity plus 0 form a representative set in F of the residue class field $F(\mathfrak{p})$.*

Proof. The elements of the set $0 \cup C$ fall in different cosets of $\mathfrak{o}(\mathfrak{p})$ modulo $\mathfrak{m}(\mathfrak{p})$ by Lemma 32:7. But the number of elements is equal to the number of cosets. **q. e. d.**

32:9. *Let E/F be a finite extension of local fields. Suppose that E is a splitting field of*

$$x^{(N\mathfrak{p})^n} - x$$

over F. Then E/F is unramified of degree n.

---

[1] It is a known result of galois theory, easily deduced from the Fundamental Theorem of Abelian Groups and the fact that $x^n = 1$ has at most n solutions in a field, that *every finite subgroup of the multiplicative group of an arbitrary field is cyclic.*

Proof. E contains all $((N\mathfrak{p})^n - 1)^{th}$ roots of unity and these are $(N\mathfrak{p})^n - 1$ in number since the polynomial $x^{(N\mathfrak{p})^n} - x$ has distinct roots over F. All these roots have period prime to $N\mathfrak{p}$, and hence to $N\mathfrak{P}$. But by Proposition 32:8 there are $N\mathfrak{P} - 1$ elements in E which have period prime to $N\mathfrak{P}$. Hence $(N\mathfrak{p})^n - 1 \leq N\mathfrak{P} - 1$. Hence $(N\mathfrak{p})^n \leq N\mathfrak{P}$.

Now every finite subgroup of $\dot{F}$ is cyclic. Hence $E = F(\zeta)$ with $\zeta^{(N\mathfrak{p})^n} = \zeta$. Proposition 32:6 tells us that E/F is unramified with residue class field $\bar{E} = \bar{F}(\bar{\zeta})$. Here $\bar{\zeta}^{(N\mathfrak{p})^n} = \bar{\zeta}$, hence every α in $\bar{E}$ satisfies the equation $\alpha^{(N\mathfrak{p})^n} = \alpha$. So $N\mathfrak{P} \leq (N\mathfrak{p})^n$. Hence $(N\mathfrak{p})^n = N\mathfrak{P}$. This gives $n = f(\mathfrak{P}|\mathfrak{p}) = [E:F]$ and proves that E/F is unramified of degree n.

q. e. d.

32:10. *Let E/F be a finite extension of local fields. Suppose the extension is unramified of degree n. Then E is a splitting field over F of the polynomial*

$$x^{(N\mathfrak{p})^n} - x\,.$$

Proof. $N\mathfrak{P} = (N\mathfrak{p})^n$. So E contains all $((N\mathfrak{p})^n - 1)^{th}$ roots of unity. Hence it contains a splitting field H of the given polynomial. But $[H:F] = n$ by Proposition 32:9. So $H = E$. **q. e. d.**

32:11. **Example.** Consider a finite extension E/F of local fields. If K_m is a subfield of E such that K_m/F is unramified of degree m, then K_m is the splitting field in E of

$$x^{(N\mathfrak{p})^m} - x$$

over F. Hence K_m is unique. Consider another such field K_r and suppose r divides m. Then every root of

$$x^{(N\mathfrak{p})^r} - x = 0$$

is also a root of

$$x^{(N\mathfrak{p})^m} = x^{(N\mathfrak{p})^r \cdots (N\mathfrak{p})^r} = x\,.$$

Hence $K_r \subseteq K_m$ whenever r divides m.

32:12. *Any unramified extension E/F of local fields at $\mathfrak{P}|\mathfrak{p}$ is galois with cyclic galois group $\mathfrak{G}(E/F)$. There is a unique $\sigma \in \mathfrak{G}(E/F)$ such that*

$$\sigma x \equiv x^{N\mathfrak{p}} \bmod \mathfrak{m}(\mathfrak{P}) \quad \forall\, x \in \mathfrak{o}(\mathfrak{P})\,.$$

This σ generates $\mathfrak{G}(E/F)$.

Proof. E/F is galois since E is a splitting field of the separable polynomial $x^{N\mathfrak{P}-1} - 1$ over F. Let $(^-, \bar{E})$ be a residue class field of E at $\mathfrak{P}$. Consider a typical $\varrho \in \mathfrak{G}(E/F)$. Then ϱ is topological by Corollary 11:18a, hence ϱ sends $\mathfrak{o}(\mathfrak{P})$ onto $\mathfrak{o}(\mathfrak{P})$ and $\mathfrak{m}(\mathfrak{P})$ onto $\mathfrak{m}(\mathfrak{P})$, hence it induces an automorphism $\bar{\varrho}$ on $\bar{E}$ that satisfies

$$\bar{\varrho}\bar{x} = \overline{\varrho x} \quad \forall\, x \in \mathfrak{o}(\mathfrak{P})\,.$$

Clearly $\bar{\varrho}$ is in $\mathfrak{G}(\bar{E}/\bar{F})$. Hence the rule $\varrho \rightarrowtail \bar{\varrho}$ provides a mapping

$$\mathfrak{G}(E/F) \rightarrowtail \mathfrak{G}(\bar{E}/\bar{F}) .$$

This mapping is clearly a group homomorphism. If $\bar{\varrho} = \bar{1}$, then

$$\varrho x - x \in \mathfrak{m}(\mathfrak{P}) \quad \forall x \in \mathfrak{o}(\mathfrak{P}) ,$$

hence $\varrho\zeta = \zeta$ for all $(N\mathfrak{P} - 1)^{th}$ roots of unity ζ in E by Corollary 32:8a, hence $\varrho = 1$. In other words, the above mapping is injective. But $\mathfrak{G}(E/F)$ and $\mathfrak{G}(\bar{E}/\bar{F})$ have the same number of elements since the given extension is unramified. Hence the above mapping is bijective.

Now the galois group of an extension of finite fields is cyclic. Hence $\mathfrak{G}(E/F)$ is cyclic. In fact $\mathfrak{G}(\bar{E}/\bar{F})$ is generated by an automorphism σ_0 such that

$$\sigma_0 x_0 = x_0^{N\mathfrak{p}} \quad \forall x_0 \in \bar{E} .$$

So if σ is chosen in $\mathfrak{G}(E/F)$ with $\bar{\sigma} = \sigma_0$, we have

$$\sigma x \equiv x^{N\mathfrak{p}} \bmod \mathfrak{m}(\mathfrak{P}) \quad \forall x \in \mathfrak{o}(\mathfrak{P}) .$$

Hence σ is the generator of $\mathfrak{G}(E/F)$ with the required property. The uniqueness of σ follows from the injectivity of the map $\varrho \rightarrow \bar{\varrho}$. **q. e. d.**

32:12a. $\sigma\zeta = \zeta^{N\mathfrak{p}}$ *for every* $(N\mathfrak{P} - 1)^{th}$ *root of unity in* E.

Proof. Apply Corollary 32:8a. **q. e. d.**

32:13. **Definition.** The automorphism σ of Proposition 32:12 is called the Frobenius automorphism.

32:14. *Let Ω/F be a finite extension of local fields with subextensions E/F and H/F. Then*

$$E/F \text{ unramified} \Rightarrow EH/H \text{ unramified.}$$

Proof. Write $E = F(\zeta)$ with ζ satisfying the equation $x^{(N\mathfrak{p})^n} - x = 0$; this is possible by Proposition 32:10. But then $EH = H(\zeta)$; and $H(\zeta)/H$ is unramified by Corollary 32:6a. **q. e. d.**

32:15. *Let E/F be an extension of local fields with spots $\mathfrak{P}|\mathfrak{p}$. Suppose $E = F(\alpha)$ with α a root of the so-called Eisenstein polynomial*

$$x^r + a_1 x^{r-1} + \cdots + a_r$$

which has all $a_i \in \mathfrak{m}(\mathfrak{p})$ and a_r a prime element at $\mathfrak{p}$. Then E/F is fully ramified of degree r, and α is a prime element at $\mathfrak{P}$.

Proof. Pick a valuation $|\ | \in \mathfrak{P}$. Then α satisfies

$$\alpha^r + a_1 \alpha^{r-1} + \cdots + a_r = 0$$

with all $|a_i| \leq |a_r| < 1$. Hence $|\alpha| < 1$ by the Principle of Domination. But then $|a_i \alpha^{r-i}| < |a_r|$ for $i = 1, 2, \ldots, r-1$. Hence $|\alpha^r| = |a_r|$, again by the Principle of Domination. So $|\alpha| = \sqrt[r]{|a_r|}$. But a_r is a prime element at $\mathfrak{p}$. Hence $e \geqq r$. However $n \leqq r$ since α satisfies an equation of degree r over F. Hence $n \leqq r \leqq e \leqq n$. Hence $r = e = n$ and the given extension is fully ramified of degree r with α a prime element at $\mathfrak{P}$. **q. e. d.**

§ 33. Global fields

§ 33 A. The normalized valuations

Consider a global field F. Let $\Omega = \Omega_F$ stand for the set of all non-trivial spots on F. If $\mathfrak{p}$ is a non-archimedean spot in Ω, then $\mathfrak{p}$ is discrete, and $F_\mathfrak{p}$ is a local field, and a normalized valuation $|\ |_\mathfrak{p}$ has been defined on F and $F_\mathfrak{p}$ (see § 32 A). What about the remaining spots in Ω? If F is an algebraic function field there are no remaining spots. If F is an algebraic number field there will be a finite number of archimedean spots in Ω, some of them real and some of them complex; for each of these a normalized valuation $|\ |_\mathfrak{p}$ has been defined in § 12.

We make the rule for this entire paragraph on Global Fields, that $|\ |_\mathfrak{p}$ will always denote the normalized valuation (discrete, real, or complex) on the global field F (or on the completion $F_\mathfrak{p}$) at the spot $\mathfrak{p} \in \Omega$.

It is easily seen that a topological isomorphism of one local field onto another is analytic with respect to the normalized valuations. The same applies to complete archimedean fields. In particular, the topological isomorphism between two completions of a global field at the same spot is analytic with respect to the normalized valuations on these completions.

Suppose we have to consider a finite separable extension E/F of global fields and let $\mathfrak{P}|\mathfrak{p}$ be two spots on E/F. Then $|\ |_\mathfrak{P}$ denotes the normalized valuation on $E_\mathfrak{P}$ at $\mathfrak{P}$, and $|\ |_\mathfrak{p}$ the normalized valuation on $F_\mathfrak{p}$ at $\mathfrak{p}$. We also let $|\ |_\mathfrak{p}$ denote the normalized valuation on $F_\mathfrak{P}$ at $\mathfrak{p}$. We claim that

$$|\alpha|_\mathfrak{P} = |N_{\mathfrak{P}|\mathfrak{p}}\, \alpha|_\mathfrak{p} \quad \forall\, \alpha \in E_\mathfrak{P}$$

where $N_{\mathfrak{P}|\mathfrak{p}}\colon E_\mathfrak{P} \to F_\mathfrak{p}$ is the norm function of § 15 B. To prove this in the discrete case we apply Proposition 32:3 and obtain

$$|\alpha|_\mathfrak{P} = |N_{E_\mathfrak{P}/F_\mathfrak{P}}\, \alpha|_\mathfrak{p} = |N_{\mathfrak{P}|\mathfrak{p}}\, \alpha|_\mathfrak{p}.$$

If $\mathfrak{p}$ and $\mathfrak{P}$ are both complex or both real the result is trivial; if $\mathfrak{p}$ is real and $\mathfrak{P}$ complex we have

$$|\alpha|_\mathfrak{P} = |\alpha|^2 = |N_{E_\mathfrak{P}/F_\mathfrak{P}}\, \alpha| = |N_{\mathfrak{P}|\mathfrak{p}}\, \alpha| = |N_{\mathfrak{P}|\mathfrak{p}}\, \alpha|_\mathfrak{p}$$

where $|\ |$ denotes the ordinary absolute value on $E_\mathfrak{P}$ and $F_\mathfrak{p}$.

§ 33B. Product formula

33:1. Theorem. *Let α be a non-zero element of the global field F. Then*

$$|\alpha|_{\mathfrak{p}} = 1 \quad \textit{for almost all } \mathfrak{p} \in \Omega\,,$$

and

$$\prod_{\mathfrak{p} \in \Omega} |\alpha|_{\mathfrak{p}} = 1\,.$$

Proof. 1) First suppose that F is a rational number field Q. Write

$$\alpha = \pm p_1^{\nu_1} p_2^{\nu_2} \cdots p_r^{\nu_r}$$

where $p_1, \ldots, p_r$ are distinct prime numbers in Q. The non-archimedean spots are the p-adic spots for all prime numbers p; now $Np = p$ by Remark 31:4; and $\operatorname{ord}_{p_i} \alpha = \nu_i$; hence

$$|\alpha|_{p_i} = \left(\frac{1}{p_i}\right)^{\nu_i}.$$

Similarly $|\alpha|_p = 1$ for all prime numbers other than $p_1, \ldots, p_r$. The only remaining spot is the one determined by the ordinary absolute value which in this case is the normalized valuation $|\ |_\infty$ at ∞. This gives $|\alpha|_\infty = p_1^{\nu_1} \cdots p_r^{\nu_r}$. Finally,

$$\prod_{\mathfrak{p} \in \Omega} |\alpha|_{\mathfrak{p}} = \left(\frac{1}{p_1}\right)^{\nu_1} \cdots \left(\frac{1}{p_r}\right)^{\nu_r} (p_1^{\nu_1} \cdots p_r^{\nu_r}) = 1\,.$$

This completes the proof of the first case.

2) A similar proof works for a rational function field Q. Write $Q = k(x)$ by choosing some suitable transcendental x over the constant field k. Write

$$\alpha = \varepsilon p_1^{\nu_1} \cdots p_r^{\nu_r}$$

where $\varepsilon \in k$ and the p_i are irreducible monic polynomials over k. The finite spots are determined by the prime polynomials p; now here $Np = q^{\deg p}$ where q is the number of elements of k, again by Remark 31:4; hence

$$|\alpha|_{p_i} = \left(\frac{1}{q}\right)^{\nu_i \deg p_i}.$$

Similarly $|\alpha|_p = 1$ for all other prime polynomials. At the infinite spot we have $N\infty = q$, and so

$$|\alpha|_\infty = \left(\frac{1}{q}\right)^{-(\deg \alpha)}$$

where

$$\deg \alpha = \nu_1 \deg p_1 + \cdots + \nu_r \deg p_r\,.$$

Hence

$$\prod_{\mathfrak{p} \in \Omega} |\alpha|_{\mathfrak{p}} = 1\,.$$

3) Finally the general case of any global field F. By Proposition 31:9 there is a rational global field Q over which F is finite and separable. Let S be the set of all finite spots on Q, ∞ the single infinite spot on Q. Let T, T' be sets of spots on F with

$$T||S\,,\quad T'||\infty\,.$$

Then S is Dedekind by Proposition 31:2; so T is Dedekind by Theorem 23:1 since F/Q is finite and separable; hence $|\alpha|_{\mathfrak{p}} = 1$ for almost all $\mathfrak{p}$ in T by the definition of a Dedekind set of spots. But T' is a finite set of spots since F/Q is finite and separable; and $\Omega_F = T \cup T'$. Hence $|\alpha|_{\mathfrak{p}} = 1$ for almost all $\mathfrak{p} \in \Omega_F$.

The product formula has already been established for Q; we shall use this fact to establish it for F. Consider a typical $\alpha \in \dot{F}$. Each $\mathfrak{p} \in \Omega_F$ induces a spot $p \in \Omega_Q$; then by § 33A and Theorem 15:3 we obtain

$$\prod_{\mathfrak{p}} |\alpha|_{\mathfrak{p}} = \prod_{p} \Big(\prod_{\mathfrak{p}|p} |\alpha|_{\mathfrak{p}} \Big) = \prod_{p} \Big(\prod_{\mathfrak{p}|p} |N_{\mathfrak{p}|p}\,\alpha|_p \Big) = \prod_{p} (|N_{F/Q}\,\alpha|_p) = 1\,.$$

q. e. d.

§ 33C. Comment on ideal theory

Suppose S is a Dedekind set of spots on the global field F. Let us follow the convention of § 22 and write $\mathfrak{o}$ for $\mathfrak{o}(S)$ and $\mathfrak{p}$ for the prime ideal $\mathfrak{p} = \mathfrak{o}(S) \cap \mathfrak{m}(\mathfrak{p})$ determined by the spot $\mathfrak{p}$. Every ideal $\mathfrak{a} \in I(S)$ can be expressed uniquely in the form

$$\mathfrak{a} = \prod_{\mathfrak{p} \in S} \mathfrak{p}^{\nu_{\mathfrak{p}}}\,.$$

We define

$$N\mathfrak{a} = \prod_{\mathfrak{p} \in S} (N\mathfrak{p})^{\nu_{\mathfrak{p}}}\,.$$

Note that $N(\mathfrak{a}\mathfrak{b}) = (N\mathfrak{a})\,(N\mathfrak{b})$, and $N\mathfrak{a}$ is a natural number if $\mathfrak{a} \subseteq \mathfrak{o}$. The significance of $N\mathfrak{a}$ is given by the following proposition.

33:2. *Let S be a Dedekind set of spots on the global field F, and let $\mathfrak{a} \subseteq \mathfrak{b}$ be fractional ideals at S. Then*

$$(\mathfrak{b}:\mathfrak{a}) = N\mathfrak{a}/N\mathfrak{b}\,.$$

Proof. The various isomorphisms used in the proof are obtained from § 22D. We can assume that $\mathfrak{b} = \mathfrak{o}$ since there is an additive isomorphism

$$\mathfrak{b}/\mathfrak{a} \rightarrowtail \mathfrak{o}/\mathfrak{b}^{-1}\mathfrak{a}$$

and since $N(\mathfrak{b}^{-1}\mathfrak{a}) = N\mathfrak{a}/N\mathfrak{b}$; we now have to prove that $(\mathfrak{o}:\mathfrak{a}) = N\mathfrak{a}$. Take the factorization $\mathfrak{a} = \mathfrak{p}_1^{\nu_1}\,\mathfrak{p}_2^{\nu_2} \ldots \mathfrak{p}_r^{\nu_r}$ into distinct prime factors; then

$$\mathfrak{o}/\mathfrak{a} \rightarrowtail (\mathfrak{o}/\mathfrak{p}_1^{\nu_1}) \oplus \cdots \oplus (\mathfrak{o}/\mathfrak{p}_r^{\nu_r})\,;$$

so this reduces things to a power of a single prime ideal; accordingly let $\mathfrak{a} = \mathfrak{p}^\nu$ with $\nu \in \mathbf{N}$. Form the chain

$$\mathfrak{o} \supset \mathfrak{p} \supset \cdots \supset \mathfrak{p}^\nu .$$

Then

$$\mathfrak{p}^i/\mathfrak{p}^{i+1} \rightarrowtail \mathfrak{o}/\mathfrak{p} \rightarrowtail F(\mathfrak{p}) .$$

Hence

$$(\mathfrak{p}^i : \mathfrak{p}^{i+1}) = N\mathfrak{p} .$$

Hence $(\mathfrak{o} : \mathfrak{p}^\nu) = (N\mathfrak{p})^\nu$. **q. e. d.**

33:3. $$N\mathfrak{a} = \prod_{\mathfrak{p} \in S} \frac{1}{|\mathfrak{a}|_\mathfrak{p}} .$$

Proof. Take the factorization into prime ideals:

$$\mathfrak{a} = \prod_{\mathfrak{p} \in S} \mathfrak{p}^{\nu_\mathfrak{p}} .$$

Then

$$N\mathfrak{a} = \prod_{\mathfrak{p} \in S} (N\mathfrak{p})^{\nu_\mathfrak{p}} = \prod_{\mathfrak{p} \in S} \frac{1}{|\mathfrak{p}|_\mathfrak{p}^{\nu_\mathfrak{p}}} = \prod_{\mathfrak{p} \in S} \frac{1}{|\mathfrak{a}|_\mathfrak{p}} .$$

q. e. d.

§ 33D. Introduction of the idèle group J_F

Ω is the set of all non-trivial spots (discrete or archimedean) on our global field F. Consider the multiplicative group

$$\prod_{\mathfrak{p} \in \Omega} \dot{F}_\mathfrak{p}$$

consisting of the direct product of all the multiplicative groups $\dot{F}_\mathfrak{p}$; a typical element of this big group is defined in terms of its $\mathfrak{p}$-coordinates, say

$$\mathfrak{i} = (\mathfrak{i}_\mathfrak{p})_{\mathfrak{p} \in \Omega} \quad (\mathfrak{i}_\mathfrak{p} \in \dot{F}_\mathfrak{p}) ;$$

and the multiplication in the direct product is, by definition, coordinatewise. An idèle is defined to be an element $\mathfrak{i}$ of the above direct product which satisfies the following extra condition:

$$|\mathfrak{i}_\mathfrak{p}|_\mathfrak{p} = 1 \quad \text{for almost all } \mathfrak{p} \in \Omega .$$

The set of all idèles is a subgroup of the direct product called the group of idèles and written J_F. We let $\mathfrak{i}_\mathfrak{p}$ denote the $\mathfrak{p}$-coordinate of any $\mathfrak{i} \in J_F$ at the spot $\mathfrak{p} \in \Omega$.

For any $\alpha \in \dot{F}$ and any $\mathfrak{i} \in J_F$ we let $\alpha\mathfrak{i}$ stand for the idèle in J_F whose $\mathfrak{p}$-coordinate at each $\mathfrak{p} \in \Omega$ is equal to $\alpha\mathfrak{i}_\mathfrak{p}$. We define the volume of an idèle $\mathfrak{i}$ to be the positive real number

$$\|\mathfrak{i}\| = \prod_{\mathfrak{p} \in \Omega} |\mathfrak{i}_\mathfrak{p}|_\mathfrak{p} ,$$

where $|\ |_{\mathfrak{p}}$ denotes the normalized valuation at the spot $\mathfrak{p}$; the product used in the definition is essentially a finite one since almost all $|\mathfrak{i}_{\mathfrak{p}}|_{\mathfrak{p}}$ are equal to 1. It is evident that

$$\|\mathfrak{i}\mathfrak{j}\| = \|\mathfrak{i}\|\,\|\mathfrak{j}\|\,.$$

And the Product Formula shows that

$$\|\alpha\mathfrak{i}\| = \|\mathfrak{i}\|\,.$$

We say that a subset X of F is bounded by the idèle $\mathfrak{i} \in J_F$ if every $x \in X$ satisfies

$$|x|_{\mathfrak{p}} \leqq |\mathfrak{i}_{\mathfrak{p}}|_{\mathfrak{p}} \quad \forall\, \mathfrak{p} \in \Omega\,.$$

As a trivial example we observe that the field element 0 is bounded by every $\mathfrak{i} \in J_F$. We let

$$M(\mathfrak{i})$$

stand for the total number of field elements bounded by the idèle $\mathfrak{i}$; then $M(\mathfrak{i})$ is either a natural number or ∞. The mapping

$$x \leftrightarrow \alpha x$$

sets up a bijection between the field elements bounded by $\mathfrak{i}$ and those bounded by $\alpha\mathfrak{i}$; hence

$$M(\mathfrak{i}) = M(\alpha\mathfrak{i})\,.$$

§ 33E. The "density" $M(\mathfrak{i})/\|\mathfrak{i}\|$

The purpose of this subparagraph is to find very rough estimates for the "density" of field elements bounded by an idèle. We shall use the Pigeon Holing Principle: if l letters are to be placed in h boxes then at least one box must receive l/h or more letters.

33:4. **Theorem.** *Let F be a global field. Then there is a constant $D > 0$ which depends only on F such that*

$$M(\mathfrak{i}) \leqq \max\,(1, D\|\mathfrak{i}\|) \quad \forall\, \mathfrak{i} \in J_F\,.$$

In particular, an idèle can bound no more than a finite number of elements of F.

Proof. 1) Take a discrete spot $\mathfrak{q} \in \Omega$ and fix it. Let ω be the number of archimedean spots in Ω; if F is an algebraic function field, this number is 0; if F is an algebraic number field it is finite since F is then a finite extension of its prime field. We claim that

$$D = 4^{\omega} N\mathfrak{q}$$

will do the job. So we must consider a typical idèle $\mathfrak{i} \in J_F$ and show that it satisfies the inequality of the theorem.

2) If we replace $\mathfrak{i}$ by $\alpha\mathfrak{i}$ with $\alpha \in \dot{F}$ then neither $M(\mathfrak{i})$ nor $\|\mathfrak{i}\|$ is altered. Hence by the Weak Approximation Theorem it is enough to consider an $\mathfrak{i}$ for which $|\mathfrak{i}_\mathfrak{q}|_\mathfrak{q} = 1$. Let us do so. If 0 is the only field element bounded by $\mathfrak{i}$, then $M(\mathfrak{i}) = 1 \leq \max(1, D\|\mathfrak{i}\|)$ and we are through. So let us assume that there is a set X which is bounded by $\mathfrak{i}$ and consists of M elements of F with $1 < M < \infty$; we will be through if we can prove that $M \leq D\|\mathfrak{i}\|$. The rest of the proof is spent on this inequality.

3) We introduce Dedekind ideal theory at the single spot $\mathfrak{q}$. Put $\mathfrak{o} = \mathfrak{o}(\mathfrak{q})$ and $\mathfrak{q} = \mathfrak{m}(\mathfrak{q})$. We have $X \subseteq \mathfrak{o}$ since X is bounded by the idèle $\mathfrak{i}$ and $|\mathfrak{i}_\mathfrak{q}|_\mathfrak{q} = 1$. Form the chain of ideals

$$\mathfrak{o} \supset \mathfrak{q} \supset \cdots \supset \mathfrak{q}^r \supset \mathfrak{q}^{r+1} \supset \cdots,$$

choose that $r \geqq 0$ for which

$$(N\mathfrak{q})^r < M \leqq (N\mathfrak{q})^{r+1}.$$

Now $(\mathfrak{o}:\mathfrak{q}^r) = (N\mathfrak{q})^r$ by Proposition 33:2. So by the Pigeon Holing Principle, there is at least one coset of $\mathfrak{o}$ modulo $\mathfrak{q}^r$ which contains two distinct elements $\alpha_1, \alpha_2 \in X$. Here $\alpha_1 - \alpha_2 \in \mathfrak{q}^r$. Hence

$$|\alpha_1 - \alpha_2|_\mathfrak{q} \leqq |\mathfrak{q}|_\mathfrak{q}^r = \left(\frac{1}{N\mathfrak{q}}\right)^r \leqq \frac{N\mathfrak{q}}{M}.$$

But

$$|\alpha_1 - \alpha_2|_\mathfrak{p} \leqq \begin{cases} |\mathfrak{i}_\mathfrak{p}|_\mathfrak{p} & \text{if } \mathfrak{p} \text{ discrete} \\ 4\,|\mathfrak{i}_\mathfrak{p}|_\mathfrak{p} & \text{if } \mathfrak{p} \text{ archimedean.} \end{cases}$$

Hence by the Product Formula,

$$\begin{aligned} 1 &= \prod_\mathfrak{p} |\alpha_1 - \alpha_2|_\mathfrak{p} \\ &\leqq 4^\omega \Big(\prod_{\mathfrak{p} \neq \mathfrak{q}} |\mathfrak{i}_\mathfrak{p}|_\mathfrak{p}\Big) N\mathfrak{q}/M \\ &= 4^\omega N\mathfrak{q}\,\|\mathfrak{i}\|/M. \end{aligned}$$

Thus $M \leqq D\|\mathfrak{i}\|$, as we asserted. **q. e. d.**

33:5. Theorem. *Let F be a global field. Then there is a constant C $(0 < C < 1)$ which depends only on F such that*

$$C\|\mathfrak{i}\| < M(\mathfrak{i}) \quad \forall\, \mathfrak{i} \in J_F.$$

In particular, an idèle with a big enough volume bounds at least one non-zero field element.

Proof. 1) Since F is a global field it is a finite separable extension of a rational global field Q. Let Z be the ring of integers of Q. Let S denote the set of all finite spots on Q, let ∞ be the infinite spot. Define $T \subseteq \Omega_F$

and $T' \subseteq \Omega_F$ by

$$T||S, T'||\infty .$$

Then T is a Dedekind set of spots on F by Proposition 31:2 and Theorem 23:1; and T' is a finite set of spots which is either all archimedean or all discrete.

Since T is a Dedekind set of spots it introduces Dedekind ideal theory into F. We put $\mathfrak{o} = \mathfrak{o}(T)$ as usual. Then $Z = \mathfrak{o}(S) \subseteq \mathfrak{o}(T) = \mathfrak{o}$.

Take a base $x_1, \ldots, x_n$ for F over Q. We can find a non-zero $m \in Z$ which is arbitrarily small at any predetermined finite subset of S, and hence at any predetermined finite subset of T; the Product Formula will ensure at the same time that m is arbitrarily large at ∞, and hence at all $\mathfrak{p} \in T'$. Replace $x_1, \ldots, x_n$ by $mx_1, \ldots, mx_n$. Then the new $x_1, \ldots, x_n$ still form a base for F/Q; but they also have the following additional properties:

$$\begin{cases} |x_1|_\mathfrak{p} \leqq 1, \ldots, |x_n|_\mathfrak{p} \leqq 1 & \forall\, \mathfrak{p} \in T, \\ |x_1|_\mathfrak{p} \geqq 1, \ldots, |x_n|_\mathfrak{p} \geqq 1 & \forall\, \mathfrak{p} \in T'. \end{cases}$$

In particular, $x_1, \ldots, x_n$ are now in $\mathfrak{o}$. We assume that the chosen base has these additional properties.

By the Weak Approximation Theorem we can find a number $B \in F$ such that

$$|B|_\mathfrak{p} \geqq n 2^{n+1} |x_i|_\mathfrak{p}$$

for $1 \leqq i \leqq n$ and for all $\mathfrak{p} \in T'$. Put

$$C = \prod_{\mathfrak{p} \in T'} \frac{1}{|B^4|_\mathfrak{p}} .$$

This will be our C. Note that $|B|_\mathfrak{p} \geqq 4 > 1$ for all $\mathfrak{p} \in T'$.

2) We have to consider a typical idèle $\mathfrak{i}$ and show that it satisfies the inequality of the theorem. By the Weak Approximation Theorem there is an $\alpha \in F$ which is arbitrarily close to $B^3/\mathfrak{i}_\mathfrak{p}$ at each $\mathfrak{p} \in T'$; if the approximation is good enough we can achieve

$$|B^2|_\mathfrak{p} < |\alpha \mathfrak{i}_\mathfrak{p}|_\mathfrak{p} < |B^4|_\mathfrak{p} \quad \forall\, \mathfrak{p} \in T' .$$

in virtue of the continuity of the map $\beta \rightarrow |\beta|$. Now pick $m \in Z$ such that

$$|\alpha\, \mathfrak{i}_\mathfrak{p} m|_\mathfrak{p} \leqq 1 \quad \forall\, \mathfrak{p} \in T .$$

If we replace $\mathfrak{i}$ by the idèle $\alpha m \mathfrak{i}$, then neither $M(\mathfrak{i})$ nor the volume $\|\mathfrak{i}\|$ is changed. So let us make this replacement. In effect, this allows us to make the following assumption about the idèle $\mathfrak{i}$ under consideration:

there is an $m \in Z$ such that

$$\begin{cases} |\mathfrak{i}_\mathfrak{p}|_\mathfrak{p} \leq 1 & \forall\, \mathfrak{p} \in T \\ |mB^2|_\mathfrak{p} < |\mathfrak{i}_\mathfrak{p}|_\mathfrak{p} < |mB^4|_\mathfrak{p} & \forall\, \mathfrak{p} \in T'. \end{cases}$$

3) We must now introduce three new idèles $\mathfrak{h}$, $\mathfrak{j}$, $\mathfrak{k}$. Once we have done so we will be able to state the idea of the proof and the reason behind the notation. Let us define these idèles by specifying their $\mathfrak{p}$-coordinates:

(i) $\mathfrak{h}_\mathfrak{p} = 1 \quad \forall\, \mathfrak{p} \in T, \qquad \mathfrak{h}_\mathfrak{p} = mB \quad \forall\, \mathfrak{p} \in T'$,

(ii) $\mathfrak{j}_\mathfrak{p} = \mathfrak{i}_\mathfrak{p} \quad \forall\, \mathfrak{p} \in T, \qquad \mathfrak{j}_\mathfrak{p} = mB^2 \quad \forall\, \mathfrak{p} \in T'$,

(iii) $\mathfrak{k}_\mathfrak{p} = \mathfrak{i}_\mathfrak{p} \quad \forall\, \mathfrak{p} \in T, \qquad \mathfrak{k}_\mathfrak{p} = mB^4 \quad \forall\, \mathfrak{p} \in T'$.

The rest of the proof now goes as follows. First get a lower estimate on $M(\mathfrak{h})$. Use this in an application of the Pigeon Holing Principle to get a lower estimate for $M(\mathfrak{j})$. This gives a lower estimate for $M(\mathfrak{i})$ since $|\mathfrak{j}_\mathfrak{p}|_\mathfrak{p} \leq |\mathfrak{i}_\mathfrak{p}|_\mathfrak{p}$ for all $\mathfrak{p} \in \Omega$ in virtue of the choice of $\mathfrak{i}$ in step 2). On the other hand $\mathfrak{k}$ provides an upper estimate on the volume of $\mathfrak{i}$ since $|\mathfrak{i}_\mathfrak{p}|_\mathfrak{p} \leq |\mathfrak{k}_\mathfrak{p}|_\mathfrak{p}$ for all $\mathfrak{p} \in \Omega$. We therefore obtain a lower estimate on the density. It will be shown that this estimate is greater than C. So much for the idea. Let us now carry out the details.

4) First we want our estimate for $M(\mathfrak{h})$. We define

$$X = \{m_1 x_1 + \cdots + m_n x_n \mid m_i \in Z \quad \text{with} \quad |m_i|_\infty \leq |m|_\infty\}.$$

Suppose we count X. In the number theoretic case we get elements of X if the m_i vary through the subset $\{0, \pm 1, \pm 2, \ldots, \pm m\}$ of Z; hence X contains strictly more than

$$|m|_\infty^n$$

elements. In the function theoretic case we do a similar thing, only this time the m_i are to be polynomials with $\deg m_i \leq \deg m$; here again we find that X contains strictly more than

$$q^{n \deg m} = |m|_\infty^n$$

elements, where q is the number of elements in the constant field of Q. An easy calculation now shows that X is bounded by the idèle $\mathfrak{h}$ (the definition of B is used in this calculation: indeed, B was specifically designed to make this calculation work). Hence we have the desired estimate:

$$M(\mathfrak{h}) > |m|_\infty^n.$$

5) Let $\mathfrak{a}$ be the ideal at T that is defined by $|\mathfrak{a}|_\mathfrak{p} = |\mathfrak{i}_\mathfrak{p}|_\mathfrak{p}$ for all $\mathfrak{p} \in T$. Thus $\mathfrak{a} \subseteq \mathfrak{o}$ and by Proposition 33:3

$$N\mathfrak{a} = \prod_{\mathfrak{p} \in T} \frac{1}{|\mathfrak{i}_\mathfrak{p}|_\mathfrak{p}}.$$

Now every field element bounded by $\mathfrak{h}$ is in $\mathfrak{o}$; hence the Pigeon Holing Principle gives at least one coset of $\mathfrak{o}$ modulo $\mathfrak{a}$ which contains t field elements

$$\alpha_1, \alpha_2, \ldots, \alpha_t$$

that are bounded by $\mathfrak{h}$ with

$$t > |m|_\infty^n \prod_{\mathfrak{p} \in T} |\mathfrak{i}_\mathfrak{p}|_\mathfrak{p} .$$

Then

$$\alpha_1 - \alpha_1, \alpha_2 - \alpha_1, \ldots, \alpha_t - \alpha_1$$

are in $\mathfrak{a}$; an easy computation then shows that these t elements are bounded by $\mathfrak{j}$; hence

$$M(\mathfrak{j}) > |m|_\infty^n \prod_{\mathfrak{p} \in T} |\mathfrak{i}_\mathfrak{p}|_\mathfrak{p} .$$

6) The final calculations. We have

$$\frac{M(\mathfrak{i})}{\|\mathfrak{i}\|} > \frac{M(\mathfrak{j})}{\|\mathfrak{k}\|} > |m|_\infty^n \prod_{\mathfrak{p} \in T'} \frac{1}{|m B^4|_\mathfrak{p}} .$$

But

$$|m|_\infty^n = \prod_{\mathfrak{p} \in T'} |m|_\infty^{n(\mathfrak{p}|\infty)} = \prod_{\mathfrak{p} \in T'} |m|_\mathfrak{p} .$$

Hence

$$\frac{M(\mathfrak{i})}{\|\mathfrak{i}\|} > \prod_{\mathfrak{p} \in T'} \frac{1}{|B^4|_\mathfrak{p}} = C .$$

q. e. d.

§ 33F. The group of S-units

Let S be any non-empty set of spots on the global field F. The set

$$\mathfrak{u}(S) = \{\alpha \in \dot{F} \mid |\alpha|_\mathfrak{p} = 1 \quad \forall\, \mathfrak{p} \in S\}$$

is a subgroup of $\dot{F}$ called the group of S-units; the elements of this group are called the S-units of F. We shall often write $\mathfrak{u}$ instead of $\mathfrak{u}(S)$. If S happens to be a Dedekind set of spots, then $\mathfrak{u}(S)$ coincides with our earlier definition of the group of units of F at S.

We shall call $\mathfrak{u}(\Omega)$ the group of absolute units of F. This group has the following simple description.

33:6. *The group of absolute units of a global field is cyclic of finite order. It consists of all the roots of unity in the given field.*

Proof. Suppose we have shown that $\mathfrak{u}(\Omega)$ is cyclic of finite order. Then every element of $\mathfrak{u}(\Omega)$ has to be a root of unity; conversely every root of unity in F must clearly also fall in $\mathfrak{u}(\Omega)$. Hence the second part of the proposition follows from the first. It therefore remains for us to prove that $\mathfrak{u}(\Omega)$ is a finite cyclic group.

Now $\mathfrak{u}(\Omega)$ is bounded by the idèle $(1)_{\mathfrak{p}\in\Omega}$. Hence by Theorem 33:4 $\mathfrak{u}(\Omega)$ is a finite subgroup of $\dot{F}$. But every finite subgroup of the multiplicative group of an arbitrary field is cyclic. Hence $\mathfrak{u}(\Omega)$ is finite cyclic. **q. e. d.**

33:7. *Let F be a global field and let K be a positive constant. Then the number of discrete spots $\mathfrak{p}$ on F such that $N\mathfrak{p} \leqq K$ is finite.*

Proof. We can assume that K is a natural number. Take $K+1$ distinct elements of F, say $\alpha_0, \alpha_1, \ldots, \alpha_K$ with $\alpha_0 = 0$. By the Product Formula there is a finite set T of spots on F such that $|\alpha_i - \alpha_j|_{\mathfrak{p}} = 1$ for $0 \leqq i < j \leqq K$ whenever $\mathfrak{p} \in \Omega_F - T$. Let us suppose that T has been chosen large enough to include all archimedean spots on F. Our proposition will be established if we can prove that $N\mathfrak{p} > K$ for all $\mathfrak{p} \in \Omega_F - T$. Suppose to the contrary that there is a spot $\mathfrak{p} \in \Omega_F - T$ such that $N\mathfrak{p} \leqq K$. Now $\mathfrak{o}(\mathfrak{p})$ has exactly $N\mathfrak{p}$ cosets modulo $\mathfrak{m}(\mathfrak{p})$; and $\alpha_0, \alpha_1, \ldots, \alpha_K$ fall in $\mathfrak{o}(\mathfrak{p})$; hence there is at least one coset of $\mathfrak{o}(\mathfrak{p})$ modulo $\mathfrak{m}(\mathfrak{p})$ which contains two distinct α's, say α_i and α_j; this implies that $\alpha_i - \alpha_j \in \mathfrak{m}(\mathfrak{p})$. But this is impossible since $|\alpha_i - \alpha_j|_{\mathfrak{p}} = 1$. So we do indeed have $N\mathfrak{p} > K$ for any $\mathfrak{p} \in \Omega_F - T$. **q. e. d.**

In practice S will be a proper subset of Ω consisting of almost all discrete spots in Ω. (For example S could be the set of all discrete spots on an algebraic number field.) When we are in this situation we shall let s stand for the number of spots in $\Omega - S$; so here $1 \leqq s < \infty$.

33:8. *Let S be a set of discrete spots on the global field F. Suppose that $\Omega - S$ consists of exactly s spots with $2 \leqq s < \infty$. Then for each $\mathfrak{q} \in \Omega - S$ there is an S-unit $\varepsilon_{\mathfrak{q}}$ such that*

$$|\varepsilon_{\mathfrak{q}}|_{\mathfrak{q}} > 1\,, \quad |\varepsilon_{\mathfrak{q}}|_{\mathfrak{p}} < 1 \quad \forall\, \mathfrak{p} \in \Omega - (S \cup \mathfrak{q})\,.$$

Proof. 1) We put $N\mathfrak{q} = 1$ if $\mathfrak{q}$ is archimedean ($N\mathfrak{q}$ is already defined if $\mathfrak{q}$ is discrete). Let C stand for the constant of Theorem 33:5: then any idèle $\mathfrak{i}$ with $C\,\|\mathfrak{i}\| \geqq 1$ will have to bound at least one non-zero field element. Let C be chosen so small that there is at least one spot $\mathfrak{p}$ in S for which $N\mathfrak{p} \leqq N\mathfrak{q}/C$. Thus the set

$$W = \{\mathfrak{p} \in S \,|\, N\mathfrak{p} \leqq N\mathfrak{q}/C\}$$

is non-empty; and this is a finite subset of S by Proposition 33:7. Put

$$X = \left\{x \in F \,\middle|\, \frac{C}{N\mathfrak{q}} \leqq |x|_{\mathfrak{p}} \leqq 1 \quad \forall\, \mathfrak{p} \in S\right\}.$$

Note that for each $x \in X$ and for each $\mathfrak{p} \in S - W$ we have

$$\frac{1}{N\mathfrak{p}} < \frac{C}{N\mathfrak{q}} \leqq |x|_{\mathfrak{p}} \leqq 1\,;$$

but $|x|_{\mathfrak{p}}$ is a power of $N\mathfrak{p}$; hence $|x|_{\mathfrak{p}} = 1$. Hence

$$X \subseteq \mathfrak{u}(S - W)\,.$$

On the other hand, at each $\mathfrak{p} \in W$ the function $|x|_\mathfrak{p}$ takes on just a finite number of values as x runs through X. Hence the set of S-tuples of the form $(|x|_\mathfrak{p})_{\mathfrak{p} \in S}$ with $x \in X$ is finite. Hence we can find $x_1, \ldots, x_r \in X$ such that

$$X = \bigcup_{1 \leq i \leq r} x_i \, \mathfrak{u}(S) .$$

2) The second step of the proof consists in defining an idèle $\mathfrak{i}$ in J_F with certain properties. The $\mathfrak{p}$-coordinate $\mathfrak{i}_\mathfrak{p}$ of $\mathfrak{i}$ can be any element of $\dot{F}_\mathfrak{p}$ which satisfies

$$|\mathfrak{i}_\mathfrak{p}|_\mathfrak{p} \begin{cases} = 1 & \text{if} \quad \mathfrak{p} \in S \\ < |x_i|_\mathfrak{p} & \text{if} \quad 1 \leq i \leq r\,, \mathfrak{p} \in \Omega - (S \cup \mathfrak{q}) . \end{cases}$$

This still leaves the $\mathfrak{q}$-coordinate unspecified; here $\mathfrak{i}_\mathfrak{q}$ can be any element of $\dot{F}_\mathfrak{q}$ which makes the volume satisfy the inequality

$$1 \leq C \, \|\mathfrak{i}\| \leq N \mathfrak{q} .$$

Why is there such an $\mathfrak{i}_\mathfrak{q}$? If $\mathfrak{q}$ is archimedean it is because the value group of $F_\mathfrak{q}$ is all of P; if $\mathfrak{q}$ is discrete one takes any $\mathfrak{i}_\mathfrak{q} \in F_\mathfrak{q}$ with $|\mathfrak{i}_\mathfrak{q}|_\mathfrak{q} = (N\mathfrak{q})^{-m}$ where m is a number in Z for which

$$(N \mathfrak{q})^m \leq C \prod_{\mathfrak{p} \neq \mathfrak{q}} |\mathfrak{i}_\mathfrak{p}|_\mathfrak{p} \leq (N \mathfrak{q})^{m+1} .$$

Thus our definition is in order and we have our idèle $\mathfrak{i}$.

3) Since $C \, \|\mathfrak{i}\| \geq 1$ there is a non-zero field element x bounded by the idèle $\mathfrak{i}$. Then

$$\prod_{\mathfrak{p} \in \Omega - S} |x|_\mathfrak{p} \leq \prod_{\mathfrak{p} \in \Omega - S} |\mathfrak{i}_\mathfrak{p}|_\mathfrak{p} = \|\mathfrak{i}\| \leq N \mathfrak{q}/C .$$

Hence for every $\mathfrak{p}' \in S$ we have

$$1 = \prod_{\mathfrak{p} \in \Omega} |x|_\mathfrak{p} \leq |x|_{\mathfrak{p}'} \, N \mathfrak{q}/C ,$$

hence

$$C/N \mathfrak{q} \leq |x|_{\mathfrak{p}'} \leq 1 ,$$

in other words $x \in X$. So by step 1) we can find an x_i and an S-unit ε such that $x = \varepsilon x_i$. We claim that ε is the S-unit required by the proposition. For

$$|\varepsilon|_\mathfrak{p} = \frac{|x|_\mathfrak{p}}{|x_i|_\mathfrak{p}} \leq \frac{|\mathfrak{i}_\mathfrak{p}|_\mathfrak{p}}{|x_i|_\mathfrak{p}} < 1$$

holds for all $\mathfrak{p} \in \Omega - (S \cup \mathfrak{q})$; and so $|\varepsilon|_\mathfrak{q} > 1$ by the Product Formula.

q. e. d.

33:9. *Let S be the set of spots given in the last proposition, and let T be a subset of $s - 1$ elements of $\Omega - S$. An S-unit $\varepsilon_\mathfrak{q}$ which satisfies*

$$|\varepsilon_\mathfrak{q}|_\mathfrak{q} > 1, \quad |\varepsilon_\mathfrak{q}|_\mathfrak{p} < 1 \quad \forall \, \mathfrak{p} \in \Omega - (S \cup \mathfrak{q})$$

is given for each $\mathfrak{q} \in T$. *Then these S-units are multiplicatively independent and they generate a subgroup of finite index*[1] *in* $\mathfrak{u}(S)$.

Proof. Let $\mathfrak{q}'$ be that spot in $\Omega - S$ which is not in T.

1) If we had a dependence relation among the $\varepsilon_\mathfrak{q}$ we could write it

$$\prod_{\mathfrak{q} \in P} \varepsilon_\mathfrak{q}^{\nu_\mathfrak{q}} = \prod_{\mathfrak{q} \in T-P} \varepsilon_\mathfrak{q}^{\nu_\mathfrak{q}}$$

with $\nu_\mathfrak{q} \geqq 0$ for all $\mathfrak{q}$ in T and $\nu_\mathfrak{q} > 0$ for all $\mathfrak{q}$ in P, where P is some non-empty subset of T. Put

$$\alpha = \prod_{\mathfrak{q} \in P} \varepsilon_\mathfrak{q}^{\nu_\mathfrak{q}} = \prod_{\mathfrak{q} \in T-P} \varepsilon_\mathfrak{q}^{\nu_\mathfrak{q}} .$$

Then $|\alpha|_\mathfrak{p} = 1$ for all $\mathfrak{p} \in S$. And for each $\mathfrak{p} \in P$ we have

$$|\alpha|_\mathfrak{p} = \Big| \prod_{\mathfrak{q} \in T-P} \varepsilon_\mathfrak{q}^{\nu_\mathfrak{q}} \Big|_\mathfrak{p} \leqq 1 .$$

And similarly $|\alpha|_\mathfrak{p} \leqq 1$ for all $\mathfrak{p} \in T - P$. But

$$|\alpha|_{\mathfrak{q}'} = \Big| \prod_{\mathfrak{q} \in P} \varepsilon_\mathfrak{q}^{\nu_\mathfrak{q}} \Big|_{\mathfrak{q}'} < 1 .$$

This contradicts the Product Formula for α. Hence there can be no dependence among the $\varepsilon_\mathfrak{q}$. This proves the first part.

2) Define an idèle $\mathfrak{i}$ in J_F by the equations

$$\mathfrak{i}_\mathfrak{p} = 1 \quad \forall\, \mathfrak{p} \in S \cup \mathfrak{q}', \qquad \mathfrak{i}_\mathfrak{q} = \varepsilon_\mathfrak{q} \quad \forall\, \mathfrak{q} \in T .$$

Let $\alpha_1, \ldots, \alpha_r$ be all the S-units that are bounded by the idèle $\mathfrak{i}$; we have $r \geqq 1$ since, for instance, 1 is bounded by $\mathfrak{i}$; and we have $r < \infty$ by Theorem 33:4. Let H denote the subgroup of $\mathfrak{u}$ that is generated by the given $\varepsilon_\mathfrak{q}$ ($\mathfrak{q} \in T$). It is enough to prove that

$$\mathfrak{u} \subseteq \bigcup_{1 \leqq i \leqq r} \alpha_i H ,$$

for then $(\mathfrak{u}:H) \leqq r < \infty$.

Now there are certainly elements η of H that are bounded by $\mathfrak{i}$ (for instance 1 is one of them); pick such an η for which $|\eta|_{\mathfrak{q}'}$ is smallest. Then we claim that $|\eta|_\mathfrak{q} > 1$ for each $\mathfrak{q} \in T$; for if we have a $\mathfrak{q} \in T$ with $|\eta|_\mathfrak{q} \leqq 1$, then $\varepsilon_\mathfrak{q} \eta$ is an element of H which is bounded by $\mathfrak{i}$ and has $|\varepsilon_\mathfrak{q} \eta|_{\mathfrak{q}'} < |\eta|_{\mathfrak{q}'}$, and this denies the minimal property in the choice of η. Hence we have exhibited an $\eta \in H$ which is bounded by $\mathfrak{i}$ and satisfies

$$|\eta|_{\mathfrak{q}'} < 1 , \quad |\eta|_\mathfrak{q} > 1 \quad \forall\, \mathfrak{q} \in T .$$

Consider a typical $\alpha \in \mathfrak{u}$. Then there is an $\eta' \in H$ such that

$$|\alpha \eta'|_\mathfrak{q} \leqq |\varepsilon_\mathfrak{q}|_\mathfrak{q} \quad \forall\, \mathfrak{q} \in T$$

[1] For this reason the construction of the $\varepsilon_\mathfrak{q}$ is sometimes called the construction of "enough" units.

(for instance, a sufficiently high power of η^{-1}). Choose an η' of the above sort for which $|\alpha\eta'|_{\mathfrak{q}'}$ is smallest (why is this possible?). We then repeat an argument already used in this proof to show that

$$|\alpha\eta'|_{\mathfrak{q}'} < 1\,, \quad |\alpha\eta'|_{\mathfrak{q}} > 1 \quad \forall\, \mathfrak{q} \in T\,.$$

But this means that $\alpha\eta'$ is bounded by the idèle $\mathfrak{i}$. Hence $\alpha\eta' = \alpha_i$ for some i. So $\alpha \in \alpha_i H$. Hence $\mathfrak{u} \subseteq \bigcup_{1 \leq i \leq r} \alpha_i H$. **q. e. d.**

33:10. **The Dirichlet Unit Theorem.** *Let S be a set of discrete spots on a global field F, and suppose that the number of spots in $\Omega - S$ is equal to s with $1 \leqq s < \infty$. Then the group of S-units $\mathfrak{u}(S)$ is a direct product of the group of absolute units with $s-1$ infinite cyclic groups.*

Proof. Let $\mathfrak{u} = \mathfrak{u}(S)$. If $s = 1$ we have $\mathfrak{u} = \mathfrak{u}(\Omega)$ by the Product Formula, and so we are through. Hence assume that $s > 1$. We can then apply the two preceding propositions. They give us a free commutative group H of rank $s-1$ which has finite index in $\mathfrak{u}$. Hence $\mathfrak{u}$ is a finitely generated abelian group. Since H contains $s-1$ independent elements we must have rank $\mathfrak{u} \geqq s-1$. Now $\mathfrak{u}^i \subseteq H$ where i is the index $(\mathfrak{u}:H) < \infty$, and furthermore any s elements of H are dependent, hence rank $\mathfrak{u} = s-1$. So $\mathfrak{u}$ is a direct product of its elements of finite order with $s-1$ infinite cyclic groups. But the elements of finite order are the roots of unity, and these are the absolute units by Proposition 33:6. **q. e. d.**

§ 33G. The very strong approximation theorem

33:11. **Theorem.** *Let T be any finite set of spots on the global field F and let $\mathfrak{q} \in \Omega - T$. Suppose that $\alpha_{\mathfrak{p}} \in F$ is given, one for each $\mathfrak{p} \in T$. Then for each $\varepsilon > 0$ there is an $A \in F$ such that*

$$\begin{cases} |A - \alpha_{\mathfrak{p}}|_{\mathfrak{p}} < \varepsilon & \forall\, \mathfrak{p} \in T \\ \qquad |A|_{\mathfrak{p}} \leqq 1 & \forall\, \mathfrak{p} \in \Omega - (T \cup \mathfrak{q})\,. \end{cases}$$

Proof. 1) We can take it that $0 < \varepsilon < 1$. Suppose we enlarge T to some finite set of spots which still does not contain $\mathfrak{q}$; and suppose we define $\alpha_{\mathfrak{p}}$ to be 0 at each of the new spots introduced; if we can prove the theorem for the new set of spots, then we shall have it for the original T. Accordingly we can assume that the given set T has the following additional properties: $T \cup \mathfrak{q}$ contains all archimedean spots, and $S = \Omega - (T \cup \mathfrak{q})$ is Dedekind.

2) Now apply the Weak Approximation Theorem. This gives an element $\alpha \in F$ which is arbitrarily close to $\alpha_{\mathfrak{p}}$ at each $\mathfrak{p} \in T$; this is still not the required A since in general there will be a finite set of spots $W \subseteq S$ at which $|\alpha|_{\mathfrak{p}} > 1$. We shall however be through if we can prove that, given any finite subset W of S, there is a $\beta \in F$ which is arbitrarily small at all $\mathfrak{p} \in W$, which is arbitrarily close to 1 at all $\mathfrak{p} \in T$, and which is integral at all $\mathfrak{p} \in S$. For if this can be done, and if we take good enough

approximations all round, then $\alpha\beta$ will be close to $\alpha_\mathfrak{p}$ at all $\mathfrak{p} \in T$, and it will have $|\alpha\beta|_\mathfrak{p} \leqq 1$ at all $\mathfrak{p} \in S$, so $A = \alpha\beta$ will be the element we are after. We therefore have to prove that such a β does in fact exist.

3) Let δ be the degree of approximation needed for β. Since S is a Dedekind set of spots on F we can introduce Dedekind ideal theory. Put $\mathfrak{o} = \mathfrak{o}(S)$ and let $\mathfrak{a}$ be an integral ideal at S for which

$$|\mathfrak{a}|_\mathfrak{p} < \delta \quad \forall\, \mathfrak{p} \in W.$$

Then

$$r = (\mathfrak{o} : \mathfrak{a}) = N\mathfrak{a} < \infty .$$

Pick a representative set $\alpha_1, \ldots, \alpha_r$ of $\mathfrak{o}$ modulo $\mathfrak{a}$. Now Proposition 33:8 provides us with an S-unit E which is arbitrarily small at all $\mathfrak{p} \in T$, say

$$|E\alpha_i|_\mathfrak{p} < \delta \quad (1 \leqq i \leqq r, \quad \mathfrak{p} \in T) .$$

Then $E\alpha_1, \ldots, E\alpha_r$ still fall in distinct cosets of $\mathfrak{o}$ modulo $\mathfrak{a}$ since E is an S-unit; hence these elements are also a representative set of $\mathfrak{o}$ modulo $\mathfrak{a}$. Let $E\alpha_1$, say, fall in the same coset as -1. Put $\beta = E\alpha_1 + 1$. Then $\beta \in \mathfrak{o}$ so that β is integral at all $\mathfrak{p} \in S$; and $\beta = E\alpha_1 + 1$ is in $\mathfrak{a}$ since $E\alpha_1$ is in the coset of -1, so

$$|\beta|_\mathfrak{p} \leqq |\mathfrak{a}|_\mathfrak{p} < \delta \qquad \forall\, \mathfrak{p} \in W;$$

and

$$|\beta - 1|_\mathfrak{p} = |E\alpha_1|_\mathfrak{p} < \delta \quad \forall\, \mathfrak{p} \in T .$$

Hence β has the required properties. **q. e. d.**

33:11a. **Corollary.** *The A of the theorem can be made to satisfy the additional inequality $|A|_\mathfrak{q} > 1/\varepsilon$.*

Proof. Fix a $\mathfrak{q}' \in \Omega - (T \cup \mathfrak{q})$. Choose $A \in F$ in such a way that

$$\begin{cases} |A - \alpha_\mathfrak{p}|_\mathfrak{p} < \frac{1}{n} & \forall\, \mathfrak{p} \in T \\ \qquad |A|_\mathfrak{p} \leqq 1 & \forall\, \mathfrak{p} \in \Omega - (T \cup \mathfrak{q} \cup \mathfrak{q}') \\ \qquad |A|_{\mathfrak{q}'} < \frac{1}{n}, \end{cases}$$

where n is a natural number. If n is big enough, the Product Formula will force $|A|_\mathfrak{q} > 1/\varepsilon$. **q. e. d.**

§ 33H. Finiteness of class number[1]

33:12. **Theorem.** *Let S be a set of discrete spots on the global field F. Then S is Dedekind if and only if $S \subset \Omega$.*

[1] Our approach to the theory of S-units and the finiteness of class number is based on the Pigeon Holing Principle and the Product Formula. This is the modern elementary approach of E. Artin and G. Whaples, *Bull. Am. Math. Soc.* (1945), pp. 469—492. For the very elegant topological approach based on fundamental properties of locally compact groups see K. Iwasawa, *Ann. Math.* 57 (1953), pp. 331—356.

Proof. If we are given $S \subset \Omega$, then S is Dedekind by the Very Strong Approximation Theorem. Conversely, let a Dedekind set S be given. Suppose if possible that $S = \Omega$. Then $\mathfrak{o}(S)$ must be $\mathfrak{u}(S) \cup 0$ by the Product Formula. So $\mathfrak{o}(S)$ is a finite set. But this is impossible since F is the quotient field of $\mathfrak{o}(S)$ by Proposition 21:3. Hence we cannot have $S = \Omega$. So $S \subset \Omega$. **q. e. d.**

33:13. **Theorem.** *Let S be a Dedekind set of spots on the global field F. Then the ideal class number $h_F(S)$ is finite.*

Proof. 1) Fix $\mathfrak{q} \in \Omega - S$. Put $N\mathfrak{q} = 1$ if $\mathfrak{q}$ is archimedean ($N\mathfrak{q}$ is already defined if $\mathfrak{q}$ is discrete). Consider the constant $N\mathfrak{q}/C$ where C is the constant of Theorem 33:5. We make the following claim: given any fractional ideal $\mathfrak{a}$ at S, there is a non-zero $\alpha \in \mathfrak{a}$ such that

$$N(\alpha\mathfrak{o}) \leqq (N\mathfrak{q}/C)\, N\mathfrak{a}\,.$$

To see this, consider an idèle $\mathfrak{i}$ in J_F whose coordinates satisfy

$$|\mathfrak{i}_\mathfrak{p}|_\mathfrak{p} = |\mathfrak{a}|_\mathfrak{p} \quad \forall\, \mathfrak{p} \in S$$

and whose volume satisfies

$$1 \leqq C\|\mathfrak{i}\| \leqq N\mathfrak{q}\,.$$

(Such an $\mathfrak{i}$ exists by an argument used in step 2) of Proposition 33:8.) Since $C\|\mathfrak{i}\| \geqq 1$ the idèle $\mathfrak{i}$ bounds at least one non-zero field element α by Theorem 33:5. By Proposition 33:3 and the Product Formula,

$$\begin{aligned} N(\alpha\mathfrak{o}) &= \prod_{\mathfrak{p} \in \Omega - S} |\alpha|_\mathfrak{p} \\ &\leqq \prod_{\mathfrak{p} \in \Omega - S} |\mathfrak{i}_\mathfrak{p}|_\mathfrak{p} \\ &= \|\mathfrak{i}\| \left(\prod_{\mathfrak{p} \in S} \frac{1}{|\mathfrak{i}_\mathfrak{p}|_\mathfrak{p}} \right) \\ &= \|\mathfrak{i}\|\, (N\mathfrak{a}) \\ &\leqq (N\mathfrak{q}/C)\, N\mathfrak{a}\,. \end{aligned}$$

But $|\alpha|_\mathfrak{p} \leqq |\mathfrak{i}_\mathfrak{p}|_\mathfrak{p} \leqq |\mathfrak{a}|_\mathfrak{p}$ for all $\mathfrak{p} \in S$. Hence $\alpha \in \mathfrak{a}$. So we have proved our claim.

2) We immediately deduce from the result of step 1) that we have a constant K (namely $K = N\mathfrak{q}/C$) which depends only on F and S and has the following property: given any fractional ideal $\mathfrak{a}$ at S, there is an integral ideal $\mathfrak{a}'$ in the same ideal class as $\mathfrak{a}$ such that $1 \leqq N\mathfrak{a}' \leqq K$.[1] But

[1] The only thing that matters here is that K is constant. But the number K is obviously of significance when calculating class numbers of specific fields since it allows you to restrict yourself to integral ideals $\mathfrak{a}$ which satisfy $1 \leqq N\mathfrak{a} \leqq K$. From this point of view the smaller the K the better. For class number computations in number fields and for Minkowski's value of K we refer the reader to E. Artin, *Theory of algebraic numbers* (Göttingen lectures, 1956).

it follows from Proposition 33:7 that the number of integral ideals $\mathfrak{b}$ with $N\mathfrak{b} \leqq K$ is finite. Hence the total number of possible $\mathfrak{a}'$ is finite. In other words, the ideal class number $h_F(S)$ is finite. **q. e. d.**

33:13a. **Corollary.** *There is a Dedekind set S_0 consisting of almost all spots on F such that $h_F(S) = 1$ whenever $S \subseteq S_0$.*

Proof. Start with a Dedekind set T consisting of almost all spots on F. Let $\mathfrak{a}_1, \ldots, \mathfrak{a}_r$ be a representative set for the ideal class group at T. Take a subset S_0 consisting of almost all spots in T such that $|\mathfrak{a}_i|_\mathfrak{p} = 1$ for $1 \leqq i \leqq r$ and for all $\mathfrak{p} \in S_0$. We say that $h_F(S) = 1$ for any $S \subseteq S_0$. To see this consider a typical fractional ideal $\mathfrak{b}$ at S. Take $\mathfrak{a} \in I(T)$ with $|\mathfrak{a}|_\mathfrak{p} = |\mathfrak{b}|_\mathfrak{p}$ for all $\mathfrak{p} \in S$. Then $\mathfrak{a} = \alpha\mathfrak{a}_i$ for some $\mathfrak{a}_i$ and some $\alpha \in \dot{F}$. Hence

$$|\mathfrak{b}|_\mathfrak{p} = |\mathfrak{a}|_\mathfrak{p} = |\alpha\mathfrak{a}_i|_\mathfrak{p} = |\alpha|_\mathfrak{p} \quad \forall\, \mathfrak{p} \in S\,.$$

Hence $\mathfrak{b} = \alpha\mathfrak{o}$. So $h_F(S) = 1$. **q. e. d.**

§ 33I. More about idèles

Let us return to a discussion of the idèle group J_F of a global field F. If α is any non-zero field element we know from the Product Formula that $|\alpha|_\mathfrak{p} = 1$ for almost all $\mathfrak{p} \in \Omega$. Hence we can associate an idèle

$$(\alpha)_{\mathfrak{p} \in \Omega}$$

with every such α. An idèle which can be expressed in this form is called a principal idèle; the set of all principal idèles is a subgroup of J_F which will be written P_F. The rule $\alpha \rightarrowtail (\alpha)_{\mathfrak{p}\in\Omega}$ provides a natural isomorphism

$$\dot{F} \rightarrowtail P_F \subseteq J_F\,.$$

Now fix a set of spots $S \subseteq \Omega$ (in practice S will be a Dedekind set consisting of almost all spots in Ω). We call the idèle $\mathfrak{i}$ an S-idèle if

$$|\mathfrak{i}_\mathfrak{p}|_\mathfrak{p} = 1 \quad \forall\, \mathfrak{p} \in S\,.$$

The set of S-idèles is a subgroup J_F^S of the full idèle group J_F. We put

$$P_F^S = J_F^S \cap P_F\,.$$

Thus restriction of the natural map $\dot{F} \rightarrowtail P_F$ produces

$$\mathfrak{u}(S) \rightarrowtail P_F^S\,.$$

33:14. **Theorem.** *Let S be a Dedekind set of spots on the global field F. Then*

$$(J_F : P_F J_F^S) = h_F(S) < \infty\,.$$

Proof. We define a mapping

$$\varphi: J_F \to I(S)$$

as follows: given an idèle $\mathfrak{i}$ let $\varphi\mathfrak{i}$ be that fractional ideal $\mathfrak{a}$ at S for which $|\mathfrak{a}|_\mathfrak{p} = |\mathfrak{i}_\mathfrak{p}|_\mathfrak{p}$ at all $\mathfrak{p} \in S$. Thus

$$|\varphi\mathfrak{i}|_\mathfrak{p} = |\mathfrak{i}_\mathfrak{p}|_\mathfrak{p} \quad \forall\, \mathfrak{p} \in S$$

is the defining equation of φ. This mapping is a homomorphism since

$$|\varphi(\mathfrak{i}\mathfrak{j})|_\mathfrak{p} = |\mathfrak{i}_\mathfrak{p}\mathfrak{j}_\mathfrak{p}|_\mathfrak{p} = |\varphi\mathfrak{i}|_\mathfrak{p}\,|\varphi\mathfrak{j}|_\mathfrak{p} = |(\varphi\mathfrak{i})\,(\varphi\mathfrak{j})|_\mathfrak{p}$$

holds for all $\mathfrak{p}$ in S. It is clearly surjective with kernel J_F^S. And the inverse image of the group of principal ideals $P(S)$ is $P_F J_F^S$. So the kernel of the composite map

$$J_F \to I(S) \to I(S)/P(S)$$

is $P_F J_F^S$. Hence

$$J_F/P_F J_F^S \rightarrowtail I(S)/P(S)\,.$$

q. e. d.

33:14a. **Corollary.** *There is a Dedekind set S_0 consisting of almost all spots on F such that $(J_F : P_F J_F^S) = 1$ whenever $S \subseteq S_0$.*

§ 33J. Note on classical ideal theory

In the classical approach to algebraic number theory one works with a subfield F of the complex numbers $\mathbf{C}$ which is finite over the rational numbers $\mathbf{Q}$. The algebraic integers are those complex numbers α which satisfy a polynomial equation

$$\alpha^m + a_1\alpha^{m-1} + \cdots + a_m = 0 \quad (\text{all } a_i \in \mathbf{Z})\,.$$

And the ring of integers R of the algebraic number field F consists of all those algebraic integers which fall in F. How is this related to the modern approach which we have given here? Of course we expect it to be a special case, and in fact it is. For instance, consider the set of all discrete spots S on $\mathbf{Q}$ and let $T||S$ on F. Then $\mathbf{Z} = \mathfrak{o}(S)$ by Proposition 31:2. Hence $R = \mathfrak{o}(T)$ by Proposition 23:2. So the ring of integers in the classical sense is that ring of integers in our sense which corresponds to the set of all discrete spots on F. In particular, the Dirichlet Unit Theorem gives the structure of the group of units of R. And the ideal class group of the fractional ideal theory determined by the ring R is finite.

Part Two

Abstract Theory of Quadratic Forms

Chapter IV

Quadratic Forms and the Orthogonal Group

We leave the arithmetic theory of fields in order to develop a different subject, the abstract theory of quadratic forms. In the latter half of the book we shall combine these two subjects into the arithmetic theory of quadratic forms. Our immediate purpose is to introduce a quadratic form and an orthogonal geometry on an arbitrary finite dimensional vector space and to study certain groups of linear transformations that leave the quadratic form invariant. We must make the assumption[1] from now on that the field of scalars F does not have characteristic 2. As we indicated, our vector spaces are assumed to be finite dimensional.

§ 41. Forms, matrices and spaces

§ 41A. Quadratic spaces

Let V be an n-dimensional vector space over an abstract field F of characteristic not 2. Consider a symmetric bilinear form[2] B on V, i. e. a mapping

$$B: V \times V \to F$$

with the following properties:

$$B(x, y+z) = B(x, y) + B(x, z)$$

$$B(\alpha x, y) = \alpha B(x, y), \quad B(x, y) = B(y, x)$$

for all $x, y, z \in V$ and all $\alpha \in F$. If we put $Q(x) = B(x, x)$ we obtain a mapping

$$Q: V \to F.$$

[1] There will be two exceptions to this assumption about the characteristic: the Wedderburn theory of §§ 51 and 52 and the abstract theory of lattices of § 81 can be over fields of any characteristic.

[2] We restrict ourselves entirely to symmetric bilinear forms over fields of characteristic not 2, and to the so-called orthogonal geometry determined by such forms. If the bilinear form is alternating instead of symmetric the geometry is called symplectic, if the bilinear form is hermitian the geometry is called unitary. There is also a theory of quadratic forms and a corresponding orthogonal geometry over fields of characteristic 2. We refer the reader to J. Dieudonné, *La géométrie des groupes classiques* (Berlin, 1955) for information about these geometries.

A quadratic map on the vector space V is any mapping $Q: V \to F$ which can be obtained from a symmetric bilinear form B by the substitution $x = y$ in the above way. (This is not the definition in characteristic 2.) The following identities hold:

$$Q(\alpha x) = \alpha^2 Q(x)$$

$$Q(x+y) = Q(x) + Q(y) + 2B(x, y)$$

$$Q\left(\sum_i \alpha_i x_i\right) = \sum_i \alpha_i^2\, Q(x_i) + 2 \sum_{i<j} \alpha_i \alpha_j B(x_i, x_j)\,.$$

The second identity shows that the symmetric bilinear form B associated with the quadratic map Q is unique. A quadratic space is a composite object consisting of a (finite dimensional) vector space V, a quadratic map Q, and the symmetric bilinear form B associated with Q. We say that a quadratic space is binary, ternary, quaternary, quinary, ..., n-ary, according as its dimension is 2, 3, 4, 5, ..., n. An n-ary space is usually assumed to be non-zero. The quadratic space V is said to represent a field element α if there is a vector x in V such that $Q(x) = \alpha$, in other words if $\alpha \in Q(V)$. We say that V is universal if $Q(V) = F$. Two subsets X and Y of V are called orthogonal if $B(X, Y) = 0$.

For any fixed scalar α we define

$$B^\alpha(x, y) = \alpha B(x, y) \quad \forall\, x, y \in V\,.$$

This provides V with a new symmetric bilinear form

$$B^\alpha: V \times V \to F\,,$$

and hence puts a new quadratic structure on V. We let Q^α denote the quadratic map associated with B^α. When this is done we say that we have scaled by α and we let V^α denote the vector space V provided with its new quadratic structure. We should not confuse scaling of the quadratic space V with scaling of a vector $x \in V$; by the latter we mean that the vector x is replaced by αx. Thus

$$Q^\alpha(x) = \alpha Q(x)\,, \quad \text{while} \quad Q(\alpha x) = \alpha^2 Q(x)\,.$$

Note that

$$Q^\alpha(V) = \alpha Q(V)\,.$$

Consider two finite dimensional vector spaces V and W over F. We know from linear algebra that the set of all linear transformations of V into W is a finite dimensional vector space over F in a natural way. This set will be written

$$L_F(V, W)$$

or simply $L(V, W)$. When $V = W$ the composition of transformations provides $L(V, V)$ with a multiplicative law which makes it into a ring with identity; we then use

$$L_F(V)$$

or $L_n(V)$ or just $L(V)$ instead of $L_F(V, V)$. Scalar and ring multiplication in $L(V)$ are related by the equations

$$\alpha(\varphi\,\psi) = (\alpha\,\varphi)\,\psi = \varphi(\alpha\,\psi)$$

for all $\alpha \in F$ and all $\varphi, \psi \in L(V)$. (See § 51B for the significance of these equations.) The invertible linear transformations in $L_F(V)$ form a group called the general linear group and written

$$GL_F(V)\,, \quad GL_n(V)\,, \quad \text{or} \quad GL(V)\,.$$

Suppose now that V and W are quadratic spaces and let Q and B denote the quadratic map and bilinear form on each of them. A linear transformation $\sigma \in L(V, W)$ is called a representation of V into W with respect to the quadratic maps Q on V and W if

$$Q(\sigma x) = Q(x) \quad \forall\, x \in V\,.$$

If σ is a representation, the identity

$$B(x, y) = \frac{1}{2}\{Q(x+y) - Q(x) - Q(y)\}$$

implies that

$$B(\sigma x, \sigma y) = B(x, y) \quad \forall\, x, y \in V\,.$$

An injective representation is called an isometry of V into W. And V and W are said to be isometric if there exists an isometry σ of V onto W. We let $V \rightarrowtail W$ denote a representation, $V \rightarrowtail W$ an isometry of V into W, and $V \twoheadrightarrow W$ or $V \cong W$ an isometry of V onto W. The set of all isometries of V into W will be written

$$O(V, W)\,.$$

If $V = W$ we write

$$O_n(V) \quad \text{or} \quad O(V)$$

instead of $O(V, V)$. In this situation every isometry is invertible, so

$$O_n(V) \subseteq GL_n(V)\,.$$

Moreover, the composition of two isometries is an isometry, so is the inverse of an isometry, hence the set $O_n(V)$ is actually a subgroup of $GL_n(V)$. This group is called the orthogonal group of V with respect to the quadratic map Q. If α is a non-zero scalar we have

$$O_n(V) = O_n(V^\alpha)\,.$$

41:1. *Let σ be a linear transformation of the quadratic space V into the quadratic space W, and let $x_1, \ldots, x_n$ be a base for V. Suppose*

$$B(\sigma x_i, \sigma x_j) = B(x_i, x_j)$$

for $1 \leqq i \leqq n$ *and* $1 \leqq j \leqq n$. *Then* σ *is a representation.*

Proof. Express a typical $x \in V$ in the given base: $x = \sum_i \alpha_i x_i$. Then

$$\begin{aligned} Q(\sigma x) &= Q\left(\sum_i \alpha_i \sigma x_i\right) \\ &= \sum_i \alpha_i^2 Q(\sigma x_i) + 2 \sum_{i<j} \alpha_i \alpha_j B(\sigma x_i, \sigma x_j) \\ &= \sum_i \alpha_i^2 Q(x_i) + 2 \sum_{i<j} \alpha_i \alpha_j B(x_i, x_j) \\ &= Q(x) . \end{aligned}$$

q. e. d.

§41B. Symmetric matrices

Let M be an $m \times m$ symmetric matrix and N an $n \times n$ symmetric matrix over F. We write

$$M \rightarrow N \quad (\text{over } F)$$

and say that M is represented by N if there is an $n \times m$ matrix T with coefficients in F such that

$$M = {}^tTNT ,$$

where ${}^tT$ stands for the transpose of T. If this can be done with an invertible T over F we say that M is equivalent to N and write

$$M \cong N \quad (\text{over } F) .$$

Equivalence of matrices is clearly an equivalence relation on the set of all $n \times n$ symmetric matrices over F.

Now consider an n-dimensional quadratic space V with bilinear form B and quadratic map Q. We can associate a symmetric matrix N with each base $x_1, \ldots, x_n$ for V by taking the $n \times n$ matrix whose i,j entry is $B(x_i, x_j)$. We call N the matrix of the quadratic space V in the base $x_1, \ldots, x_n$ and write

$$V \cong N \quad \text{in} \quad x_1, \ldots, x_n .$$

If there is at least one base $x_1, \ldots, x_n$ for which this holds, then we say that V has the matrix N and we write

$$V \cong N .$$

What happens to N under a change of base? Suppose that $V \cong N'$ in the base $x_1', \ldots, x_n'$ and let $T = (t_{ij})$ be the matrix that carries the first base to the second, i. e. let

$$x_j' = \sum_\lambda t_{\lambda j} x_\lambda .$$

Then

$$B(x_i', x_j') = B\left(\sum_\mu t_{\mu i} x_\mu , \sum_\lambda t_{\lambda j} x_\lambda\right) = \sum_{\mu, \lambda} t_{\mu i} B(x_\mu, x_\lambda) t_{\lambda j} .$$

Hence

$$N' = {}^tTNT\,.$$

In other words, N' and N are equivalent over F. And it is now clear that if N'' is equivalent to N, then there is a base $x_1'', \ldots, x_n''$ for V in which $V \cong N''$.

If V is just a vector space with base $x_1, \ldots, x_n$, and if N is a given $n \times n$ symmetric matrix, is there a bilinear form B or V with respect to which V has the matrix N in $x_1, \ldots, x_n$? Yes. Just define

$$B\left(\sum_\lambda \alpha_\lambda x_\lambda, \sum_\mu \beta_\mu x_\mu\right) = \sum_{\lambda,\mu} \alpha_\lambda \beta_\mu a_{\lambda\mu}$$

for typical vectors $\sum \alpha_\lambda x_\lambda$ and $\sum \beta_\mu x_\mu$ of V, where $a_{\lambda\mu}$ is the λ, μ entry of N.

Given a symmetric $n \times n$ matrix N we let $< N >$ stand for an n-dimensional quadratic space which has the matrix N. For example, if α is a given field element then $< \alpha >$ denotes a line which contains a non-zero vector x for which $Q(x) = \alpha$. We shall find it convenient to use $< M >$ for different quadratic spaces having the same matrix M.

41:2. *Let U and V be quadratic spaces with matrices M and N respectively. Then*

(1) *$U \rightarrowtail V$ if and only if $M \rightarrowtail N$*
(2) *$U \cong V$ if and only if $M \cong N$.*

Proof. (1) Put $M = (a_{ij})$ and $N = (b_{ij})$. Take a base $x_1, \ldots, x_m$ for U in which $B(x_i, x_j) = a_{ij}$ and a base $y_1, \ldots, y_n$ for V in which $B(y_i, y_j) = b_{ij}$.

First suppose that we have a representation $\sigma: U \rightarrowtail V$. Then $B(z_i, z_j) = a_{ij}$, where $z_i = \sigma x_i$. Write $z_i = \sum_\mu t_{\mu i} y_\mu$ and let T be the $n \times m$ matrix $T = (t_{\mu i})$. Then

$$a_{ij} = B(z_i, z_j) = B\left(\sum_\mu t_{\mu i} y_\mu, \sum_\lambda t_{\lambda j} y_\lambda\right) = \sum_{\mu,\lambda} t_{\mu i} b_{\mu\lambda} t_{\lambda j}\,.$$

Hence $M = {}^tTNT$ as asserted.

Conversely, suppose that we are given an $n \times m$ matrix T such that $M = {}^tTNT$. Define $z_i = \sum_\mu t_{\mu i} y_\mu$ and use the preceding calculation to observe that $B(z_i, z_j)$ can now be proved equal to a_{ij}. Construct the linear transformation $\sigma \in L(U, V)$ given by $\sigma x_i = z_i$ for $1 \leqq i \leqq m$. Then

$$B(\sigma x_i, \sigma x_j) = a_{ij} = B(x_i, x_j)\,.$$

Hence σ is a representation by Proposition 41:1. So $U \rightarrowtail V$.

(2) The second part of the proposition is done in the same way.

q. e. d.

Suppose we take m vectors $z_1, \ldots, z_m$ in the quadratic space V. The determinant

$$\det(B(z_i, z_j))$$

of the $m \times m$ matrix $(B(z_i, z_j))$ is called the discriminant of the vectors $z_1, \ldots, z_m$ and is written

$$d_B(z_1, \ldots, z_m)$$

or simply $d(z_1, \ldots, z_m)$. In particular, if $x_1, \ldots, x_n$ is a base for V and if $V \cong N$ in this base, then

$$d(x_1, \ldots, x_n) = \det N .$$

If we take another base $x_1', \ldots, x_n'$ the equation $N' = {}^tT N T$ shows that

$$d(x_1', \ldots, x_n') = \alpha^2 d(x_1, \ldots, x_n)$$

for some $\alpha \in \dot{F}$. Hence the canonical image of $d(x_1, \ldots, x_n)$ in $0 \cup (\dot{F}/\dot{F}^2)$ is independent of the base, it is called the discriminant of the quadratic space V, and it is written

$$d_B V \quad \text{or} \quad dV .$$

If V consists just of the single point 0, the discriminant is defined to be

$$dV = 1 .$$

The set $0 \cup (\dot{F}/\dot{F}^2)$ used in the definition of the discriminant is formed in the obvious way: take the quotient group $\dot{F}/\dot{F}^2$, adjoin 0 to it, and define 0 times anything to be 0. We shall often write $dV = \alpha$ with α in F; this will really mean that dV is the canonical image of α in $0 \cup (\dot{F}/\dot{F}^2)$. It is equivalent to saying that V has a base $V = Fx_1 + \cdots + Fx_n$ in which

$$d_B(x_1, \ldots, x_n) = \alpha .$$

41:3. **Example.** If $V \cong N$ and $\alpha \in F$, then $V^\alpha \cong \alpha N$. Hence $dV^\alpha = \alpha^n dV$.

41:4. *Any m vectors $x_1, \ldots, x_m$ in the quadratic space V with $d(x_1, \ldots, x_m) \neq 0$ are independent.*

Proof. A dependence $\sum \alpha_i x_i = 0$ yields $\sum \alpha_i B(x_i, x_j) = 0$ for $1 \leq j \leq m$. This is a dependence among the rows of the matrix $(B(x_i, x_j))$; such a dependence is impossible since $d(x_1, \ldots, x_m) \neq 0$. **q. e. d.**

§ 41 C. Quadratic forms

Let us briefly indicate the connection between quadratic spaces and the general definition of a quadratic form. In general a linear, quadratic, cubic, ..., d-ic form over an arbitrary field is a homogeneous polynomial of degree d in a polynomial ring in sufficiently many variables, say in

$$F[X_1, \ldots, X_n, \ldots] .$$

For instance,

$$3X_1 + 5X_2 + 2X_3$$

is a ternary linear form. And

$$f(X_1, X_2) = X_1^2 + 3X_1X_2 + 7X_2^2$$

is a binary quadratic form. Now if the characteristic is not 2, every n-ary quadratic form can be expressed with symmetric cross terms, for instance here we have

$$f(X_1, X_2) = X_1^2 + \frac{3}{2}X_1X_2 + \frac{3}{2}X_2X_1 + 7X_2^2.$$

In this way we can associate a symmetric matrix, here it would be

$$\begin{pmatrix} 1 & \frac{3}{2} \\ \frac{3}{2} & 7 \end{pmatrix},$$

with a given quadratic form. Hence we can associate a quadratic space with a given quadratic form. The quadratic space V associated with $f(X_1, X_2)$ would have a base x_1, x_2 in which V had the above matrix, and

$$Q(\alpha_1 x_1 + \alpha_2 x_2) = \alpha_1^2 + 3\alpha_1\alpha_2 + 7\alpha_2^2;$$

in other words the quadratic map Q is obtained from f by substituting α_1, α_2 for X_1, X_2. This brings out the connection between quadratic forms and quadratic maps and it also shows why we must not expect the definitions of § 41A to work when the characteristic is equal to 2.

One could define the representation and equivalence of quadratic forms using polynomials. These definitions, when correctly interpreted, correspond to the ones given here in terms of matrices and spaces. From now on the term quadratic form will be used for the quadratic map Q on a quadratic space V.

§ 42. Quadratic spaces

§ 42A. Orthogonal splittings

Consider the quadratic space V with its symmetric bilinear form B and associated quadratic form Q. We say that V has the orthogonal splitting

$$V = V_1 \perp \cdots \perp V_r$$

into subspaces $V_1, \ldots, V_r$ if V is the direct sum

$$V = V_1 \oplus \cdots \oplus V_r$$

with the V_i pair-wise orthogonal, i. e. with

$$B(V_i, V_j) = 0 \quad \text{for} \quad 1 \leq i < j \leq r.$$

We call the V_i the components of the splitting. We also use the notation

$$\mathop{\perp}_{1}^{r} V_i \quad \text{and} \quad \mathop{\oplus}_{1}^{r} V_i$$

for orthogonal splitting and direct sum respectively. We formally define

$$\mathop{\perp}_{\emptyset} V_i = 0 \quad \text{and} \quad \mathop{\oplus}_{\emptyset} V_i = 0\,,$$

where $\emptyset$ denotes the empty set. We say that a subspace U splits V, or that it is a component of V, if there exists a subspace W such that

$$V = U \perp W\,.$$

By taking a base for each component we see that the discriminant satisfies

$$d(V_1 \perp \cdots \perp V_r) = dV_1 \ldots dV_r,$$

the multiplication being performed of course in $0 \cup (\dot{F}/\dot{F}^2)$. If $X_1, \ldots, X_r$ are quadratic spaces which are not necessarily subspaces of V we write

$$V \cong X_1 \perp \cdots \perp X_r$$

to signify that V has an orthogonal splitting

$$V = V_1 \perp \cdots \perp V_r$$

in which each subspace V_i is isometric to X_i for $1 \leqq i \leqq r$.

Suppose that V is just an abstract vector space, but that it is also a direct sum

$$V = V_1 \oplus \cdots \oplus V_r$$

of quadratic subspaces $V_1, \ldots, V_r$. Then there is a unique symmetric bilinear form on V which induces the given bilinear forms on the V_i and under which

$$V = V_1 \perp \cdots \perp V_r\,.$$

For if $B_1, \ldots, B_r$ are the respective given bilinear forms, define

$$B(\Sigma x_\lambda, \Sigma y_\lambda) = \Sigma B_\lambda(x_\lambda, y_\lambda)$$

for typical vectors $\Sigma x_\lambda, \Sigma y_\lambda$ of $\mathop{\oplus}_{\lambda} V_\lambda$; it is easily seen that B has the required properties. Incidentally, this shows that if we are given any quadratic spaces $V_1, \ldots, V_r$, not necessarily contained in a given space, there exists a quadratic space V such that

$$V \cong V_1 \perp \cdots \perp V_r\,.$$

As an example of the notation consider the equation

$$V \cong \langle M \rangle \perp \langle N \rangle$$

with M and N symmetric matrices over F. This means that V is a quadratic space with a base in which

$$V \cong \left(\begin{array}{c|c} M & 0 \\ \hline 0 & N \end{array}\right).$$

Similarly

$$V \cong \langle\alpha_1\rangle \perp \cdots \perp \langle\alpha_n\rangle$$

with all α_i in F means that V has a base $x_1, \ldots, x_n$ in which $Q(x_i) = \alpha_i$ for $1 \leqq i \leqq n$ and $B(x_i, x_j) = 0$ for $1 \leqq i < j \leqq n$. This prompts us to make the following definition: a base $x_1, \ldots, x_n$ for the quadratic space V is said to be an orthogonal base if

$$B(x_i, x_j) = 0 \quad \text{for} \quad 1 \leqq i < j \leqq n.$$

42:1. *Every non-zero quadratic space has an orthogonal base.*

Proof. If $Q(V) = 0$ we are through. Otherwise we can pick $x \in V$ with $Q(x) \neq 0$. Take a base $x, x_2, \ldots, x_n$ for V. Then

$$x,\ x_2 - \frac{B(x, x_2)}{Q(x)}\, x,\ \ldots,\ x_n - \frac{B(x, x_n)}{Q(x)}\, x$$

is still a base for V, and the last $n-1$ vectors span an $(n-1)$-dimensional subspace W for which $B(x, W) = 0$. Now apply induction to W. **q. e. d.**

Consider two quadratic spaces $V = V_1 \perp \ldots \perp V_r$ and $W = W_1 + \ldots + W_r$ with the subspaces W_i pair-wise orthogonal (do not necessarily assume that the W_i give an orthogonal splitting for W). Suppose that a representation $\sigma_i : V_i \rightarrowtail W_i$ is given for each i. Then we know from linear algebra that there is a unique linear transformation σ of V into W which agrees with each σ_i on V_i. In fact the reader may easily verify that σ is also a representation $\sigma : V \rightarrowtail W$. We write this representation in the form

$$\sigma = \sigma_1 \perp \cdots \perp \sigma_r.$$

The important case is where $V = W$, all $V_i = W_i$, and all σ_i are in $O(V_i)$; in this event

$$\sigma_1 \perp \cdots \perp \sigma_r \in O(V);$$

if we also take $\tau = \tau_1 \perp \cdots \perp \tau_r$ with all $\tau_i \in O(V_i)$ we obtain the following rules:

$$\sigma\tau = (\sigma_1\tau_1) \perp \cdots \perp (\sigma_r\tau_r)$$

$$\sigma^{-1} = \sigma_1^{-1} \perp \cdots \perp \sigma_r^{-1}$$

$$\det\sigma = \det\sigma_1 \det\sigma_2 \ldots \det\sigma_r$$

where det is the determinant of the linear transformation in question.

§ 42B. Orthogonal complements

Let U be a subspace of the quadratic space V. By the orthogonal complement U^* of U in V we mean the subspace

$$U^* = \{x \in V \mid B(x, U) = 0\}.$$

We define the radical of V as the subspace

$$\operatorname{rad} V = \{x \in V \mid B(x, V) = 0\}.$$

Thus $\operatorname{rad} V = V^*$. We shall say that V is a regular quadratic space if $\operatorname{rad} V = 0$.

42:2. *Let the quadratic space V be a sum of pair-wise orthogonal subspaces, i. e. $V = V_1 + \cdots + V_r$ with $B(V_i, V_j) = 0$ for $1 \leqq i < j \leqq r$. Then*

$$\operatorname{rad} V = \operatorname{rad} V_1 + \cdots + \operatorname{rad} V_r.$$

Proof. Take a typical $x \in \operatorname{rad} V$ and express it as $x = \Sigma x_\lambda$ with each $x_\lambda \in V_\lambda$. Then for each $i (1 \leqq i \leqq r)$ we have

$$B(x_i, V_i) = B(\Sigma x_\lambda, V_i) \subseteq B(x, V) = 0,$$

so each x_i is in $\operatorname{rad} V_i$, hence $x \in \Sigma \operatorname{rad} V_i$. Conversely, if we take $x = \Sigma x_\lambda$ with each $x_\lambda \in \operatorname{rad} V_\lambda$ we have

$$B(x, V) \subseteq B(x_1, V_1) + \cdots + B(x_r, V_r) = 0,$$

so $x \in \operatorname{rad} V$. **q. e. d.**

42:2a. *V is regular if and only if all the V_i are regular.*

42:2b. *If V is regular, then $V = V_1 \perp \cdots \perp V_r$.*

Proof. It is only necessary to show that the sum is direct. So let us take $x_1 + \cdots + x_r = 0$ with $x_i \in V_i$ for $1 \leqq i \leqq r$. Then

$$0 = B(x_1 + \cdots + x_r, V_i) = B(x_i, V_i).$$

Hence $x_i \in \operatorname{rad} V_i = 0$. **q. e. d.**

42:3. *The quadratic space V is regular if and only if $dV \neq 0$.*

Proof. Take an orthogonal base

$$V = (Fx_1) \perp \cdots \perp (Fx_n).$$

Then V is regular if and only if all Fx_i are regular. But the line Fx_i is regular if and only if $Q(x_i) \neq 0$. Hence V is regular if and only if

$$Q(x_1)\, Q(x_2) \ldots Q(x_n) \neq 0.$$

This quantity is simply the discriminant $d(x_1, \ldots, x_n)$. **q. e. d.**

42:4. *Let U be a regular subspace of the quadratic space V. Then U splits V, in fact $V = U \perp U^*$. And if $V = U \perp W$ is any other splitting, then $W = U^*$.*

Proof. Take an orthogonal base for U:

$$U = (Fx_1) \perp \cdots \perp (Fx_p).$$

Since U is regular, all $Q(x_i) \neq 0$. A typical $z \in V$ can be written $z = y + w$ with

$$y = \frac{B(z, x_1)}{Q(x_1)} x_1 + \cdots + \frac{B(z, x_p)}{Q(x_p)} x_p,$$

$$w = z - y.$$

Here y is clearly in U; and an easy computation gives $B(w, x_i) = 0$ for $1 \leqq i \leqq p$, hence $w \in U^*$. This proves that $V = U + U^*$. Now $U \cap U^* = \operatorname{rad} U = 0$. So $V = U \oplus U^*$. Hence $V = U \perp U^*$.

If $V = U \perp W$, then $W \subseteq U^*$. But we have just proved that $V = U \perp U^*$. Hence $\dim W = \dim U^*$. Hence $W = U^*$. **q. e. d.**

Let us show how the given bilinear form B naturally determines an isomorphism Φ of V onto the dual space V', i. e. onto the set of linear functionals on V, when V is regular under B. Fix a vector x in V for a moment. Then

$$\varphi_x(y) = B(x, y)$$

defines a linear functional $\varphi_x\colon V \rightarrowtail F$ since B is bilinear. Hence φ_x is in the dual space V'. Now free x. This associates a linear functional φ_x with each $x \in V$, so we have a mapping

$$\Phi\colon V \rightarrowtail V'$$

defined by $\Phi(x) = \varphi_x$. The bilinearity of B again shows that Φ is linear. And the regularity of V shows that Φ is injective and hence bijective. Hence we have defined the Φ we were after.

We know from linear algebra that every base $x_1, \ldots, x_n$ for V has a dual base in V'. If we let Φ carry the dual back to V we obtain a base $y_1, \ldots, y_n$ for V with the property that

$$B(x_i, y_j) = \delta_{ij} \quad \text{(Kronecker delta).}$$

Moreover this base is unique. We shall call it the dual of the base $x_1, \ldots, x_n$ with respect to the bilinear form B.

42:5. **Example.** Let V be a regular quadratic space, let N be the matrix of V in the base $x_1, \ldots, x_n$, and let M be the matrix of V in the dual base $y_1, \ldots, y_n$. We claim that $M = N^{-1}$. For by § 41B we have $M = {}^tTNT$ where $T = (t_{\lambda j})$ is the matrix determined by $y_j = \sum_\lambda t_{\lambda j} x_\lambda$. On the other hand we have

$$\sum_\lambda B(x_i, x_\lambda)\, t_{\lambda j} = B(x_i, y_j) = \delta_{ij},$$

so that NT is the identity matrix. Hence $M = {}^tT$. Hence $M = MNM$. So $M = N^{-1}$ as asserted.

42:6. *U is an arbitrary subspace of a regular quadratic space V. Then*

$$\dim U + \dim U^* = \dim V\,, \quad U^{**} = U\,.$$

Proof. Consider the isomorphism Φ of V onto the dual V'. It is easily seen from the definition of Φ that ΦU^* is the annihilator of U in V'. Hence

$$\dim U^* = \dim \Phi U^* = \dim V - \dim U\,.$$

As for the second part, we have $U \subseteq U^{**}$ by definition of U^*, hence $U = U^{**}$ by a dimension argument. **q. e. d.**

§ 42C. Radical splittings

Consider the radical $\operatorname{rad} V$ of the quadratic space V and let U be any subspace of V for which $V = U \oplus \operatorname{rad} V$. Then clearly $V = U \perp \operatorname{rad} V$ and we call this a radical splitting of V. Obviously U is not unique unless V is regular or $V = \operatorname{rad} V$, but we shall see in Proposition 42:8 that it is always unique up to isometry. The equations

$$\operatorname{rad} V = \operatorname{rad} U \perp \operatorname{rad}(\operatorname{rad} V) = \operatorname{rad} U \perp \operatorname{rad} V$$

imply that $\operatorname{rad} U = 0$, and so U is regular.

42:7. *Let $\sigma\colon V \rightarrowtail W$ be a representation of quadratic spaces. If V is regular, then σ is an isometry.*

Proof. Take x in the kernel of σ. Then

$$B(x, V) = B(\sigma x, \sigma V) = 0\,.$$

Hence $x \in \operatorname{rad} V$. Hence $x = 0$. **q. e. d.**

42:8. *Let $V = U \perp \operatorname{rad} V$ and $V_1 = U_1 \perp \operatorname{rad} V_1$ be radical splittings of the quadratic spaces V and V_1. Then*

(1) $V \rightarrowtail V_1$ *if and only if* $U \rightarrowtail U_1$,

(2) $V \cong V_1$ *if and only if* $U \cong U_1$ *with* $\operatorname{rad} V \cong \operatorname{rad} V_1$.

Proof. (1) Suppose we are given a representation $\sigma\colon U \rightarrowtail U_1$. Define a representation $\tau\colon \operatorname{rad} V \rightarrowtail \operatorname{rad} V_1$ by putting $\tau x = 0$ for all x in $\operatorname{rad} V$. Then $\sigma \perp \tau$ is a representation of V into V_1.

Conversely let $\sigma\colon V \rightarrowtail V_1$ be given. Define the linear transformation $\varphi\colon U \rightarrowtail U_1$ by putting

$$\sigma x = \varphi x + z$$

with $x \in U$, $\varphi x \in U_1$, $z \in \operatorname{rad} V_1$. Then

$$Q(x) = Q(\sigma x) = Q(\varphi x)$$

and so φ is a representation.

(2) Let $\sigma\colon V \rightarrowtail\!\!\!\rightarrow V_1$ be the given isometry. Clearly σ carries $\operatorname{rad} V$ to $\operatorname{rad} V_1$ so that $\operatorname{rad} V \cong \operatorname{rad} V_1$. And we know from the first part that $U \rightarrowtail U_1$. But every representation of a regular space is an isometry by Proposition 42:7. Hence $U \rightarrowtail U_1$. Hence $U \rightarrowtail\!\!\!\rightarrow U_1$. The converse is clear. **q. e. d.**

One of the main problems in the theory of quadratic forms is the problem of finding invariants that will fully describe the equivalence class of a given form. The interpretation of this in the language of vector spaces is the following: find invariants that will determine whether or not two quadratic spaces over the same field are isometric. This theory has been fully developed over certain fields, notably over the field of real numbers and over the local and global fields of number theory. We shall take this up in Chapter VI. For the present we mention that by Proposition 42:8 it is enough to consider regular quadratic spaces only.

For this reason we shall usually assume that all given quadratic spaces are regular. Of course this does not eliminate all mention of spaces that are not regular since, for instance, the subspaces of a regular space are not necessarily regular.

§ 42D. Isotropy

Let x be a non-zero vector in the quadratic space V: we call x isotropic if $Q(x) = 0$, we call it anisotropic if $Q(x) \neq 0$. Let V be a non-zero quadratic space: we call V isotropic if it contains an isotropic vector, we call it anisotropic if it does not contain an isotropic vector, we call it totally isotropic if each of its non-zero vectors is isotropic. All these definitions will apply to non-zero vectors and non-zero spaces only. We have V totally isotropic if and only if V is non-zero with $Q(V) = 0$. For any quadratic space V we have

$$Q(V) = 0 \Leftrightarrow B(V, V) = 0 ,$$

because of the identity $B(x, y) = \frac{1}{2}\{Q(x+y) - Q(x) - Q(y)\}$. Do isotropic spaces exist? Yes. For instance, any space having a matrix with a zero on the diagonal is isotropic. The simplest and most important example of a regular isotropic space is the hyperbolic plane: a space V is called a hyperbolic plane if it has the matrix $V \cong \begin{pmatrix} 0 & 1 \\ 1 & 0 \end{pmatrix}$ in one of its bases. Thus all hyperbolic planes are isometric, regular, and with discriminant -1. Suppose the hyperbolic plane V is written $V = Fx + Fy$ with $Q(x) = Q(y) = 0$. Then the lines Fx and Fy are isotropic. And the equation

$$Q(\alpha x + \beta y) = 2\alpha\beta B(x, y)$$

shows that these are the only isotropic lines in V. This equation also shows that $Q(V) = F$, in other words every hyperbolic plane is universal.

42:9. *The following assertions are equivalent for a binary quadratic space V:*

(1) *V is a hyperbolic plane,*

(2) *V is isotropic and regular,*

(3) $dV = -1$.

Proof. (1) $\Rightarrow$ (2). This follows by inspection of the defining matrix of a hyperbolic plane.

(2) $\Rightarrow$ (3). Take an isotropic vector x in V and extend it to a base for V. The matrix of V in this base has the form

$$\begin{pmatrix} 0 & \beta \\ \beta & \gamma \end{pmatrix} \qquad \beta, \gamma \in F.$$

So $dV = -\beta^2$. But V is given regular, hence $\beta \in \dot{F}$. But $-\beta^2$ and -1 are equal in $\dot{F}/\dot{F}^2$. Hence $dV = -1$.

(3) $\Rightarrow$ (1). $Q(V) \neq 0$ since V is regular. Take a non-zero field element α in $Q(V)$ and pick $x \in V$ with $Q(x) = \alpha$. Then Fx splits V since it is regular, hence $V = (Fx) \perp (Fy)$ for some $y \in V$. Now the information $dV = -1$ implies that $-Q(x)\,Q(y)$ is a non-zero square, so we can assume after a suitable scaling of y that $Q(y) = -\alpha$. Then

$$V = F\left(\frac{x+y}{2}\right) + F\left(\frac{x-y}{\alpha}\right).$$

The matrix of V in this base is easily seen to be $\begin{pmatrix} 0 & 1 \\ 1 & 0 \end{pmatrix}$. Hence V is a hyperbolic plane. **q. e. d.**

42:10. *Every regular isotropic quadratic space is split by a hyperbolic plane. Hence it is universal.*

Proof. Let x be an isotropic vector in the given space V. Since V is regular there is a $y \in V$ with $B(x, y) \neq 0$. Then $U = Fx + Fy$ is a regular binary isotropic space, hence it is a hyperbolic plane. Being regular it must split V. **q. e. d.**

42:11. *Let V be a regular quadratic space, let α be a scalar. Then $\alpha \in Q(V)$ if and only if $<-\alpha> \perp V$ is isotropic.*

Proof. We have to consider $W = Fz \perp V$ with $Q(z) = -\alpha$. If there is an x in V with $Q(x) = \alpha$, then $Q(z + x) = 0$ and W is isotropic. Conversely suppose that W is isotropic. If V is isotropic it is universal by the last proposition and we are through. If V is not isotropic we must have a non-zero scalar β and a non-zero vector $y \in V$ such that $Q(\beta z + y) = 0$. But then $-\beta^2\alpha + Q(y) = 0$. So $\alpha = Q(y)/\beta^2 \in Q(V)$. **q. e. d.**

42:12. *Let U be a regular ternary subspace of a regular quaternary space V. Suppose V has discriminant* 1. *Then V is isotropic if and only if U is isotropic.*

Proof. We must take an isotropic V and deduce that U is isotropic. Write $V = U \perp <\alpha>$ with $\alpha \in \dot{F}$. Then $-\alpha \in Q(U)$ by Proposition 42:11. Hence U has a splitting $U = P \perp <-\alpha>$ with P a plane. So

$$V = P \perp <-\alpha> \perp <\alpha>,$$

hence $1 = dV = -dP$. So $dP = -1$ and P is a hyperbolic plane by Proposition 42:9. Hence U is isotropic. **q. e. d.**

42:13. *V is a regular quadratic space, U is a subspace with $Q(U) = 0$, and $x_1, \ldots, x_r$ is a base for U. Then there is a subspace $H_1 \perp \cdots \perp H_r$ of V in which each H_i is a hyperbolic plane with $x_i \in H_i$.*

Proof. If $r = 1$ we take $y_1 \in V$ with $B(x_1, y_1) \neq 0$ and put $H_1 = Fx_1 + Fy_1$. Then H_1 is a hyperbolic plane with the desired property. Let us proceed by induction to any $r > 1$. Put $U_{r-1} = Fx_1 + \cdots + Fx_{r-1}$ and $U_r = U$.

Then $U_{r-1} \subset U_r$, so $U_r^* \subset U_{r-1}^*$. Pick $y_r \in U_{r-1}^* - U_r^*$ and put $H_r = Fx_r + Fy_r$. Then $B(x_i, y_r) = 0$ for $1 \leq i \leq r-1$, hence $B(x_r, y_r) \neq 0$. Hence H_r is a hyperbolic plane containing x_r.

Write $V = H_r \perp H_r^*$. Then $H_r \subseteq U_{r-1}^*$ since $x_r \in U_{r-1}^*$ and $y_r \in U_{r-1}^*$, hence $U_{r-1} \subseteq H_r^*$. Apply the inductive assumption to U_{r-1} regarded as a subspace of H_r^*. This gives

$$H_1 \perp \cdots \perp H_{r-1} \subseteq H_r^*$$

with $x_i \in H_i$, for $1 \leq i \leq r-1$. So $H_1 \perp \cdots \perp H_{r-1} \perp H_r$ has the desired properties. **q. e. d.**

§ 42E. Involutions and symmetries

The identity 1_V of the ring $L_n(V)$ is by definition that linear transformation which leaves every vector of V fixed, i. e. $1_V(x) = x$ for all x in V. And the negative -1_V of 1_V reverses every vector of V, i. e. $-1_V(x) = -x$ for all $x \in V$. Clearly both $\pm 1_V$ are in $O_n(V)$ since $Q(\pm x) = Q(x)$ for all x in V. Recall from linear algebra that a linear transformation σ is called an involution if $\sigma^2 = 1_V$.

42:14. *V is a quadratic space and $\sigma \in O_n(V)$. Then σ is an involution if and only if there is a splitting $V = U \perp W$ for which $\sigma = -1_U \perp 1_W$.*

Proof. If $\sigma = -1_U \perp 1_W$, then $\sigma^2 = 1_U \perp 1_W = 1_V$ and σ is an involution. Conversely suppose that σ is an involution. Then $\sigma^2 = 1 (= 1_V)$. Consider the linear transformations $\sigma - 1$ and $\sigma + 1$ and put $U = (\sigma - 1)V$ and $W = (\sigma + 1)V$. A typical $x \in V$ has the form

$$x = -(\sigma - 1)\frac{x}{2} + (\sigma + 1)\frac{x}{2} \in U + W\,.$$

So $V = U + W$. A typical y in U has the form $(\sigma - 1)x$, so

$$\sigma y = \sigma^2 x - \sigma x = x - \sigma x = -y\,;$$

hence $\sigma y = -y$ for all y in U; similarly $\sigma z = z$ for all z in W. This shows first of all that $U \cap W = 0$, so $V = U \oplus W$. And secondly it implies that $B(U, W) = 0$ since for typical $y \in U$ and $z \in W$ we have

$$B(y, z) = B(\sigma y, \sigma z) = B(-y, z)\,,$$

i. e. $B(y, z) = 0$. Hence $V = U \perp W$. And thirdly it says that σ is -1_U on U and 1_W on W, so $\sigma = -1_U \perp 1_W$. **q. e. d.**

42:14a. *W is the set of vectors left fixed by σ, U the set of vectors reversed by σ.*

The most important involutions on a quadratic space V are the symmetries which we now define. Fix a vector $y \in V$ with $Q(y) \neq 0$. Define a mapping $\tau_y : V \rightarrow V$ by the formula

$$\tau_y x = x - \frac{2B(x, y)}{Q(y)}\, y\,.$$

Then the following facts can be verified directly: τ_y is linear, it is an involution, it is a representation and hence an element of $O_n(V)$, it reverses every vector in the line Fy, it leaves every vector in the hyperplane $(Fy)^*$ fixed. We call τ_y the symmetry with respect to the vector y or with respect to the line Fy; note that $\tau_y = \tau_{y'}$ if and only if Fy and Fy' are the same line. In particular there are exactly as many symmetries as there are anisotropic lines in V. It is easily seen that

$$\sigma\tau_y\sigma^{-1} = \tau_{\sigma y}$$

whenever $\sigma \in O_n(V)$. And every symmetry, being an involution, is its own inverse.

42:15. **Example.** Let $H = Fx + Fy$ be a hyperbolic plane with $Q(x) = Q(y) = 0$. Consider a vector $z = x - \alpha y$ for some $\alpha \neq 0$. Then z is anisotropic, so we can form the symmetry τ_z. An easy computation gives

$$\tau_z x = \alpha y, \quad \tau_z y = \alpha^{-1} x.$$

Now every anisotropic z falls in the line $F(x - \alpha y)$ for some $\alpha \in \dot{F}$. Hence every symmetry has the above action on x and y. In particular, every symmetry on a hyperbolic plane interchanges the two isotropic lines. Conversely, if $\sigma \in O_2(V)$ interchanges the two isotropic lines it must be a symmetry: $\sigma x = \alpha y$ with $\alpha \in \dot{F}$, hence the equation $B(x, y) = B(\sigma x, \sigma y)$ yields $\sigma y = \alpha^{-1} x$, hence $\sigma = \tau_{x-\alpha y}$. Finally we note that the action of a σ in $O_2(V)$ which is not a symmetry is described by

$$\sigma x = \alpha x, \quad \sigma y = \alpha^{-1} y;$$

hence every such σ is of the form

$$\sigma = \tau_{x-y}\tau_{x-\alpha y}.$$

§ 42F. Witt's theorem

42:16. **Theorem.** *Let U and W be isometric regular subspaces of a quadratic space V. Then U^* and W^* are isometric.*

Proof. 1) First suppose that U and W are lines, say $U = Fx$ and $W = Fy$ with $Q(x) = Q(y) \neq 0$. Then

$$Q(x+y) + Q(x-y) = 2Q(x) + 2Q(y) = 4Q(x),$$

hence either $Q(x+y)$ or $Q(x-y)$ is not zero. Replacing y by $-y$ if necessary allows us to assume that $Q(x-y) \neq 0$. We may therefore form the symmetry τ_{x-y}. We have

$$\tau_{x-y} x = x - \frac{2B(x, x-y)}{Q(x-y)}(x-y).$$

But

$$\begin{aligned} Q(x-y) &= Q(x) + Q(y) - 2B(x,y) \\ &= 2Q(x) - 2B(x,y) \\ &= 2B(x, x-y) . \end{aligned}$$

Hence $\tau_{x-y}x = y$. Hence $\tau_{x-y}U^* = W^*$. In other words U^* and W^* are isometric when U and W are lines.

2) Now the general case by induction to $\dim U$. Since U and W are given isometric we can take non-trivial splittings $U = U_1 \perp U_2$ and $W = W_1 \perp W_2$ with $U_1 \cong W_1$ and $U_2 \cong W_2$. Then the inductive assumption says that $U_2 \perp U^*$ is isometric to $W_2 \perp W^*$, hence there is a splitting $U_2 \perp U^* = X \perp Y$ with $X \cong W_2$ and $Y \cong W^*$. But then $X \cong W_2 \cong U_2$, and the inductive assumption again says that $Y \cong U^*$. Hence $W^* \cong U^*$.

q. e. d.

42:17. **Theorem.** *Let V and V' be regular isometric quadratic spaces, let U be any subspace of V, and let σ be an isometry of U into V'. Then there is a prolongation of σ to an isometry of V onto V'.*

Proof. Write $U = W \perp \operatorname{rad} U$ and let $x_1, \ldots, x_r$ be a base for $\operatorname{rad} U$. By Proposition 42:13 there is a subspace

$$H = H_1 \perp \cdots \perp H_r$$

of the quadratic space W^* in which each H_i is a hyperbolic plane such that $x_i \in H_i$. Since H is regular it splits W^*, hence there is a subspace S of W^* such that

$$V = H \perp S \perp W .$$

Put $U' = \sigma U$, $W' = \sigma W$, and $x_i' = \sigma x_i$ for $1 \leq i \leq r$. So

$$\operatorname{rad} U' = \sigma(\operatorname{rad} U) = F x_1' + \cdots + F x_r' .$$

And

$$U' = W' \perp \operatorname{rad} U' .$$

We can repeat the preceding construction on this arrangement to obtain a splitting

$$V' = H' \perp S' \perp W'$$

in which

$$H' = H_1' \perp \cdots \perp H_r'$$

where the H_i' are hyperbolic planes in which $x_i' \in H_i'$. Now there is clearly an isometry of H onto H' which agrees with σ on each x_i, hence on rad U. Also the given σ carries W to W'. Hence there is a prolongation of σ to an isometry σ of $H \perp W$ onto $H' \perp W'$. An easy application of Theorem 42:16 now says that S is isometric to S'. Hence there is a prolongation of σ to an isometry of V onto V'. **q. e. d.**

One can use the last two theorems to attach an invariant called the index to a regular quadratic space V. Consider a maximal subspace M of V with the property $Q(M) = 0$, let M' be another such subspace, and suppose for the sake of argument that $\dim M \leqq \dim M'$. Then there is an isomorphism σ of M into M', and this isomorphism is in fact an isometry since $Q(M) = Q(M') = 0$, hence there is a prolongation of σ to an isometry σ of V onto V by Theorem 42:17. Then $M \subseteq \sigma^{-1}M'$. But $Q(\sigma^{-1}M') = 0$. Hence $M = \sigma^{-1}M'$ and so $\dim M = \dim M'$. We have therefore proved that all maximal subspaces M of V with the property $Q(M) = 0$ have the same dimension. This dimension is called the index of V and is written $\operatorname{ind} V$.

There is another way of looking at the index. If we keep splitting off hyperbolic planes in V we ultimately obtain a splitting

$$V = H_1 \perp \cdots \perp H_r \perp V_0$$

with $0 \leqq 2r \leqq \dim V$ in which each H_i is a hyperbolic plane and V_0 is either 0 or anisotropic. By Theorem 42:16 we see first that r does not depend on how the splitting is performed, and then that V_0 is unique up to isometry. An easy application of Proposition 42:13 shows that r is actually the index of V. In particular this proves that $r = \operatorname{ind} V$ satisfies the inequality

$$0 \leqq 2 \operatorname{ind} V \leqq \dim V .$$

We call V a hyperbolic space if $0 < 2r = \dim V$. Thus V is hyperbolic if and only if it has a splitting

$$V = H_1 \perp \cdots \perp H_r$$

in which $r \geqq 1$ with all the H_i hyperbolic planes.

§ 42G. Effective orthogonalization

A good practical way of specifying a quadratic space is by giving a symmetric matrix. Suppose a space V is determined by a given symmetric matrix M. The fact that V always has an orthogonal base is the same as saying that there is a diagonal matrix equivalent to M. Is there an effective procedure for finding this diagonal matrix? Such a procedure is implicit in the proof of the existence of an orthogonal base (Proposition 42:1) and we formalize it here.

Start with a symmetric matrix M and consider the following transformations on M:

(a) multiply column i by a non-zero scalar α, then multiply the resulting row i by the same α,

(b) interchange columns i and j, then interchange the resulting rows i and j,

(c) add α times column i to column j, then α times the resulting row i to row j (α any scalar, $i \neq j$).

Call each of these three transformations an elementary transformation on the matrix M. An elementary transformation can always be interpreted in terms of a change of base on the space V. To see this let us consider a base $x_1, \ldots, x_n$ for V in which the associated matrix is M. Then the matrix obtained by the first elementary transformation is actually the matrix of V in the base obtained by replacing x_i by αx_i and leaving all other x_j fixed. Similarly the second elementary transformation corresponds to interchanging x_i and x_j. And the third corresponds to replacing x_j by $x_j + \alpha x_i$. Hence the matrix obtained by an elementary transformation, or by any series of elementary transformations, is equivalent to M. We leave it to the reader to show that a series of elementary transformations can always be explicitly found which will transform M into a matrix

$$\operatorname{diag}(a_1, \ldots, a_{n-r}, 0, \ldots, 0)$$

with all a_i non-zero. Incidentally the number r is the dimension of the radical of V.

§ 43. Special subgroups of $O_n(V)$

The given quadratic space V is non-zero and regular throughout this paragraph. We show that the symmetries of V generate $O_n(V)$, and we make a preliminary study of certain subgroups O_n^+, Ω_n and Z_n of $O_n(V)$.

§ 43A. Rotations and reflexions

Consider the orthogonal group $O_n(V)$ of the regular non-zero quadratic space V. Fix a base $x_1, \ldots, x_n$ for V and let M denote the matrix of V in this base. Consider an isometry σ of V and let $T = (t_{ij})$ be the matrix of σ in the above base, i. e.

$$\sigma x_j = \sum_i t_{ij} x_i .$$

Then $\sigma x_1, \ldots, \sigma x_n$ is also a base for V and the matrix of V in this base is M since σ is an isometry. Hence by § 41B we have $M = {}^tTMT$. Now the determinant of the linear transformation σ is equal to $\det T$, hence

$$\det \sigma = \pm 1 \quad \text{if} \quad \sigma \in O_n(V)$$

since V is regular. We call σ a rotation if $\det \sigma = +1$, and we let $O_n^+(V)$ or $O^+(V)$ or O_n^+ denote the set of rotations of V. We call σ a reflexion if $\det \sigma = -1$ and we let $O_n^-(V)$, etc. stand for the set of reflexions of V. It is clear that $O_n^+(V)$ is a subgroup of $O_n(V)$, and that $O_n^-(V)$ is not. We have

$$O_n = O_n^+ \cup O_n^- , \quad O_n^+ \cap O_n^- = \emptyset .$$

O_n^+ is not empty since, for instance, the square of any isometry is a rotation. O_n^- is not empty since, for instance, every symmetry is a reflexion (see § 42E). The map

$$\det : O_n \to \pm 1$$

is a surjective group homomorphism with kernel O_n^+, hence O_n^+ is a normal subgroup of O_n with

$$(O_n : O_n^+) = 2 .$$

The preceding discussion shows that the base $x_1, \ldots, x_n$ determines a group isomorphism of $O_n(V)$ onto the group of matrices $\{T\}$ with the property that

$$M = {}^tTMT .$$

The matrices T with this property are called the automorphs of M. Since $\det M \neq 0$, every automorph has determinant ± 1. The automorphs of determinant $+1$ are called the proper automorphs of M; they correspond to the rotations of V under the above group isomorphism.

Since every symmetry is a reflexion, and since there are as many symmetries as there are anisotropic lines, the number of reflexions is at least equal to the number of anisotropic lines. Now there are just as many rotations as there are reflexions. Hence the number of rotations is at least equal to the number of anisotropic lines.

43:1. *V is a hyperbolic space and σ is an isometry of V onto V which is identity on a maximal totally isotropic subspace of V. Then σ is a rotation.*

Proof. Let M be the given maximal totally isotropic subspace of V, let r be its dimension, so $2r$ is the dimension of V. By Proposition 42:13 there is another maximal totally isotropic subspace N of V such that $V = M \oplus N$. For any $x \in M$, $y \in N$ we have $\sigma x = x$ and

$$B(x, \sigma y - y) = B(x, \sigma y) - B(x, y) = B(x, \sigma y) - B(x, \sigma y) = 0 ,$$

so $\sigma y - y \in M^*$. But $M \subseteq M^*$, so by comparing dimensions $M = M^*$. Hence $\sigma y - y \in M$ for all $y \in N$. Take a base $x_1, \ldots, x_r$ for M and a base $y_1, \ldots, y_r$ for N. Computing $\det\sigma$ in the resulting base for V gives $\det\sigma = 1$. Hence σ is a rotation. **q. e. d.**

43:2. *Let U be a hyperplane of the regular quadratic space V and let σ be an isometry of U into V. If U is regular there are exactly two prolongations of σ to O_n and the one is a symmetry times the other. If U is not regular there is exactly one prolongation of σ to O_n.*

Proof. (1) First do the case of a regular hyperplane U. By Witt's theorem there is a prolongation σ_1 of σ to O_n. Let σ_2 be any element of O_n. Then σ_2 is a prolongation of σ if and only if $\sigma_2^{-1}\sigma_1$ is identity on U. This means that σ_1 and $\sigma_1\tau_y$ are the only prolongations of σ, where Fy is the line orthogonal to the hyperplane U.

(2) Now let us suppose that U is not regular. We know that there is at least one prolongation of σ to O_n. If there were another we would have a ϱ in O_n which is identity on U but not on V. Let us prove that this is impossible. U^* is a line since U is a hyperplane, hence $\operatorname{rad} U = U \cap U^*$ is a line Fx say; a radical splitting of U therefore leads to a splitting

$$V = W \perp (Fx + Fy) , \quad Q(x) = Q(y) = 0 , \quad W \subseteq U .$$

Now ϱ leaves W fixed. Hence it leaves $Fx + Fy$ fixed. But $\varrho x = x$. Hence $\varrho y = \alpha y$ for some scalar α. But

$$B(x, y) = B(\varrho x, \varrho y) = B(x, \alpha y) = \alpha B(x, y) .$$

Hence $\alpha = 1$. Hence ϱ is identity on V. Hence every ϱ which is identity on U is identity on V. So σ has just one prolongation. **q. e. d.**

§43B. Generation of O_n by symmetries

43:3. **Theorem.** *Every isometry σ of a regular n-ary quadratic space onto itself is a product of at most n symmetries.*

Proof.[1] 1) The first step is in fact a lemma. We suppose that the isometry σ on the space V satisfies the condition

$$x \text{ anisotropic in } \dot{V} \Rightarrow \sigma x - x \text{ isotropic in } \dot{V}, \qquad (*)$$

and we deduce that $n \geqq 4$, n is even, and σ is a rotation. It is evident that we cannot have $n = 1$. If we have $n = 2$ we pick an anisotropic x in V; then the condition (*) implies that x and σx are independent with discriminant $d(x, \sigma x)$ equal to 0; this denies the regularity of V. We therefore must have $n \geqq 3$. We shall now deduce from condition (*) that n is even and σ is a rotation.

We claim that $Q(\sigma x - x) = 0$ holds for all $x \in V$, not just for the anisotropic ones. To see this we consider an isotropic y in V; then there is a hyperbolic plane which contains y and splits V, hence there is a vector z with $Q(z) \neq 0$ and $B(y, z) = 0$. Hence $Q(y + \varepsilon z) \neq 0$, hence

$$Q(\sigma(y + \varepsilon z) - (y + \varepsilon z)) = 0, \quad Q(\sigma z - z) = 0 ,$$

by the condition (*), hence

$$Q(\sigma y - y) + 2\varepsilon\, B(\sigma y - y, \sigma z - z) = 0 .$$

All this holds for any $\varepsilon \in \dot{F}$, in particular for $\varepsilon = \pm 1$. Do it for $\varepsilon = +1$, then for $\varepsilon = -1$, then add. We obtain $Q(\sigma y - y) = 0$ as we asserted.

Hence the space $W = (\sigma - 1) V$ satisfies $Q(W) = 0$. Now for any x in V and any $y \in W^*$ we have

$$\begin{aligned} B(x, \sigma y - y) &= B(\sigma x, \sigma y - y) - B(\sigma x - x, \sigma y - y) \\ &= B(\sigma x, \sigma y - y) \\ &= B(\sigma x, \sigma y) - B(\sigma x, y) \\ &= B(x, y) - B(\sigma x, y) \\ &= -B(\sigma x - x, y) \\ &= 0 . \end{aligned}$$

[1] A two or three line proof will show that σ is a product of symmetries, even that it is a product of at most $2n - 1$ symmetries. The difficulty is in showing that σ is a product of at most n symmetries.

Hence $\sigma y - y$ is in $\operatorname{rad} V$. But V is regular. Hence $\sigma y = y$ for all y in W^*. Hence $Q(W^*) = 0$ by condition(*). Hence

$$W \subseteq W^* \subseteq W^{**} = W .$$

Hence $W = W^*$. Hence n is even and $\dim W = n/2$, in other words V is a hyperbolic space and W is a maximal totally isotropic subspace. But σ is identity on $W^* = W$. Hence σ is a rotation by Proposition 43:1.

2) Now we can prove the theorem. The proof is by induction on n. For $n = 1$ the result is trivial, so let $n > 1$.

Suppose that there is an anisotropic vector x in V such that $\sigma x = x$. Then the restriction of σ to the hyperplane U orthogonal to Fx is an element of $O_{n-1}(U)$. By the inductive assumption this restriction is a product of at most $n - 1$ symmetries taken with respect to lines in U. Now each of these symmetries has a natural prolongation obtained by taking the symmetry on V with respect to the original line in U. The product of the prolongations taken in the original order agrees with σ on U, and also on Fx where both σ and the product are identity, hence they agree on V. Hence σ is a product of at most $n - 1$ symmetries when it leaves an anisotropic vector fixed.

Next suppose that there is an anisotropic vector x such that $Q(\sigma x - x) \neq 0$. Form the symmetry $\tau_{\sigma x - x}$. Now $\tau_{\sigma x - x}\sigma$ is easily seen to leave x fixed. And x is anisotropic. Hence $\tau_{\sigma x - x}\sigma$ is a product of at most $n - 1$ symmetries. Hence σ is a product of at most n symmetries.

Hence any σ which does not satisfy condition (*) of step 1) is a product of at most n symmetries. Consider a σ which satisfies (*). Then n is even and σ is a rotation. Fix a symmetry τ of V. Then $\tau\sigma$ is a reflexion, hence it cannot satisfy (*), hence $\tau\sigma$ is a product of at most n symmetries, hence σ is a product of at most $n + 1$ symmetries. But σ cannot be a product of $n + 1$ symmetries since σ is a rotation and $n + 1$ is odd. Hence σ is a product of at most n symmetries. **q. e. d.**

If σ is any isometry of V onto itself we call the subspace

$$\{x \in V \mid \sigma x = x\}$$

the fixed space of σ. For example 0 is the fixed space of -1_V and V is the fixed space of 1_V. The fixed space of a symmetry is the hyperplane orthogonal to the line used in defining the symmetry. We must take care not to confuse the fixed space of σ with the subspaces left fixed by σ; for instance V is always left fixed by σ, but it is the fixed space of σ only when σ is the identity.

43:3a. **Corollary.** *If σ is a product of r symmetries, then the dimension of its fixed space is at least $n - r$.*

Proof. Express σ as a product $\sigma = \tau_1 \ldots \tau_r$ of symmetries and let U_i be the fixed space of τ_i for $1 \leqq i \leqq r$. Then $U_1 \cap \cdots \cap U_r$ is contained in the fixed space of σ. It is therefore enough to prove that if $U_1, \ldots, U_r$ are any r hyperplanes in a vector space V, then

$$\dim (U_1 \cap \cdots \cap U_r) \geqq n - r .$$

We do this by induction to r. For $r = 1$ it is the definition of a hyperplane. For $r > 1$ we have

$$\begin{aligned} \dim (U_1 \cap \cdots \cap U_r) &= \dim (U_1 \cap \cdots \cap U_{r-1}) + \dim U_r \\ &\quad - \dim((U_1 \cap \cdots \cap U_{r-1}) + U_r) \\ &\geqq (n - r + 1) + (n - 1) - (n) \\ &= n - r . \end{aligned}$$

q. e. d.

43:3b. **Corollary.** *Suppose σ can be expressed as a product of n symmetries. Then it can be expressed as a product of n symmetries with the first (or last) symmetry chosen arbitrarily.*

Proof. Write σ as a product of n symmetries, say $\sigma = \tau_1 \ldots \tau_n$. Let τ be an arbitrary given symmetry. Then by the theorem we can express $\tau\sigma$ as a product of at most n symmetries, hence

$$\sigma = \tau\tau_2' \ldots \tau_r'$$

with $r \leqq n + 1$. Here $\det\sigma$ will be $(-1)^r$. On the other hand $\det\sigma$ is $(-1)^n$ by hypothesis. Hence r and n have the same parity. In particular, $r \leqq n$. If $r < n$, put an even number of τ's at the end of the above expression for σ to obtain

$$\sigma = \tau\tau_2' \ldots \tau_n' .$$

This allows us to choose the first symmetry in an arbitrary way. Similarly with the last. **q. e. d.**

§ 43C. Binary and ternary spaces

The orthogonal groups of binary and ternary spaces have certain special properties that can be used in the general theory. (Incidentally the theory of the binary orthogonal group is quite different from the theory in higher dimensions.) For instance if V is a regular binary space every reflexion is a symmetry by Theorem 43:3; so the number of rotations, which is always equal to the number of reflexions, is also equal to the number of symmetries and hence to the number of anisotropic lines in V. As another example let us show that $O_2^+(V)$ is commutative. To see this first consider a typical symmetry τ and a typical rotation σ. Then $\tau\sigma = \tau_1$ is a reflexion and hence a symmetry. So

$$\tau\sigma\tau = \tau\tau\tau_1\tau = \tau_1\tau = \sigma^{-1}.$$

Hence for any other rotation ϱ we have

$$\sigma\varrho\sigma^{-1} = \tau\tau_1\varrho\,\tau_1\tau = \tau\varrho^{-1}\tau = \varrho .$$

So O_2^+ is commutative. Later we shall see that O_n^+ is not commutative when $n \geqq 3$.

43:4. *V is a regular quadratic space with $1 \leqq \dim V \leqq 3$. Let d be the dimension of the fixed space of an isometry σ of V onto V. Then σ is a product of $n - d$, but not of less than $n - d$, symmetries of V.*

Proof. It is clear that σ is not a product of $r < n - d$ symmetries for if it were, the fixed space of σ would have dimension $\geqq n - r > d$. We must therefore prove that σ has at least one expression as a product of $n - d$ symmetries. For $n = 1$ the result is clear.

Next consider $n = 2$. If $d = 0$, σ is neither the identity nor a symmetry, hence it is a product of two symmetries as required. If $d = 1$ the fixed space is a line; it cannot be an isotropic line by Proposition 43:2; hence the fixed space is a regular line, hence σ is the symmetry with respect to the line orthogonal to this fixed space. If $d = 2$, σ is the identity. This finishes the case $n = 2$.

Finally $n = 3$. If $d = 0$, σ cannot be a product of less than three symmetries. If $d = 1$ the fixed space is a line; here it is enough to prove that σ is a rotation; suppose if possible that σ is not a rotation; then σ is a reflexion and so $-\sigma$ is a rotation; hence $-\sigma$ keeps a vector fixed, hence σ reverses a vector; so σ keeps one vector fixed and it reverses another; these two vectors are clearly independent; hence σ^2 keeps every vector in a certain plane fixed; but σ^2 is also a rotation; hence $\sigma^2 = 1_V$ by Proposition 43:2; therefore σ is both a reflexion and an involution; by Proposition 42:14 σ will either be -1_V or a symmetry; neither is possible since the fixed space of σ is a line; hence σ is indeed a rotation, hence a product of two symmetries. If $d = 2$ the fixed space is a plane, hence a regular plane by Proposition 43:2; hence σ is a symmetry with respect to the line orthogonal to this plane. The case $d = 3$ is trivial. **q. e. d.**

The last proposition shows that for ternary spaces the fixed space of any rotation other than the identity is a line. This line is called the axis of the rotation. We regard every line as the axis of the identity rotation. The set of all rotations with given axis is a subgroup of $O_3^+(V)$. If the axis in question is an anisotropic line L, then there is a natural isomorphism (obtained by restriction) of the group of rotations having L as axis onto the group $O_2^+(L^*)$.

43:5. *An isotropic line is given in a regular ternary space. Then the multiplicative group of rotations having this line as axis is isomorphic to the additive group of the field of scalars.*

Proof. Let Fx be the axis in question. The quadratic space V is split by a hyperbolic plane containing x, hence we can find vectors such that

$$V = (Fx + Fy) \perp Fz$$

with $Q(x) = Q(y) = 0$ and $B(x, y) = 1$. Let U denote the plane $U = Fx \perp Fz$; this plane has radical Fx.

In order to establish the isomorphism of F onto the group of rotations with axis Fx we consider a typical α in F and we define the linear map $\varrho: U \rightarrowtail V$ by the equations

$$\varrho x = x\,,\quad \varrho z = \alpha x + z\,.$$

This map is an isometry by Proposition 41:1; hence there is a prolongation of ϱ to an isometry ϱ of V onto V by Witt's theorem. Moreover this prolongation is unique by Proposition 43:2. We have therefore associated a unique isometry of V onto V with the field element α; we write this isometry as ϱ_α; its defining equations are

$$\varrho_\alpha x = x\,,\quad \varrho_\alpha z = \alpha x + z\,.$$

Clearly $\varrho_{\alpha+\beta} = \varrho_\alpha \varrho_\beta$. In particular $\varrho_\alpha = \varrho_{\alpha/2}^2$ so that ϱ_α is always a rotation; and Fx is the axis of this rotation. Hence the map $\alpha \rightarrow \varrho_\alpha$ gives an isomorphism of F into the group of rotations with axis Fx. It remains for us to prove that this isomorphism is surjective. Consider a typical $\sigma \in O_3^+(V)$ with axis Fx. Write

$$\begin{aligned}\sigma x &= x\\ \sigma y &= ax + by + cz\\ \sigma z &= dx + ey + fz\,.\end{aligned}$$

Then $B(\sigma x, \sigma y) = 1$ so that $b = 1$. And $B(\sigma x, \sigma z) = 0$ so that $e = 0$. But $\det\sigma = 1$. Hence $f = 1$. Hence σ is a rotation whose action on U is given by

$$\sigma x = x\,,\quad \sigma z = dx + z\,.$$

Hence σ and ϱ_d agree on U. Hence they are equal by Proposition 43:2.

q. e. d.

43:5a. *Every rotation on an isotropic axis is the square of a rotation on the same axis. There is at least one rotation on this axis which is not the identity.*

§ 43D. The commutator subgroup Ω_n of O_n

We let $\Omega_n(V)$ or simply Ω_n stand for the commutator subgroup of the orthogonal group $O_n(V)$ of a regular quadratic space V. Clearly

$$\Omega_n \subseteq O_n^+\,.$$

The groups O_n^+, Ω_n have been defined for non-zero regular spaces only; any reference to these groups will carry the implicit understanding that the underlying space is non-zero and regular; the same understanding will be made with the group Z_n in § 43E.

43:6. *Ω_n contains the square of every element of O_n. It is generated by the commutators of the form $\tau_x\tau_y\tau_x\tau_y = \tau_x\tau_y\tau_x^{-1}\tau_y^{-1}$ where τ_x and τ_y are symmetries. In particular it is generated by the squares of elements of O_n^+.*

Proof. Let G be the subgroup of O_n that is generated by the commutators $\tau_x\tau_y\tau_x\tau_y$. For any σ in O_n we have $\sigma\tau_x\sigma^{-1} = \tau_{\sigma x}$, hence

$$\sigma(\tau_x\tau_y\tau_x\tau_y)\,\sigma^{-1} = \tau_{\sigma x}\tau_{\sigma y}\tau_{\sigma x}\tau_{\sigma y}\,,$$

hence G is a normal subgroup of O_n. We may therefore form the quotient group O_n/G. Let bar denote the natural map of O_n onto O_n/G. Then for typical symmetries τ_x and τ_y we have $\bar\tau_x\bar\tau_y\bar\tau_x\bar\tau_y = 1$, hence $\bar\tau_x\bar\tau_y = \bar\tau_y\bar\tau_x$. So if we consider typical $\sigma, \varrho \in O_n$ and express them in terms of symmetries by the equations

$$\sigma = \tau_1 \ldots \tau_m\,, \quad \varrho = \tau_{m+1} \ldots \tau_r\,,$$

we have

$$\bar\sigma\bar\varrho = \bar\tau_1 \ldots \bar\tau_m\bar\tau_{m+1} \ldots \bar\tau_r = \bar\tau_{m+1} \ldots \bar\tau_r\bar\tau_1 \ldots \bar\tau_m = \bar\varrho\bar\sigma\,.$$

Hence O_n/G is commutative. Hence $\Omega_n \subseteq G$. But $G \subseteq \Omega_n$ by the definition of G. Hence $G = \Omega_n$.

Finally we have

$$\bar\sigma^2 = \bar\tau_1 \ldots \bar\tau_m\bar\tau_1 \ldots \bar\tau_m = \overline{\tau_1\tau_1} \ldots \overline{\tau_m\tau_m} = 1\,.$$

Hence $\sigma^2 \in G = \Omega_n$. Thus Ω_n contains the square of every element of O_n. But Ω_n is generated by the squares $(\tau_x\tau_y)^2$ by definition of $G = \Omega_n$. Hence Ω_n is generated by the squares of all elements of O_n^+. **q. e. d.**

43:7. *Ω_n is also the commutator subgroup of O_n^+ when $n \geqq 3$.*

Proof. Let Ω_n^+ be the commutator subgroup of O_n^+, at least until we have proved that it is equal to Ω_n. Every commutator of O_n^+ is a commutator of O_n so $\Omega_n^+ \subseteq \Omega_n$. We must reverse this inclusion relation. Since we proved in Proposition 43:6 that Ω_n is generated by commutators of the form $\tau_x\tau_y\tau_x\tau_y$ where τ_x and τ_y are symmetries of V, it is enough to prove that every such element $\tau_x\tau_y\tau_x\tau_y$ is in Ω_n^+.

We claim that $Fx + Fy$ is contained in a regular ternary subspace T of V. If $Fx + Fy$ is regular this fact is clearly true. If $Fx + Fy$ is not regular its radical is a line Fz, hence $Fx + Fy = Fx + Fz$. Then there is a hyperbolic plane H which is orthogonal to Fx and contains Fz. So $T = Fx \perp H$ is the required ternary space. In either case we have our T, hence a splitting

$$V = T \perp W\,.$$

For the rest of the proof we let $\bar\tau_x$ and $\bar\tau_y$ be the symmetries of T with respect to x and y, and we let 1 denote the identity of $O_{n-3}(W)$.

Then

$$\tau_x\tau_y\tau_x\tau_y = (\bar\tau_x\bar\tau_y\bar\tau_x\bar\tau_y)\perp 1$$
$$= (-\bar\tau_x\perp 1)(-\bar\tau_y\perp 1)(-\bar\tau_x\perp 1)(-\bar\tau_y\perp 1)\,.$$

Now $-\bar\tau_x$ is a rotation on T since $\bar\tau_x$ is a symmetry, hence $(-\bar\tau_x\perp 1)$ is a rotation on V. And $(-\bar\tau_x\perp 1)^{-1} = (-\bar\tau_x\perp 1)$. The same with $(-\bar\tau_y\perp 1)$. Hence $\tau_x\tau_y\tau_x\tau_y$ is a commutator of elements of O_n^+, hence it is in Ω_n^+.

q. e. d.

43:8. **Remarks.** When $n = 1$ the group O_1 is a group of two elements so its entire structure is trivial. If $n = 2$ we know that O_2^+ is commutative so that its commutator subgroup is 1_V; we shall see in Corollary 43:12a that $\Omega_2 = 1$ if and only if V is a hyperbolic plane over a field of three elements; in particular, Proposition 43:7 is only exceptionally true for binary spaces. We know that the commutator subgroup Ω_2 of O_2 is generated by the set of squares of all rotations; but this set is a group since O_2^+ is commutative; hence Ω_2 is the set of squares of all rotations; symbolically, $\Omega_2 = (O_2^+)^2$.

§43E. The center Z_n of O_n

The symbol $Z_n(V)$ or simply Z_n will denote the center of the orthogonal group of a regular quadratic space. In this subparagraph we assume that a regular n-ary quadratic space V is given as the underlying space and that O_n, O_n^+, Ω_n, Z_n refer to the orthogonal group $O_n(V)$ of V.

43:9. *Let σ be an isometry of the regular space V onto itself which leaves every line fixed. Then $\sigma = \pm 1_V$.*

Proof. Fix a vector x in V with $Q(x) \neq 0$. Then $\sigma x = \alpha x$ by hypothesis, and $\alpha^2 Q(x) = Q(\sigma x) = Q(x)$ so that $\alpha = \pm 1$. If y falls in Fx we have $\sigma y = \alpha y$. Otherwise x and y will be independent. We have $\sigma y = \beta y$ for some $\beta \in F$ and we have to prove that $\beta = \alpha$. Now $\sigma(x+y) = \gamma(x+y)$ for some $\gamma \in F$. Hence

$$\alpha x + \beta y = \sigma x + \sigma y = \sigma(x+y) = \gamma x + \gamma y\,.$$

Hence $\alpha = \gamma = \beta$. **q. e. d.**

43:10. *Let V be a regular n-ary isotropic space with $n \geqq 3$ and let σ be an isometry of V onto itself which leaves all isotropic lines fixed. Then $\sigma = \pm 1_V$.*

Proof. Since V is isotropic it is split by a hyperbolic plane, say $V = H \perp W$ with H the hyperbolic plane. Note that H, indeed that any hyperbolic plane in V is left fixed by σ since it is spanned by two isotropic lines each of which is left fixed. Our first claim is that σ leaves every line in W fixed. We need only consider the line Fz in W with $Q(z) \neq 0$. Now H is universal, so it represents $-Q(z)$, hence there is a hyperbolic plane containing the vector z and contained in the space $H \perp Fz$; this plane is left fixed by σ, hence σz falls in it, hence σz falls in $H \perp Fz$. But

$\sigma H = H$. Hence σz, being orthogonal to $\sigma H = H$, falls in Fz. Thus our first claim is established. So Proposition 43:9 tells us that σ is either $\pm 1_W$ on W. We can replace the given σ by $-\sigma$ if necessary, so we can assume that $\sigma z = z$ for all z in W.

Let $\bar{\sigma}$ denote the restriction of σ to H. It is enough to prove that $\bar{\sigma} = 1_H$. Now $\bar{\sigma}$ is in $O_2(H)$; and $\bar{\sigma}$ does not interchange the two isotropic lines of H, so $\bar{\sigma}$ is a rotation (see Example 42:15). It therefore suffices to find a single non-zero vector of H which is left fixed by $\bar{\sigma}$, since the fixed space of a rotation of H is either 0 or H itself. Pick an anisotropic vector z in W and then an x in H with $Q(x) = -Q(z)$. Then $x + z$ is isotropic, so $\sigma(x+z) = \alpha(x+z)$ for some α in F. Hence

$$\alpha x + \alpha z = \sigma(x+z) = \sigma x + \sigma z = \sigma x + z\,.$$

Then $\alpha x + (\alpha - 1)\, z = \sigma x \in H$, hence $\alpha = 1$ and $\sigma x = x$. **q. e. d.**

43:11. *Let V be any regular n-ary quadratic space, other than a hyperbolic plane over a field of three elements, and let σ be an isometry of V onto itself which leaves all anisotropic lines fixed. Then $\sigma = \pm 1_V$.*

Proof. The case $n = 1$ is trivial so assume $n \geqq 2$. If V is anisotropic we are through by Proposition 43:9. Hence assume that V is isotropic.

First we do the case $n = 2$. Here V is a hyperbolic plane. If σ leaves the two isotropic lines of V fixed, then it leaves all lines fixed and we are through. Suppose if possible that σ interchanges the isotropic lines of V. Then σ is a symmetry by Example 42:15, say with respect to the line Fx. Let Fy be the line orthogonal to Fx. Consider any anisotropic vector of the form $x + \alpha y$ with $\alpha \in F$. Then

$$\sigma x = -x\,, \quad \sigma y = y\,, \quad \sigma(x + \alpha y) = \gamma(x + \alpha y)$$

for some $\gamma \in F$. Hence

$$\gamma x + \gamma\alpha y = \sigma(x + \alpha y) = -x + \alpha y\,.$$

Hence by comparing coefficients $\alpha = 0$. In other words Fx and Fy are the only anisotropic lines in V. But there are at least four anisotropic lines in a hyperbolic plane over a field of five or more elements. Hence our assumption that σ interchanges the isotropic lines of V is untenable. Hence $\sigma = \pm 1_V$.

There remains the general case with $n \geqq 3$. Every regular plane in V is spanned by two anisotropic lines, hence it is left fixed by σ. If we can show that every isotropic vector x falls in two distinct hyperbolic planes, then σx will fall in each of the planes, hence in the line Fx; this will imply that every isotropic line, hence every line, is left fixed by σ. So we will be through by Proposition 43:9. So let us find two hyperbolic planes containing x. There is at least one hyperbolic plane $H = Fx + Fy$

containing x since V is regular. Take a non-zero vector z orthogonal to H. Then $H' = Fx + F(y+z)$ is a hyperbolic plane containing x and distinct from H. So our assertion is established. **q. e. d.**

43:12. *$Z_n = \{\pm 1_V\}$ with one exception, when V is a hyperbolic plane over a field of three elements. In the exceptional case $Z_2 = O_2$.*

Proof. Clearly $\pm 1_V$ are in Z_n. We must prove the converse, so we consider a typical σ in Z_n. Let Fx be any anisotropic line in V. Then

$$\tau_x = \sigma \tau_x \sigma^{-1} = \tau_{\sigma x},$$

hence σx is in the line Fx. So σ leaves all anisotropic lines fixed. Hence $\sigma = \pm 1_V$ by Proposition 43:11.

There remains the exceptional case of a hyperbolic plane V over a field of three elements. Here V contains exactly two distinct anisotropic lines. Hence O_2^- consists of two distinct symmetries, hence O_2 is of order 4. Every group of order 4 is commutative. **q. e. d.**

43:12a. *O_n is commutative in exactly two exceptional cases, when $n = 1$ and when V is a hyperbolic plane over a field of three elements. Otherwise it is not commutative.*

Proof. The exceptional cases are already known to have commutative groups. If $n > 1$, V has two or more anisotropic lines (for instance the lines of an orthogonal base) and so O_n has order at least 4. In particular O_n cannot be $\{\pm 1_V\}$. **q. e. d.**

43:12b. *O_n^+ is commutative when n is 1 or 2. It is not commutative when $n \geqq 3$.*

Proof. The commutativity of O_1^+ is trivial, for O_2^+ it was proved in § 43C. If $n \geqq 3$ we have $\Omega_n \neq 1$ since O_n is not commutative, but Ω_n is the commutator subgroup of O_n^+, hence O_n^+ is not commutative. **q. e. d.**

43:13. *The centralizer of Ω_n in O_n is Z_n when $n \geqq 3$.*

Proof. We must take a typical σ in O_n that commutes with every element of Ω_n and we must prove that σ is in Z_n.

First suppose that V is anisotropic. It is enough to prove that every plane is left fixed by σ, for then every line is left fixed by σ, and so $\sigma = \pm 1_V$ by Proposition 43:9. So we consider a typical plane U in V and prove $\sigma U = U$. Let $V = U \perp W$ be the corresponding splitting. Since U is anisotropic it contains at least three distinct anisotropic lines, hence $O_2^+ = O_2^+(U)$ has at least three rotations; now a binary space has exactly two involutions that are rotations, namely $\pm 1_U$; hence there is a rotation ϱ in O_2^+ with $\varrho^2 \neq 1$. Put $\bar{\varrho} = \varrho \perp 1_W$. Then W is the fixed space of $\bar{\varrho}^2$ since 0 is the fixed space of ϱ^2. Hence σW is the fixed space of $\sigma \bar{\varrho}^2 \sigma^{-1}$. But $\bar{\varrho}^2$ is in Ω_n, hence $\sigma \bar{\varrho}^2 \sigma^{-1} = \bar{\varrho}^2$ by hypothesis. So σW is also the fixed space of $\bar{\varrho}^2$. Hence $\sigma W = W$. Hence $\alpha U = U$ as asserted.

We are left with the case of an isotropic space V. By Proposition 43:10 it is enough to prove that σ leaves a typical isotropic line Fx fixed. Now there is a hyperbolic plane, and hence a regular ternary subspace U of V, which contains Fx. Take the splitting $V = U \perp W$. By Corollary 43:5a there is a rotation ϱ in $O_3^+ = O_3^+(U)$ which has axis Fx and whose square is not the identity. Put $\bar{\varrho} = \varrho \perp 1_W$. The fixed space of $\bar{\varrho}^2$ is $Fx \perp W$ since Fx is the fixed space of ϱ^2. Hence the fixed space of $\sigma\bar{\varrho}^2\sigma^{-1}$ is $F(\sigma x) \perp (\sigma W)$. But $\bar{\varrho}^2$ is in Ω_n, hence $\sigma\bar{\varrho}^2\sigma^{-1} = \bar{\varrho}^2$ by hypothesis. So $F(\sigma x) \perp \sigma W$ is also the fixed space of $\bar{\varrho}^2$. Hence

$$F(\sigma x) \perp (\sigma W) = Fx \perp W.$$

But $F(\sigma x)$ and Fx are the radicals in this equation, hence $F(\sigma x) = Fx$. Thus σ keeps the typical isotropic line Fx fixed and this is what we asserted. **q. e. d.**

43:13a. *Suppose $n \geqq 3$. Then the center of O_n^+ is $O_n^+ \cap Z_n$. And the center of Ω_n is $\Omega_n \cap Z_n$.*

§43F. Irreducibility of Ω_n for $n \geqq 3$

43:14. *V is a regular n-ary quadratic space with $n \geqq 3$, and U is a subspace with $0 \subset U \subset V$. Then there is a σ in Ω_n such that $\sigma U \neq U$.*

Proof. First suppose that U is a regular subspace of V. Take the splitting $V = U \perp W$ and consider the involution $-1_U \perp 1_W$. There is a σ in Ω_n which does not commute with this involution since the centralizer of Ω_n in O_n is $\{\pm 1_V\}$. If we had $\sigma U = U$ we could write $\sigma = \tau \perp \varrho$ with $\tau \in O(U)$ and $\varrho \in O(W)$; but then σ would commute with $-1_U \perp 1_W$. Hence $\sigma U \neq U$.

Now suppose that U is not regular. Replacing U by $\operatorname{rad} U$ if necessary allows us to assume that $Q(U) = 0$. Fix a line Fx inside U. If we can move it out of U using an element of Ω_n we shall be through. To this end we consider a regular ternary subspace T of V which contains x. Then $T \cap U = Fx$ since Fx is a maximal totally isotropic subspace of T. Take the splitting $V = T \perp W$ and fix a hyperbolic plane $Fx + Fy$ inside T with $Q(y) = 0$. By Corollary 43:5a there is a σ in $O_3^+(T)$ which is the square of a rotation on T and whose fixed space is the line Fy. Put $\bar{\sigma} = \sigma \perp 1_W$. Then $\bar{\sigma}$ is the square of a rotation and hence it is in Ω_n. We claim that $\sigma x = \bar{\sigma} x$ is not in U. Now σx is in T and $T \cap U = Fx$, so it is enough to show that σ moves the line Fx. If σ did not move the line Fx, then σ would leave both the hyperbolic plane $Fx + Fy$ and the vector y fixed, hence it would induce an isometry of this plane onto itself which left y fixed, hence it would be the identity on $Fx + Fy$ by Example 42:15, and this is impossible since Fy is the fixed space of σ.

q. e. d.

43:14a. $\Omega_n \cap Z_n \subset \Omega_n$ *if* $n \geqq 3$.

43:15. *Let* V *be any regular* n*-ary quadratic space other than a hyperbolic plane over a field of three elements, and let* α *be a non-zero element of* $Q(V)$. *Then there is a base* $V = Fx_1 + \cdots + Fx_n$ *in which* $Q(x_i) = \alpha$ *for* $1 \leqq i \leqq n$. *This holds with* $\alpha = 0$ *if* V *is isotropic.*

Proof. For $n = 1$ the result is trivial. If $n = 2$ we can assume that α is not zero. Take x in V with $Q(x) = \alpha$. There are at least three anisotropic lines in V since V is not a hyperbolic plane over a field of three elements. Then the symmetry τ with respect to a line which is neither equal to Fx nor orthogonal to it shifts Fx. So x, τx is a base with the desired property.

If $n \geqq 3$ we fix x in V with $Q(x) = \alpha$. If α is 0 we assume that $x \neq 0$. Let U be the subspace of V that is spanned by σx as σ runs through Ω_n. Then $\sigma U = U$ for all σ in Ω_n. Hence $U = V$ by Proposition 43:14. We can therefore find $\sigma_1, \ldots, \sigma_n$ in Ω_n such that $\sigma_1 x, \ldots, \sigma_n x$ is a base for V.

q. e. d.

Chapter V

The Algebras of Quadratic Forms

Our purpose in this chapter is to introduce three algebras of importance in the theory of quadratic forms, the Clifford algebra, the quaternion algebra, and the Hasse algebra. The Clifford algebra will be developed from first principles and its main use for us will be in the definition of an invariant called the spinor norm. The quaternion algebra and the Hasse algebra play an important role in the arithmetic theory of quadratic forms. The definition of the Hasse algebra depends on some of the structure theory of central simple algebras, in particular it needs Wedderburn's theorem and the theory of similarity of algebras that is normally used in defining the Brauer group. We have therefore included a proof of Wedderburn's theorem and some of its consequences. Also included as a convenience to the reader is a brief discussion of the tensor product of finite dimensional vector spaces[1].

The general assumption that the characteristic of the underlying field F is not 2 will not be used in the first two paragraphs of this chapter.

---

[1] Our presentation of the tensor product, of the method of extending the field of scalars, and of the Clifford algebra, is strictly for finite dimensional vector spaces over fields. These concepts can also be developed using invariant methods for modules over commutative rings. For further information we refer the reader to C. Chevalley, *Fundamental concepts of algebra* (New York, 1956).

§ 51. Tensor products

§ 51 A. Abstract vector spaces

Consider finite dimensional vector spaces T, U, V, W over an arbitrary field F. By a bilinear mapping t of $U \times V$ into T we mean a mapping

$$t\colon U \times V \rightarrowtail T$$

which has the following properties:

$$t(u+u', v+v') = t(u, v) + t(u', v) + t(u, v') + t(u', v'),$$
$$t(\alpha u, \beta v) = \alpha\beta t(u, v)$$

whenever $u, u' \in U$ and $v, v' \in V$ and $\alpha, \beta \in F$. A tensor product of U and V is a composite object (t, T) consisting of a vector space T and a bilinear mapping $t\colon U \times V \rightarrowtail T$ which satisfies the following universal mapping property: given any bilinear mapping w of $U \times V$ into a vector space W, there is exactly one F-linear map φ such that $\varphi \circ t = w$.

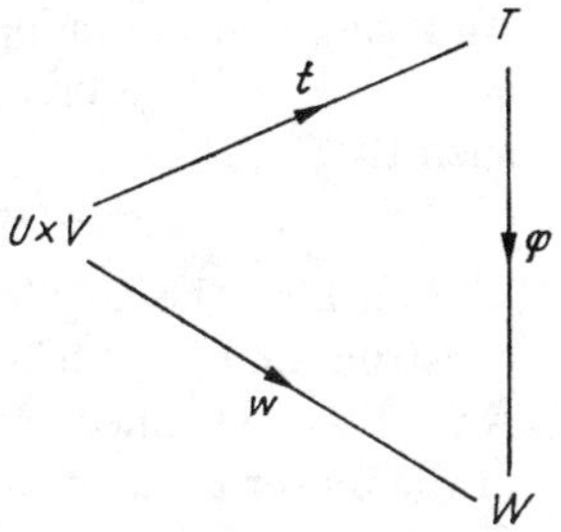

It is easy to see that tensor products exist, particularly for the finite dimensional vector spaces under discussion here. Fix a base $x_1, \ldots, x_m$ for U and a base $y_1, \ldots, y_n$ for V, then take an mn-dimensional vector space T over F with a base $\{z_{\lambda\mu}\}$ where $1 \leqq \lambda \leqq m$, $1 \leqq \mu \leqq n$. Define a bilinear mapping t: $U \times V \rightarrowtail T$ by the equation

$$t\Big(\sum_{\lambda} \alpha_\lambda x_\lambda, \sum_{\mu} \beta_\mu y_\mu\Big) = \sum_{\lambda,\mu} \alpha_\lambda \beta_\mu z_{\lambda\mu}.$$

Then it is easily seen that the composite object (t, T) satisfies the universal mapping property stated above. Hence *tensor products exist.*

Consider two tensor products (t, T) and (t', T') of U and V. We claim that there is exactly one isomorphism φ of T onto T' such that $\varphi \circ t = t'$. The mapping φ required in this assertion is already at hand: it has to be the unique linear map φ of T into T' which satisfies the equation $\varphi \circ t = t'$, and whose existence is assured by the universal mapping property of tensor products. We just have to prove that φ is bijective. To this end consider the linear map $\varphi'\colon T' \rightarrowtail T$ for which $\varphi' \circ t' = t$. Then $\varphi' \circ \varphi$ is a linear map of T into T such that $(\varphi' \circ \varphi) \circ t = t$. By the definition of T as a tensor product there can be just one such mapping $\varphi' \circ \varphi$ of T into T; now the identity 1_T on T is such a mapping; hence $\varphi' \circ \varphi = 1_T$. Similarly $\varphi \circ \varphi' = 1_{T'}$. These two equations imply, by a simple set-

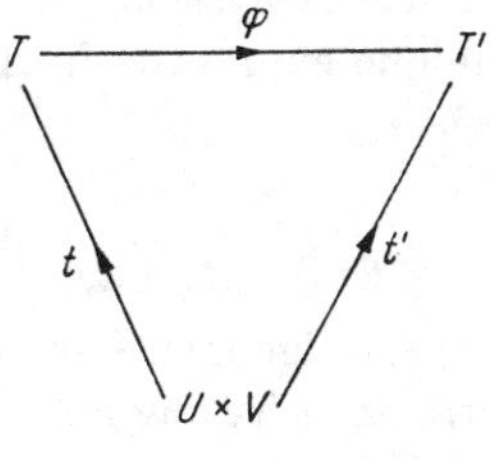

theoretic argument, that φ and φ' are bijective. Hence φ is an isomorphism of T onto T'. We have therefore proved our claim. So *tensor products are unique.*

The existence and uniqueness of tensor products allows us to talk of *the* tensor product of two vector spaces. Instead of the arbitrary symbols (t, T) for the tensor product we use $\otimes$ for the bilinear map t and $U\otimes V$ for the space T. The image of $(u, v) \in U \times V$ under the bilinear map $\otimes$ is written $u\otimes v$. Thus the tensor product $(\otimes, U\otimes V)$ is simply a composite object consisting of a bilinear map $\otimes$ and a vector space $U\otimes V$. The bilinearity of $\otimes$ now reads

$$(u+u')\otimes(v+v') = u\otimes v + u'\otimes v + u\otimes v' + u'\otimes v',$$

$$(\alpha u)\otimes(\beta v) = \alpha\,\beta\,(u\otimes v)\,.$$

And the universal mapping property says that whenever a bilinear map $w\colon U\times V \rightarrowtail W$ is given, there exists a unique linear map $\varphi\colon U\otimes V \rightarrowtail W$ such that

$$\varphi(u\otimes v) = w(u, v) \quad \forall\, (u, v) \in U \times V\,.$$

Consider the explicit tensor product (t, T) defined earlier in the construction of the tensor product of U and V. This had dimension mn. And it was spanned by vectors of the form $t(u, v)$. So the uniqueness of the tensor product says that all tensor products have these properties. Hence

$$\dim U\otimes V = \dim U \cdot \dim V\,.$$

And $U\otimes V$ is spanned by the vectors $u\otimes v$ as u, v run through U, V. So if bases $x_1, \ldots, x_m$ and $y_1, \ldots, y_n$ are chosen for U and V, the mn vectors $x_\lambda\otimes y_\mu$ will form a base for $U\otimes V$. Hence every vector of $U\otimes V$ can be put in the form

$$\sum_{\mu=1}^{n} u_\mu\otimes y_\mu \quad \text{with all } u_\mu \text{ in } U.$$

The reader may easily verify using linear methods that the u_μ in the above expression are unique. Similarly every vector in $U\otimes V$ has a unique expression in the form

$$\sum_{\lambda=1}^{m} x_\lambda\otimes v_\lambda \quad \text{with all } v_\lambda \text{ in } V\,.$$

§ 51B. Algebras

An algebra[1] A over a field F is a vector space provided with a ring structure having an identity 1_A in which scalar and ring multiplication are related by the equations

$$\alpha(xy) = (\alpha x)\,y = x(\alpha y)$$

[1] Strictly speaking this should be called an associative algebra with identity.

for all $\alpha \in F$ and all $x, y \in A$. The algebra is called commutative if it is commutative under its ring structure. It is called a division algebra if it is a skew field under its ring structure. Note the equation

$$(\alpha 1_A)\, x = \alpha x = x(\alpha 1_A)$$

for all α in F and all x in A.

We make the general assumption that every algebra is finite dimensional over its field of scalars.

51:1. **Example.** The set of linear transformations $L_F(V)$ of a vector space V into itself is both a ring and a vector space over F. In fact it is an algebra over F. (See § 41A.)

A mapping φ of the algebra A into an algebra B is called an algebra homomorphism if it is an F-linear ring homomorphism such that $\varphi(1_A)=1_B$.

Here is a convenient way of checking whether a given mapping is an algebra homomorphism. Let $x_1, \ldots, x_n$ be a base for the algebra A and let φ be a mapping of A into an algebra B over the same field F. Suppose that φ if F-linear and that it preserves multiplication on the given base for A, i. e. that

$$\varphi(x_i x_j) = \varphi(x_i)\, \varphi(x_j) \quad \text{for} \quad 1 \leqq i \leqq n\,, 1 \leqq j \leqq n\,.$$

Then an easy calculation involving linearity gives

$$\varphi(\Sigma\, \alpha_\lambda x_\lambda \cdot \Sigma\, \beta_\mu x_\mu) = \varphi(\Sigma\, \alpha_\lambda x_\lambda) \cdot \varphi(\Sigma\, \beta_\mu x_\mu)\,.$$

Hence φ is a ring homomorphism. So a map $\varphi\colon A \rightarrow B$ which is F-linear, which preserves multiplication on a given base for A, and which sends the identity to the identity, will have to be an algebra homomorphism of A into B.

The following rule is useful in the explicit construction or definition of an algebra. Let a vector space V with a base $x_1, \ldots, x_n$ be given over a field F. For each i, j ($1 \leqq i \leqq n$, $1 \leqq j \leqq n$) a vector of V is specified, and this vector is formally denoted as a product $x_i x_j$. (In practice this can be done by specifying n^3 scalars α_{ij}^k and then taking $x_i x_j$ to be the vector $\sum_{k=1}^{n} \alpha_{ij}^k x_k$.) We can extend these products by linearity to a law of multiplication on V, i. e. we define multiplication by the formula

$$\Big(\sum_\lambda \alpha_\lambda x_\lambda\Big) \Big(\sum_\mu \beta_\mu x_\mu\Big) = \sum_{\lambda,\mu} \alpha_\lambda\, \beta_\mu (x_\lambda x_\mu)\,.$$

This law is clearly distributive with respect to addition on V. And it satisfies

$$\alpha(xy) = (\alpha x)\, y = x(\alpha y)$$

for all α in F and all x, y in V. Suppose further that the law on V satisfies

$$(x_i x_j)\, x_k = x_i (x_j x_k)$$

for all relevant i, j, k. A linear argument then shows that multiplication is associative and so V is a ring. If there is also a vector $1 \in V$ such that $1 x_i = x_i = x_i 1$ for $1 \leq i \leq n$, then 1 is an identity for the ring V and so V is actually an algebra over F. In other words, whenever we define an algebra in this way we have to check two things: associativity among the defining basis vectors, and the existence of an element which acts as the identity on the defining base. It is clear that the multiplication obtained on V is uniquely determined by the specified values of the $x_i x_j$.

Let us extend the concept of a tensor product to algebras. We consider algebras A, B, C over the same field F. We call a mapping

$$w: A \times B \rightarrow C$$

a multiplicative bilinear mapping if it is bilinear and satisfies the equations

$$w(1_A, 1_B) = 1_C,$$

$$w(x, y) \cdot w(x', y') = w(xx', yy')$$

for all $x, x' \in A$ and all $y, y' \in B$. Now consider a tensor product (t, T) of A and B regarded just as vector spaces over F. Then T is a vector space and t is bilinear. We claim that there is a unique law of multiplication on T which makes T into an algebra and makes t multiplicative. For consider bases $x_1, \ldots, x_m$ for A and $y_1, \ldots, y_n$ for B. Then the mn vectors $t(x_\lambda, y_\mu)$ form a base for T. Define

$$t(x_i, y_j) \cdot t(x_k, y_l) = t(x_i x_k, y_j y_l)$$

for all relevant i, j, k, l. Extend this by linearity to a law of multiplication on T. A linear argument shows that

$$t(x, y)\, t(x', y') = t(xx', yy')$$

holds for all $x, x' \in A$ and for all $y, y' \in B$. Hence the new multiplication is associative on the above basis vectors for T, and $t(1_A, 1_B)$ acts as the identity on these basis vectors. So we have made T into an algebra with identity $t(1_A, 1_B)$. Furthermore, any law which makes T into an algebra and makes t multiplicative must agree with the law defined above on the basis vectors of T, hence the two laws must be identical. We have therefore proved our claim: *there is a unique law of multiplication which makes T into an algebra and makes t multiplicative.*

We now agree that a tensor product (t, T) of algebras A, B shall always be their tensor product as vector spaces in which T has been made into an algebra with the above property. In the standard notation for tensor products this reads as follows: $A \otimes B$ is made into an algebra by a uniquely determined multiplication having the property that

$$(x \otimes y) \cdot (x' \otimes y') = xx' \otimes yy'$$

for all $x, x' \in A$ and all $y, y' \in B$; and $1_A \otimes 1_B$ is the identity of $A \otimes B$.

Consider the tensor product $A \otimes B$ of the algebras A, B and let

$$w: A \times B \rightarrowtail C$$

be a multiplicative bilinear map into some third algebra C. We know that there is a unique F-linear map

$$\varphi: A \otimes B \rightarrowtail C$$

such that

$$\varphi(x \otimes y) = w(x,y) \quad \forall\ (x,y) \in A \times B .$$

In fact φ must be an algebra homomorphism. For if we take bases $x_1, \ldots, x_m$ for A and $y_1, \ldots, y_n$ for B we have

$$\begin{aligned} \varphi((x_i \otimes y_j)(x_k \otimes y_l)) &= \varphi(x_i x_k \otimes y_j y_l) \\ &= w(x_i x_k, y_j y_l) \\ &= \varphi(x_i \otimes y_j) \cdot \varphi(x_k \otimes y_l) , \end{aligned}$$

so that φ preserves multiplication on a base for $A \otimes B$; and furthermore,

$$\varphi(1_A \otimes 1_B) = w(1_A, 1_B) = 1_C ,$$

so that φ preserves the identity. Hence φ is indeed an algebra homomorphism as we asserted. Incidentally this shows that the uniqueness map between two tensor products of A and B is actually an algebra isomorphism.

If we have algebra isomorphisms $\varphi: A \rightarrowtail A'$ and $\psi: B \rightarrowtail B'$, then there is an algebra isomorphism

$$A \otimes B \rightarrowtail A' \otimes B' .$$

For $w(x, y) = (\varphi x) \otimes (\psi y)$ defines a multiplicative bilinear map

$$w: A \times B \rightarrowtail A' \otimes B' ,$$

hence there is an algebra homomorphism

$$\varrho: A \otimes B \rightarrowtail A' \otimes B'$$

such that $\varrho(x \otimes y) = (\varphi x) \otimes (\psi y)$ for all $(x, y) \in A \times B$. Now ϱ is clearly surjective, hence it is bijective since $A \otimes B$ and $A' \otimes B'$ have the same dimension.

Let us prove the following algebra isomorphisms:

$$F \otimes A \cong A , \quad A \otimes B \cong B \otimes A ,$$

$$A \otimes (B \otimes C) \cong (A \otimes B) \otimes C .$$

In the first instance we introduce a multiplicative bilinear map w: $F \times A \rightarrowtail A$ by the equation $w(\alpha, x) = \alpha x$. This gives us an algebra homomorphism $\varphi: F \otimes A \rightarrowtail A$ such that $\varphi(\alpha \otimes x) = \alpha x$ for all $\alpha \in F$ and all $x \in A$. This map φ is surjective, hence bijective by a comparison of dimensions. Hence $F \otimes A$ is indeed isomorphic to A. Similarly $A \otimes B$

is proved isomorphic to $B\otimes A$ by using the multiplicative bilinear map $w(x, y) = y\otimes x$ of $A \times B$ into $B\otimes A$.

The best way for us to prove the third isomorphism without getting too involved with the general theory of tensor products is to use bases $x_1, \ldots, x_m$ for A, and $y_1, \ldots, y_n$ for B, and $z_1, \ldots, z_r$ for C. The $m\,n\,r$ vectors of the form $x_i\otimes(y_j\otimes z_k)$ now form a base for $A\otimes(B\otimes C)$. Make up a multiplication table for this base in terms of the multiplication tables of the three given bases. Now do the same thing with the $(x_i\otimes y_j)\otimes z_k$ in $(A\otimes B)\otimes C$. It turns out that the two multiplication tables obtained are the same. Hence the F-linear isomorphism determined by

$$x_i\otimes(y_j\otimes z_k) \rightarrow (x_i\otimes y_j)\otimes z_k$$

preserves multiplication among these basis elements, hence it is a ring isomorphism, hence an algebra isomorphism of $A\otimes(B\otimes C)$ onto $(A\otimes B)\otimes C$ as required.

Consider algebras $A_1, \ldots, A_r$ over F with $r \geqq 2$. We use the symbol

$$A_1\otimes\cdots\otimes A_r$$

to denote an algebra that is defined inductively by the equation

$$A_1\otimes\cdots\otimes A_r = (A_1\otimes\cdots\otimes A_{r-1})\otimes A_r.$$

This algebra is uniquely determined up to an algebra isomorphism. It will also be written

$$\bigotimes_{1\leqq i\leqq r} A_i\,.$$

It follows easily by induction, using the associativity formula for $A\otimes(B\otimes C)$, that

$$(A_1\otimes\cdots\otimes A_s)\otimes(A_{s+1}\otimes\cdots\otimes A_r)$$

is algebra isomorphic to $A_1\otimes\cdots\otimes A_r$. Similarly the commutativity formula $A\otimes B\cong B\otimes A$ shows that

$$A_1\otimes\cdots\otimes A_r$$

is independent of the order of the factors.

§ 52. Wedderburn's theorem on central simple algebras

§ 52A. Central simple algebras

A is an algebra over the field F. The mapping $\alpha \rightarrow \alpha 1_A$ determines an algebra isomorphism

$$F \rightarrowtail F1_A \subseteq A\,,$$

so A naturally contains a subfield $F1_A$ isomorphic to the field of scalars F. This subfield is part of the center of A since

$$(\alpha 1)\,x = \alpha x = x(\alpha 1)$$

for all $\alpha \in F$ and all $x \in A$. If $F 1_A$ is actually the entire center of A we call A a central algebra over F.

By a left ideal of A we mean a left ideal of the ring A. The equation $\alpha x = (\alpha 1) x$ shows that every left ideal of A is actually a subspace of the vector space A. Similarly with right ideals and with two-sided ideals. We call A a simple algebra if it contains no two-sided ideals other that 0 and A itself. For example, every division algebra is simple.

A central simple algebra over a field F is an algebra which is central and simple over F.

52:1. *If A and B are central simple algebras over F, then so is $A \otimes_F B$.*

Proof. 1) First we prove that $A \otimes B$ is central, i. e. that a typical element z in the center of $A \otimes B$ is in $F(1_A \otimes 1_B)$. Fix a base $y_1, \ldots, y_n$ for B and express z in the form

$$z = \sum_\lambda u_\lambda \otimes y_\lambda \quad (u_\lambda \in A) .$$

Then $z(x \otimes 1) = (x \otimes 1) z$ holds for all x in A, hence

$$\sum_\lambda (u_\lambda x) \otimes y_\lambda = \sum_\lambda (x u_\lambda) \otimes y_\lambda .$$

Now this representation is unique by § 51A, so $u_\lambda x = x u_\lambda$ for all x in A, hence each u_λ is in the center of A, so $u_\lambda \in F 1_A$ for $1 \leqq \lambda \leqq n$. Hence z has the form $z = 1_A \otimes v$ for some $v \in B$. Now repeat the argument on B instead of A. This gives $v \in F 1_B$. Hence z is in $F(1_A \otimes 1_B)$. So $A \otimes B$ is central.

2) Here we shall prove that $A \otimes B$ is simple. We must consider a non-zero two-sided ideal $\mathfrak{a}$ of $A \otimes B$ and prove that it is all of $A \otimes B$. First an observation. Suppose

$$u_1 \otimes v_1 + \cdots + u_p \otimes v_p \quad (u_i \in A, v_i \in B)$$

is an element of $\mathfrak{a}$ in which u_1 is non-zero; the ideal generated by u_1 in A is all of A since A is simple; hence we can find $u_1', \ldots, u_p'$ in A such that $u_1 = 1_A$ and

$$u_1' \otimes v_1 + \cdots + u_p' \otimes v_p \in \mathfrak{a} .$$

Similarly we can arrange to have

$$u_1 \otimes v_1' + \cdots + u_p \otimes v_p' \in \mathfrak{a}$$

with $v_1' = 1_B$ when v_1 is non-zero.

So much for the observation. Now we can prove that $\mathfrak{a} = A \otimes B$. Of all non-zero z in $\mathfrak{a}$ and of all bases $y_1, \ldots, y_n$ for B, make a choice in which k is minimal in the representation

$$z = u_1 \otimes y_1 + \cdots + u_k \otimes y_k \quad (u_i \in A) .$$

We can assume that we actually have

$$z = 1_A \otimes y_1 + u_2 \otimes y_2 + \cdots + u_k \otimes y_k$$

by the above observation. If $k > 1$, then $u_k \notin F 1_A$ since if it were we could change the base and produce a smaller k. We can therefore take $x \in A$ with $x u_k \neq u_k x$ since $F 1_A$ is the center of A. But then

$$(x \otimes 1_B)\, z - z\,(x \otimes 1_B)$$

is a non-zero element of $\mathfrak{a}$ with a smaller k. This is impossible. Hence $k = 1$. Therefore $z = 1_A \otimes y_1$. Now go back to the above observation. This gives us a z in $\mathfrak{a}$ of the form $z = 1_A \otimes 1_B$. Hence $\mathfrak{a} = A \otimes B$. Hence $A \otimes B$ is simple. q. e. d.

52:2. *Let A be any algebra over F and let B, C be two central simple subalgebras of A which contain the identity of A. If B and C commute element-wise, then $\Sigma\, b c$ with $b \in B, c \in C$ is a typical element of the subalgebra generated by B and C. And this algebra is isomorphic to $B \otimes_F C$.*

Proof. Only the second part really needs proof. We can suppose that A is actually the algebra generated by B and C. Define a multiplicative bilinear map

$$w : B \times C \to A$$

by the equation $w(b, c) = bc$. Then by § 51B there will be an algebra homomorphism $\varphi : B \otimes C \to A$ such that $\varphi(b \otimes c) = bc$ for all $b \in B, c \in C$. This map is clearly onto A. And its kernel is 0 since $B \otimes C$ is simple. Hence $\varphi : B \otimes C \rightarrowtail A$ is an algebra isomorphism. q. e. d.

§ 52B. The algebra $R_A(\mathfrak{a})$.

This is a preamble to the proof of Wedderburn's theorem. A is an algebra over F, and $\mathfrak{a}, \mathfrak{b}$ are left ideals in A. A mapping $\varphi : \mathfrak{a} \to \mathfrak{b}$ is called A-linear if it satisfies the equations

$$(x + y)\, \varphi = (x \varphi) + (y \varphi)\,, \quad (a x)\, \varphi = a (x \varphi)$$

for all $x, y \in \mathfrak{a}$ and all $a \in A$. (All of a sudden the mapping φ appears on the right! This is our only exception to the rule that mappings are always to be written on the left.) We know that $\mathfrak{a}$ and $\mathfrak{b}$ can also be regarded as vector spaces over F, and if this is done the equations

$$(\alpha x)\, \varphi = ((\alpha 1_A)\, x)\, \varphi = (\alpha 1_A)\, (x \varphi) = \alpha (x \varphi)$$

show that every A-linear mapping is F-linear.

We let $R_A(\mathfrak{a})$ denote the set of A-linear maps of the left ideal $\mathfrak{a}$ into itself. Actually we shall regard $R_A(\mathfrak{a})$ as an algebra over F where the laws are provided in the following way. Given $\varphi, \psi \in R_A(\mathfrak{a})$ and $\alpha \in F$ we define $\varphi + \psi$, $\varphi \cdot \psi$ and $\alpha \varphi$ by the formulas

$$\begin{aligned} x(\varphi + \psi) &= (x \varphi) + (x \psi) \\ x(\varphi \cdot \psi) &= (x \varphi)\, \psi \\ x(\alpha \varphi) &= \alpha (x \varphi)\,, \end{aligned}$$

for each x in $\mathfrak{a}$. Each of these three mappings is clearly an A-linear mapping of $\mathfrak{a}$ into $\mathfrak{a}$, hence each of them is an element of $R_A(\mathfrak{a})$. If we now define zero and identity by the equations

$$(x)\,0 = 0, \quad (x)\,1 = x\,,$$

then an easy verification shows that we have made $R_A(\mathfrak{a})$ into an algebra over F. (As a matter of fact $R_A(\mathfrak{a})$ is a vector subspace of the algebra $L_F(\mathfrak{a})$ of all F-linear maps of $\mathfrak{a}$ into itself; it is not a subring of $L_F(\mathfrak{a})$ since multiplication has been twisted by writing mappings on the right; it is precisely in order to obtain an algebra having this twisted multiplication that the A-linear mappings were written on the right.)

Our interest is really in the minimal left ideals of A. As the name implies, a minimal left ideal is a left ideal which properly contains exactly one left ideal, the zero ideal. Minimal left ideals always exist since every left ideal is a subspace of A, and A is finite dimensional.

52:3. *If $\mathfrak{a}$ is a minimal left ideal in an algebra A over F, then $R_A(\mathfrak{a})$ is a division algebra over F.*

Proof. All that we have to do is prove that a typical non-zero φ in $R_A(\mathfrak{a})$ is invertible. Now $(\mathfrak{a})\,\varphi$ is a left ideal contained in $\mathfrak{a}$ and $\mathfrak{a}$ is minimal, hence $(\mathfrak{a})\,\varphi = \mathfrak{a}$, hence φ is surjective. And the kernel of φ is also a left ideal contained in $\mathfrak{a}$, so it is 0. Hence $\varphi\colon \mathfrak{a} \rightarrowtail \mathfrak{a}$ is bijective. Let ψ be the inverse mapping of φ. Then ψ is A-linear, hence in $R_A(\mathfrak{a})$. And $\varphi\,\psi = 1 = \psi\,\varphi$. Hence φ is invertible. **q. e. d.**

52:4. *Let $\mathfrak{a}$ and $\mathfrak{b}$ be minimal left ideals in a simple algebra A. Then there is an A-linear bijection φ of $\mathfrak{a}$ onto $\mathfrak{b}$. And $R_A(\mathfrak{a})$ is algebra isomorphic to $R_A(\mathfrak{b})$.*

Proof. 1) The set of points

$$\{a \in A \mid a x = 0 \quad \forall\, x \in \mathfrak{b}\}$$

is a two-sided ideal in A, hence it is 0 since A is simple. So there is a $b \in \mathfrak{b}$ with $0 \subset \mathfrak{a} b \subseteq \mathfrak{b}$. But $\mathfrak{a} b$ is a left ideal of A, hence $\mathfrak{a} b = \mathfrak{b}$ since $\mathfrak{b}$ is minimal. Define the surjection $\varphi\colon \mathfrak{a} \twoheadrightarrow \mathfrak{b}$ by the equation $(a)\,\varphi = a b$ for each $a \in \mathfrak{a}$. This is clearly A-linear. And its kernel is a left ideal contained in $\mathfrak{a}$, hence it is 0. So we do indeed have an A-linear bijection φ of $\mathfrak{a}$ onto $\mathfrak{b}$.

2) Take a typical $\sigma \in R_A(\mathfrak{a})$ and define

$$\Psi(\sigma) = \varphi^{-1} \sigma\, \varphi$$

where the A-linear map $\varphi^{-1}\colon \mathfrak{b} \rightarrowtail \mathfrak{a}$ denotes the inverse of φ. Then $\varphi^{-1} \sigma\, \varphi$ maps $\mathfrak{b}$ into $\mathfrak{b}$ and it is clearly A-linear, hence it is in $R_A(\mathfrak{b})$. So we have constructed a mapping

$$\Psi\colon R_A(\mathfrak{a}) \rightarrowtail R_A(\mathfrak{b})\,.$$

This mapping is clearly bijective. We leave it to the reader to check that it is an algebra isomorphism. **q. e. d.**

§ 52C. The algebra $M_n(D)$

D is a division algebra over the field F. We let $M_n(D)$ denote the set of all $n \times n$ matrices with coefficients in D, and we define addition and multiplication of matrices in the usual way. This makes $M_n(D)$ into a ring with identity. For each $\alpha \in F$ and each $(d_{ij}) \in M_n(D)$ we define the scalar multiple

$$\alpha(d_{ij}) = (\alpha d_{ij}) \, .$$

This makes $M_n(D)$ into an algebra over F. (To prove finite dimensionality use the finite dimensionality of D over F.)

We let e_{ij} denote the matrix with 1_D in the (i, j) position and 0 everywhere else. There are n^2 of these matrices and they are called the defining matrices of $M_n(D)$. They satisfy the equations

$$e_{ij} e_{kl} = \begin{cases} e_{il} & \text{if } j = k \\ 0 & \text{if } j \neq k \, . \end{cases}$$

Also $e_{11} + \cdots + e_{nn}$ is the identity of $M_n(D)$.

52:5. *$M_n(F)$ is central simple over F.*

Proof. 1) First we prove it is central. We must consider a typical matrix $\sum\limits_{\lambda,\mu} \alpha_{\lambda\mu} e_{\lambda\mu}$ in the center and prove that it is diagonal with all diagonal entries equal. Fix $i \neq j$ and take e_{ij}. Then

$$e_{ij}\Big(\sum_{\lambda,\mu} \alpha_{\lambda\mu} e_{\lambda\mu}\Big) = \sum_{\mu} \alpha_{j\mu} e_{i\mu} \, ,$$

and

$$\Big(\sum_{\lambda,\mu} \alpha_{\lambda\mu} e_{\lambda\mu}\Big) e_{ij} = \sum_{\lambda} \alpha_{\lambda i} e_{\lambda j} \, ,$$

hence

$$\sum_{\mu} \alpha_{j\mu}\, e_{i\mu} = \sum_{\lambda} \alpha_{\lambda i}\, e_{\lambda j} \, .$$

Comparing coefficients gives $\alpha_{j\mu} = 0$ whenever $\mu \neq j$; and also $\alpha_{jj} = \alpha_{ii}$. Hence the given matrix is of the required type.

2) In order to prove simplicity we must show that a non-zero two-sided ideal $\mathfrak{a}$ contains the identity matrix. Take an element $\sum\limits_{\lambda,\mu} \alpha_{\lambda\mu}\, e_{\lambda\mu}$ in $\mathfrak{a}$ with α_{ij}, say, non-zero. Then

$$\alpha_{ij}^{-1}\, e_{pi}\Big(\sum_{\lambda,\mu} \alpha_{\lambda\mu}\, e_{\lambda\mu}\Big)\, e_{jp} = e_{pp}$$

is in $\mathfrak{a}$ for $1 \leqq p \leqq n$. Hence the identity $e_{11} + \cdots + e_{nn}$ is in $\mathfrak{a}$. q. e. d.

52:6. *There is an algebra isomorphism*

$$M_n(F) \otimes_F M_r(F) \cong M_{nr}(F) \, .$$

Proof. Let e_{ij} $(1 \leqq i \leqq n,\ 1 \leqq j \leqq n)$ be the defining matrices of $M_n(F)$, let f_{kl} $(1 \leqq k \leqq r,\ 1 \leqq l \leqq r)$ be the defining matrices of $M_r(F)$.

Each number $\lambda (1 \leqq \lambda \leqq nr)$ can be expressed uniquely in the form

$$\lambda = (i-1)\, r + k \qquad (1 \leqq i \leqq n,\, 1 \leqq k \leqq r)\,.$$

And each $\mu (1 \leqq \mu \leqq nr)$ can be expressed uniquely in the form

$$\mu = (j-1)\, r + l \qquad (1 \leqq j \leqq n,\, 1 \leqq l \leqq r)\,.$$

Put

$$E_{\lambda\mu} = e_{ij} \otimes f_{kl}\,.$$

These $(nr)^2$ vectors form a base for $M_n(F) \otimes M_r(F)$ since the above defining matrices form bases for $M_n(F)$ and $M_r(F)$ respectively. An easy computation gives

$$E_{\lambda\mu} E_{\nu\varrho} = \begin{cases} E_{\lambda\varrho} & \text{if} \quad \mu = \nu \\ 0 & \text{if} \quad \mu \neq \nu\,. \end{cases}$$

But this means that $M_n(F) \otimes M_r(F)$ has a base whose multiplication table is the same as the multiplication table of the defining matrices of $M_{nr}(F)$; and these defining matrices form a base for $M_{nr}(F)$. Hence $M_n(F) \otimes M_r(F)$ is algebra isomorphic to $M_{nr}(F)$. **q. e. d.**

52:7. *Suppose D is a central division algebra over F. Then there is an algebra isomorphism*

$$M_n(F) \otimes_F D \cong M_n(D)\,.$$

Proof. Let $\bar{D}$ be the subalgebra of $M_n(D)$ consisting of the diagonal matrices of the form $\operatorname{diag}(d, \ldots, d)$ with $d \in D$. This subalgebra is an algebra isomorphic to D. Let A be the subalgebra of $M_n(D)$ consisting of all matrices of the form (d_{ij}) with $d_{ij} \in F\,1_D$. This subalgebra is algebra isomorphic to $M_n(F)$. Now $\bar{D}$ and A contain the identity of $M_n(D)$, they commute element-wise, and they are both central simple over F. Furthermore, $M_n(D)$ is the algebra generated by $\bar{D}$ and A. Apply Proposition 52:2. This says that $\bar{D} \otimes A$ is algebra isomorphic to $M_n(D)$. Hence $D \otimes M_n(F)$ is algebra isomorphic to $M_n(D)$. **q. e. d.**

52:8. *Let A be a central simple algebra over the field F. Suppose there is a division algebra D over F such that A is algebra isomorphic to the matrix algebra $M_n(D)$. Then D is central, n is unique, and D is unique up to an algebra isomorphism.*

Proof. 1) So $M_n(D)$ is central. Take d in the center of D. Then the matrix $\bar{d} = \operatorname{diag}(d, \ldots, d)$ is in the center of $M_n(D)$, hence it is of the form $\bar{d} = \overline{\alpha\, 1_D}$ for some α in F, hence $d = \alpha\, 1_D$, hence $d \in F\, 1_D$. So D is central.

2) Put $A' = M_n(D)$. The given isomorphism carries A to A', it carries a typical minimal left ideal $\mathfrak{a}$ to a minimal left ideal $\mathfrak{a}'$ of A', and it induces an algebra isomorphism

$$R_A(\mathfrak{a}) \rightarrowtail R_{A'}(\mathfrak{a}')\,.$$

If we can prove that $R_{A'}(\mathfrak{a}')$ is algebra isomorphic to D for some minimal left ideal $\mathfrak{a}'$ of A', then it will be isomorphic to D for all minimal left ideals $\mathfrak{a}'$ of A' by Proposition 52:4, hence $R_A(\mathfrak{a})$ will be isomorphic to D for every minimal left ideal $\mathfrak{a}$ of A. This clearly will prove that D is unique up to isomorphism. And the isomorphism

$$A \cong M_n(F) \otimes_F D$$

will imply that n is unique. So the whole proof boils down to this: show that $R_{A'}(\mathfrak{a}')$ is algebra isomorphic to D for at least one minimal left ideal $\mathfrak{a}'$ of $A' = M_n(D)$.

3) Let $\bar{d}$ stand for the matrix $\operatorname{diag}(d, \ldots, d)$ whenever d is in D. Note that $\bar{d} e_{ij} = e_{ij} \bar{d}$ for all i, j. A typical element of $M_n(D)$ now has the form

$$\sum \bar{d}_{ij} e_{ij} = \sum e_{ij} \bar{d}_{ij} \quad (d_{ij} \in D) .$$

Let $\mathfrak{a}'$ be the set of all matrices of the form

$$\bar{d}_1 e_{11} + \cdots + \bar{d}_n e_{n1} \quad \text{with} \quad d_i \in D ,$$

i. e. the set of all matrices in which every column but the first is identically zero. Then $\mathfrak{a}'$ is clearly a left ideal in $A' = M_n(D)$. We claim it is minimal. For suppose that

$$\bar{d}_1 e_{11} + \cdots + \bar{d}_n e_{n1}$$

is an element of a left ideal $\mathfrak{b}$ contained in $\mathfrak{a}'$ with $d_i \neq 0$, say. Then

$$\bar{d}\, e_{j1} = \bar{d}\, \bar{d}_i^{-1} e_{ji} (\bar{d}_1 e_{11} + \cdots + \bar{d}_n e_{n1})$$

is in $\mathfrak{b}$ for any $d \in D$ and $1 \leq j \leq n$. Hence $\mathfrak{b} = \mathfrak{a}'$. So $\mathfrak{a}'$ is indeed a minimal left ideal of A'.

We prove $R_{A'}(\mathfrak{a}') \cong D$ for this minimal left ideal $\mathfrak{a}'$. Each $d \in D$ defines an A'-linear map $\varphi_d \colon \mathfrak{a}' \to \mathfrak{a}'$ by the equation

$$(x)\, \varphi_d = x \bar{d} \quad \forall\, x \in \mathfrak{a}' .$$

So we have an algebra isomorphism

$$D \rightarrowtail R_{A'}(\mathfrak{a}')$$

defined by $d \to \varphi_d$. It remains for us to prove this isomorphism is surjective. So consider a typical $\psi \in R_{A'}(\mathfrak{a}')$. Write

$$(e_{11})\, \psi = \bar{d}_1 e_{11} + \cdots + \bar{d}_n e_{n1} .$$

Then for $1 < i \leq n$,

$$\bar{d}_i e_{11} = e_{1i} (\bar{d}_1 e_{11} + \cdots + \bar{d}_n e_{n1}) = (e_{1i} e_{11})\, \psi = 0 .$$

Hence $(e_{11})\, \psi = \bar{d}_1 e_{11}$. And

$$(e_{i1})\, \psi = (e_{i1} e_{11})\, \psi = e_{i1} (e_{11}\, \psi) = \bar{d}_1 e_{i1} .$$

Hence $(x)\, \psi = x \bar{d}_1$ for all $x \in \mathfrak{a}'$. Hence $\psi = \varphi_{d_1}$. **q. e. d.**

§ 52D. The algebra $L_D(V)$

As a final preparation for the proof of Wedderburn's theorem let us recall some facts about vector spaces over skew fields. Let D be a division algebra over F, and let V be an n-dimensional right vector space over D. We use $L_D(V)$ to denote the set of all D-linear maps of V into V. We return to the practice of writing mappings on the left. Then $L_D(V)$ is a ring in a natural way, multiplication being the composition of maps. It is also a vector space over F under the following scalar multiplication:

$$(\alpha\varphi)\,x = (\varphi x)\,(\alpha 1_D)$$

for all $\alpha \in F$, $\varphi \in L_D(V)$, $x \in V$. And

$$\alpha(\varphi\,\psi) = (\alpha\varphi)\,\psi = \varphi(\alpha\,\psi)$$

for all $\alpha \in F$ and $\varphi, \psi \in L_D(V)$. So $L_D(V)$ is actually an algebra over F. (Finite dimensionality can be proved in several ways; for instance it follows from the isomorphism $L_D(V) \rightarrowtail M_n(D)$ below.) If we fix a base $x_1, \ldots, x_n$ for V over D, then we can associate an $n \times n$ matrix with each $\sigma \in L_D(V)$ in the usual way, i. e. we write

$$\sigma x_j = \sum_{i=1}^{n} x_i\, d_{ij} \quad \text{with} \quad d_{ij} \in D\,,$$

and we send σ to the matrix (d_{ij}). This gives us an algebra isomorphism

$$L_D(V) \rightarrowtail M_n(D)\,.$$

§ 52E. Wedderburn's theorem

52:9. Theorem. *Let A be a central simple algebra*[1] *over F. Then there is a central simple division algebra D over F such that A is isomorphic to the algebra*

$$M_n(F) \otimes_F D\,.$$

Here n is unique and D is unique up to isomorphism.

Proof. 1) Let $\mathfrak{a}$ be a minimal left ideal in A and put $D = R_A(\mathfrak{a})$. Then D is an algebra over F and in fact it is a division algebra by Proposition 52:3. This will be the division algebra of the theorem.

Now by the definition of the laws of composition on $D = R_A(\mathfrak{a})$ we have

$$(x+y)\,\varphi = x\varphi + y\varphi\,, \quad x(\varphi+\psi) = x\varphi + x\psi\,,$$
$$x(\varphi\,\psi) = (x\varphi)\,\psi\,, \quad (x)\,1_D = x$$

---

[1] We have limited our discussion of the Wedderburn theorem to central simple algebras, but essentially the same proof will show that "a simple ring with identity which satisfies the minimum condition on left ideals is isomorphic to a full matrix ring $M_n(D)$ over a division ring D". For a different proof, and for further references to the literature, we refer the reader to the revised edition of B. L. VAN DER WAERDEN's *Algebra* vol. 2 (Berlin, 1959).

for all $x, y \in \mathfrak{a}$ and all $\varphi, \psi \in D$. This simply means that $\mathfrak{a}$ can be regarded as a right vector space over the division algebra D. In order to prove that $\mathfrak{a}$ is finite dimensional over D we restrict the scalars to $F\,1_D$; it is enough to prove that $\mathfrak{a}$ is finite dimensional over $F\,1_D$; but

$$\alpha x = \alpha(x\,1_D) = x(\alpha\,1_D)\,,$$

and $\mathfrak{a}$ is finite dimensional over F, hence it is over $F\,1_D$. So $\mathfrak{a}$ is indeed finite dimensional over D. We now form the algebra $L_D(\mathfrak{a})$ of D-linear maps of $\mathfrak{a}$ into $\mathfrak{a}$ (see § 52D).

2) Our first task is to find a natural isomorphism of the algebra A into the algebra $L_D(\mathfrak{a})$. Given $a \in A$ let $\sigma_a : \mathfrak{a} \rightarrowtail \mathfrak{a}$ be defined by the equation

$$\sigma_a(x) = ax \quad \forall\, x \in \mathfrak{a}\,.$$

Then σ_a is clearly additive; and

$$\sigma_a(x\varphi) = a(x\varphi) = (ax)\,\varphi = (\sigma_a x)\,\varphi\,,$$

by definition of $D = R_A(\mathfrak{a})$; hence σ_a is D-linear. So we have a mapping

$$A \rightarrowtail L_D(\mathfrak{a})$$

defined by $a \rightarrowtail \sigma_a$. It is easily verified that this mapping is an algebra homomorphism. Its kernel is a two-sided ideal in A, hence it is 0 since A is simple. So we have found our algebra isomorphism

$$A \rightarrowtail L_D(\mathfrak{a})\,.$$

3) The difficult step in the proof is showing that the above map is surjective. Before we attempt this we must establish some facts about annihilators in the algebra A. Given a D-subspace U of $\mathfrak{a}$ we put

$$U^0 = \{a \in A \mid au = 0 \quad \forall\, u \in U\}\,;$$

this is a left ideal of A. Conversely, given a left ideal $\mathfrak{b}$ in A we put

$$\mathfrak{b}^0 = \{x \in \mathfrak{a} \mid bx = 0 \quad \forall\, b \in \mathfrak{b}\}\,;$$

this is a D-subspace of $\mathfrak{a}$. We can therefore form the D-subspace U^{00}. Clearly $U \subseteq U^{00}$. Our purpose here in step 3) is to prove that $U = U^{00}$. We do this by induction on $\dim_D U$. If $U = 0$ we have $U^0 = A$ and then $U^{00} = A^0 = 0$. Now consider a U with $0 \subset U \subseteq \mathfrak{a}$, and assume the result for all subspaces of lower dimension. We must prove it for U. Write $U = W + zD$ with W a hyperplane of U and $z \in U - W$. Consider $y \in U^{00}$. We must prove that $y \in U$. If $y \in W$ we are through, so suppose y is not in W. Then $W^0 y$ is not 0, it is a left ideal contained in $\mathfrak{a}$, hence $W^0 y = \mathfrak{a}$. Similarly $W^0 z = \mathfrak{a}$. We define a map $\mathfrak{a} \rightarrowtail \mathfrak{a}$ as follows: take typical $l\,z \in \mathfrak{a}$ with $l \in W^0$ and send it to $l\,y \in \mathfrak{a}$. Why is this well-defined? If $l\,z = l'z$, then

$$(l - l')\,U = (l - l')\,(W + zD) = 0\,,$$

since $l - l' \in W^0$, hence $(l - l') \in U^0$, hence $(l - l')\, y = 0$ since $y \in U^{00}$, hence $ly = l'y$. So the map

$$lz \rightarrow ly \quad (l \in W^0)$$

is a well-defined map of $\mathfrak{a}$ into $\mathfrak{a}$. It is clearly A-linear. Hence there is a $\varphi \in R_A(\mathfrak{a}) = D$ such that

$$(lz)\, \varphi = ly \quad \forall\, l \in W^0 .$$

Then

$$l(y - z\varphi) = 0 \quad \forall\, l \in W^0 ,$$

hence $y - z\varphi \in W^{00} = W$, hence

$$y \in W + zD = U .$$

We have therefore proved that $U^{00} = U$.

4) Now take a base for the right D-space $\mathfrak{a}$:

$$\mathfrak{a} = z_1 D + \cdots + z_n D .$$

Consider a typical $\sigma \in L_D(\mathfrak{a})$. We must find an $a \in A$ such that σ is equal to the map σ_a of step 2). Once this is done we will have established the surjectivity of the map $A \rightarrowtail L_D(\mathfrak{a})$. Let W_i be the hyperplane spanned by all the vectors $z_1, \ldots, z_n$ except z_i. If we had $W_i^0 z_i = 0$ we would have $W_i^0\, \mathfrak{a} = 0$ and so $W_i = W_i^{00} = \mathfrak{a}$, which is absurd. Hence $W_i^0\, z_i \neq 0$, hence $W_i^0\, z_i = \mathfrak{a}$. Pick $a_i \in W_i^0$ such that $a_i z_i = \sigma\, z_i$; then $a_i z_j = 0$ for $j \neq i$. Do this for $1 \leqq i \leqq n$, and put $a = a_1 + \cdots + a_n$. Then

$$\sigma_a z_i = a z_i = a_i\, z_i = \sigma z_i .$$

Hence σ_a and σ agree on the base $z_1, \ldots, z_n$. Hence $\sigma_a = \sigma$. And we have established our algebra isomorphism

$$A \rightarrowtail L_D(\mathfrak{a}) .$$

5) We therefore have an algebra isomorphism $A \rightarrowtail M_n(D)$. Then D is central by Proposition 52:8. It is simple since it is a division algebra. Hence

$$A \cong M_n(F) \otimes_F D$$

by Proposition 52:7. And the uniqueness follows from the same two propositions. This completes the proof of one of the most remarkable results in modern algebra. **q. e. d.**

§ 52F. Similarity of algebras

Consider two central simple algebras A and A' over F. By Wedderburn's theorem there are natural numbers r and r', and central simple division algebras D and D' over F, such that

$$A \cong D \otimes M_r(F) \quad \text{and} \quad A' \cong D' \otimes M_{r'}(F) .$$

We also know that D and D' are essentially unique. We say that A is similar to A' and we write

$$A \sim A'$$

if A and A' have isomorphic division algebra components, i. e. if D is algebra isomorphic to D'. It is easily checked that similarity is an equivalence relation. Of course isomorphic algebras are similar. And the concept of similarity coincides with the concept of isomorphism for algebras of the same dimension. Note that $A \sim A'$ if and only if

$$A \otimes M_p(F) \cong A' \otimes M_q(F)$$

for some p and q. From this we can deduce the following result: if A_i and A_i' are central simple algebras over F with $A_i \sim A_i'$ for $1 \leqq i \leqq t$, then

$$A_1 \otimes \cdots \otimes A_t \sim A_1' \otimes \cdots \otimes A_t' .$$

We write $A \sim 1$ and say that A splits if $A \cong M_p(F)$ for some $p \geqq 1$; this is the same as saying that the division algebra component of A is isomorphic to F. Obviously,

$$A \sim 1 \Rightarrow A \otimes B \sim B \sim B \otimes A$$

for all central simple B. There is also the cancellation law[1]

$$A \otimes B \sim A' \otimes B \Rightarrow A \sim A',$$

but the proof of this must wait until we have made a brief study of the reciprocal of a central simple algebra.

The reciprocal of the central simple algebra A is defined in the following way: leave the vector space structure of A unchanged and provide A with the new ring structure determined by the twisted multiplication

$$a * b = ba \quad (a, b \in A) .$$

This new ring is again a central simple algebra over F, it is called the reciprocal algebra of A, and it will be written A^*.

52:10. *$A \otimes_F A^*$ splits.*

Proof. It is enough to prove that $A \otimes_F A^*$ is algebra isomorphic to $L_F(A)$ since $L_F(A)$ is algebra isomorphic to the matrix algebra $M_n(F)$ where $n = \dim_F A$.

For each $(a, b) \in A \times A^*$ we define a mapping $\varphi_{a,b}$ of A into itself by writing

$$\varphi_{a,b}(x) = a\, x\, b \quad \forall\, x \in A .$$

---

[1] The reader who is prepared to talk about the set of similarity classes of central simple algebras over F will see that the tensor product $\otimes$ induces a law of multiplication on this set, and that the resulting object is a group. This group is called the Brauer group of F.

Clearly $\varphi_{a,b}$ is an F-linear map of A into A, i. e. $\varphi_{a,b} \in L_F(A)$. We therefore have a multiplicative bilinear map

$$w: A \times A^* \to L_F(A)$$

given by $w(a, b) = \varphi_{a,b}$. Hence there is an algebra homomorphism

$$\lambda: A \otimes A^* \to L_F(A)$$

such that $\lambda(a \otimes b) = \varphi_{a,b}$ for all $(a, b) \in A \times A^*$. The kernel of λ is a two-sided ideal in the central simple algebra $A \otimes A^*$, hence it is 0. So λ is injective. But

$$\dim A \otimes A^* = n^2 = \dim L_F(A) .$$

So λ is bijective. Then

$$\lambda: A \otimes A^* \rightarrowtail\!\!\!\to L_F(A)$$

is the desired algebra isomorphism. **q. e. d.**

Now we can prove the cancellation result

$$A \otimes B \sim A' \otimes B \Rightarrow A \sim A' .$$

Namely,

$$\begin{aligned} A \otimes B \sim A' \otimes B &\Rightarrow A \otimes B \otimes B^* \sim A' \otimes B \otimes B^* \\ &\Rightarrow A \otimes M_q(F) \sim A' \otimes M_q(F) \\ &\Rightarrow A \sim A' . \end{aligned}$$

§ 53. Extending the field of scalars

E/F is an arbitrary extension of fields.

§ 53 A. Abstract vector spaces

Consider a vector space V over F and a vector space T over E. As usual we assume that V and T are finite dimensional over F and E respectively. If we say that V is contained in T we understand, of course, that V is a subset of T and we also tacitly assume that the laws induced by T on V agree with the given laws on V. We say that an E-space T is an E-ification of the F-space V if V is contained in T, and if every base for V over F is also a base for T over E. It is clear that if T contains V, and if there is at least one base for V that is a base for T, then every base for V is a base for T and so T is an E-ification of V.

Every space V over F has an E-ification. The construction is almost trivial. Take an E-space T' which has no points in common with V, and such that $\dim_E T' = \dim_F V$. Let $x'_1, \ldots, x'_n$ be a base for T' over E and define the F-space

$$V' = F x'_1 + \cdots + F x'_n .$$

Then T' is an E-ification of V'. Now V and V' are isomorphic F-spaces. Hence there is an E-space T containing V which is an E-ification of V.

The last step in the above construction is an application of what we shall call *the identification process*. This is our first direct use of this process and there might be some point in elaborating on it this once. So let us go back to the isomorphic F-spaces V and V', and the E-ification T' of V' with T' disjoint from V. Define a new set

$$T = (T' - V') \cup V$$

and prolongate the F-isomorphism $\sigma\colon V \rightarrowtail V'$ to a bijection $\sigma\colon T \rightarrowtail T'$ by the formula

$$\sigma x = \begin{cases} \sigma x & \text{if} \quad x \in V \\ x & \text{if} \quad x \in T' - V'. \end{cases}$$

There is a unique E-space structure on T which makes $\sigma^{-1}\colon T' \rightarrowtail T$ an E-isomorphism, namely the one carried from T' to T by σ^{-1}. Then T' is an E-ification of $\sigma^{-1}V'$. But the F-space structure on V is identical to the F-space structure on $\sigma^{-1}V' = V$ since σ was chosen as an F-isomorphism. Hence T is an E-ification of V.

If T' is any E-space containing V, and if T is an E-ification of V, then it is easy to see that there is exactly one E-linear map of T into T' which is the identity on V. This immediately implies the following uniqueness theorem for E-ifications: if T and T' are E-ifications of V, there is exactly one E-linear isomorphism $\varphi\colon T \rightarrowtail T'$ such that $\varphi x = x$ for all $x \in V$.

In view of the existence and uniqueness of E-ifications we use the symbol EV to denote an E-ification of V. We shall often refer to EV as the E-ification of V. Note that

$$\dim_E EV = \dim_F V .$$

And if U is any subspace of V, the subspace of EV that is spanned by U over E is an E-ification of U. If we have

$$V = U_1 \oplus \cdots \oplus U_r$$

and we take $EU_i \subseteq EV$ in the above way, then

$$EV = EU_1 \oplus \cdots \oplus EU_r.$$

§ 53B. Algebras

Now consider an algebra A over F and let the vector space $B \supseteq A$ be an E-ification of the vector space A. By choosing a base for A it becomes clear that there is a unique multiplication on B which agrees with the given multiplication on A and makes B into an algebra over E. Here B will have the same identity as A. The algebra B is then called an E-ification of the algebra A. It is written EA. The E-linear uniqueness isomorphism between two E-ifications is actually an algebra isomorphism.

And if C is an algebra over E which contains A and which has the same identity element as A, there is a unique algebra homomorphism of EA into C which is the identity map on A.

By taking bases we can readily see that isomorphic algebras have isomorphic E-ifications. And we can obtain algebra isomorphisms

$$E(A_1 \otimes_F A_2) \rightarrowtail (EA_1) \otimes_E (EA_2)\,,$$

and hence

$$E\Big(\bigotimes_{1 \leqq i \leqq r} A_i\Big) \rightarrowtail \bigotimes_{1 \leqq i \leqq r} (EA_i)\,.$$

§ 53C. Quadratic spaces

Now suppose V is a quadratic space over F, and let B and Q be the corresponding bilinear and quadratic forms on V. Let T be an E-ification of V. By considering a base for V we see that there is a unique symmetric bilinear form

$$B : T \times T \rightarrow E$$

which agrees with the given B on V. We then have an associated quadratic form Q on T; and T has been made into a quadratic space in a unique way. This quadratic space is called an E-ification of V and is written EV. Clearly V and EV have the same matrix in any given base for V. So the uniqueness map between any two E-ifications is actually an isometry. And isometric spaces have isometric E-ifications. If we have a splitting $V = U_1 \perp \cdots \perp U_r$ and take E-ifications $EU_i \subseteq EV$, then

$$EV = EU_1 \perp \cdots \perp EU_r\,.$$

We can easily show that $\operatorname{rad} EV = E \operatorname{rad} V$. As for discriminants, we have

$$dV = dV' \Rightarrow d(EV) = d(EV')\,.$$

§ 54. The Clifford algebra

V will denote a regular n-ary quadratic space over the field F, B will be the associated symmetric bilinear form, Q the associated quadratic form. An orthogonal base in which

$$V = Fx_1 \perp \cdots \perp Fx_n$$

is fixed for V.

We say that an algebra A is compatible with the quadratic space V if V is a subspace of A such that

$$x^2 = Q(x)\, 1_A \quad \forall\, x \in V\,,$$

where 1_A denotes the identity of A. If this holds, then the equation

$$Q(x+y) = Q(x) + Q(y) + 2B(x, y)$$

implies that

$$xy + yx = 2B(x, y)\,1_A \qquad \forall\, x, y \in V\,.$$

In particular $xy = -yx$ whenever x and y are orthogonal vectors in V. So the given basis vectors satisfy the relations

$$x_i x_j = \begin{cases} -x_j x_i & \text{if } i \neq j \\ Q(x_i)\,1_A & \text{if } i = j\,. \end{cases}$$

If a vector $x \in V$ has an inverse $x^{-1} \in A$, then the equations

$$Q(x)\,x^{-1} = x x x^{-1} = x$$

show that x must be anisotropic. Conversely every anisotropic vector $x \in V$ has an inverse in A, namely

$$x^{-1} = (Qx)^{-1}\,x\,.$$

Let C be an algebra compatible with V. We call C a Clifford algebra of the quadratic space V if it satisfies the following universal mapping property: given any algebra A compatible with V, there is exactly one algebra homomorphism $\varphi\colon C \to A$ such that $\varphi x = x$ for all $x \in V$. We shall prove later that V always has a Clifford algebra. In the meantime let us settle the question of uniqueness.

54:1. *Let C and C' be Clifford algebras of the regular quadratic space V. Then there is exactly one algebra isomorphism $\varphi\colon C \rightarrowtail\!\!\rightarrow C'$ which is the identity map on V.*

Proof. By definition of the Clifford algebra there is exactly one algebra homomorphism $\varphi\colon C \to C'$ which is the identity on V. Similarly we have a $\varphi'\colon C' \to C$. We have to prove that φ is bijective. Now $\varphi' \circ \varphi\colon C \to C$ is an algebra homomorphism which is the identity on V; but there is exactly one such map; hence $\varphi' \circ \varphi$ is the identity map on C; similarly $\varphi \circ \varphi'$ is the identity map on C'; a simple set-theoretic argument then shows that φ must be bijective. **q. e. d.**

54:2. *V is a regular quadratic space with an orthogonal base $x_1, \ldots, x_n$, and A is an algebra compatible with V. Then the subring A' generated by V in A is a subalgebra containing the identity element of A. It is spanned by all products of the form*

$$x_1^{e_1} \ldots x_n^{e_n} \quad \textit{with} \quad e_i = 0, 1\,.$$

Proof. A' consists of all finite sums of the form

$$\Sigma\,(xy\ldots)$$

with $x, y, \ldots \in V$. This is clearly a subspace of A. It also contains

$$(Qx_1)^{-1} x_1^2 = 1_A\,.$$

Hence A' is a subalgebra containing the identity element of A.

Each product in the above sum is a linear combination of products of the given basis vectors, hence a linear combination of products of the form $x_1^{m_1} \dots x_n^{m_n}$ with $m_i \geqq 0$ in virtue of the relation $x_i x_j = -x_j x_i$ for $i \neq j$, hence a linear combination of elements $x_1^{e_1} \dots x_n^{e_n}$ with $e_i = 0,1$ in virtue of the relation $x_i^2 = Q(x_i)\, 1_A$. **q. e. d.**

54:2a. $\dim_F A' \leqq 2^n$.

54:2b. *If* $\dim A' = 2^n$, *then* A' *is a Clifford algebra of* V.

Proof. 1) We have the formula

$$(x_1^{d_1} \dots x_n^{d_n})\,(x_1^{e_1} \dots x_n^{e_n}) = \prod_{1 \leqq j < i \leqq n} (-1)^{d_i e_j}\,(x_1^{e_1+d_1} \dots x_n^{e_n+d_n})$$

whenever $d_i = 0,1$ and $e_i = 0,1$ for $1 \leqq i \leqq n$. The proof of the formula follows from the relation $x_i x_j = -x_j x_i$ for $i \neq j$ in this way: move $x_1^{e_1}$ step by step to $x_1^{d_1}$ and count the number of changes of sign, then move $x_2^{e_2}$ to $x_2^{d_2}$, and so on, then count the total number of changes of sign.

2) We now prove that A' is a Clifford algebra of V under the assumption that $\dim_F A' = 2^n$. This assumption guarantees that the products

$$x_1^{e_1} \dots x_n^{e_n} \quad (e_i = 0,1)$$

form a base for A'. Consider a typical algebra A^* compatible with V and let $*$ denote its law of multiplication. We can define an F-linear map $\varphi\colon A' \to A^*$ such that

$$x_1^{e_1} \dots x_n^{e_n} \to x_1^{e_1} * \dots * x_n^{e_n}.$$

This map preserves the identity; and it is easily seen to preserve multiplication on the above base for A' because of the formula of step 1); hence it is an algebra homomorphism; clearly it is the identity map on V. Furthermore, any F-linear map with all these properties must be φ itself. Hence A' is indeed a Clifford algebra of V. **q. e. d.**

54:3. *Every regular quadratic space has a Clifford algebra.*

Proof. 1) V is the space in question, $x_1, \dots, x_n$ a fixed orthogonal base for V. We must introduce some notation. (The usefulness of this new notation is confined to the actual construction of the Clifford algebra and it should be forgotten once the construction has been completed and the proposition has been proved.) $K_2 = \{0,1\}$ is a fixed finite field of two elements and $\delta, \varepsilon, \zeta$ are typical elements in K_2. For each $\alpha \in \dot{F}$ and each $\varepsilon \in K_2$ we define

$$\alpha^\varepsilon = \begin{cases} 1 & \text{if } \varepsilon = 0 \\ \alpha & \text{if } \varepsilon = 1\,. \end{cases}$$

We then have the following rules:

$$(\alpha\,\beta)^{\varepsilon} = \alpha^{\varepsilon}\,\beta^{\varepsilon}, \qquad 1^{\varepsilon} = 1\,,$$

$$\alpha^{\varepsilon+\delta} = \alpha^{\varepsilon}\,\alpha^{\delta} \quad \text{if} \quad \alpha = \pm 1\,,$$

$$\alpha^{\delta\varepsilon}\,\alpha^{\zeta\delta+\varepsilon\zeta} = \alpha^{\delta\varepsilon+\zeta\delta}\,\alpha^{\varepsilon\zeta}.$$

We let K be the cartesian product

$$K = K_2 \times \cdots \times K_2 \quad (n \text{ times})$$

made into an additive group by component-wise addition: thus for typical

$$\varDelta = (\delta_1, \ldots, \delta_n)\,, \quad \mathsf{E} = (\varepsilon_1, \ldots, \varepsilon_n)$$

in K we have

$$\varDelta + \mathsf{E} = (\delta_1 + \varepsilon_1, \ldots, \delta_n + \varepsilon_n)\,.$$

2) Take a vector space C' of dimension 2^n over F which is disjoint from V. Fix a base for C' and label it with elements of K. So $X_\varDelta$ denotes a typical vector in this base. Define

$$X_\varDelta X_{\mathsf{E}} = \Big(\prod_{1 \leq j < i \leq n} (-1)^{\delta_i \varepsilon_j}\Big)\Big(\prod_{1 \leq i \leq n} q_i^{\delta_i \varepsilon_i}\Big)\, X_{\varDelta + \mathsf{E}}$$

where q_i is the scalar $q_i = Q(x_i)$. Extend this to a law of multiplication on C' in the manner of § 51B. A straightforward computation using the rules given in step 1) gives the associativity relation

$$X_\varDelta(X_{\mathsf{E}} X_{\mathsf{Z}}) = (X_\varDelta X_{\mathsf{E}})\,X_{\mathsf{Z}}$$

on these basis vectors; clearly the identity relation

$$X_\varDelta X_0 = X_\varDelta = X_0 X_\varDelta$$

is true. Hence (§ 51B) C' is an algebra over F. And by definition of C',

$$\dim C' = 2^n.$$

3) Put

$$x_1' = X_{(1,0,\ldots,0)}\,, \ldots, \; x_n' = X_{(0,\ldots,0,1)}\,,$$

then let V' be the n-dimensional subspace

$$V' = F x_1' + \cdots + F x_n'$$

of C'. It follows by inspection from the definition of multiplication that

$$x_i'\,x_j' + x_j'\,x_i' = \begin{cases} 0 & \text{if} \quad j \neq i \\ 2q_i 1_{C'} & \text{if} \quad j = i\,. \end{cases}$$

Hence $xy + yx \in F\,1_{C'}$ for all $x, y \in V'$. So for each $x, y \in V'$ we can define $B(x,y)$ such that

$$B(x,y)\,1_{C'} = \frac{1}{2}\,(xy + yx)\,.$$

Then $B: V' \times V' \rightarrow F$ is a symmetric bilinear form on V'. So V' is naturally a quadratic space with quadratic form Q, say. We have

$$Q(x)\, 1_{C'} = x^2 \quad \forall\, x \in V';$$

and C' is clearly the subring generated by V'; and $\dim C' = 2^n$, hence C' is a Clifford algebra of V'. Now

$$V' = (F x_1') \perp \cdots \perp (F x_n')$$

with $Q(x_i') = q_i = Q(x_i)$ for $1 \leqq i \leqq n$. Hence V is isometric to V'. The identification process (see § 53A for a detailed application of the identification process) then gives us a Clifford algebra C containing V.

q. e. d.

54:3a. *The Clifford algebra C of a regular n-ary quadratic space V is generated by V. And* $\dim C = 2^n$.

Proof. This is true for at least one Clifford algebra of V, namely the one just constructed in the last proof. Hence by Proposition 54:1 it is true for C. **q. e. d.**

Since by Proposition 54:2 the 2^n vectors

$$x_1^{e_1} \ldots x_n^{e_n} \quad (e_i = 0,1)$$

of the Clifford algebra C span C, and since we have just proved that $\dim C = 2^n$, it follows that these 2^n vectors form a base for C. We shall call this base the base derived from $x_1, \ldots, x_n$, or simply the derived base of C.

Any element of the Clifford algebra C of V which can be expressed in the form

$$\sum_{\text{finite}} (xy \ldots)$$

with each term $(xy\ldots)$ a product of an even number of elements $x, y, \ldots$ of V, is called an even element of C. The set C^+ of all even elements of C is a subalgebra containing the identity of C. Clearly C^+ is spanned by all even products of the basis elements $x_1, \ldots, x_n$ of V, hence it follows from the relations $x_i x_j = -x_j x_i$ (for $i \neq j$) and $x_i^2 = Q(x_i)\, 1_C$ that C^+ is spanned by the derived basis vectors of the form

$$x_1^{e_1} \ldots x_n^{e_n}$$

with $e_i = 0,1$ and $e_1 + \cdots + e_n \equiv 0 \bmod 2$. Now there are 2^{n-1} basis vectors of this sort. Hence

$$\dim C^+ = 2^{n-1}.$$

54:4. *a is an even element of the Clifford algebra C of the regular quadratic space V. If a commutes with every element of V, then a is in $F 1_C$.*

Proof. Express a as a linear combination

$$a = \sum_{\lambda} \alpha_\lambda X_\lambda,$$

where the X_λ are distinct even vectors of the base for C that is derived from the given orthogonal base $x_1, \ldots, x_n$ for V. For any such X_λ we have $x_i X_\lambda = \pm X_\lambda x_i$ for $1 \leq i \leq n$. Furthermore, if we consider a particular X_{λ_0} in which

$$X_{\lambda_0} = x_1^{e_1} \ldots x_i^{e_i} \ldots x_n^{e_n}$$

with $e_i = 1$, say, then $X_{\lambda_0} x_i = -x_i X_{\lambda_0}$. Now $a x_i = x_i a$. So

$$\begin{aligned}(\alpha_{\lambda_0} X_{\lambda_0} + \sum_{\lambda \neq \lambda_0} \alpha_\lambda X_\lambda)\, x_i &= x_i (\alpha_{\lambda_0} X_{\lambda_0} + \sum_{\lambda \neq \lambda_0} \alpha_\lambda X_\lambda) \\ &= \Big(-\alpha_{\lambda_0} X_{\lambda_0} + \sum_{\lambda \neq \lambda_0} \pm \alpha_\lambda X_\lambda\Big)\, x_i\,.\end{aligned}$$

Hence $\alpha_{\lambda_0} = 0$. Hence $a \in F 1_C$. **q. e. d.**

54:5. *C is a Clifford algebra of a regular quadratic space V. Then there is exactly one algebra anti-isomorphism $C \rightarrowtail C$ which is the identity map on V.*

Proof. Let C^* be the vector space C provided with the new twisted multiplication

$$a * b = ba \quad \forall\, a, b \in C\,.$$

Then C^* is again an algebra over F. It is easily seen to be a Clifford algebra of V. We therefore have exactly one algebra isomorphism $a \rightarrowtail \bar{a}$ of C onto C^* which is the identity on V. But to say that the map $a \rightarrowtail \bar{a}$ of C onto C^* is an isomorphism is the same as saying that it is an anti-isomorphism of C onto C in virtue of the equations

$$\overline{ab} = \bar{a} * \bar{b} = \bar{b}\bar{a}\,.$$

q. e. d.

54:6. *V is a regular quadratic space and $\tau_{u_1}, \ldots, \tau_{u_r}$ are symmetries such that $\tau_{u_1} \ldots \tau_{u_r} = 1_V$. Then $Q(u_1) \ldots Q(u_r) \in \dot{F}^2$.*

Proof. Take a Clifford algebra C of V. Then for any anisotropic $u \in V$ and for all $x \in V$ we have

$$\begin{aligned}\tau_u x &= x - \frac{2B(x,u)}{Q(u)}\, u \\ &= x - (Qu)^{-1}(xu + ux)\, u \\ &= x - x - u\, x\, u^{-1} \\ &= -\, u\, x\, u^{-1}.\end{aligned}$$

Now r is even since 1_V is a rotation, hence the given equation $\tau_{u_1} \ldots \tau_{u_r} = 1_V$ implies that

$$(u_1 \ldots u_r)\, x\, (u_1 \ldots u_r)^{-1} = x \quad \forall\, x \in V\,.$$

So $(u_1 \ldots u_r)$ commutes with x for all $x \in V$, hence $u_1 \ldots u_r = \alpha 1_C$ for some $\alpha \in F$ by Proposition 54:4. Apply the anti-isomorphism bar of Proposition 54:5:

$$\begin{aligned} Q(u_1)\dots Q(u_r)\,1_C &= (u_1\dots u_r)\,(u_r\dots u_1)\\ &= (u_1\dots u_r)\,(\bar u_r\dots \bar u_1)\\ &= (u_1\dots u_r)\,\overline{(u_1\dots u_r)}\\ &= (\alpha 1_C)\,\overline{(\alpha 1_C)}\\ &= \alpha^2 1_C\,. \end{aligned}$$

q. e. d.

§ 55. The spinor norm

We consider a non-zero regular n-ary quadratic space V with its corresponding symmetric bilinear form B and the associated quadratic form Q. Consider any $\sigma \in O_n(V)$. By Theorem **43:3** we can express σ as a product of symmetries, say

$$\sigma = \tau_{u_1}\dots\tau_{u_r}\,.$$

Suppose this is done in some other way, say

$$\sigma = \tau_{v_1}\dots\tau_{v_s}\,.$$

Then $\sigma\sigma^{-1} = 1_V$ so that

$$\tau_{u_1}\dots\tau_{u_r}\tau_{v_s}\dots\tau_{v_1} = 1_V\,,$$

hence by Proposition 54:6

$$Q(u_1)\dots Q(u_r) \in Q(v_1)\dots Q(v_s)\,\dot F^2\,.$$

We can therefore attach a well-defined invariant[1] to any $\sigma \in O_n$, namely the canonical image of $Q(u_1)\dots Q(u_r)$ in $\dot F/\dot F^2$. This invariant is called the spinor norm of σ and is written

$$\theta(\sigma) \in \dot F/\dot F^2\,.$$

Clearly

$$\theta(\sigma\tau) = \theta(\sigma)\,\theta(\tau)\,,$$

so we have a group homomorphism

$$\theta\colon O_n \to \dot F/\dot F^2\,.$$

Our interest in the spinor norm of a reflexion is just a passing one and we shall usually confine ourselves to the restriction

$$\theta\colon O_n^+ \to \dot F/\dot F^2\,.$$

The kernel of this restriction is written $O_n'(V)$. Thus

$$O_n'(V) = \{\sigma \in O_n^+(V) \mid \theta(\sigma) = 1\}\,.$$

---

[1] Professor ZASSENHAUS has shown me some of his unpublished results on the spinor norm in which well-definedness is proved without making use of the Clifford algebra. As a by-product he obtains the explicit formula $\theta(\sigma) = \det\left(\frac{1_V+\sigma}{2}\right)$ for the spinor norm of a rotation σ which satisfies $\det(1_V+\sigma) \neq 0$.

A commutator clearly has spinor norm 1, so

$$\Omega_n \subseteq O'_n \subseteq O^+_n \subseteq O_n .$$

In fact each of these groups is a normal subgroup of O_n.

Suppose that the quadratic space V is scaled by a non-zero α in F. Let us consider the isometries of V^α. We have already mentioned the obvious fact that $O_n(V) = O_n(V^\alpha)$. The anisotropic lines of V and V^α are the same, and the symmetries τ_u are the same linear transformations whether u is regarded as an element of V or as an element of V^α since

$$x - \frac{2B(x, u)}{Q(u)} u = x - \frac{2B^\alpha(x, u)}{Q^\alpha(u)} u .$$

However the spinor norm of τ_u is changed by the factor α on passing from V to V^α. On the other hand a rotation, being a product of an even number of symmetries, will have its spinor norm changed by an even power of α and hence not at all. In particular,

$$O'_n(V) = O'_n(V^\alpha) .$$

The other subgroups Z_n, Ω_n, O^+_n of O_n are clearly unchanged on scaling V by α.

We are not going to be too rigid in our use of the θ notation, although it will always be clear from the context just what we have in mind. The equation $\theta\sigma = \alpha$ with $\alpha \in \dot{F}$ will often appear; this really means that $\theta\sigma$ is the canonical image of α in $\dot{F}/\dot{F}^2$. Occasionally we regard $\theta\sigma$ as the full coset $\alpha\dot{F}^2$ taken as a subset of $\dot{F}$. More generally, consider any subset X of O_n; strictly speaking the symbol θX is the image of X in $\dot{F}/\dot{F}^2$ under θ; but we shall also regard it as the full set of images

$$\theta X = \bigcup_{\sigma \in X} \theta(\sigma) \dot{F}^2$$

of X in $\dot{F}$; if X is a subgroup of O_n, then θX (in the relaxed notation) is a subgroup of $\dot{F}$.

55:1. **Example.** Let us go back to the hyperbolic plane $H = Fx + Fy$ of Example 42:15. In fact suppose

$$H \cong \begin{pmatrix} 0 & \gamma \\ \gamma & 0 \end{pmatrix} \text{ in the base } x, y.$$

If τ is a symmetry it must interchange the two isotropic lines, in fact it must have the form

$$\tau x = \alpha y , \quad \tau y = \alpha^{-1} x .$$

What is the spinor norm of the symmetry described by these equations? We have seen that τ is actually the symmetry with respect to the line $x - \alpha y$, i.e. $\tau = \tau_{x - \alpha y}$. Hence

$$\theta(\tau) = Q(x - \alpha y) = -2\alpha\gamma .$$

From this we can derive the spinor norm of the rotation described by the equations

$$\sigma x = \alpha x\,, \quad \sigma y = \alpha^{-1} y\,.$$

For σ has the form $\sigma = \tau_{x-y}\tau_{x-\alpha y}$, hence $\theta(\sigma) = (-2\,\gamma)\,(-2\,\alpha\,\gamma)$, hence

$$\theta(\sigma) = \alpha\,.$$

55:2. *V is a regular n-ary quadratic space with $n \geqq 2$. Then $\theta(O_n^+)$ is the subgroup of $\dot{F}$ consisting of all non-zero scalars of the form*

$$\alpha_1 \ldots \alpha_{2r}$$

with $2r \leqq n$ and $\alpha_1, \ldots, \alpha_{2r}$ in $Q(V)$.

Proof. The proof is immediate. **q. e. d.**

55:2a. *If V is isotropic, then $\theta(O_n^+) = \dot{F}$.*

55:3. *Let $u_1, \ldots, u_r$ and $v_1, \ldots, v_r$ be anisotropic vectors in a regular quadratic space V, and suppose that $\{Q(u_1), \ldots, Q(u_r)\}$ is a permutation of $\{Q(v_1), \ldots, Q(v_r)\}$. Then*

$$\tau_{u_1} \ldots \tau_{u_r} \in \tau_{v_1} \ldots \tau_{v_r}\,\Omega_n\,.$$

Proof. Let $w_1, \ldots, w_r$ be a rearrangement of $v_1, \ldots, v_r$ in which $Q(w_i) = Q(u_i)$ for $1 \leqq i \leqq r$. Let bar be the natural homomorphism of O_n onto O_n/Ω_n. Then O_n/Ω_n is commutative, so

$$\bar{\tau}_{v_r} \ldots \bar{\tau}_{v_1}\,\bar{\tau}_{u_1} \ldots \bar{\tau}_{u_r} = (\bar{\tau}_{w_1}\,\bar{\tau}_{u_1}) \ldots (\bar{\tau}_{w_r}\,\bar{\tau}_{u_r})\,.$$

Hence

$$\tau_{v_r} \ldots \tau_{v_1}\,\tau_{u_1} \ldots \tau_{u_r} \in (\tau_{w_1}\,\tau_{u_1}) \ldots (\tau_{w_r}\,\tau_{u_r})\,\Omega_n\,.$$

It therefore suffices to prove that $\tau_w\tau_u \in \Omega_n$ whenever $Q(w) = Q(u) \neq 0$. By Theorem 42:17 there is a $\sigma \in O_n$ such that $\sigma w = u$. Now

$$\sigma\tau_w\,\sigma^{-1} = \tau_{\sigma w} = \tau_u\,,$$

hence

$$\tau_w\tau_u = \tau_w\sigma\tau_w^{-1}\,\sigma^{-1} \in \Omega_n\,.$$

q. e. d.

55:4. *$V = U \perp W$ is a splitting of the regular quadratic space V, and σ is an element of $O(V)$ with the splitting $\sigma = \tau \perp \varrho$ where $\tau \in O(U)$ and $\varrho \in O(W)$. Then $\theta(\sigma) = \theta(\tau)\,\theta(\varrho)$.*

Proof. Express τ as a product of symmetries $\tau = \tau_1 \ldots \tau_r$ with $\tau_i \in O(U)$. Then

$$\tau \perp 1_W = (\tau_1 \perp 1_W) \ldots (\tau_r \perp 1_W)\,,$$

and so

$$\theta(\tau \perp 1_W) = \prod_i \theta(\tau_i \perp 1_W) = \prod_i \theta(\tau_i) = \theta(\tau)\,.$$

Similarly $\theta(1_U \perp \varrho) = \theta(\varrho)$. Then

$$\theta(\sigma) = \theta(\tau \perp 1_W)\,\theta(1_U \perp \varrho) = \theta(\tau)\,\theta(\varrho)\,.$$

q. e. d.

55:5. *V is a regular n-ary space with* $1 \leqq n \leqq 3$. *Then*

$$(O_n^+)^2 = \Omega_n = O'_n .$$

Proof. Here $(O_n^+)^2$ stands for the set of squares of rotations of V. We saw in Proposition 43:6 that Ω_n is generated by $(O_n^+)^2$, in particular

$$(O_n^+)^2 \subseteq \Omega_n \subseteq O'_n .$$

First we prove that $\Omega_n = O'_n$. Let us consider a typical $\sigma \in O'_n$. Since σ is a rotation and $n \leqq 3$, we can write σ as a product of two symmetries, $\sigma = \tau_x \tau_y$ say. Then $\theta(\sigma) = 1$, so $Q(x)\,Q(y) \in \dot{F}^2$, so $Q(x) \in Q(y)\,\dot{F}^2$, so we can assume after suitably scaling the vector y that $Q(x) = Q(y)$. But then $\tau_y \in \tau_x \Omega_n$ by Proposition 55:3. Hence $\sigma = \tau_x \tau_y \in \Omega_n$.

Now we show that $(O_n^+)^2 = \Omega_n$. If $n = 1$ this is obvious. If $n = 2$ then O_n^+ is commutative, so $(O_n^+)^2$ is a group, so this group must be Ω_n (see Remark 43:8). Consider the case $n = 3$. Take a typical $\sigma \in O'_3$. If σ is a rotation with an isotropic axis, then σ is the square of a rotation by Corollary 43:5a, i. e. $\sigma \in (O_3^+)^2$. Otherwise σ will be a rotation with an anisotropic line L as axis. We can then write

$$\sigma = 1_L \perp \varrho$$

where ϱ is a rotation on the plane L^* orthogonal to L. But $\theta(\varrho) = \theta(\sigma) = 1$ by Proposition 55:4 since σ is in O'_3. Hence $\varrho \in (O_2^+(L^*))^2$. Hence $\sigma \in (O_3^+)^2$. So in general $O'_3 \subseteq (O_3^+)^2$. **q. e. d.**

55:6. *U is a regular subspace of a regular quadratic space V. If* $Q(U) = Q(V)$ *and* $\Omega(U) = O'(U)$, *then* $\Omega(V) = O'(V)$.

Proof. We have to consider a typical $\sigma \in O'(V)$ and prove that it is in $\Omega(V)$. Express σ as a product of symmetries:

$$\sigma = \tau_{v_1} \ldots \tau_{v_r} .$$

Take $u_1, \ldots, u_r \in U$ with $Q(u_i) = Q(v_i)$ for $1 \leqq i \leqq r$, and put

$$\varrho = \tau_{u_1} \ldots \tau_{u_r} \in O'(V) .$$

Then $\sigma \in \varrho\,\Omega(V)$ by Proposition 55:3, so it is enough to prove that $\varrho \in \Omega(V)$. If we take the splitting $V = U \perp W$, then $\varrho = \bar{\varrho} \perp 1_W$ with $\bar{\varrho} \in O(U)$ since each u_i is in U. But

$$\theta(\bar{\varrho}) = \theta(\varrho) = 1 ,$$

so $\bar{\varrho} \in O'(U)$, hence $\bar{\varrho} \in \Omega(U)$, hence $\varrho \in \Omega(V)$. **q. e. d.**

55:6a. *If V is isotropic, then* $\Omega(V) = O'(V)$. *And we have group isomorphisms*

$$O^+/O' = O^+/\Omega \rightarrowtail \dot{F}/\dot{F}^2 .$$

Proof. Take a hyperbolic plane $H \subseteq V$. Then $Q(H) = \dot{F} = Q(V)$; and $\Omega(H) = O'(H)$ by Proposition 55:5. Hence $\Omega(V) = O'(V)$.

We have $\theta(O_n^+) = \dot{F}$ by Corollary 55:2a. Hence the spinor norm $\theta: O_n^+ \to \dot{F}/\dot{F}^2$ is surjective. Its kernel is O'_n by definition of O'_n. Hence $O_n^+/O'_n \rightarrowtail \dot{F}/\dot{F}^2$. **q. e. d.**

55:7. *Let V be a regular n-ary quadratic space. Then $\theta(-1_V)$ is equal to the discriminant dV. And -1_V is in O'_n if and only if n is even with $dV = 1$.*

Proof. 1) Take an orthogonal base $x_1, \ldots, x_n$ for V. Then $\tau_{x_1} \ldots \tau_{x_n} = -1_V$, hence

$$\theta(-1_V) = Q(x_1) \ldots Q(x_n)\,\dot{F}^2 = dV\,.$$

2) If -1_V is in O'_n it is a rotation with $\theta(-1_V) = 1$, hence n is even with $dV = 1$. Similarly with the converse. **q. e. d.**

§ 56. Special subgroups of $O_n(V)$

The subgroups $O_n^+, O'_n, \Omega_n, Z_n$ of the orthogonal group O_n of a regular n-ary quadratic space V give rise to a normal series

$$1 \lhd \Omega_n \cap Z_n \lhd \Omega_n \lhd O'_n \lhd O_n^+ \lhd O_n$$

and it is natural to ask for a description of the factors of this series, say in terms of F and V. A study of the factor

$$\Omega_n/\Omega_n \cap Z_n$$

would take us too far afield so here we must content ourselves with a statement of one of the known basic results: the group $\Omega_n/\Omega_n \cap Z_n$ is simple when V is isotropic and $n \geqq 5$.[1] However we can conveniently study the remaining factors of the series. First the obvious remark that O_n/O_n^+ is a cyclic group of order 2 allows us to concentrate on the factors

$$\Omega_n \cap Z_n\,, \quad O'_n/\Omega_n\,, \quad O_n^+/O'_n\,.$$

We can use the results of §§ 43E and 55 to obtain the following descriptions when V is arbitrary with $1 \leqq n \leqq 3$, or when V is isotropic with $n \geqq 2$:

1. $\Omega_n \cap Z_n$ is $\{\pm 1_V\}$ if n is even with $dV = 1$, otherwise it is 1_V,
2. $O'_n/\Omega_n = 1$,
3. O_n^+/O'_n is isomorphic to the group $\{\alpha\beta \in \dot{F} \mid \alpha, \beta \in Q(V)\}$ modulo $\dot{F}^2$ when $1 \leqq n \leqq 3$,

3'. O_n^+/O'_n is isomorphic to $\dot{F}/\dot{F}^2$ in the isotropic case.

We are therefore left with the factors $\Omega_n \cap Z_n$, O'_n/Ω_n, O_n^+/O'_n for an anisotropic V with $n \geqq 4$. The structure of these groups depends strongly on the nature of F and V. We shall discuss this part of the theory over certain fields of interest in subsequent chapters of the book.

---

[1] See E. Artin, *Geometric algebra* (New York, 1957) for a proof of this result and for results on the structure of $O_n(V)$ when $1 \leqq n \leqq 4$.

§ 57. Quaternion algebras

§ 57A. The quaternion algebra $\left(\frac{\alpha, \beta}{F}\right)$.

Let a field F and two scalars α, β in $\dot{F}$ be given. Take a 4-dimensional space V and a base $1, x_1, x_2, x_3$ for V. Thus

$$V = F1 + Fx_1 + Fx_2 + Fx_3 .$$

Define a multiplication on these basis elements by the multiplication table

	1	x_1	x_2	x_3
1	1	x_1	x_2	x_3
x_1	x_1	$\alpha 1$	x_3	αx_2
x_2	x_2	$-x_3$	$\beta 1$	$-\beta x_1$
x_3	x_3	$-\alpha x_2$	βx_1	$-\alpha\beta 1$

and extend this by linearity to a multiplication on V by the method of § 51B. By inspecting the table we find that the basis vectors satisfy the associativity relations, and the element 1 satisfies the identity relation, hence V is an algebra.

The vector space V and the base $1, x_1, x_2, x_3$ used in the above construction can, of course, be quite arbitrary. In order to be more specific we assume that, for each pair α, β in $\dot{F}$, a space V and a base $1, x_1, x_2, x_3$ are taken in some way and then fixed for the rest of the discussion. The algebra obtained by the preceding construction is then called the quaternion algebra

$$\left(\frac{\alpha, \beta}{F}\right),$$

and the base $1, x_1, x_2, x_3$ is called the defining base of the algebra. The elements of a quaternion algebra are called quaternions, the elements of $F1$ are called scalar quaternions, the elements of $Fx_1 + Fx_2 + Fx_3$ are called pure quaternions, the set of pure quaternions is written

$$\left(\frac{\alpha, \beta}{F}\right)^0 = Fx_1 + Fx_2 + Fx_3 .$$

If E is any extension field of F, then the E-ification $E\left(\frac{\alpha, \beta}{F}\right)$ has a base with the same multiplication table as $\left(\frac{\alpha, \beta}{E}\right)$, hence there is an algebra isomorphism

$$E\left(\frac{\alpha, \beta}{F}\right) \rightarrowtail \left(\frac{\alpha, \beta}{E}\right).$$

When there is no danger of confusion we write (α, β) instead of $\left(\frac{\alpha, \beta}{F}\right)$. It is convenient to remember that

$$x_i x_j = -x_j x_i \in \dot{F} x_k, \quad x_i^2 \in \dot{F} 1$$

for any permutation i, j, k of the digits 1, 2, 3.

Consider an arbitrary quaternion

$$x = \xi_0 1 + \xi_1 x_1 + \xi_2 x_2 + \xi_3 x_3 \quad (\xi_i \in F) .$$

We define the conjugate of x to be the quaternion

$$\bar{x} = \xi_0 1 - \xi_1 x_1 - \xi_2 x_2 - \xi_3 x_3 .$$

A direct computation gives

$$\overline{\eta x} = \eta \bar{x} , \quad \overline{x+y} = \bar{x} + \bar{y} , \quad \overline{xy} = \bar{y}\bar{x} ,$$

for all $x, y \in \left(\frac{\alpha, \beta}{F}\right)$ and all $\eta \in F$. We define the norm N and trace T of x by the equations

$$N x = x\bar{x} , \quad T x = x + \bar{x} .$$

An easy computation gives

$$N x = (\xi_0^2 - \xi_1^2 \alpha - \xi_2^2 \beta + \xi_3^2 \alpha\beta)\, 1 , \quad T x = 2\, \xi_0 1 .$$

In particular this shows that both Nx and Tx are scalar quaternions satisfying

$$N\bar{x} = N x , \quad T\bar{x} = T x .$$

For any quaternions x and y we have

$$N(xy) = Nx \cdot Ny , \quad T(x+y) = Tx + Ty .$$

We see that x is pure if and only if $\bar{x} = -x$, and x is scalar if and only if $\bar{x} = x$. We see that

$$x^2 = \begin{cases} -Nx & \text{if } x \text{ is pure} \\ Nx & \text{if } x \text{ is scalar.} \end{cases}$$

Hence x^2 is a scalar quaternion whenever x is either a scalar quaternion or a pure one. From this it follows that x^2 is not a scalar quaternion if x is neither scalar nor pure. Hence $x^2 \in F1$ if and only if x is either scalar or pure.

57:1. *x is any element in a quaternion algebra. Then x is invertible if and only if $Nx \in \dot{F}\, 1$. If this condition is satisfied, then $x^{-1} = (Nx)^{-1}\, \bar{x}$.*

Proof. Suppose x^{-1} exists. Then

$$x^{-1}(Nx) = x^{-1}\, x\bar{x} = \bar{x} \neq 0 ,$$

hence $Nx \neq 0$, hence $Nx \in \dot{F}\, 1$. Conversely, suppose $Nx \in \dot{F}\, 1$. Then

$$((Nx)^{-1}\, \bar{x})\, x = (Nx)^{-1}\, N\bar{x} = (Nx)^{-1}\, Nx = 1 ,$$

and similarly $x((Nx)^{-1}\, \bar{x}) = 1$. **q. e. d.**

57:2. $\left(\frac{\alpha, \beta}{F}\right)$ *is central simple.*

Proof. 1) First let us prove that the given algebra is central. We must consider a typical

$$x = \xi_0 1 + \xi_1 x_1 + \xi_2 x_2 + \xi_3 x_3$$

in the center and prove that it is scalar. Let i, j, k be any permutation of the digits 1, 2, 3. Then

$$0 = x x_k - x_k x = 2\,(\xi_i x_i + \xi_j x_j)\, x_k\,.$$

Multiplying by x_k^{-1} we get $\xi_i x_i + \xi_j x_j = 0$. Hence $\xi_i = 0$ and $\xi_j = 0$. Similarly $\xi_k = 0$. Hence x is scalar. So $\left(\frac{\alpha,\,\beta}{F}\right)$ is central.

2) We have to show that a non-zero two-sided ideal $\mathfrak{a}$ is the entire quaternion algebra. It is enough to find an element of $\dot{F}\,1$ in $\mathfrak{a}$. Take a non-zero quaternion

$$x = \xi_0 1 + \xi_1 x_1 + \xi_2 x_2 + \xi_3 x_3$$

in $\mathfrak{a}$. If ξ_1, ξ_2, ξ_3 are all 0, then $x \in \dot{F}\,1$ and we are through. Hence we may assume that $\xi_j \neq 0$ for some j with $1 \leq j \leq 3$. Let i, j, k be a permutation of the digits 1, 2, 3. Then

$$x x_k - x_k x = 2\,(\xi_i x_i + \xi_j x_j)\, x_k$$

is in $\mathfrak{a}$, hence

$$y = \xi_i x_i + \xi_j x_j \in \mathfrak{a}\,,$$

so

$$y x_j + x_j y = 2 \xi_j x_j^2$$

is in $\mathfrak{a}$. But this element is in $\dot{F}\,1$. Hence $\left(\frac{\alpha,\,\beta}{F}\right)$ is simple. **q. e. d.**

57:3. *y and z are two elements of an arbitrary algebra A over F such that*

$$y^2 = \alpha 1_A\,,\quad z^2 = \beta\, 1_A\,,\quad yz = -zy$$

with α, β in $\dot{F}$. Then the subspace

$$F\,1_A + Fy + Fz + Fyz$$

is an algebra isomorphic to $\left(\frac{\alpha,\,\beta}{F}\right)$.

Proof. Let C denote the given subspace. A linear argument using the given relations shows that C is actually a subring, hence a subalgebra of A. Take the F-linear surjection

$$\varphi : \left(\frac{\alpha,\,\beta}{F}\right) \to C$$

defined by

$$\varphi 1 = 1_A\,,\quad \varphi x_1 = y\,,\quad \varphi x_2 = z\,,\quad \varphi x_3 = yz\,.$$

A simple argument involving the associativity of multiplication shows that φ preserves products. Hence φ is an algebra homomorphism. Its kernel is a two-sided ideal in $\left(\frac{\alpha,\,\beta}{F}\right)$, hence it is 0, hence φ is a bijection, hence an algebra isomorphism. **q. e. d.**

57:4. **Example.** Let us prove that $\left(\frac{1,-1}{F}\right)$ is algebra isomorphic to the matrix algebra $M_2(F)$. Consider the defining matrices $e_{11}, e_{12}, e_{21}, e_{22}$ of $M_2(F)$. Then $e_{11}+e_{22}$ is the identity matrix 1 of $M_2(F)$. Put $y=e_{21}+e_{12}$, $z=e_{21}-e_{12}$, so that

$$y^2=1\,,\quad z^2=-1\,,\quad yz=-zy\,.$$

Apply Proposition 57:3. Then

$$M_2(F)=F1+Fy+Fz+Fyz\cong\left(\frac{1,-1}{F}\right).$$

57:5. **Example.** The Clifford algebra of the quadratic space $V\cong\langle\alpha\rangle\perp\langle\beta\rangle$ is isomorphic to the quaternion algebra (α,β).

57:6. *An algebra isomorphism φ of one quaternion algebra onto another sends the pure quaternions to the pure quaternions and satisfies*

$$\overline{\varphi x}=\varphi\bar{x}\,,\quad N(\varphi x)=\varphi(Nx)\,,\quad T(\varphi x)=\varphi(Tx)$$

for all x.

Proof. 1) It is enough to characterize pure quaternions in a form that is invariant under isomorphism. But we have such a characterization on hand: a non-zero x is pure if and only if x^2 is in the center $F1$ and x is not.

2) Write $x=\xi_0 1+y$ with $\xi_0\in F$ and y pure. Then φy is pure. Hence

$$\overline{\varphi x}=\overline{\xi_0 1+\varphi y}=\xi_0 1-\varphi y=\varphi(\xi_0 1-y)=\varphi\bar{x}\,.$$

And so

$$N(\varphi x)=(\varphi x)\,(\overline{\varphi x})=(\varphi x)\,(\varphi\bar{x})=\varphi(Nx)\,.$$

Similarly $T(\varphi x)=\varphi(Tx)$. **q. e. d.**

§ 57B. The quadratic form of a quaternion algebra

Consider the quaternion algebra (α,β) over F. Then

$$\frac{1}{2}\,(x\bar{y}+y\bar{x})=\frac{1}{2}\,T(x\bar{y})\in F\,1\,.$$

For each pair of quaternions x, y we let $B(x,y)$ denote that uniquely determined element of F for which

$$B(x,y)\,1=\frac{1}{2}\,T(x\bar{y})\,.$$

This provides (α,β) with a symmetric bilinear form B so that (α,β) can be regarded as a quadratic space in a natural way. The associated quadratic form satisfies

$$Q(x)\,1=Nx\quad\text{for all } x\,.$$

If x is a scalar quaternion and y is pure, then $x\bar{y}$ is pure so

$$B(x,y)\,1=\frac{1}{2}\,T(x\bar{y})=0\,.$$

Hence

$$(\alpha, \beta) = (F\,1) \perp (\alpha, \beta)^0.$$

If x and y are both pure, then

$$B(x, y)\,1 = \frac{1}{2}(x\bar{y} + y\bar{x}) = -\frac{1}{2}(xy + yx)\,,$$

hence

$$B(x, y) = 0 \Leftrightarrow xy = -yx \quad (\text{if } x, y \text{ pure})\,.$$

If x is scalar then $Q(x)\,1 = x^2$; if x is pure then $Q(x)\,1 = -x^2$. In particular,

$$(\alpha, \beta) \cong <1> \perp <-\alpha> \perp <-\beta> \perp <\alpha\beta>\,.$$

Hence the discriminant of (α, β) is equal to

$$d(\alpha, \beta) = 1\,.$$

57:7. **Example.** Let C be a quaternion algebra and let V be a binary subspace of the quadratic space C^0 of pure quaternions such that $V \cong <-\alpha> \perp <-\beta>$ with α, β in $\dot{F}$. So there are x, y in C^0 such that

$$x^2 = \alpha 1\,, \quad y^2 = \beta 1\,, \quad xy = -yx\,.$$

Hence by Proposition 57:3 there is an algebra isomorphism

$$C \cong (\alpha, \beta)\,.$$

57:8. *Let C and D be quaternion algebras over the field F. Then the following assertions are equivalent:*

(1) *C is algebra isomorphic to D*

(2) *C is isometric to D*

(3) *C^0 is isometric to D^0.*

Proof. Let C be the quaternion algebra (α, β), let D be the quaternion algebra (γ, δ). If C and D are isomorphic as algebras they will be isometric as quadratic spaces since an algebra isomorphism preserves norms by Proposition 57:6. If C is isometric to D, then C^0 is isometric to D^0 by Witt's theorem (§ 42F). If C^0 is isometric to D^0, then D^0 contains a binary subspace of the form $<-\alpha> \perp <-\beta>$, hence D is algebra isomorphic to C by Example 57:7. **q. e. d.**

57:9. *Let α and β be non-zero elements of a field F. Then the following assertions are equivalent:*

(1) *(α, β) is algebra isomorphic to $(1, -1)$*

(2) *(α, β) is not a division algebra*

(3) *(α, β) is isotropic*

(4) *$(\alpha, \beta)^0$ is isotropic*

(5) *$<\alpha> \perp <\beta>$ represents 1*

(6) *$\alpha \in N_{E/F}E$ where $E = F(\sqrt{\beta})$.*

Proof. (1) *implies* (2). Let $1, x_1, x_2, x_3$ be the defining base of the quaternion algebra $(1, -1)$. Then $(x_1 + x_2)^2 = 0$. So $(1, -1)$ is not a division algebra. Hence (α, β) is not a division algebra.

(2) *implies* (3). If (α, β) is not a division algebra, then there is at least one element $x \neq 0$ in (α, β) which does not have an inverse. By Proposition 57:1, Nx must be 0; hence $Q(x) = 0$; hence (α, β) is isotropic.

(3) *implies* (4). Apply Proposition 42:12.

(4) *implies* (5). We are given that

$$<-\alpha> \perp <-\beta> \perp <\alpha\,\beta>$$

is isotropic, hence

$$<\alpha^2\beta> \perp <\alpha\,\beta^2> \perp <-\alpha^2\,\beta^2>$$

is isotropic, i. e.

$$<\beta> \perp <\alpha> \perp <-1>$$

is isotropic. So $<\alpha> \perp <\beta>$ represents 1 by Proposition 42:11.

(5) *implies* (6). By Proposition 42:11 the given information implies that

$$<\alpha> \perp <\beta> \perp <-1>$$

is isotropic, hence

$$<-\alpha> \perp <-\beta> \perp <1>$$

is isotropic, hence $<1> \perp <-\beta>$ represents α by Proposition 42:11. So there exist $\xi, \eta \in F$ such that $\alpha = \xi^2 - \beta\eta^2$. Then in $E = F(\sqrt{\beta})$ we have

$$\alpha = (\xi + \eta\sqrt{\beta})\,(\xi - \eta\sqrt{\beta}) = N_{E/F}\,(\xi + \eta\sqrt{\beta}) \in N_{E/F}(E)\,.$$

(6) *implies* (1). We have $\xi, \eta \in F$ such that

$$\alpha = N_{E/F}(\xi + \eta\sqrt{\beta})\,.$$

So $\alpha = \xi^2 - \eta^2\,\beta$, hence

$$<-\alpha> \perp <-\beta> \perp <1>$$

is isotropic, hence

$$<-\beta> \perp <-\alpha> \perp <\alpha\,\beta>$$

is isotropic. So there is an x in $(\alpha, \beta)^0$ with $x \neq 0$ and $Q(x) = 0$; this x has $Nx = 0$ and is therefore not invertible. So (α, β) cannot be a division algebra. Hence by Wedderburn's theorem (α, β) is algebra isomorphic to $M_2(F)$, hence it is algebra isomorphic to $(1, -1)$ by Example 57:4.

q. e. d.

57:10. *Given* $\alpha, \beta, \gamma, \lambda, \mu \in \dot{F}$. *Then we have the following algebra isomorphisms*:

(1) $(1, \alpha) \cong (1, -1) \cong (\alpha, -\alpha) \cong (\alpha, 1-\alpha)$,

(2) $(\beta, \alpha) \cong (\alpha, \beta) \cong (\alpha\lambda^2, \beta\,\mu^2)$,

(3) $(\alpha, \alpha\,\beta) \cong (\alpha, -\beta)$

(4) $(\alpha, \beta) \otimes_F (\alpha, \gamma) \cong (\alpha, \beta\,\gamma) \otimes_F (1, -1)$.

Proof. The first three parts follow immediately from Propositions 57:8 and 57:9. The fourth part is a little harder. Let $1, x_1, x_2, x_3$ be the defining base of (α, β), and let $1, y_1, y_2, y_3$ be the defining base of (α, γ). Consider the resulting base of sixteen elements of the form $x_i \otimes y_j$ (with $x_0 = 1, y_0 = 1$) for $A = (\alpha, \beta) \otimes (\alpha, \gamma)$. Define

$$X = F(1 \otimes 1) + F(x_1 \otimes 1) + F(x_2 \otimes y_2) + F(x_3 \otimes y_2)$$
$$Y = F(1 \otimes 1) + F(1 \otimes y_2) + F(x_1 \otimes y_3) + F(-\gamma(x_1 \otimes y_1)) .$$

An easy application of Proposition 57:3 gives

$$X \cong (\alpha, \beta\gamma) , \quad Y \cong (\gamma, -\alpha^2 \gamma) \cong (1, -1) .$$

The basis elements used in defining X commute with those used in defining Y, hence X commutes element-wise with Y. Also X and Y contain the identity $1 \otimes 1$ of A. And X and Y are central simple since they are quaternion algebras. By Proposition 52:2 the algebra generated by X and Y is isomorphic to $X \otimes Y$. Comparing dimensions gives $A \cong X \otimes Y$ as desired. **q. e. d.**

57:11. **Remark.** For trivial reasons we can replace the isomorphism symbol $\cong$ by the similarity symbol $\sim$ throughout the statement of the last proposition. We also have

$$(\alpha, \beta) \otimes (\alpha, \gamma) \sim (\alpha, \beta\gamma)$$

since $(1, -1)$ splits by Example 57:4. If we take $\beta = \gamma$ we obtain

$$(\alpha, \beta) \otimes (\alpha, \beta) \sim 1$$

so that (α, β) is isomorphic to its own reciprocal.

§ 57C. The rotations of $(\alpha, \beta)^0$

57:12. *Let u be an anisotropic pure quaternion in the quaternion algebra (α, β) over F. Then u is invertible and*

$$\tau_u x = -u x u^{-1} \quad \forall x \in (\alpha, \beta)^0 .$$

Proof. u is invertible by Proposition 57:1. Then for all $x \in (\alpha, \beta)^0$ we have

$$\begin{aligned} \tau_u x &= x - \frac{2B(x, u)}{Q(u)} u \\ &= x - (x\bar{u} + u\bar{x})(Nu)^{-1} u \\ &= x - (xu + ux)(Nu)^{-1} \bar{u} \\ &= -uxu^{-1} . \end{aligned}$$

q. e. d.

57:13. *Let y be an invertible quaternion in (α, β). Then the map σ_y defined by*

$$\sigma_y x = y x y^{-1} \quad \forall x \in (\alpha, \beta)^0$$

is a rotation of $(\alpha, \beta)^0$. *Every rotation of* $(\alpha, \beta)^0$ *has this form. And* $\theta(\sigma_y) = Q(y)$.[1]

Proof. 1) Clearly σ_y is an F-linear injection, hence bijection, of $(\alpha, \beta)^0$ onto itself. It is an isometry since

$$N(\sigma_y x) = N(yxy^{-1}) = (Ny)(Nx)(Ny)^{-1} = Nx$$

for all x in $(\alpha, \beta)^0$.

2) Next we prove that σ_y is a rotation. Suppose not. Then σ_y is a product of three symmetries of $(\alpha, \beta)^0$, hence by Proposition 57:12 there is an invertible quaternion v such that

$$yxy^{-1} = -vxv^{-1} \quad \forall\, x \in (\alpha, \beta)^0 .$$

Hence there is an invertible quaternion w such that

$$wx = -xw \quad \forall\, x \in (\alpha, \beta)^0 .$$

Write $w = \xi_0 1 + w_0$ with $\xi_0 \in F$ and w_0 pure. Then taking $x = w_0$ we get $ww_0 = -w_0 w$. But

$$ww_0 = (\xi_0 1 + w_0)\, w_0 = w_0 w .$$

Hence $ww_0 = 0$. But w is invertible. Hence $w_0 = 0$. Hence $w = \xi_0 1$. Hence $(\xi_0 1)\, x = -x(\xi_0 1)$ for all $x \in (\alpha, \beta)^0$. Hence $\xi_0 = 0$. Hence $w = 0$. And this is absurd since w is invertible.

3) A typical rotation σ of $(\alpha, \beta)^0$ is a product of two symmetries $\sigma = \tau_{u_1}\tau_{u_2}$ with u_1, u_2 pure. Then by Proposition 57:12 we have $\sigma x = (u_1 u_2)\, x\, (u_1 u_2)^{-1}$ for all $x \in (\alpha, \beta)^0$. Hence $\sigma = \sigma_{u_1 u_2}$. So every rotation of $(\alpha, \beta)^0$ has the desired form.

4) Finally we must compute the spinor norm $\theta(\sigma_y)$. Write $\sigma_y = \tau_{u_1}\tau_{u_2}$ with u_1, u_2 pure. Then

$$yxy^{-1} = (u_1 u_2)\, x\, (u_1 u_2)^{-1} \quad \forall\, x \in (\alpha, \beta)^0 .$$

Hence $(u_1 u_2)^{-1} y$ is in the center $F1$. Hence $y = \alpha\, u_1 u_2$. So

$$Q(y)\, 1 = Ny = \alpha^2 N u_1\, N u_2 = \alpha^2\, Q(u_1)\, Q(u_2)\, 1 .$$

But $\theta(\sigma_y) = Q(u_1)\, Q(u_2)$. Hence $\theta(\sigma_y) = Q(y)$. **q. e. d.**

§ 58. The Hasse algebra

Our purpose in this paragraph is to study an invariant called the Hasse algebra of a regular quadratic space. So let us consider a regular n-ary space V over the field F.

Take an orthogonal base for V and fix it for the moment. Suppose that

$$V \cong \langle\alpha_1\rangle \perp \cdots \perp \langle\alpha_n\rangle$$

---

[1] This result is the basis of the 3-dimensional part of the proof of the Strong Approximation Theorem for Rotations (§ 104).

in this base. We define the Hasse algebra

$$S_F V = \bigotimes_{1 \leqq i \leqq n} \left(\frac{\alpha_i, d_i}{F}\right)$$

where $d_i = \alpha_1 \ldots \alpha_i$. When there is just one field under discussion we write SV instead of $S_F V$. The Hasse algebra is central simple since it is a tensor product of quaternion algebras. We can therefore apply the theory of similarity of algebras to it. But before we proceed further we must be sure that SV is uniquely determined, at least up to an algebra isomorphism.

58:1. **Lemma.** *Let $\mathfrak{X}_0$ and $\mathfrak{X}_*$ be two orthogonal bases for V. Then there is a chain of orthogonal bases*

$$\mathfrak{X}_0 \cdots \rightarrowtail \mathfrak{X}_{i-1} \rightarrowtail \mathfrak{X}_i \cdots \rightarrowtail \mathfrak{X}_*$$

in which $\mathfrak{X}_i$ is obtained by altering at most two adjacent basis vectors of $\mathfrak{X}_{i-1}$.

Proof. We can assume that $n \geqq 3$. Put $\mathfrak{X}_* = \{y_1, \ldots, y_n\}$.

1) First we prove the following: there is a chain

$$\mathfrak{X}_0 \rightarrowtail \cdots \rightarrowtail \mathfrak{X}$$

of the required type in which y_1 is the first basis vector of $\mathfrak{X}$. Of all bases $\mathfrak{X}$ which can be obtained by such a chain we choose an $\mathfrak{X}$ in which y_1 has most coordinates 0. In fact we can take $\mathfrak{X} = \{x_1, \ldots, x_n\}$ with

$$y_1 = \alpha_1 x_1 + \cdots + \alpha_p x_p$$

where $\alpha_i \neq 0$ for $1 \leqq i \leqq p$. If $p = 1$ we are through. We will derive a contradiction for $p > 1$. If $p \geqq 3$ we cannot have

$$\begin{cases} Q(\alpha_1 x_1) + Q(\alpha_2 x_2) = 0 \\ Q(\alpha_2 x_2) + Q(\alpha_3 x_3) = 0 \\ Q(\alpha_3 x_3) + Q(\alpha_1 x_1) = 0\,, \end{cases}$$

for then we would have $Q(\alpha_3 x_3) = 0$ after a suitable elimination. We can therefore assume that $Q(\alpha_1 x_1) + Q(\alpha_2 x_2) \neq 0$ when $p \geqq 3$. For $p = 2$ this is automatically satisfied since $Q(y_1) \neq 0$. Define

$$\bar{x}_1 = \alpha_1 x_1 + \alpha_2 x_2\,, \quad \bar{x}_2 = x_2 - \frac{B(\bar{x}_1, x_2)}{Q(\bar{x}_1)} \bar{x}_1\,.$$

This gives a new base in a chain of the required type with a smaller p. This is a contradiction. Hence $p = 1$.

2) The lemma can now be proved by induction to n. Run a chain of the required type from $\mathfrak{X}_0$ to a base $\mathfrak{X} = \{y_1, z_2, \ldots, z_n\}$. Then

$$(Fz_2) \perp \cdots \perp (Fz_n) = (Fy_2) \perp \cdots \perp (Fy_n)\,,$$

so we can run a chain of the required type from $\{z_2, \ldots, z_n\}$ to $\{y_2, \ldots, y_n\}$, hence from $\mathfrak{X}$ to $\mathfrak{X}_*$, hence from $\mathfrak{X}_0$ to $\mathfrak{X}_*$. **q. e. d.**

58:2. *SV is well-defined, at least up to an algebra isomorphism.*

Proof. We shall use the rules established in Proposition 57:10 throughout the proof. If $n = 1$ the result is immediate. So assume $n > 1$. The lemma shows that it is enough to compare SV in two orthogonal bases $x_1, \ldots, x_n$ and $x'_1, \ldots, x'_n$ in which

$$x_\lambda = x'_\lambda \quad \text{for} \quad \lambda \neq i, i+1$$

where i is some fixed integer with $1 \leqq i \leqq n-1$. So here we will have

$$\alpha_\lambda = \alpha'_\lambda \quad \text{for} \quad \lambda \neq i, i+1$$

where $\alpha_\lambda = Q(x_\lambda)$ and $\alpha'_\lambda = Q(x'_\lambda)$. We also have

$$F x_i \perp F x_{i+1} = F x'_i \perp F x'_{i+1}$$

so that $d_\lambda = d'_\lambda$ for $1 \leqq \lambda < i$ and $d_\lambda \in d'_\lambda \dot{F}^2$ for $i+1 \leqq \lambda \leqq n$. We have to show that

$$\bigotimes_{1 \leqq i \leqq n} (\alpha_i, d_i) \cong \bigotimes_{1 \leqq i \leqq n} (\alpha'_i, d'_i) .$$

This reduces to proving

$$(\alpha_i, d_i) \otimes (\alpha_{i+1}, d_{i+1}) \cong (\alpha'_i, d'_i) \otimes (\alpha'_{i+1}, d'_{i+1}) .$$

A simple calculation using the rules shows that the left hand side is similar to

$$(\alpha_i \alpha_{i+1}, -d_{i-1}) \otimes (\alpha_i, \alpha_{i+1}) .$$

Similarly with the right hand side. So we are reduced to proving

$$(\alpha_i, \alpha_{i+1}) \cong (\alpha'_i, \alpha'_{i+1}) .$$

But the Clifford algebra of $F x_i \perp F x_{i+1} = F x'_i \perp F x'_{i+1}$ is isomorphic to both (α_i, α_{i+1}) and $(\alpha'_i, \alpha'_{i+1})$ by Example 57:5. Hence $(\alpha_i, \alpha_{i+1}) \cong (\alpha'_i, \alpha'_{i+1})$. Hence SV is well-defined up to an isomorphism. **q. e. d.**

58:3. **Remark.** Now that the Hasse algebra is well-defined we should give some rules for operating with it[1]. So consider our regular n-ary space V over F. If V has a splitting

$$V \cong \langle \alpha_1 \rangle \perp \cdots \perp \langle \alpha_n \rangle \quad (\alpha_i \in F) ,$$

then

$$SV \sim \bigotimes_{1 \leqq i \leqq j \leqq n} (\alpha_i, \alpha_j) .$$

For a non-trivial splitting $V = U \perp W$ we obtain

$$SV \sim SU \otimes (dU, dW) \otimes SW ;$$

here dU denotes the discriminant of U or, strictly speaking, a representative of the discriminant in $\dot{F}$; similarly with dW. If we scale V by a

[1] See E. Witt's paper on quadratic forms in *Crelle's J.* **176** (1937), pp. 31—44, for the connection between the Hasse algebra and the Clifford algebra.

non-zero α in F we find

$$SV^\alpha \sim (\alpha, (-1)^{n(n+1)/2}(dV)^{n+1}) \otimes SV\,.$$

The behavior under E-ification is described by

$$S(EV) \cong ES(V)\,.$$

If V is ternary we have

$$(-1,-1) \otimes SV \sim (-\alpha_1\alpha_2, -\alpha_1\alpha_3)\,.$$

If V is quaternary we have

$$(-1,-1) \otimes SV \sim (-\alpha_1\alpha_2, -\alpha_1\alpha_3) \otimes (\alpha_4, dV)\,.$$

58:4. **Theorem.** *V and W are regular n-ary quadratic spaces with $1 \leqq n \leqq 3$. Then V is isometric to W if and only if*

$$\dim V = \dim W\,, \quad dV = dW\,, \quad SV \sim SW\,.$$

Proof. We need only prove the sufficiency. For $n = 1$ it is trivial, in fact the discriminant is enough. Next let us do the ternary case. Let $dV = \alpha$ with $\alpha \in \dot{F}$. Then $SV^\alpha \sim SW^\alpha$ by Remark 58:3, and $dV^\alpha = dW^\alpha = 1$. In effect this allows us to assume that $dV = dW = 1$. Hence we have splittings

$$V \cong \langle -\alpha\rangle \perp \langle -\beta\rangle \perp \langle \alpha\beta\rangle$$

$$W \cong \langle -\gamma\rangle \perp \langle -\delta\rangle \perp \langle \gamma\delta\rangle$$

for suitable $\alpha, \beta, \gamma, \delta \in \dot{F}$. Hence

$$V \cong (\alpha, \beta)^0\,, \quad \text{and} \quad W \cong (\gamma, \delta)^0.$$

On the other hand, if we apply Remark 58:3 to the above splitting for V we obtain

$$(-1,-1) \otimes SV \sim (-\alpha\beta, \beta) \sim (\alpha, \beta)\,,$$

and similarly

$$(-1,-1) \otimes SW \sim (\gamma, \delta)\,.$$

Hence $(\alpha, \beta) \sim (\gamma, \delta)$. Hence $(\alpha, \beta)^0$ and $(\gamma, \delta)^0$ are isometric by Proposition 57:8. So V is isometric to W.

The binary case follows from the ternary case. If V and W have the same invariants, then so do $V \perp \langle 1\rangle$ and $W \perp \langle 1\rangle$. Hence $V \perp \langle 1\rangle$ and $W \perp \langle 1\rangle$ are isometric by the ternary case just proved. Hence $V \cong W$ by Witt's theorem. **q. e. d.**

58:5. **Example.** Over certain fields the invariants $\dim V$, dV, SV are enough to characterize V under isometry even when $n \geqq 4$. However this is not true in general. For instance the real quadratic spaces

$$V \cong \perp_1^4 \langle 1\rangle \quad \text{and} \quad W \cong \perp_1^4 \langle -1\rangle$$

have the same invariants but are not isometric.

58:6. *A regular ternary space V is isotropic if and only if $SV \sim (-1,-1)$.*

Proof. V is isotropic if and only if it is split by a hyperbolic plane, i.e. if and only if $V \cong \langle 1\rangle \perp \langle -1\rangle \perp \langle -dV\rangle$. By Theorem 58:4 this is true if and only if $SV \sim (-1,-1)$. **q. e. d.**

58:7. *V is a regular quaternary space with discriminant d, and E is the field $E = F(\sqrt{d})$. Then V is isotropic if and only if EV is isotropic.*

Proof. We must consider an isotropic space EV and deduce that V is isotropic. So suppose that V is not isotropic. Then d must be a non-square in F and E/F is quadratic. Every element of EV has the form $x + \sqrt{d}\, y$ with x, y in V; take an isotropic vector of this form. Then

$$Q(x) + dQ(y) + 2\sqrt{d}\, B(x, y) = 0\,.$$

So $Q(x) = -dQ(y)$ and $B(x,y) = 0$ since the extension is quadratic. If $Q(y) = 0$ we have $Q(x) = 0$, hence x and y are both 0 since there are no isotropic vectors in V; but then $x + \sqrt{d}\, y$ is not isotropic. So in fact $Q(x)$ and $Q(y)$ are non-zero. Let us write

$$Q(y) = \varepsilon\,, \quad B(x,y) = 0\,, \quad Q(x) = -d\varepsilon\,.$$

Then

$$V \cong \langle \varepsilon\rangle \perp \langle -d\varepsilon\rangle \perp P$$

where P is a plane contained in V. If we use this expression to compute the discriminant of V we find that P must have discriminant -1. Hence P is a hyperbolic plane. Hence V is isotropic. **q. e. d.**

58:7a. *V is isotropic if and only if $E(SV) \sim \left(\frac{-1,-1}{E}\right)$.*

Proof. Take a regular ternary subspace U of V. By Propositions 42:12 and 58:7 V is isotropic if and only if EU is isotropic. By Proposition 58:6 EU is isotropic if and only if $S(EU) \sim \left(\frac{-1,-1}{E}\right)$. But $E(SV) \cong S(EV) \sim S(EU)$. **q. e. d.**

58:8. **Theorem.** *Let F be a field with the property that every regular quinary space over it is isotropic. Then two regular quadratic spaces U and V over F are isometric if and only if*

$$\dim U = \dim V\,, \quad dU = dV\,, \quad SU \sim SV\,.$$

Proof. We need only do the sufficiency. Let n be the common dimension. For $1 \leqq n \leqq 3$ the result is true over any field by Theorem 58:4. So assume that $n \geqq 4$. Then $U \perp \langle -1\rangle$ is isotropic by hypothesis, hence $1 \in Q(U)$ by Proposition 42:11. Similarly $1 \in Q(V)$. So we have splittings

$$U \cong U' \perp \langle 1\rangle\,, \quad V \cong V' \perp \langle 1\rangle\,.$$

But U' and V' have the same invariants. An inductive argument then gives $U' \cong V'$. Hence $U \cong V$. **q. e. d.**

Part Three

Arithmetic Theory of Quadratic Forms over Fields

Chapter VI

The Equivalence of Quadratic Forms

One of the major accomplishments in the theory of quadratic forms is the classification of the equivalence class of a quadratic form over arithmetic fields. We are ready to present this part of the theory. Roughly speaking it goes as follows: the global solution is completely described by local and archimedean solutions, the local solution involves the dimension, the discriminant, and an invariant called the Hasse symbol, the complex archimedean solution is trivial, and the real archimedean solution is the well-known law of inertia of Sylvester.

§ 61. Complete archimedean fields

Let us consider the theory of quadratic forms over a complete archimedean field F, i. e. over a field which is complete at the archimedean spot $\mathfrak{p}$. We know from § 12 that there is a topological isomorphism of F onto either the real field R or the complex field C; in the first instance $\mathfrak{p}$ is called real, in the second complex. It is best to treat the real and complex cases separately.

§ 61A. The real case

If F is a real complete field, then $(\dot{F}:\dot{F}^2) = 2$ and ± 1 are representatives of the cosets of $\dot{F}$ modulo $\dot{F}^2$. So a vector x in a regular n-ary quadratic space V over F will satisfy exactly one of the conditions

$$Q(x) \in \dot{F}^2, \quad Q(x) = 0, \quad Q(x) \in -\dot{F}^2.$$

We call V a positive definite quadratic space over F if

$$Q(x) \in \dot{F}^2 \quad \forall\, x \in \dot{V};$$

we call it negative definite if

$$Q(x) \in -\dot{F}^2 \quad \forall\, x \in \dot{V};$$

we call it definite if it is either positive or negative definite; we call it indefinite if it is not definite.

We can refine an arbitrary orthogonal base to a base

$$V = (Fx_1) \perp \cdots \perp (Fx_p) \perp (Fy_1) \perp \cdots \perp (Fy_q)$$

in which $Q(x_i) = 1$ for $1 \leqq i \leqq p$ and $Q(y_j) = -1$ for $1 \leqq j \leqq g$, i. e. in which

$$V \cong <1> \perp \cdots \perp <1> \perp <-1> \perp \cdots \perp <-1> .$$

Here we have $0 \leqq p \leqq n$ and $0 \leqq g \leqq n$. Since a sum of elements of $\dot{F}^2$ is again in $\dot{F}^2$ we can conclude that V is positive definite if and only if $p = n$; it is negative definite if and only if $g = n$; it is indefinite if and only if $0 < p < n$. Hence V is indefinite if and only if it is isotropic. We have $Q(\dot{V})$ equal to

$$\dot{F}^2, F, -\dot{F}^2$$

according as V is positive definite, indefinite, or negative definite. Every subspace of a positive definite space is regular and positive definite; similarly with negative definite spaces. The only space which is both positive and negative definite is the trivial space 0.

Suppose P is a maximal positive definite subspace of V. Then P is regular and

$$P \cong <1> \perp \cdots \perp <1> .$$

Let P' be some other maximal positive definite subspace of V, say with $\dim P \leqq \dim P'$. Then there is an isometry σ of P into P'. By Witt's theorem there is a prolongation of σ to an isometry σ of V onto V. So $\sigma^{-1} P'$ will be a positive definite space containing P. Hence $\dim P = \dim P'$. We have therefore proved: all maximal positive definite subspaces of V have the same dimension. We call this dimension the positive index of V and write it $\text{ind}^+ V$. Similarly define the negative index $\text{ind}^- V$.

Let us return to our orthogonal base $x_1, \ldots, x_p, y_1, \ldots, y_g$. It is easily seen that $Fx_1 \perp \cdots \perp Fx_p$ is a maximal positive definite subspace of V. And $Fy_1 \perp \cdots \perp Fy_g$ is a maximal negative definite subspace of V. Hence

$$\text{ind}^+ V = p\,, \quad \text{ind}^- V = g\,.$$

In particular,

$$\text{ind}^+ V + \text{ind}^- V = \dim V\,.$$

If $p \geqq g$ we have a splitting

$$V = H_1 \perp \cdots \perp H_g \perp V_0$$

with H_i a hyperbolic plane for $1 \leqq i \leqq g$, and V_0 positive definite and therefore 0 or anisotropic. Hence in this case the index of V in the sense of § 42F satisfies $\text{ind}\, V = g$. Similarly we have $\text{ind}\, V = p$ when $p \leqq g$. In other words,

$$\text{ind}\, V = \min\,(\text{ind}^+ V, \text{ind}^- V)\,.$$

61:1. **Theorem.** *Let U and V be regular quadratic spaces over a real complete field F. Then U is represented by V if and only if*

$$\operatorname{ind}^+ U \leqq \operatorname{ind}^+ V, \quad \operatorname{ind}^- U \leqq \operatorname{ind}^- V.$$

For isometry the conditions read

$$\operatorname{ind}^+ U = \operatorname{ind}^+ V, \quad \operatorname{ind}^- U = \operatorname{ind}^- V.$$

Proof. The proof is almost trivial and the details are omitted.

q. e. d.

61:2. **Remark.** $<-1> \perp <-1>$ does not represent 1. Hence the quaternion algebra $(-1, -1)$ is a division algebra by Proposition 57:9. On the other hand an arbitrary quaternion algebra will be isomorphic to $(1, -1)$ or $(-1, -1)$. Hence there are essentially two distinct quaternion algebras over a real complete field F, namely

$$\left(\frac{1, -1}{F}\right) \text{ and } \left(\frac{-1, -1}{F}\right).$$

§ 61B. The complex case

Here everything is trivial. Since F is topologically isomorphic to C we have $\dot{F} = \dot{F}^2$. Hence every regular n-ary quadratic space V has a splitting

$$V \cong <1> \perp \cdots \perp <1>.$$

And V is isotropic when $n \geqq 2$. Also

$$U \rightarrowtail V \quad \text{if and only if} \quad \dim U \leqq \dim V,$$

and

$$U \cong V \quad \text{if and only if} \quad \dim U = \dim V.$$

There is essentially one quaternion algebra, namely

$$\left(\frac{1, -1}{F}\right).$$

§ 61C. Special subgroups of $O_n(V)$

We conclude this paragraph by giving the structure of the groups

$$\Omega_n \cap Z_n, \quad O'_n/\Omega_n, \quad O^+_n/O'_n$$

of a regular n-ary quadratic space V over a complete archimedean field F. Recall that we first raised this question over a general field in § 56.

In the complex case we can apply the results stated in § 56 and we find that

$$\Omega_n = O'_n = O^+_n$$

with

$$\Omega_n \cap Z_n = \begin{cases} \{\pm 1_V\} & \text{if } n \text{ is even} \\ 1_V & \text{if } n \text{ is odd.} \end{cases}$$

Now suppose that F is real. Then by Proposition 55:2 we have

$$(O_n^+ : O_n') = \begin{cases} 2 \text{ if } V \text{ is indefinite} \\ 1 \text{ if } V \text{ is definite.} \end{cases}$$

By Propositions 55:5 and 55:6 we have

$$O_n' = \Omega_n .$$

Hence by Proposition 55:7,

$$\Omega_n \cap Z_n = \begin{cases} \{\pm 1_V\} \text{ if } dV = 1 \text{ with } n \text{ even} \\ \quad 1_V \text{ otherwise.} \end{cases}$$

§ 62. Finite fields

Next we consider quadratic spaces over finite fields. Let F be a finite field of q elements. Consider the multiplicative homomorphism

$$\varphi : \dot{F} \to \dot{F}^2$$

defined by the equation $\varphi x = x^2$. The kernel of φ consists of the elements ± 1 since the equation $x^2 - 1 = 0$ has precisely these roots in the field F, and these two roots are distinct since F does not have characteristic 2. Hence $\dot{F}^2$ is a group of $\frac{1}{2}(q-1)$ elements. Hence $\dot{F}/\dot{F}^2$ is a group of 2 elements. Thus every element of $\dot{F}$ is either a square or a square times a fixed non-square.

62:1. *A regular n-ary quadratic space over a finite field is universal if $n \geq 2$.*

Proof. It is enough to prove this for binary spaces. By scaling the space we can reduce things to the following: prove that a typical binary space V represents at least one non-square. This we now do. Write $V \cong \langle\varepsilon\rangle \perp \langle\delta\rangle$ with $\varepsilon, \delta \in \dot{F}$. If ε or δ is a non-square we are through. Hence we can assume that $V \cong \langle 1\rangle \perp \langle 1\rangle$. If -1 is a square in F, then V is a hyperbolic plane and we are through. Hence we can assume that -1 is a non-square.

$\dot{F}^2$ and $1 + \dot{F}^2$ are finite sets containing the same number of elements. These two sets are not equal since 1 is in the first set but not in the second. Hence there is an element α in $\dot{F}$ such that $1 + \alpha^2$ is not in $\dot{F}^2$. This element $1 + \alpha^2$ cannot be 0, hence it is a non-square in F, and this non-square is clearly represented by $V \cong \langle 1\rangle \perp \langle 1\rangle$. **q. e. d.**

62:1a. *A regular quadratic space V over a finite field F has a splitting*

$$V \cong \langle 1\rangle \perp \cdots \perp \langle 1\rangle \perp \langle dV\rangle .$$

So there are essentially two regular quadratic spaces of given dimension over F.

62:1b. *V is isotropic if $n \geqq 3$.*[1]

62:2. **Theorem.** *U and V are regular quadratic spaces over a finite field F. Then U is isometric to V if and only if*

$$\dim U = \dim V\,, \quad dU = dV\,.$$

Proof. Apply Corollary 62:1a. **q. e. d.**

62:3. **Remark.** $<\alpha> \perp <\beta>$ represents 1 whenever $\alpha, \beta \in \dot{F}$. Hence every quaternion algebra is isomorphic to $(1, -1)$ by Proposition 57:9. So there is essentially one quaternion algebra over a finite field F, namely

$$\left(\frac{1, -1}{F}\right).$$

62:4. **Remark.** The factor groups

$$\Omega_n \cap Z_n\,, \quad O'_n/\Omega_n\,, \quad O^+_n/O'_n$$

of a regular n-ary quadratic space V $(n \geqq 2)$ over a finite field are described by the equations

$$(O^+_n : O'_n) = 2\,, \quad O'_n = \Omega_n\,,$$

and

$$\Omega_n \cap Z_n = \begin{cases} \{\pm 1_V\} \text{ if } dV = 1 \text{ with } n \text{ even} \\ \quad 1_V \text{ otherwise.} \end{cases}$$

§ 63. Local fields

Now consider a local field F. Let us recall some of the basic definitions and notation of §§ 13, 16 and 32. F is a field with a complete and discrete prime spot $\mathfrak{p}$ and the residue class field $F(\mathfrak{p})$ is a finite field of $N\mathfrak{p}$ elements. We let $\mathfrak{o}$ stand for the ring of integers $\mathfrak{o}(\mathfrak{p})$, $\mathfrak{u}$ for the group of units $\mathfrak{u}(\mathfrak{p})$, $\mathfrak{p}$ for the maximal ideal $\mathfrak{m}(\mathfrak{p})$, π for a prime element of F at $\mathfrak{p}$, ord for the order function $\text{ord}_\mathfrak{p}$, and $|\ |$ for the normalized valuation $|\ |_\mathfrak{p}$. We know from § 22E that the fractional ideals of F at $\mathfrak{p}$ are the powers

$$\mathfrak{p}^\nu = \pi^\nu \mathfrak{o} \quad (\nu \in \mathsf{Z})\,.$$

Remember that in this part of the book we are making the general assumption that the characteristic of F is not 2. However it is still possible for the residue class field to have characteristic 2. This is what happens for instance in the case of the 2-adic numbers. We shall call F a dyadic local field if its residue class field has characteristic 2; thus for a dyadic field we have

$$\chi(F) = 0 \quad \text{and} \quad \chi(F(\mathfrak{p})) = 2\,.$$

We call F non-dyadic if it is not dyadic; here we have $\chi(F(\mathfrak{p})) > 2$ and

$$\chi(F) = 0 \quad \text{or} \quad \chi(F) = \chi(F(\mathfrak{p}))\,.$$

[1] This was originally proved by DICKSON and generalized by C. CHEVALLEY, *Abh. Math. Sem. Hamburg* (1935), pp. 73—75, to forms of any degree: *every form of degree d in $d+1$ variables over a finite field has a non-trivial zero.*

Note that F is dyadic if and only if

$$0 < |2| < 1 \quad (\text{or } 0 < \operatorname{ord} 2 < \infty)\,;$$

it is non-dyadic if and only if

$$|2| = 1 \quad (\text{or } \operatorname{ord} 2 = 0)\,.$$

We saw in § 62 that exactly half the non-zero elements of a finite field of characteristic not 2 are squares; in particular this is true of the residue class field $F(\mathfrak{p})$ of a non-dyadic local field. On the other hand if F is a dyadic local field, then its residue class field is a finite field of characteristic 2; since all finite fields are perfect we can conclude that every element of $F(\mathfrak{p})$ is a square, i. e. that

$$(F(\mathfrak{p}))^2 = F(\mathfrak{p}) \quad \text{if } \mathfrak{p} \text{ dyadic.}$$

This has the following important consequence in F: if $\varepsilon, \varepsilon'$ are units in a dyadic local field, there is a unit δ such that

$$\varepsilon' \equiv \varepsilon\,\delta^2 \mod \pi.$$

In particular, in the dyadic case every unit is a square mod $\mathfrak{p}$.

§ 63 A. Quadratic defect

63:1. **Local Square Theorem.** *Let α be an integer in the local field F. Then there is an integer β such that*

$$1 + 4\pi\,\alpha = (1 + 2\pi\,\beta)^2.$$

Proof. The polynomial $\pi x^2 + x - \alpha$ is reducible by the Reducibility Criterion of Proposition 13:9. Hence we have $\beta, \beta' \in F$ such that

$$x^2 + \pi^{-1}x - \pi^{-1}\alpha = (x - \beta)\,(x - \beta')\,.$$

Then the product of the roots is equal to $-\pi^{-1}\alpha$, hence one of the roots, say β, must be in $\mathfrak{o}$. But

$$\beta = \frac{1}{2}\left(-\pi^{-1} \pm \sqrt{\pi^{-2} + 4\pi^{-1}\alpha}\right)$$

by the quadratic formula. Hence

$$1 + 4\pi\,\alpha = (1 + 2\pi\,\beta)^2.$$

q. e. d.

63:1a. **Corollary.** *Suppose ε, δ are units in F such that $\varepsilon \equiv \delta \mod 4\pi$. Then $\varepsilon \in \delta\,\mathfrak{u}^2$.*

63:1b. **Corollary.** *$\dot{F}^2$ is an open subset of F.*

Consider any element ξ of the local field F. Then ξ has at least one expression in the form $\xi = \eta^2 + \alpha$ with η, α in F. We write

$$\xi = \eta^2 + \alpha \quad (\eta, \alpha \in F)$$

in all possible ways and take the intersection

$$\mathfrak{d}(\xi) = \bigcap_{\alpha} \alpha\mathfrak{o}\,.$$

Then $\mathfrak{d}(\xi)$ is either a fractional ideal or 0. We call $\mathfrak{d}(\xi)$ the quadratic defect of ξ. Clearly

$$\mathfrak{d}(\xi) \subseteq \xi\mathfrak{o}\,.$$

If ξ is a square in F, then $\mathfrak{d}(\xi) = 0$. Using the Local Square Theorem one can easily show that the converse is also true. Hence

$$\xi \in F^2 \Leftrightarrow \mathfrak{d}(\xi) = 0\,.$$

In particular ξ always has an expression

$$\xi = \eta^2 + \alpha \quad \text{with} \quad \alpha\mathfrak{o} = \mathfrak{d}(\xi)\,.$$

From this it follows that

$$\mathfrak{d}(a^2\xi) = a^2\mathfrak{d}(\xi) \quad \forall\, a, \xi \in F\,.$$

We have

$$\mathfrak{d}(\xi) = \xi\mathfrak{o} \quad \text{if } \operatorname{ord}_{\mathfrak{p}}\xi \text{ is odd.}$$

When $\operatorname{ord}_{\mathfrak{p}}\xi$ is even we can write $\xi = \pi^{2r}\varepsilon$ with ε a unit, and then $\mathfrak{d}(\xi) = \pi^{2r}\mathfrak{d}(\varepsilon)$. So it is enough to study the quadratic defect on the group of units $\mathfrak{u}$. For a unit ε we can write $\varepsilon = \delta^2 + \alpha$ with $\delta \in \mathfrak{u}$ and $\alpha\mathfrak{o} = \mathfrak{d}(\varepsilon) \subseteq \mathfrak{o}$.

What is the intuitive meaning of the quadratic defect? We have just seen that having defect 0 is equivalent to being a square. Consider a non-square unit ε with defect $\mathfrak{d}(\varepsilon) = \mathfrak{p}^d \subseteq \mathfrak{o}$. Then we can write $\varepsilon = \delta^2 + \alpha$ with $\delta \in \mathfrak{u}$, $\alpha \in \mathfrak{p}^d$, while such an expression is impossible with an α in $\mathfrak{p}^{d+1}$. So here the quadratic defect is that ideal $\mathfrak{p}^d$ with the property that ε is congruent to a square $\bmod \mathfrak{p}^d$ but not $\bmod \mathfrak{p}^{d+1}$.

63:2. *Let ε be a unit in the local field F. If F is non-dyadic, then $\mathfrak{d}(\varepsilon)$ is 0 or $\mathfrak{o}$; if F is dyadic, then $\mathfrak{d}(\varepsilon)$ is one of the ideals*

$$0 \subset 4\mathfrak{o} \subset 4\mathfrak{p}^{-1} \subset 4\mathfrak{p}^{-3} \subset \cdots \subset \mathfrak{p}^3 \subset \mathfrak{p}\,.$$

Proof. 1) If F is non-dyadic, then it follows from the Local Square Theorem that $\mathfrak{d}(\varepsilon)$ is 0 whenever it is not $\mathfrak{o}$, i. e. whenever $\mathfrak{d}(\varepsilon) \subseteq \mathfrak{p}$.

2) Now consider the case of a dyadic field F. If $\mathfrak{d}(\varepsilon) \subset 4\mathfrak{o}$ then $\mathfrak{d}(\varepsilon) \subseteq 4\mathfrak{p}$ and so ε is a square by the Local Square Theorem, hence $\mathfrak{d}(\varepsilon) = 0$. It remains for us to consider an ε with $\mathfrak{d}(\varepsilon) = \mathfrak{p}^d$ and $4\mathfrak{o} \subset \mathfrak{p}^d \subseteq \mathfrak{o}$, and to prove that d must then be odd. Suppose if possible that we have such an ε with an even d. Put $d = 2r$. We can write $\varepsilon = \delta^2 + \alpha$ with $\delta \in \mathfrak{u}$ and $\alpha\mathfrak{o} = \mathfrak{p}^d$. Replacing ε by ε/δ^2 gives us a new ε of the form

$$\varepsilon = 1 + \varepsilon_1\pi^{2r}$$

with $\varepsilon_1 \in \mathfrak{u}$, $\mathfrak{d}(\varepsilon) = \mathfrak{p}^{2r}$, and $2\mathfrak{o} \subset \mathfrak{p}^r \subsetneq \mathfrak{o}$. By the perfectness of the residue class field there is a unit δ_1 such that $\delta_1^2 \equiv \varepsilon_1 \bmod \pi$. Then

$$1 + \varepsilon_1 \pi^{2r} \equiv 1 + \delta_1^2 \pi^{2r} \equiv (1 + \delta_1 \pi^r)^2$$

modulo π^{2r+1}. Hence we have an expression

$$\varepsilon = (1 + \delta_1 \pi^r)^2 + \alpha_1 \quad \text{with } \alpha_1 \in \mathfrak{p}^{2r+1}.$$

This contradicts the fact that $\mathfrak{d}(\varepsilon) = \mathfrak{p}^{2r}$. Hence d must be odd.

q. e. d.

63:3. *Let ε be a unit in the local field F. Then $\mathfrak{d}(\varepsilon) = 4\mathfrak{o}$ if and only if $F(\sqrt{\varepsilon})/F$ is quadratic unramified.*

Proof. Here it is understood that $F(\sqrt{\varepsilon})$ is provided with that unique spot which divides the given spot $\mathfrak{p}$ on F. By Proposition 32:3, $F(\sqrt{\varepsilon})$ is also a local field.

1) First let us be given $\mathfrak{d}(\varepsilon) = 4\mathfrak{o}$. This certainly makes the given extension quadratic. Multiplying ε by the square of a unit in F allows us to assume that $\varepsilon \equiv 1 \bmod 4$, hence that $\frac{1}{4}(\varepsilon - 1) \in \mathfrak{o}$. Now $F(\sqrt{\varepsilon})$ can be obtained by adjoining a root $\alpha = \frac{1}{2}(-1 + \sqrt{\varepsilon})$ of the polynomial

$$x^2 + x + \frac{1}{4}(1 - \varepsilon)$$

to F. But Proposition 32:6 applies to this situation. Hence $F(\sqrt{\varepsilon})/F$ is unramified.

2) Now suppose $F(\sqrt{\varepsilon})/F$ is quadratic unramified. We can assume that ε is given in the form $\varepsilon = 1 + \alpha$ with $\mathfrak{d}(\varepsilon) = \alpha\mathfrak{o}$. Since the given extension is quadratic we know that $4\mathfrak{o} \subseteq \alpha\mathfrak{o} \subsetneq \mathfrak{o}$. This finishes the proof for the non-dyadic case. Now assume that F is dyadic and prove that $\alpha\mathfrak{o} = 4\mathfrak{o}$. Write $A = -1 + \sqrt{\varepsilon}$. Then

$$A(A + 2) = \alpha.$$

Let $|\ |$ be the prolongation of the given valuation on F to $F(\sqrt{\varepsilon})$. If we had $|A| > |2|$ we would have

$$|\alpha| = |A|^2 > |4|,$$

hence α would have even order in $F(\sqrt{\varepsilon})$, hence it would have even order in F since the given extension is unramified; so $\mathfrak{d}(\varepsilon)$ would be equal to $\mathfrak{p}^{2r}$ with $\mathfrak{p}^{2r} \supset 4\mathfrak{o}$ and this is impossible by Proposition 63:2. Hence $|A| \leqq |2|$. This implies that $|\alpha| \leqq |4|$. Hence $\alpha\mathfrak{o} \subseteq 4\mathfrak{o}$ and so $\alpha\mathfrak{o} = 4\mathfrak{o}$ as required. **q. e. d.**

63:4. *There is a unit ε in the local field F with $\mathfrak{d}(\varepsilon) = 4\mathfrak{o}$. If ε' is any other such unit, then $\varepsilon' \in \varepsilon\mathfrak{u}^2$.*

Proof. By Proposition 32:9 there is an unramified quadratic extension E/F. Since $\chi(F) \neq 2$ we can obtain E by adjoining a square root to F;

since E/F is unramified this will have to be a square root of an element of even order. Hence we can write $E = F(\sqrt{\varepsilon})$ with ε a unit in F. Then $\mathfrak{d}(\varepsilon) = 4\mathfrak{o}$ by Proposition 63:3.

Now consider ε'. Then $E' = F(\sqrt{\varepsilon'})$ is quadratic unramified over F by Proposition 63:3. By Proposition 32:10 the two fields E and E' are splitting fields of the same polynomial over F, hence $\sqrt{\varepsilon'} \in F(\sqrt{\varepsilon})$, hence $\sqrt{\varepsilon'} \in \sqrt{\varepsilon}\, F$. So $\varepsilon' \in \varepsilon \mathfrak{u}^2$. **q. e. d.**

63:5. *Let* $\varepsilon = 1 + \alpha$ *be a unit in a dyadic field with* $|4| < |\alpha| < 1$ *and* $\operatorname{ord} \alpha$ *odd. Then* $\mathfrak{d}(\varepsilon) = \alpha\mathfrak{o}$.

Proof. Put $\alpha\mathfrak{o} = \mathfrak{p}^d$. Clearly $\mathfrak{d}(\varepsilon) \subseteq \mathfrak{p}^d$. Suppose if possible that $\mathfrak{d}(\varepsilon) \subseteq \mathfrak{p}^{d+1}$. Then there is a $\gamma \in \mathfrak{o}$ such that

$$1 + \alpha \equiv (1 + \gamma)^2 \bmod \pi^{d+1}.$$

So $|\gamma(\gamma + 2)| = |\alpha|$ by the Principle of Domination since $|\alpha| = |\pi^d|$. If we had $|\gamma| \leqq |2|$ we would have $|\alpha| \leqq |4|$, contrary to the assumptions; if we had $|\gamma| > |2|$ we would have $|\alpha| = |\gamma|^2$, contrary to the assumptions. Hence we cannot have $\mathfrak{d}(\varepsilon) \subseteq \mathfrak{p}^{d+1}$. So $\mathfrak{d}(\varepsilon)$ is indeed equal to $\alpha\mathfrak{o}$. **q. e. d.**

63:6. **Remark.** Each of the ideals in Proposition 63:2 will actually appear as the quadratic defect of some unit ε. To get defect 0 take $\varepsilon = 1$, to get $4\mathfrak{o}$ apply Proposition 63:4, to get $\mathfrak{p}^d$ with d odd and $4\mathfrak{o} \subset \mathfrak{p}^d \subset \mathfrak{o}$ take $\varepsilon = 1 + \pi^d$ and apply Proposition 63:5.

We conclude this subparagraph with local index computations that will be needed later in the global theory. For any fractional ideal $\mathfrak{p}^r$ with $r > 0$ the set

$$1 + \mathfrak{p}^r = \{1 + \alpha\pi^r \mid \alpha \in \mathfrak{o}\}$$

is a neighborhood of the identity 1 under the $\mathfrak{p}$-adic topology on F. Clearly $1 + \mathfrak{p}^r$ is a subgroup of the group of units $\mathfrak{u}$ and we have

$$1 + \mathfrak{p}^{r+1} \subset 1 + \mathfrak{p}^r \subseteq \mathfrak{u} \qquad (\text{if } r > 0)\,.$$

63:7. **Lemma.** *Let* θ *be a homomorphism of a commutative group* G *into some other group, let* θG *be the image of* G *and* G_θ *the kernel of* θ. *Then for any subgroup* H *of* G *we have*

$$(G : H) = (\theta G : \theta H)\,(G_\theta : H_\theta)\,,$$

where the left hand side is finite if and only if the right hand side is.

Proof. By the isomorphism theorems of group theory we obtain

$$\begin{aligned}(G : H) &= (G : G_\theta H)\,(G_\theta H : H)\\ &= (\theta G : \theta(G_\theta H))\,(G_\theta : G_\theta \cap H)\\ &= (\theta G : \theta H)\,(G_\theta : H_\theta)\,.\end{aligned}$$

q. e. d.

63:8. *F is a local field at* $\mathfrak{p}$, $\mathfrak{u}$ *is the group of units, and* $1+\mathfrak{p}^r$ *is a neighborhood of the identity with* $r > 0$. *Then*

(1) $(\mathfrak{u}: 1+\mathfrak{p}^r) = (N\mathfrak{p}-1)(N\mathfrak{p})^{r-1}$,

(2) $(1+\mathfrak{p}^r)^2 = 1+2\mathfrak{p}^r$ *if* $\mathfrak{p}^r \subseteq 2\mathfrak{p}$.

Proof. (1) Consider a residue class field $(^-, \bar{F})$ of F at $\mathfrak{p}$. The restriction of the bar map is a multiplicative homomorphism of $\mathfrak{u}$ onto the non-zero elements of $\bar{F}$ with kernel $1+\mathfrak{p}$. Hence $(\mathfrak{u}: 1+\mathfrak{p}) = N\mathfrak{p}-1$. Now the mapping $\varphi(1+\alpha\pi^r) = \bar{\alpha}$ is easily seen to be a homomorphism of the multiplicative group $1+\mathfrak{p}^r$ onto the additive group $\bar{F}$ with kernel $1+\mathfrak{p}^{r+1}$. Hence $(1+\mathfrak{p}^r: 1+\mathfrak{p}^{r+1}) = N\mathfrak{p}$. By taking the tower

$$\mathfrak{u} \supseteq 1+\mathfrak{p} \supset 1+\mathfrak{p}^2 \supset \cdots \supset 1+\mathfrak{p}^r$$

we obtain

$$(\mathfrak{u}: 1+\mathfrak{p}^r) = (N\mathfrak{p}-1)(N\mathfrak{p})^{r-1}.$$

(2) Clearly $(1+\mathfrak{p}^r)^2 \subseteq 1+2\mathfrak{p}^r$ when $\mathfrak{p}^r \subseteq 2\mathfrak{p}$. We must reverse the inclusion. So consider a typical element $1+2\alpha\pi^r$ of $1+2\mathfrak{p}^r$ with $\alpha \in \mathfrak{o}$. Then

$$1+2\alpha\pi^r = (1+\alpha\pi^r)^2 - \alpha^2\pi^{2r} = (1+\alpha\pi^r)^2(1+\beta\pi^{2r})$$

for some $\beta \in \mathfrak{o}$. It is enough to prove that $1+\beta\pi^{2r} \in (1+\mathfrak{p}^r)^2$ whenever $\beta \in \mathfrak{o}$ and $\mathfrak{p}^r \subseteq 2\mathfrak{p}$. By the Local Square Theorem there is a $\gamma \in 2\mathfrak{p}$ such that $(1+\gamma)^2 = 1+\beta\pi^{2r}$. Then $|\gamma+2| = |2|$ and so

$$|2\gamma| = |\gamma(\gamma+2)| = |\beta\pi^{2r}| \leq |2\pi^{r+1}|,$$

hence $\gamma \in \mathfrak{p}^{r+1} \subseteq \mathfrak{p}^r$. **q. e. d.**

63:9. *F is a local field at* $\mathfrak{p}$ *and* $\mathfrak{u}$ *is the group of units. Then*

$$(\dot{F}:\dot{F}^2) = 2(\mathfrak{u}:\mathfrak{u}^2) = 4(N\mathfrak{p})^{\operatorname{ord} 2}.$$

Proof. 1) To prove the first equality apply Lemma 63:7 with $G = \dot{F}$, $H = \dot{F}^2$, and $\theta\alpha = |\alpha|$. Then

$$(\dot{F}:\dot{F}^2) = (|\dot{F}| : |\dot{F}|^2)(\mathfrak{u}:\mathfrak{u}^2) = 2(\mathfrak{u}:\mathfrak{u}^2).$$

2) To prove the second equality apply Lemma 63:7 and Proposition 63:8. This time take $G = \mathfrak{u}$, $H = 1+\mathfrak{p}^r$ for any $r > 0$ such that $\mathfrak{p}^r \subseteq 2\mathfrak{p}$, and $\theta\alpha = \alpha^2$. Then

$$\begin{aligned}(\mathfrak{u}: 1+\mathfrak{p}^r) &= 2(\mathfrak{u}^2 : (1+\mathfrak{p}^r)^2)\\ &= 2(\mathfrak{u}^2 : 1+2\mathfrak{p}^r)\\ &= 2(\mathfrak{u}^2 : 1+\mathfrak{p}^{r+\operatorname{ord} 2}).\end{aligned}$$

Hence

$$(\mathfrak{u}:\mathfrak{u}^2)(\mathfrak{u}: 1+\mathfrak{p}^r) = 2(\mathfrak{u}: 1+\mathfrak{p}^{r+\operatorname{ord} 2}).$$

Hence

$$(\mathfrak{u}:\mathfrak{u}^2) = 2(N\mathfrak{p})^{\operatorname{ord} 2}$$

q. e. d.

§ 63B. The Hilbert symbol and the Hasse symbol

In this subparagraph F can either be the local field under discussion or any complete archimedean field. So F is either a local field at $\mathfrak{p}$, or $\mathfrak{p}$ is real and complete, or $\mathfrak{p}$ is complex and complete. In any one of these situations it is possible to replace the Hasse algebra by a simpler invariant called the Hasse symbol. In the definition of the Hasse symbol the quaternion algebras are replaced by Hilbert symbols which we now define. Given non-zero scalars α, β in an arithmetic field of the above type the Hilbert symbol

$$\left(\frac{\alpha, \beta}{\mathfrak{p}}\right),$$

or simply (α, β), is defined to be $+1$ if $\alpha\xi^2 + \beta\eta^2 = 1$ has a solution $\xi, \eta \in F$; otherwise the symbol is defined to be -1.

63:10. **Example.** Put $E = F(\sqrt{\beta})$. So E/F is of degree 1 or 2. Then

$$\alpha \in N_{E/F}E \text{ if and only if } \left(\frac{\alpha, \beta}{\mathfrak{p}}\right) = 1 .$$

Our first results refer to the local case only.

63:11. *Let V be a binary quadratic space over a local field F and let the discriminant dV be a prime element of F. Let Δ denote a fixed unit of quadratic defect $4\mathfrak{o}$. If γ is any non-zero scalar, then V represents γ or $\gamma\Delta$ but not both.*

Proof. By scaling V we can assume that $\gamma = 1$.

1) Our first task is to prove that V represents either 1 or Δ. Since dV is a prime element there is a splitting $V \cong \langle\varepsilon\rangle \perp \langle\delta\pi\rangle$ in which ε and δ are units in F. If F is non-dyadic, ε will either be a square or a square times Δ by Propositions 63:2 and 63:4. We may therefore restrict ourselves to the dyadic case.

The above unit ε can actually be any unit represented by V. In fact let it be a unit of smallest quadratic defect in $Q(V)$. We can assume that this ε has the form $\varepsilon = 1 + \beta$ with $\mathfrak{d}(\varepsilon) = \beta\mathfrak{o} \subset \mathfrak{o}$. Thus

$$V \cong \langle 1 + \beta\rangle \perp \langle\delta\pi\rangle .$$

One of three things can happen. (i) If $\mathfrak{d}(\varepsilon) = 0$ we have $\beta = 0$ and so $\varepsilon = 1$ and V represents 1 as desired. (ii) If $\mathfrak{d}(\varepsilon) = 4\mathfrak{o}$ then V represents a unit of quadratic defect $4\mathfrak{o}$ and so it represents Δ by Proposition 63:4. (iii) The one remaining possibility is $4\mathfrak{o} \subset \mathfrak{d}(\varepsilon) \subset \mathfrak{o}$. Let us prove that this is impossible. If $4\mathfrak{o} \subset \beta\mathfrak{o} \subset \mathfrak{o}$ we write $\beta = \varepsilon_1\pi^k$ with ε_1 a unit and $\mathfrak{d}(\varepsilon) = \mathfrak{p}^k$. Here k is odd. By the perfectness of the residue class field we can choose a unit λ such that $\lambda^2\delta \equiv -\varepsilon_1 \bmod \pi$. Then V represents

$$1 + \varepsilon_1\pi^k + \lambda^2\delta\pi^k \equiv 1 \bmod \pi^{k+1}$$

since k is odd. In other words V represents a unit whose quadratic defect is contained in $\mathfrak{p}^{k+1}$. This is contrary to the choice of ε.

2) Finally we have to prove that V cannot represent both 1 and Δ. If it did we would have

$$<1> \perp <\delta\pi> \cong <\Delta> \perp <\Delta\delta\pi>$$

for some $\delta \in \mathfrak{u}$. Hence $\Delta = \xi^2 + \eta^2\delta\pi$ for some ξ, η in F. By the Principle of Domination ξ has to be a unit and η has to be an integer. Hence there is an element of the form

$$\Delta/\xi^2 = 1 + \beta^2\delta\pi \quad (\beta \in \mathfrak{o})$$

with quadratic defect $4\mathfrak{o}$. This is impossible by Proposition 63:5.

q. e. d.

63:11a. *Let ε be any unit in F. Then*

$$\left(\frac{\pi, \Delta}{\mathfrak{p}}\right) = -1 \quad \textit{and} \quad \left(\frac{\varepsilon, \Delta}{\mathfrak{p}}\right) = +1 .$$

Proof. The first equation is a direct consequence of the proposition. Let us do the second. In the non-dyadic case we can find ξ, η in $\mathfrak{o}$ such that

$$\varepsilon\xi^2 + \Delta\eta^2 \equiv 1 \bmod \pi$$

since a binary quadratic form over a finite field of characteristic not 2 is universal by Proposition 62:1. Here ξ has to be a unit since Δ is a non-square, hence we have $\lambda \in \mathfrak{o}$ such that

$$\varepsilon\xi^2(1 + \lambda\pi) + \Delta\eta^2 = 1 .$$

Then $1 + \lambda\pi$ is a square by the Local Square Theorem. Hence $<\varepsilon> \perp <\Delta>$ represents 1, hence $\left(\frac{\varepsilon, \Delta}{\mathfrak{p}}\right) = 1$. In the dyadic case use the fact that $<\Delta>$ represents 1 modulo $4\mathfrak{o}$ together with the perfectness of the residue class field to show that $<\varepsilon> \perp <\Delta>$ represents 1 modulo $4\mathfrak{p}$. Then apply the Local Square Theorem. **q. e. d.**

63:11b. *The quaternion algebra (π, Δ) is a division algebra. All quaternion division algebras over F are isomorphic to it.*

Proof. The quadratic space $<\pi> \perp <\Delta>$ does not represent 1, by the proposition. Hence (π, Δ) is a division algebra by Proposition 57:9.

It is easily seen that every quaternion algebra over F is similar to a tensor product of quaternion algebras of the form $(\varepsilon, \delta\pi)$ where ε and δ are units. It therefore suffices to prove that this quaternion algebra is isomorphic to (π, Δ_0) where Δ_0 is one of the elements 1 and Δ. But

$$<\varepsilon> \perp <\delta\pi> \cong <\Delta_0> \perp <\varepsilon\delta\pi\Delta_0> .$$

Computing Hasse algebras gives

$$(\varepsilon, \delta\pi) \sim (\Delta_0, \Delta_0\varepsilon\delta) \otimes (\Delta_0, \pi) .$$

But $(\Delta_0, \Delta_0 \varepsilon\, \delta) \sim 1$ by Corollary 63:11a and Proposition 57:9. Hence $(\varepsilon, \delta\pi) \cong (\Delta_0, \pi)$. **q. e. d.**

63:12. **Example.** Corollary 63:11a shows the very simple nature of the Hilbert symbol over non-dyadic fields. For let ε, δ be units in a given non-dyadic local field. Then $\left(\frac{\varepsilon, \delta}{\mathfrak{p}}\right) = 1$ always. And $\left(\frac{\pi, \delta}{\mathfrak{p}}\right) = 1$ if and only if δ is a square.

Now consider the local field or real complete field F. There are essentially two quaternion algebras over F. Hence

$$\bigotimes_{1 \leq i \leq n} \left(\frac{\alpha_i, \beta_i}{F}\right) \sim 1$$

if and only if the number of division algebras appearing in this tensor product is even. But $\left(\frac{\alpha_i, \beta_i}{F}\right)$ is a division algebra if and only if $\left(\frac{\alpha_i, \beta_i}{\mathfrak{p}}\right) = -1$. Hence the given condition is equivalent to

$$\prod_{1 \leq i \leq n} \left(\frac{\alpha_i, \beta_i}{\mathfrak{p}}\right) = 1 .$$

So we have

$$\bigotimes_{1 \leq i \leq n} \left(\frac{\alpha_i, \beta_i}{F}\right) \sim \bigotimes_{1 \leq j \leq m} \left(\frac{\gamma_j, \delta_j}{F}\right)$$

if and only if

$$\prod_{1 \leq i \leq n} \left(\frac{\alpha_i, \beta_i}{\mathfrak{p}}\right) = \prod_{1 \leq j \leq m} \left(\frac{\gamma_j, \delta_j}{\mathfrak{p}}\right) .$$

These results are of course trivial when F is a complex complete field.

We can carry over to the Hilbert symbol the various formulas of Proposition 57:10. The first two of these formulas are trivial for the Hilbert symbol, the third now reads

$$\left(\frac{\alpha, \alpha\beta}{\mathfrak{p}}\right) = \left(\frac{\alpha, -\beta}{\mathfrak{p}}\right) ,$$

and the fourth,

$$\left(\frac{\alpha, \beta}{\mathfrak{p}}\right)\left(\frac{\alpha, \gamma}{\mathfrak{p}}\right) = \left(\frac{\alpha, \beta\gamma}{\mathfrak{p}}\right) .$$

63:13. *Let F be either a local field or a complete archimedean field at $\mathfrak{p}$. Given any non-square β in $\dot{F}$ there is an α in $\dot{F}$ such that*

$$\left(\frac{\alpha, \beta}{\mathfrak{p}}\right) = -1 .$$

Proof. The complex case cannot arise. In the real case take $\alpha = -1$. Now the local case. We can suppose that β is either a prime element or a unit of the form $\beta = 1 + \gamma$ with $\mathfrak{d}(\beta) = \gamma\mathfrak{o}$. Let $\Delta \equiv 1 \bmod 4$ be a unit of quadratic defect $4\mathfrak{o}$. If β is a prime element take $\alpha = \Delta$. If $\gamma\mathfrak{o} = 4\mathfrak{o}$

take $\alpha = \pi$. If $4\mathfrak{o} \subset \gamma\mathfrak{o} \subset \mathfrak{o}$ take $\alpha = \Delta - \beta$; then

$$<\alpha> \perp <\beta> \cong <\Delta> \perp <\Delta\,\alpha\,\beta>;$$

but

$$\mathrm{ord}_{\mathfrak{p}}\,\alpha\,\beta = \mathrm{ord}_{\mathfrak{p}}\,\alpha = \mathrm{ord}_{\mathfrak{p}}\,\gamma$$

is odd; hence $<\alpha> \perp <\beta>$ does not represent 1 by Proposition 63:11.

q. e. d.

63:13a. *Let E be a quadratic extension of F. Then*

$$(\dot{F} : N_{E/F}\dot{E}) = 2\,.$$

Proof. Since the characteristic is not 2 there is a non-square $\beta \in F$ such that $E = F(\sqrt{\beta})$. Consider the multiplicative homomorphism $\dot{F} \rightarrowtail (\pm 1)$ defined by the equation

$$\alpha \rightarrowtail \left(\frac{\alpha,\,\beta}{\mathfrak{p}}\right).$$

This mapping is surjective by the proposition; its kernel consists of those $\alpha \in \dot{F}$ for which $\left(\frac{\alpha,\,\beta}{\mathfrak{p}}\right) = 1$, i. e. for which $\alpha \in N_{E/F}\dot{E}$. Hence

$$(\dot{F} : N_{E/F}\dot{E}) = 2\,.$$

q. e. d.

Now let V be a regular n-ary quadratic space over F (here F is either a local field or a complete archimedean field at $\mathfrak{p}$). Take a splitting

$$V \cong <\alpha_1> \perp \cdots \perp <\alpha_n>$$

and define the Hasse symbol

$$S_{\mathfrak{p}}\,V = \prod_{1 \leqq i \leqq n} \left(\frac{\alpha_i,\,d_i}{\mathfrak{p}}\right)$$

where $d_i = \alpha_1 \ldots \alpha_i$. This is independent of the orthogonal splitting chosen for V since the same is true of the Hasse algebra. We clearly have

$$S_{\mathfrak{p}}\,U = S_{\mathfrak{p}}\,V \iff SU \sim SV\,.$$

From Remark 58:3 we get the formulas

$$S_{\mathfrak{p}}\,V = \prod_{1 \leqq i \leqq j \leqq n} (\alpha_i,\,\alpha_j)\,,$$
$$S_{\mathfrak{p}}(U \perp W) = S_{\mathfrak{p}}\,U \cdot (dU,\,dW) \cdot S_{\mathfrak{p}}\,W\,,$$
$$S_{\mathfrak{p}}\,V^{\alpha} = (\alpha,\,(-1)^{n(n+1)/2}(dV)^{n+1})\,S_{\mathfrak{p}}\,V\,.$$

63:14. **Example.** Let V be a quadratic space over a non-dyadic local field with

$$V \cong <\varepsilon_1> \perp \cdots \perp <\varepsilon_n>$$

where all ε_i are units and $n \geqq 3$. Then V is isotropic. For we can assume that $n = 3$. Then

$$S_{\mathfrak{p}} V = 1 = \left(\frac{-1, -1}{\mathfrak{p}}\right).$$

Now apply Proposition 58:6.

§ 63C. Quadratic forms over local fields

We are ready to characterize quadratic forms over local fields. The invariants will be the dimension, the discriminant, and the Hasse symbol. The representation problem will be solved by reducing it to the problem of isometry.

By Proposition 63:4 there is a unit of quadratic defect $4\mathfrak{o}$; we fix one such unit and write it Δ. Let V denote a regular n-ary quadratic space over F.

We start the discussion by gathering together some important information about binary spaces.

63:15. **Example**[1]. In this example V will be a regular binary quadratic space over the local field F.

(i) If $dV = -\Delta$, then $Q(\dot{V})$ is either $\mathfrak{u}\dot{F}^2$ or $\pi\mathfrak{u}\dot{F}^2$. For we can assume by scaling that $1 \in Q(V)$. Then $V \cong \langle 1\rangle \perp \langle -\Delta\rangle$. So for any unit ε the quadratic space

$$\langle -\varepsilon\rangle \perp V \cong \langle -\varepsilon\rangle \perp \langle 1\rangle \perp \langle -\Delta\rangle$$

is isotropic by the criterion of Proposition 58:6. Hence V represents ε. The same sort of argument will show that V represents no prime elements at all. Hence $Q(\dot{V})$ is precisely $\mathfrak{u}\dot{F}^2$.

(ii) If V is not isotropic and $1 \in Q(V)$, then $Q(\dot{V})$ is a subgroup of $\dot{F}$ with

$$(\dot{F} : Q(\dot{V})) = 2.$$

To prove this we write $V \cong \langle 1\rangle \perp \langle -\alpha\rangle$ and construct the group homomorphism

$$\gamma \to \left(\frac{\alpha, \gamma}{\mathfrak{p}}\right)$$

of $\dot{F}$ into $\{\pm 1\}$. It is easily seen that $Q(\dot{V})$ is the kernel of this homomorphism. And the scalar α is a non-square since V is not isotropic, so the homomorphism is surjective. Hence $(\dot{F} : Q(\dot{V})) = 2$ as asserted.

(iii) If W is another regular binary space over F, then

$$V \cong W \quad \text{if and only if} \quad Q(V) = Q(W).$$

We have to show that $V \cong W$ if $Q(V) = Q(W)$. We can assume that neither V nor W is isotropic. By scaling we can assume that V and W both represent 1. Then

$$V \cong \langle 1\rangle \perp \langle -\alpha\rangle, \quad W \cong \langle 1\rangle \perp \langle -\beta\rangle.$$

---

[1] The results of § 63, particularly of this example, are closely related to the local class field theory. For a discussion of local class field theory we refer the reader to O. F. G. SCHILLING, *The theory of valuations* (New York, 1950).

Now γ is in $Q(\dot{V})$ if and only if $\left(\frac{\alpha, \gamma}{\mathfrak{p}}\right) = 1$, and the same with W, hence

$$\left(\frac{\alpha, \gamma}{\mathfrak{p}}\right) = \left(\frac{\beta, \gamma}{\mathfrak{p}}\right) \qquad \forall\, \gamma \in \dot{F}\,,$$

hence $\alpha\beta$ is a square by Proposition 63:13, therefore $V \cong W$.

(iv) Suppose neither V nor W is isotropic. Then

$$Q(V) \subseteq Q(W) \Rightarrow V \cong W\,.$$

For we can assume by scaling that $1 \in Q(V) \subseteq Q(W)$. Hence $(\dot{F}:Q(\dot{V}))=2$ and $(\dot{F}:Q(\dot{W})) = 2$. But $Q(\dot{V}) \subseteq Q(\dot{W})$, so that $Q(\dot{V}) = Q(\dot{W})$. Hence V is isometric to W.

63:16. **Example.** Let E be a quadratic unramified extension of the local field F. We claim that

$$N_{E/F}\dot{E} = \mathfrak{u}\dot{F}^2.$$

Since E/F is a quadratic extension of a field of characteristic not 2 it can be obtained by the adjunction of a square root; since E/F is unramified there is actually a unit Δ in F such that $E = F(\sqrt{\Delta})$ and this unit must have quadratic defect $4\mathfrak{o}$ by Proposition 63:3. Now $N_{E/F}\dot{E}$ is the set of non-zero numbers represented by the quadratic space $\langle 1\rangle \perp \langle -\Delta\rangle$. This set is equal to $\mathfrak{u}\dot{F}^2$ by Example 63:15, so we have proved our claim. Note that by Theorem 14:1 the norm $N_{E/F}\alpha$ of an element α of E is a unit of F if and only if α is a unit of E. Hence

$$N_{E/F}\,\mathfrak{U} = \mathfrak{u}\,,$$

where $\mathfrak{U}$ is the group of units of E.

63:17. *Let V be an anisotropic quaternary space over a local field. Then*

$$V \cong \langle 1\rangle \perp \langle -\Delta\rangle \perp \langle \pi\rangle \perp \langle -\pi\Delta\rangle$$

where Δ is any given unit of quadratic defect $4\mathfrak{o}$.

Proof. 1) Suppose we can find a binary subspace U of V with $Q(\dot{U}) \subseteq \mathfrak{u}\dot{F}^2$; then $Q(\dot{U}) = \mathfrak{u}\dot{F}^2$ by Example 63:15, hence $Q(\dot{U}^*) \subseteq \pi\mathfrak{u}\dot{F}^2$ since V is anisotropic, hence V has the desired form by Example 63:15 again. The same holds if we can find a U with $Q(\dot{U}) \subseteq \pi\,\mathfrak{u}\dot{F}^2$.

2) Take an arbitrary splitting $V = W \perp W^*$ into binary subspaces. We can assume that W represents some units and some prime elements, else we are through by step 1). The same with W^*. From this we can conclude that V has a splitting of the form

$$V \cong \langle\varepsilon_1\rangle \perp \langle\varepsilon_2\rangle \perp \langle\varepsilon_3\pi\rangle \perp \cdots$$

with $\varepsilon_1, \varepsilon_2, \varepsilon_3$ in $\mathfrak{u}$. A double application of Proposition 63:11 gives a splitting

$$V \cong \langle\Delta_1\rangle \perp \langle -\Delta_2\rangle \perp \cdots$$

in which Δ_1 is one of the elements Δ and 1, and the same with Δ_2. This gives us a binary subspace U with discriminant $-\Delta$, hence with $Q(\dot{U}) = \mathfrak{u}\dot{F}^2$ by Example 63:15; we have therefore found a binary subspace of the type mentioned in step 1) and we are through. **q. e. d.**

63:18. **Remark.** Thus we have essentially one anisotropic quaternary space over a local field, its discriminant is 1, it represents all units and all prime elements, and so it is universal. Hence all regular quaternary spaces over a local field are universal. There are two regular quaternary spaces with discriminant 1, the hyperbolic space and the anisotropic space.

63:19. *A quadratic space V over a local field is isotropic if* $\dim V \geqq 5$.[1]

Proof. We can assume that V is regular with $\dim V = 5$. Take a splitting $V = \dot{U} \perp W$ in which U is a line and W is quaternary. Then $-Q(U) \subseteq Q(W)$ since W is universal by Remark 63:18. Hence V is isotropic. **q. e. d.**

63:20. **Theorem.** *Let F be a local field at the prime spot $\mathfrak{p}$. Then two regular quadratic spaces U and V over F are isometric if and only if*

$$\dim U = \dim V, \quad dU = dV, \quad S_{\mathfrak{p}} U = S_{\mathfrak{p}} V .$$

Proof. The equality of the Hasse symbols implies the similarity of the Hasse algebras. Now every regular quinary space over F is isotropic by Proposition 63:19. Hence Theorem 58:8 applies. Hence U is isometric to V. **q. e. d.**

63:21. **Theorem.** *Let U and V be regular quadratic spaces over a local field with* $\nu = \dim V - \dim U \geqq 0$. *Then U is represented by V if and only if*

$$\begin{aligned} &U \cong V && \text{when } \nu = 0 , \\ &U \perp \langle dU \cdot dV \rangle \cong V && \text{when } \nu = 1 , \\ &U \perp H \cong V && \text{when } \nu = 2 , \quad dU = -dV. \end{aligned}$$

Here H denotes a hyperbolic plane.

Proof. The necessity is obvious. So is the sufficiency when $\nu = 0$ or $\nu = 1$. Next consider $\nu \geqq 3$. Take splittings

$$U \cong \perp_1^m \langle \alpha_\lambda \rangle , \quad V \cong \perp_1^n \langle \beta_\mu \rangle$$

with $\alpha_\lambda, \beta_\mu \in \dot{F}$. Here $n - m \geqq 3$. Then the quaternary space

$$\langle \beta_1 \rangle \perp \langle \beta_{m+1} \rangle \perp \langle \beta_{m+2} \rangle \perp \langle \beta_{m+3} \rangle$$

represents α_1 since all regular quaternary spaces over a local field are universal by Remark 63:18. Hence V has a splitting with $\beta_1 = \alpha_1$.

[1] It is conjectured that every form of degree d in $d^2 + 1$ variables over a local field has a non-trivial zero. See S. LANG, *Ann. Math.* **55** (1952), pp. 373—390.

Repeat this m times. We obtain a splitting of the form

$$V \cong \langle\alpha_1\rangle \perp \cdots \perp \langle\alpha_m\rangle \perp \langle\beta'_{m+1}\rangle \perp \cdots \perp \langle\beta'_n\rangle .$$

Hence $U \rightarrowtail V$. To conclude we must consider $\nu = 2$. If $dU = -dV$ we are through by hypothesis. Hence assume that $dU \neq -dV$. In virtue of the case $\nu \geqq 3$ we have $U \rightarrowtail V \perp H$ where H is a hyperbolic plane. Hence

$$U \perp W \cong V \perp H$$

for some quaternary space W. The assumption about the discriminants gives $dW \neq 1$. Hence W is isotropic by Proposition 63:17. Hence we have a splitting $W = W_1 \perp H_1$ with W_1 binary and $H_1 \cong H$. But this implies that $U \perp W_1 \cong V$ by Witt's theorem. Hence $U \rightarrowtail V$ as asserted.

q. e. d.

63:22. **Theorem.** *Let V be a regular quadratic space over a local field F. Then there exists a quadratic space V' over F with*

$$\dim V' = \dim V\,, \quad dV' = dV\,, \quad S_\mathfrak{p} V' = -S_\mathfrak{p} V$$

if and only if V is neither a line nor a hyperbolic plane.

Proof. Clearly all lines with given discriminant are isometric and hence have the same Hasse symbol. This also applies to hyperbolic planes. So we must prove the converse: given that V is neither a line nor a hyperbolic plane, try to change the Hasse symbol without changing the dimension or the discriminant. Define

$$U \cong \langle 1\rangle \perp \langle\pi\rangle\,, \quad U' \cong \langle\Delta\rangle \perp \langle\pi\Delta\rangle$$

where Δ is a unit with quadratic defect $4\mathfrak{o}$. Then there is a representation

$$U' \rightarrowtail U \perp V$$

by Theorem 63:21. Hence there is a quadratic space V' with $U' \perp V' \cong U \perp V$. Clearly V' has the same discriminant and dimension as V. An easy computation with Hasse symbols shows that $S_\mathfrak{p} V' = -S_\mathfrak{p} V$.

q. e. d.

63:23. **Theorem.** *The necessary and sufficient condition that there exist a regular quadratic space V having invariants*

$$n_0 = \dim V\,, \quad d_0 = dV\,, \quad s_0 = S_\mathfrak{p} V$$

over a local field F, is that these quantities satisfy the relation

$$s_0 = \left(\frac{d_0, -1}{\mathfrak{p}}\right)$$

when $n_0 = 1$, and when $n_0 = 2$ with $d_0 \in -\dot{F}^2$. (Here it is assumed that $n_0 \in \mathsf{N}, d_0 \in \dot{F}, s_0 = \pm 1$).

Proof. The necessity follows from the definition of the Hasse symbol. Let us do the sufficiency. The spaces $<d_0>$ and $<1> \perp <-1>$ will work in the exceptional cases. Otherwise, we take any space V with the required dimension and discriminant, say

$$V \cong <1> \perp \cdots \perp <1> \perp <d_0> .$$

Now by Theorem 63:22 we have a space V' with the same dimension and discriminant, but with opposite Hasse symbol to V. Then either V or V' does the job. **q. e. d.**

63:24. **Example.** The following result will be used later in the global theory: let V be a regular quaternary space over a non-dyadic local field with dV a unit and $S_\mathfrak{p} V = 1$, and let U be any regular ternary subspace of V, then $Q(U) \supseteq \mathfrak{u}$. Scaling V by a unit shows that it is enough to prove that $1 \in Q(U)$. By Theorem 63:21 we are sure that U will represent 1 whenever $dU \neq -1$. In the exceptional case we can write $V \cong <-\delta> \perp U$ where δ is a unit for which $dV = \delta$; a computation of Hasse symbols the gives $S_\mathfrak{p} U = \left(\frac{-1,-1}{\mathfrak{p}}\right)$; hence U is isotropic, hence it is universal, hence it represents 1.

§ 64. Global notation

More definitions are needed before we can proceed with the global theory. Let F be a global field and let $\Omega = \Omega_F$ be its set of non-trivial spots. Recall from § 33D that J_F denotes the group of idèles of the field F. In § 33I we introduced the subgroup of principal idèles P_F, the subgroup of S-idèles J_F^S, and the subgroup $P_F^S = J_F^S \cap P_F$. Here S can be any subset of Ω but in practice it is usually taken to be a Dedekind set consisting of almost all spots in Ω. We define K_F^S to be the set of idèles $\mathfrak{i}$ such that

$$\begin{cases} |\mathfrak{i}_\mathfrak{p}|_\mathfrak{p} = 1 & \forall\, \mathfrak{p} \in S \\ \mathfrak{i}_\mathfrak{p} = 1 & \forall\, \mathfrak{p} \in \Omega - S . \end{cases}$$

Clearly K_F^S is a group, in fact a subgroup of J_F^S. (Here $|\ |_\mathfrak{p}$ is the normalized valuation at $\mathfrak{p}$, as usual.)

Given a spot $\mathfrak{p} \in \Omega$ we define $I_F^\mathfrak{p}$ to be the set of idèles $\mathfrak{i}$ such that

$$\mathfrak{i}_\mathfrak{q} = 1 \quad \forall\, \mathfrak{q} \in \Omega - \mathfrak{p} .$$

Again $I_F^\mathfrak{p}$ is a group. It is a subgroup of J_F^S if and only if $\mathfrak{p} \notin S$. There is a natural isomorphism

$$\dot{F}_\mathfrak{p} \rightarrowtail I_F^\mathfrak{p}$$

obtained by sending a typical $\alpha \in \dot{F}_\mathfrak{p}$ to the idèle in $I_F^\mathfrak{p}$ with $\mathfrak{p}$-coordinate equal to α and all other coordinates equal to 1.

Now let E/F be a finite separable extension. For each pair of spots $\mathfrak{P}|\mathfrak{p}$ on E/F we have the norm map

$$N_{\mathfrak{P}|\mathfrak{p}}: E_{\mathfrak{P}} \rightarrow F_{\mathfrak{p}}$$

defined in § 15B. Consider a typical idèle $\mathfrak{J}$ in J_E. Then by § 33A we have

$$|N_{\mathfrak{P}|\mathfrak{p}}\, \mathfrak{J}_{\mathfrak{P}}|_{\mathfrak{p}} = |\mathfrak{J}_{\mathfrak{P}}|_{\mathfrak{P}}\,.$$

Hence

$$\Big(\prod_{\mathfrak{P}|\mathfrak{p}} N_{\mathfrak{P}|\mathfrak{p}}\, \mathfrak{J}_{\mathfrak{P}}\Big)_{\mathfrak{p}\in\Omega_F}$$

is an idèle in J_F. We call this idèle the norm of $\mathfrak{J}$ and write it $N_{E/F}\mathfrak{J}$. This gives a group homomorphism

$$N_{E/F}: J_E \rightarrow J_F\,.$$

Similarly we obtain the following: let S be a set of spots on F and let the same letter S also denote the set of spots on E for which $S||S$; then

$$N_{E/F} J_E^S \subseteq J_F^S\,.$$

It is an immediate consequence of the definition of the norm $N_{E/F}$ and of Theorem 15:3 that the diagram

$$\begin{array}{ccc} \dot{E} & \xrightarrow{\text{nat}} & J_E \\ {\scriptstyle N_{E/F}}\downarrow & & \downarrow{\scriptstyle N_{E/F}} \\ \dot{F} & \xrightarrow[\text{nat}]{} & J_F \end{array}$$

is commutative. In other words, the norm of a field element is essentially the same as the norm of the principal idèle which it determines. In particular, $N_{E/F} P_E \subseteq P_F$.

For a tower of finite separable extensions $F \subseteq E \subseteq H$ of the global field F we find (after looking at a suitably elaborate commutative diagram) that

$$N_{H/F}\,\mathfrak{J} = N_{E/F}(N_{H/E}\mathfrak{J}) \quad \text{for all} \quad \mathfrak{J} \in J_H,$$

hence

$$N_{H/F} J_H \subseteq N_{E/F} J_E\,.$$

§ 65. Squares and norms in global fields

The purpose of this paragraph is to obtain information about squares and norms in global fields. This will be done by specializing the methods and results of global class field theory to quadratic extensions of global fields[1]. The proofs are of a technical nature involving several auxiliary computations and it would be too cumbersome to repeat our assumptions

[1] In the development of the global class field theory one obtains the main results of § 65 (with considerably more effort) for certain extensions of degree $n \geq 2$. See C. Chevalley, *Ann. Math.* **41** (1940), pp. 394—418.

in the enunciation of each proposition. Instead we shall list in the following introductory subparagraph some assumptions and conventions that will remain in force throughout § 65.

§ 65 A. Notation and conventions

F is a global field, $\Omega = \Omega_F$ is its set of non-trivial spots. Completions $F_\mathfrak{p}$ with normalized valuations $|\ |_\mathfrak{p}$ are taken and fixed at each $\mathfrak{p} \in \Omega$. S will denote a set of spots on F which satisfies the following properties:

(i) every $\mathfrak{p}$ in S is discrete and non-dyadic,
(ii) S consists of almost all spots on F,
(iii) S is not all of Ω,
(iv) $J_F = P_F J_F^S$.

Every global field possesses a set S with these four properties in virtue of Corollary 33:14a. The number of spots in $\Omega - S$ will be written s; so

$$1 \leqq s < \infty .$$

The group J_F^S is generated by the following subgroups in the following way:

$$J_F^S = K_F^S \Big(\prod_{\mathfrak{p} \in \Omega - S} I_F^\mathfrak{p} \Big) .$$

$\mathfrak{u}$ will denote the group of S-units $\mathfrak{u}(S)$ defined in § 33F, in other words

$$\mathfrak{u} = \{\alpha \in \dot{F} \mid |\alpha|_\mathfrak{p} = 1 \quad \text{for all} \quad \mathfrak{p} \in S\} .$$

For each discrete spot $\mathfrak{p} \in \Omega$ we let $\mathfrak{o}_\mathfrak{p}$ be the ring of integers and $\mathfrak{u}_\mathfrak{p}$ be the group of units of $F_\mathfrak{p}$ at $\mathfrak{p}$ (see § 13B). An element α of F (or of $F_\mathfrak{p}$) is said to be a square, or an integer, or a unit, etc., at $\mathfrak{p}$ if it is a square, integer, unit, etc., in $F_\mathfrak{p}$. An idèle $\mathfrak{i}$ is said to be a square, or an integer, or a unit, etc., at a spot $\mathfrak{p}$ if $\mathfrak{i}_\mathfrak{p}$ is a square, integer, unit, etc., at $\mathfrak{p}$. For example J_F^S consists of those elements of J_F which are units at all $\mathfrak{p}$ in S.

$J_F^{S,2}$ is defined to be the set of S-idèles which are squares at all $\mathfrak{p} \in \Omega - S$. This is a subgroup of J_F^S, and in fact

$$J_F^{S,2} = K_F^S (J_F^S)^2 .$$

We shall often use the fact that $\dot{F}_\mathfrak{p}^2$ is an open subset of $F_\mathfrak{p}$. At an archimedean $\mathfrak{p}$ this follows from the ordinary properties of the real and complex numbers, at a discrete $\mathfrak{p}$ it is Corollary 63:1b.

θ will denote an arbitrary non-square in F and E will be the quadratic extension $E = F(\sqrt{\theta})$ of F. This extension is separable since the characteristic is not 2, so the theory of § 15 can be applied to E/F. Bar will stand for conjugation in E:

$$\overline{\alpha + \beta \sqrt{\theta}} = \alpha - \beta \sqrt{\theta} \quad \forall\ \alpha, \beta \in F .$$

The mapping defined by $\Gamma \rightarrowtail \bar{\Gamma}$ is an automorphism of E, and this automorphism plus the identity automorphism of E are the two elements of the galois group of E/F. Thus

$$N_{E/F}\Gamma = \Gamma\bar{\Gamma} \quad \text{for all} \quad \Gamma \in E\,.$$

E is a global field with its own set of spots Ω_E. The valuations $|\ |_{\mathfrak{P}}$ will be normalized valuations on the fixed completions $E_{\mathfrak{P}}$ at each $\mathfrak{P} \in \Omega_E$. And S' will be the set of spots on E such that $S' || S$. We shall also use S to denote the set S'.

$\mathfrak{U}$ denotes the group of S-units on E, i. e. $\mathfrak{U} = \mathfrak{u}(S')$. For each discrete spot $\mathfrak{P} \in \Omega_E$ we let $\mathfrak{O}_{\mathfrak{P}}$ be the ring of integers and $\mathfrak{U}_{\mathfrak{P}}$ be the group of units of $E_{\mathfrak{P}}$ at $\mathfrak{P}$. For any α in $\mathfrak{U}$ we have

$$|N_{E/F}\,\alpha|_{\mathfrak{p}} = \prod_{\mathfrak{P}|\mathfrak{p}} |N_{\mathfrak{P}|\mathfrak{p}}\,\alpha|_{\mathfrak{p}} = \prod_{\mathfrak{P}|\mathfrak{p}} |\alpha|_{\mathfrak{P}} = 1$$

for each $\mathfrak{p}$ in S, hence

$$N_{E/F}\,\mathfrak{U} \subseteq \mathfrak{u} \subseteq \mathfrak{U}$$

(compare Example 15:6). If we consider the commutative diagram of § 64 we find

$$N_{E/F}\,P_E^S \subseteq P_F^S\,,$$

hence we have the commutative diagram

$$\begin{array}{ccc} \mathfrak{U} & \rightarrowtail & P_E^S \\ {\scriptstyle N_{E/F}}\downarrow & & \downarrow{\scriptstyle N_{E/F}} \\ \mathfrak{u} & \rightarrowtail & P_F^S \end{array}$$

Consider the spot $\mathfrak{p}$ on F. The extension E/F is certainly abelian, hence by § 15C the local degree $n(\mathfrak{P}|\mathfrak{p})$ is the same for all $\mathfrak{P}$ on E which divide $\mathfrak{p}$, and the common value of these local degrees is written $n_{\mathfrak{p}}$. If θ is a square at $\mathfrak{p}$, then there are exactly two distinct spots $\mathfrak{P}_1$ and $\mathfrak{P}_2$ on E which divide $\mathfrak{p}$ and we have

$$n(\mathfrak{P}_1|\mathfrak{p}) = n(\mathfrak{P}_2|\mathfrak{p}) = n_{\mathfrak{p}} = 1\,.$$

If θ is a non-square at $\mathfrak{p}$, then there is exactly one spot $\mathfrak{P}$ on E which divides $\mathfrak{p}$ and we have

$$n(\mathfrak{P}|\mathfrak{p}) = n_{\mathfrak{p}} = 2\,.$$

Similarly if $\mathfrak{p}$ is discrete we know that the ramification index $e(\mathfrak{P}|\mathfrak{p})$ depends only on $\mathfrak{p}$ and not on $\mathfrak{P}$, and it is written $e_{\mathfrak{p}}$. Of course $e_{\mathfrak{p}}$ is also the ramification index of the extension $E_{\mathfrak{P}}/F_{\mathfrak{P}}$. Similar remarks apply to the degree of inertia $f_{\mathfrak{p}} = f(\mathfrak{P}|\mathfrak{p})$. So if $n_{\mathfrak{p}} = 1$ we have

$$e_{\mathfrak{p}} = f_{\mathfrak{p}} = n_{\mathfrak{p}} = 1\,,$$

while if $n_{\mathfrak{p}} = 2$ we have

$$e_{\mathfrak{p}} = 1\,, \quad f_{\mathfrak{p}} = n_{\mathfrak{p}} = 2\,, \quad \text{or} \quad e_{\mathfrak{p}} = n_{\mathfrak{p}} = 2\,, \quad f_{\mathfrak{p}} = 1\,.$$

We say that E/F is unramified at the discrete spot $\mathfrak{p}$ if

$$e_{\mathfrak{p}} = 1, \quad f_{\mathfrak{p}} = n_{\mathfrak{p}} .$$

Thus E/F is unramified at $\mathfrak{p}$ if and only if $E_{\mathfrak{P}}/F_{\mathfrak{P}}$ is an unramified extension of local fields at each $\mathfrak{P}$ dividing $\mathfrak{p}$. Note that E/F is unramified at $\mathfrak{p}$ whenever the local degree $n_{\mathfrak{p}}$ is 1.

If $n_{\mathfrak{p}} = 2$ there is a single spot $\mathfrak{P}$ dividing $\mathfrak{p}$, so for any $\Gamma \in E$ we have $|\Gamma|_{\mathfrak{P}} = |N_{\mathfrak{P}|\mathfrak{p}}\, \Gamma|_{\mathfrak{p}} = |N_{E/F}\Gamma|_{\mathfrak{p}}$, hence

$$|\Gamma|_{\mathfrak{P}} = |\bar{\Gamma}|_{\mathfrak{P}} \quad \forall\, \Gamma \in E$$

On the other hand if $n_{\mathfrak{p}} = 1$ there will be two spots $\mathfrak{P}_1, \mathfrak{P}_2$ dividing $\mathfrak{p}$, and here it can be verified that

$$|\Gamma|_{\mathfrak{P}_1} = |\bar{\Gamma}|_{\mathfrak{P}_2} \quad \forall\, \Gamma \in E .$$

The set of spots $\Omega - S$ on F is broken up into two disjoint subsets A and B as follows:

$$A = \{\mathfrak{p} \in \Omega - S \mid n_{\mathfrak{p}} = 1\}$$
$$B = \{\mathfrak{p} \in \Omega - S \mid n_{\mathfrak{p}} = 2\} .$$

Then Ω is the disjoint union $\Omega = S \cup A \cup B$. We let a denote the number of spots in A, and b the number in B. We have

$$a \geqq 0, \quad b \geqq 0, \quad a + b = s .$$

We say that an element α of F (or of $F_{\mathfrak{p}}$) is a local norm of the extension E/F at the spot $\mathfrak{p}$ if

$$\alpha \in N_{\mathfrak{P}|\mathfrak{p}}\, E_{\mathfrak{P}} \quad \forall\, \mathfrak{P}|\mathfrak{p} .$$

Thus every element of $F_{\mathfrak{p}}$ is a local norm at $\mathfrak{p}$ when $n_{\mathfrak{p}} = 1$. In general we have the formula

$$(\dot{F}_{\mathfrak{p}} : N_{\mathfrak{P}|\mathfrak{p}}\, \dot{E}_{\mathfrak{P}}) = \begin{cases} 1 & \text{if } n_{\mathfrak{p}} = 1 \\ 2 & \text{if } n_{\mathfrak{p}} = 2 \end{cases}$$

for any $\mathfrak{P}|\mathfrak{p}$, in virtue of Corollary 63:13a. We easily see that $\alpha \in \dot{F}_{\mathfrak{p}}$ is a local norm at $\mathfrak{p}$ if and only if the Hilbert symbol on $F_{\mathfrak{p}}$ satisfies

$$\left(\frac{\alpha, \theta}{\mathfrak{p}}\right) = 1 .$$

We say that the idèle $\mathfrak{i}$ is a local norm at $\mathfrak{p}$ if $\mathfrak{i}_{\mathfrak{p}}$ is a local norm at $\mathfrak{p}$. Thus $\mathfrak{i}$ is a local norm at $\mathfrak{p}$ if and only if

$$\left(\frac{\mathfrak{i}_{\mathfrak{p}}, \theta}{\mathfrak{p}}\right) = 1 .$$

And we have the formula

$$(I_F^{\mathfrak{p}} : I_F^{\mathfrak{p}} \cap N_{E/F}\, J_E) = \begin{cases} 1 & \text{if } n_{\mathfrak{p}} = 1 \\ 2 & \text{if } n_{\mathfrak{p}} = 2. \end{cases}$$

65:1. **Example.** Let $\mathfrak{p}$ be a discrete non-dyadic spot at which θ is a unit. Then E/F is unramified at $\mathfrak{p}$ by Proposition 32:6. And by Example 63:16,

$$N_{\mathfrak{P}|\mathfrak{p}}\,\mathfrak{U}_{\mathfrak{P}} = \mathfrak{u}_{\mathfrak{p}}\ .$$

65:2. **Example.** Let $\mathfrak{i}$ be a given idèle in J_F. We claim that $\mathfrak{i}$ is in $N_{E/F}J_E$ if and only if it is a local norm at all $\mathfrak{p}$ for which $n_{\mathfrak{p}}=2$. The condition is clearly necessary by the definition of the norm of an idèle. Conversely let it be given that $\mathfrak{i}$ is a local norm at all $\mathfrak{p}$ for which $n_{\mathfrak{p}}=2$. At each of these $\mathfrak{p}$ take $\mathfrak{P}|\mathfrak{p}$ on E and $\mathfrak{J}_{\mathfrak{P}}\in E_{\mathfrak{P}}$ such that $N_{\mathfrak{P}|\mathfrak{p}}\,\mathfrak{J}_{\mathfrak{P}}=\mathfrak{i}_{\mathfrak{p}}$; then $|\mathfrak{J}_{\mathfrak{P}}|_{\mathfrak{P}}=|N_{\mathfrak{P}|\mathfrak{p}}\,\mathfrak{J}_{\mathfrak{P}}|_{\mathfrak{p}}=|\mathfrak{i}_{\mathfrak{p}}|_{\mathfrak{p}}$, so that $|\mathfrak{J}_{\mathfrak{P}}|_{\mathfrak{P}}=1$ for almost all these $\mathfrak{P}$. The remaining $\mathfrak{p}$ are the ones at which $n_{\mathfrak{p}}=1$; here there will be exactly two spots $\mathfrak{P},\mathfrak{P}'$ dividing $\mathfrak{p}$; take $\mathfrak{J}_{\mathfrak{P}}\in E_{\mathfrak{P}}$ such that $N_{\mathfrak{P}|\mathfrak{p}}\,\mathfrak{J}_{\mathfrak{P}}=\mathfrak{i}_{\mathfrak{p}}$ and $\mathfrak{J}_{\mathfrak{P}'}=1$; again we have $|\mathfrak{J}_{\mathfrak{P}}|_{\mathfrak{P}}=1$ for almost all these $\mathfrak{P}$. Hence $\mathfrak{J}=(\mathfrak{J}_{\mathfrak{P}})_{\mathfrak{P}\in\Omega_E}$ is an idèle of E, and $N_{E/F}\,\mathfrak{J}=\mathfrak{i}$. This proves our assertion. Incidentally note that the $\mathfrak{J}$ just constructed has the following property: $|\mathfrak{J}_{\mathfrak{P}}|_{\mathfrak{P}}$ is either $|\mathfrak{i}_{\mathfrak{p}}|_{\mathfrak{p}}$ or 1 whenever $\mathfrak{P}|\mathfrak{p}$.

65:3. **Example.** Next observe that $N_{E/F}\,J_E^S$ is the set of idèles $\mathfrak{i}$ in J_F^S such that $\mathfrak{i}$ is a local norm wherever $n_{\mathfrak{p}}=2$. For every element of $N_{E/F}J_E^S$ is an element of this sort by § 64 and Example 65:2. Conversely if $\mathfrak{i}$ is an idèle of this sort, then the idèle $\mathfrak{J}$ constructed in Example 65:2 is in J_E^S and has $N_{E/F}\,\mathfrak{J}=\mathfrak{i}$.

65:4. **Example.** Suppose θ is a unit at all $\mathfrak{p}$ in S. In this event $N_{E/F}J_E^S$ is the set of idèles $\mathfrak{i}$ in J_F^S such that $\mathfrak{i}$ is a local norm at all $\mathfrak{p}\in\Omega-S$ for which $n_{\mathfrak{p}}=2$. (Use Examples 65:1 and 65:3.)

§ 65 B. Index computations

65:5. *Let G be a subgroup of $\dot F$. Suppose that G is a finitely generated abelian group of rank r. Then $(G:G^2)$ is 2^{r+1} when -1 is in G, otherwise it is 2^r.*

Proof. By the Fundamental Theorem of Abelian Groups and one of the isomorphism theorems of group theory we find

$$(G:G^2)=(G_0:G_0^2)\,(G_1:G_1^2)\,\ldots\,(G_r:G_r^2)$$

where G_0 is a finite group consisting of the elements of finite order in G and where the remaining G_i are infinite cyclic groups. Hence

$$(G:G^2)=2^r\,(G_0:G_0^2)\ .$$

It remains for us to compute $(G_0:G_0^2)$. In order to do this we consider the surjective homomorphism $\varphi\colon\dot F\to\dot F^2$ defined by $\varphi x=x^2$. This has kernel (±1) since F is a field, and these two elements are distinct since the characteristic is not 2. Hence the kernel of the restriction $\varphi\colon G_0\to G_0^2$ is (±1) if -1 is in G, otherwise it is 1. In the first instance we obtain

$$(G_0:1)=2(G_0:(\pm1))=2(G_0^2:1)\ ,$$

and so $(G_0 : G_0^2) = 2$; in the second instance this argument gives $(G_0 : G_0^2) = 1$. Hence $(G : G^2)$ has the value stated. **q. e. d.**

65:6. $(P_F^S : (P_F^S)^2) = (\mathfrak{u} : \mathfrak{u}^2) = 2^s$.

Proof. The first equation is simply the fact that the natural isomorphism $\dot{F} \rightarrowtail P_F$ carries $\mathfrak{u}$ to P_F^S. The second equation follows from the Dirichlet Unit Theorem and Proposition 65:5. **q. e. d.**

65:7. $(J_F^S : J_F^{S,2}) = 4^s$.

Proof. 1) Consider the surjective homomorphism

$$J_F^S \to \prod_{\mathfrak{p} \in \Omega - S} \dot{F}_\mathfrak{p}$$

defined by suppressing the coordinates of an idèle at all spots in S. The kernel of this homomorphism is the group K_F^S, and $K_F^S \subseteq J_F^{S,2} \subseteq J_F^S$. Hence by the isomorphism theorems of group theory $(J_F^S : J_F^{S,2})$ is the product $\prod_{\mathfrak{p} \in \Omega - S} (\dot{F}_\mathfrak{p} : \dot{F}_\mathfrak{p}^2)$. We must compute this product.

2) In the function theoretic case all spots $\mathfrak{p} \in \Omega - S$ are discrete and non-dyadic, hence $(\dot{F}_\mathfrak{p} : \dot{F}_\mathfrak{p}^2) = 4$ by Proposition 63:9, hence the above product is 4^s as required.

3) In the number theoretic case if we let d, n, r, c denote the number of spots in $\Omega - S$ that are respectively dyadic, non-dyadic, real, complex, we obtain

$$\begin{aligned}
\prod_{\mathfrak{p} \in \Omega - S} (\dot{F}_\mathfrak{p} : \dot{F}_\mathfrak{p}^2) &= 4^d 4^n 2^r \prod_{\text{dyadic}} (N\mathfrak{p})^{\operatorname{ord}_\mathfrak{p} 2} \\
&= 4^{d+n} 2^r \prod_{\text{dyadic}} |2|_\mathfrak{p}^{-1} \\
&= 4^{d+n} 2^r \prod_{\text{arch}} |2|_\mathfrak{p} \\
&= 4^{d+n} 2^r 2^r 2^{2c} \\
&= 4^{d+n+r+c} \\
&= 4^s.
\end{aligned}$$

q. e. d.

65:8. *If θ is a unit at all spots of S, then* $(J_F^S : N_{E/F} J_E^S) = 2^b$.

Proof. Again consider the surjective homomorphism

$$J_F^S \to \prod_{\mathfrak{p} \in \Omega - S} \dot{F}_\mathfrak{p}$$

with kernel K_F^S. By Example 65:4 we have $K_F^S \subseteq N_{E/F} J_E^S \subseteq J_F^S$. Again by Example 65:4 we find that the image of $N_{E/F} J_E^S$ under the above homomorphism is

$$\prod_{\mathfrak{p} \in A} \dot{F}_\mathfrak{p} \times \prod_{\mathfrak{p} \in B} N_{\mathfrak{P}|\mathfrak{p}} \dot{E}_\mathfrak{P} .$$

So by the isomorphism theorems of group theory we obtain

$$(J_F^S : N_{E/F} J_E^S) = \prod_{\mathfrak{p} \in B} (\dot{F}_\mathfrak{p} : N_{\mathfrak{P}|\mathfrak{p}} \dot{E}_\mathfrak{P}) = 2^b .$$

q. e. d.

65:9. **Herbrand's lemma.** *Let Φ be a subgroup of finite index in the commutative group G. Consider homomorphisms N, T of G into G such that*

$$NTG = TNG = 1 , \quad N\Phi \subseteq \Phi , \quad T\Phi \subseteq \Phi .$$

Then

$$\frac{(G_T : NG)}{(G_N : TG)} = \frac{(\Phi_T : N\Phi)}{(\Phi_N : T\Phi)}$$

provided the indices on the right are finite. Here TG is the image of G, and G_T is the kernel of T.

Proof. Clearly $NG \subseteq G_T$ and $N\Phi \subseteq \Phi_T$. By Lemma 63:7 we find

$$\begin{aligned}(G : \Phi)\,(\Phi_T : N\Phi) &= (TG : T\Phi)\,(G_T : \Phi_T)\,(\Phi_T : N\Phi)\\ &= (TG : T\Phi)\,(G_T : N\Phi)\\ &= (TG : T\Phi)\,(G_T : NG)\,(NG : N\Phi) .\end{aligned}$$

Interchange N, T to obtain another such equation. Divide. **q. e. d.**

65:10. *Suppose θ is a unit at all $\mathfrak{p}$ in S. Then $b \geqq 1$ and*

$$(P_F^S : N_{E/F} P_E^S) = (\mathfrak{u} : N_{E/F} \mathfrak{U}) = 2^{b-1} .$$

Proof. 1) The first equation follows from the commutative diagram in § 65A. We therefore have to prove the second equation and the fact that $b \geqq 1$. Let $T : \dot{E} \rightarrowtail \dot{E}$ be the homomorphism defined by

$$T\Gamma = \Gamma / \bar{\Gamma} \quad \forall\, \Gamma \in \dot{E} .$$

Throughout this proof we use N for $N_{E/F}$. Then $N : \dot{E} \rightarrowtail \dot{E}$ is a homomorphism with

$$N\Gamma = \Gamma\bar{\Gamma} \quad \forall\, \Gamma \in \dot{E} .$$

We have

$$\begin{cases} Nx = x^2, \quad Tx = 1 & \forall\, x \in \dot{F} \\ Tx = x^2, \quad Nx = 1 & \forall\, x \in T\dot{E} , \end{cases}$$

hence

$$NT\dot{E} = TN\dot{E} = 1 .$$

Also

$$T\mathfrak{U} \subseteq \mathfrak{U} , \quad N\mathfrak{U} \subseteq \mathfrak{u} .$$

2) By the Dirichlet Unit Theorem we can express $\mathfrak{u}$ as a direct product of a finite group $\mathfrak{u}_0$ with a free commutative group $\mathfrak{u}_1$ of rank

$s-1$. We define the subgroup

$$\Phi = \mathfrak{u}_1 (T\,\mathfrak{U})$$

of $\mathfrak{U}$. It is easily verified that $\mathfrak{u}_1 \cap T\,\mathfrak{U} = 1$, hence Φ is the direct product of its subgroups $\mathfrak{u}_1$ and $T\,\mathfrak{U}$.

For any $\Gamma \in \mathfrak{U}$ we have $\Gamma^2 = (N\Gamma)\,(T\Gamma)$, hence

$$\mathfrak{U}^2 \subseteq (N\,\mathfrak{U})\,(T\,\mathfrak{U}) \subseteq \mathfrak{u}\,(T\,\mathfrak{U}) = \mathfrak{u}_0 \Phi\,.$$

Then

$$(\mathfrak{U} : \Phi) = (\mathfrak{U} : \mathfrak{u}_0 \Phi)\,(\mathfrak{u}_0 \Phi : \Phi) \leqq (\mathfrak{U} : \mathfrak{U}^2)\,(\mathfrak{u}_0 : \mathfrak{u}_0 \cap \Phi)\,.$$

Hence by Proposition 65:5 and the fact that $\mathfrak{u}_0$ is finite we obtain

$$(\mathfrak{U} : \Phi) < \infty\,.$$

3) So Φ is a finitely generated abelian group of finite index in the finitely generated abelian group $\mathfrak{U}$; from this it follows easily (using the fact that $\mathfrak{U}^i \subseteq \Phi \subseteq \mathfrak{U}$ for some i) that rank $\Phi =$ rank $\mathfrak{U}$. But $\mathfrak{U}$ has rank $a + s - 1$ by the Dirichlet Unit Theorem. Hence rank $\Phi = a + s - 1$. But rank $\mathfrak{u}_1 = s - 1$. Hence rank $T\,\mathfrak{U} = a$.

We are going to apply the Herbrand Lemma to the group $\mathfrak{U}$ and its subgroup of finite index Φ. All but the finiteness conditions are evident; so we have

$$\frac{(\mathfrak{U}_T : N\mathfrak{U})}{(\mathfrak{U}_N : T\mathfrak{U})} = \frac{(\Phi_T : N\Phi)}{(\Phi_N : T\Phi)}$$

provided the indices on the right turn out to be finite. Now $T\mathfrak{u}_1 = 1$; and if x is a typical element of $T\,\mathfrak{U}$ we have $Tx = x^2$, so $Tx = 1$ if and only if $x = \pm 1$; but $-1 = T(\sqrt{\theta}) \in T\,\mathfrak{U}$; hence $\Phi_T = (\pm 1)\,\mathfrak{u}_1$. Clearly $N\Phi = N\mathfrak{u}_1 = \mathfrak{u}_1^2 = \Phi_T^2$. Hence by Proposition 65:5 we have

$$(\Phi_T : N\Phi) = 2^s.$$

A similar argument gives $\Phi_N = T\,\mathfrak{U}$, $T\Phi = (T\,\mathfrak{U})^2$, and then

$$(\Phi_N : T\Phi) = 2^{a+1}\,.$$

Hence we do indeed have

$$(\mathfrak{U}_T : N\,\mathfrak{U}) = 2^{b-1}(\mathfrak{U}_N : T\,\mathfrak{U})\,.$$

But $\mathfrak{U}_T$ is clearly equal to $\mathfrak{u}$. So if we can prove that $(\mathfrak{U}_N : T\,\mathfrak{U}) = 1$ we will have $(\mathfrak{u} : N\,\mathfrak{U}) = 2^{b-1}$ and hence $b \geqq 1$.

4) It remains for us to prove that $\mathfrak{U}_N \subseteq T\,\mathfrak{U}$. So we must show that a typical H in $\mathfrak{U}_N$ is of the form $H = TZ$ for some Z in $\mathfrak{U}$. If $H \neq -1$ we have

$$T(1+H) = \frac{1+H}{1+\bar{H}} = H \cdot \frac{1+H}{H+1} = H$$

with $1 + H \in \dot{E}$, since $H\bar{H} = NH = 1$; if $H = -1$ we have $T(\sqrt{\theta}) = -1 = H$. So we do indeed have $Z \in \dot{E}$ with $H = TZ$, but this Z need not be in $\mathfrak{U}$ and a further adjustment is necessary.

For each $\mathfrak{p}$ in S we let $\mathfrak{P}'$ denote one of the spots on E which divide $\mathfrak{p}$; if there is another spot on E which divides $\mathfrak{p}$, let it be written $\mathfrak{P}''$; otherwise let $\mathfrak{P}''$ be $\mathfrak{P}'$. Now E/F is unramified at each $\mathfrak{p} \in S$ by Example 65:1, hence $|E|_{\mathfrak{P}'} = |F|_{\mathfrak{P}'}$, hence there is an $\alpha_\mathfrak{p}$ in F such that $|\alpha_\mathfrak{p}|_{\mathfrak{P}'} = |Z|_{\mathfrak{P}'}$. Now $J_F = P_F J_F^S$ by the general assumption being made throughout this paragraph, from which it follows easily that there is an α in F such that $|\alpha|_\mathfrak{p} = |\alpha_\mathfrak{p}|_\mathfrak{p}$ for all $\mathfrak{p}$ in S, and so $|\alpha|_{\mathfrak{P}'} = |\alpha_\mathfrak{p}|_{\mathfrak{P}'} = |Z|_{\mathfrak{P}'}$ for all $\mathfrak{P}'$. Put $Z_\alpha = Z/\alpha$. Then $|Z_\alpha|_{\mathfrak{P}'} = 1$ for all $\mathfrak{P}'$. For each $\mathfrak{P}''$ we have

$$|Z_\alpha|_{\mathfrak{P}''} = |\bar{Z}_\alpha|_{\mathfrak{P}'} = |\bar{Z}|_{\mathfrak{P}'}/|\alpha|_{\mathfrak{P}'};$$

but $TZ = H$ so $Z = H\bar{Z}$, so $|\bar{Z}|_{\mathfrak{P}'} = |Z|_{\mathfrak{P}'}$ since $H \in \mathfrak{U}$; hence $|Z_\alpha|_{\mathfrak{P}''} = 1$. Hence $Z_\alpha \in \mathfrak{U}$ and $TZ_\alpha = TZ = H$. **q. e. d.**

65:11[1]. *Let $\alpha \in F$ be a square at all spots in $\Omega - S$ and a unit at all spots in S. Then α is a square in F.*

Proof. Let θ denote any S-unit which is a non-square in F. Then our theory applies to the extension $E = F(\sqrt{\theta})$. So by Proposition 65:10 the set of spots $\mathfrak{p}$ in $\Omega - S$ at which $n_\mathfrak{p} = 2$ is non-empty. In other words, every S-unit θ which is a non-square in F is a non-square at one of the spots of Ω – S. Hence the given α must be a square in F. **q. e. d.**

65:11a. $P_F^S \cap J_F^{S,2} = (P_F^S)^2$.

65:12. $(J_F : P_F J_F^{S,2}) = (J_F^S : P_F^S J_F^{S,2}) = 2^s$.

Proof. $(J_F : P_F J_F^{S,2}) = (P_F J_F^S : P_F J_F^{S,2})$

$$\begin{aligned}
&= (J_F^S (P_F J_F^{S,2}) : P_F J_F^{S,2}) \\
&= (J_F^S : J_F^S \cap P_F J_F^{S,2}) \\
&= (J_F^S : P_F^S J_F^{S,2}) \\
&= (J_F^S : J_F^{S,2}) \div (P_F^S J_F^{S,2} : J_F^{S,2}) \\
&= 4^s \div (P_F^S : P_F^S \cap J_F^{S,2}) \\
&= 4^s \div (P_F^S : (P_F^S)^2) \\
&= 2^s.
\end{aligned}$$

q. e. d.

65:13. *Suppose that θ is a unit at all spots in S, and that $J_E = P_E J_E^S$. Then*

$$(J_F : P_F N_{E/F} J_E) = 2(P_F^S \cap N_{E/F} J_E^S : N_{E/F} P_E^S) < \infty .$$

---

[1] This result can be obtained from Corollary 65:18a.

Proof. Let N denote the norm $N_{E/F}$. We know from Proposition 65:8 that $(J_F^S : N J_E^S) = 2^b < \infty$. Hence

$$\begin{aligned}(J_F : P_F N J_E) &= (P_F J_F^S : P_F N(P_E J_E^S))\\ &= (P_F J_F^S : P_F N J_E^S)\\ &= (J_F^S (P_F N J_E^S) : P_F N J_E^S)\\ &= (J_F^S : J_F^S \cap P_F N J_E^S)\\ &= (J_F^S : P_F^S N J_E^S)\\ &= (J_F^S : N J_E^S) \div (P_F^S N J_E^S : N J_E^S)\\ &= 2^b \div (P_F^S : P_F^S \cap N J_E^S)\,.\end{aligned}$$

But

$$2^{b-1} = (P_F^S : N P_E^S) = (P_F^S : P_F^S \cap N J_E^S)\,(P_F^S \cap N J_E^S : N P_E^S).$$

Hence $(J_F : P_F N J_E)$ has the desired value. **q. e. d.**

65:14. $2 \leqq (J_F : P_F N_{E/F} J_E) < \infty$.

Proof. Take a small enough set S satisfying the general conditions of this paragraph and the special conditions of Proposition 65:13. **q. e. d.**

§ 65C. Squares

65:15. **Global Square Theorem.** *If an element of a global field F is a square at almost all spots on F, then it is a square in F.*

Proof. Suppose if possible that there is a non-square θ in F which is a square at almost all spots on F. Let S be a set of spots which satisfies all the conditions of this paragraph and is so small that θ is both a unit and a square at each of its spots. Let E be the quadratic extension $E = F(\sqrt{\theta})$. We shall deduce from these suppositions that $J_F \subseteq P_F N_{E/F} J_E$. This is of course impossible since $(J_F : P_F N_{E/F} J_E) \geqq 2$ by Proposition 65:14.

So consider a typical $\mathfrak{i}$ in J_F. By the Weak Approximation Theorem there is an α in F such that $\alpha \mathfrak{i}_{\mathfrak{p}}^{-1}$ is close to 1 at all $\mathfrak{p} \in \Omega - S$, and if the approximations are good enough we obtain $\alpha\, \mathfrak{i}_{\mathfrak{p}}^{-1} \in \dot{F}_{\mathfrak{p}}^2$ for all $\mathfrak{p} \in \Omega - S$ since $\dot{F}_{\mathfrak{p}}^2$ is an open subset of $F_{\mathfrak{p}}$. Then the idèle $(\alpha)^{-1}\mathfrak{i}$ is a local square, hence a local norm, at all $\mathfrak{p} \in \Omega - S$. But $n_{\mathfrak{p}} = 1$ for all $\mathfrak{p}$ in S by choice of S. Hence $(\alpha)^{-1}\mathfrak{i}$ is in $N_{E/F} J_E$ by Example 65:2. Hence $J_F \subseteq P_F N_{E/F} J_E$. This is the contradiction we are after. **q. e. d.**

We shall call the S-units $\varepsilon_1, \ldots, \varepsilon_r$ a set of generators of $\mathfrak{u}$ mod $\mathfrak{u}^2$ if every element of $\mathfrak{u}$ can be expressed in the form

$$\varepsilon_1^{\nu_1} \ldots \varepsilon_r^{\nu_r} \delta^2 \quad (\nu_i \in \mathsf{Z},\ \ \delta \in \mathfrak{u})\,;$$

clearly the condition $\nu_i \in \mathsf{Z}$ of this definition can be replaced by the condition $\nu_i = 0,1$. If $\varepsilon_1, \ldots, \varepsilon_r$ is a set of generators of $\mathfrak{u}$ mod $\mathfrak{u}^2$ we must have $r \geqq s$ since $(\mathfrak{u} : \mathfrak{u}^2) = 2^s$. And by the Dirichlet Unit Theorem there is always a set of generators consisting of exactly s elements. Suppose $\varepsilon_1, \ldots, \varepsilon_s$ is such a set. Then every element ε of $\mathfrak{u}$ can be expressed in the form

$$\varepsilon = \varepsilon_1^{\nu_1} \ldots \varepsilon_s^{\nu_s} \delta^2 \quad (\nu_i = 0,1; \quad \delta \in \mathfrak{u}) .$$

In fact this representation is unique (when the set of generators of $\mathfrak{u}$ mod $\mathfrak{u}^2$ consists of s elements). For if not we would have a representation $\varepsilon_1^{\mu_1} \ldots \varepsilon_{s-1}^{\mu_{s-1}} \varepsilon_s \in \mathfrak{u}^2$, say, and this would imply that the $s-1$ elements $\varepsilon_1, \ldots, \varepsilon_{s-1}$ formed a set of generators of $\mathfrak{u}$ mod $\mathfrak{u}^2$.

Consider a non-square ε in $\mathfrak{u}$. If $\varepsilon_1, \ldots, \varepsilon_s$ is a set of generators of $\mathfrak{u}$ mod $\mathfrak{u}^2$, then after a slight adjustment to the ε_i we can write $\varepsilon = \varepsilon_1 \varepsilon_2 \ldots \varepsilon_j$ for some $j \geqq 1$. So $\varepsilon, \varepsilon_2, \ldots, \varepsilon_s$ is also a set of generators of $\mathfrak{u}$ mod $\mathfrak{u}^2$. In other words, every non-square S-unit can be extended to a set of generators of $\mathfrak{u}$ mod $\mathfrak{u}^2$ which consists of exactly s elements.

We shall use $F(\sqrt{\mathfrak{u}})$ to denote the field obtained by adjoining the square roots of all S-units to F. Clearly $F(\sqrt{\mathfrak{u}}) = F(\sqrt{\varepsilon_1}, \ldots, \sqrt{\varepsilon_s})$ whenever $\varepsilon_1, \ldots, \varepsilon_s$ is a set of generators of $\mathfrak{u}$ mod $\mathfrak{u}^2$.

65:16. $[F(\sqrt{\mathfrak{u}}) : F] = 2^s$.

Proof. The proposition is a special case of the following abstract result: let K be any field of characteristic not 2, and let $\beta_1, \ldots, \beta_t$ be non-squares in K such that any product of distinct β_i is again a non-square in K; then

$$[K(\sqrt{\beta_1}, \ldots, \sqrt{\beta_t}) : K] = 2^t .$$

For $t = 1$ this result is trivial. Let us prove it by induction to t. If we can prove that any product of distinct $\beta_i (2 \leqq i \leqq t)$ is a non-square in $K(\sqrt{\beta_1})$ we shall have

$$[K(\sqrt{\beta_1}, \ldots, \sqrt{\beta_t}) : K(\sqrt{\beta_1})] = 2^{t-1}$$

and we shall be through. So consider an element x of the form

$$x = \beta_2^{\nu_2} \ldots \beta_t^{\nu_t} \quad (\nu_i = 0,1)$$

which is a square in $K(\sqrt{\beta_1})$. Then

$$x = (\alpha + \beta \sqrt{\beta_1})^2 = \alpha^2 + \beta^2 \beta_1 + 2 \alpha \beta \sqrt{\beta_1} \cdot$$

for some α, β in K. Then $\alpha \beta = 0$ since x is in K, hence either x is in K^2 or $x \beta_1$ is in K^2. In either event we have

$$\beta_1^{\mu_1} \beta_2^{\nu_2} \ldots \beta_t^{\nu_t} \in K^2$$

with $\mu_1 = 0$ or 1. Then $\nu_2 = \cdots = \nu_t = 0$ by hypothesis. Hence any product of distinct $\beta_i (2 \leqq i \leqq t)$ is a non-square in $K(\sqrt{\beta_1})$, as required. **q. e. d.**

65:17. *Let $\varepsilon_1, \ldots, \varepsilon_s$ be a set of generators of $\mathfrak{u}$ mod $\mathfrak{u}^2$. Then there are infinitely many spots $\mathfrak{p}$ in S such that*

$$\varepsilon_1 \notin F_{\mathfrak{p}}^2, \quad \varepsilon_i \in F_{\mathfrak{p}}^2 \quad (2 \leqq i \leqq s) .$$

Proof. If $s = 1$ apply the Global Square Theorem. Therefore assume that $s > 1$. Let H be the subfield $H = F(\sqrt{\varepsilon_2}, \ldots, \sqrt{\varepsilon_s})$ of $F(\sqrt{\mathfrak{u}})$. Then H is a global field and ε_1 is a non-square in H. By the Global Square Theorem there is an infinite number of spots $\mathfrak{P}$ on H such that $\varepsilon_1 \notin H_{\mathfrak{P}}^2$. These spots will induce an infinite number of spots on F. Consider a spot $\mathfrak{P}$ on H for which the induced spot $\mathfrak{p}$ on F is in S. Then $\varepsilon_1 \notin F_{\mathfrak{p}}^2$ since $\varepsilon_1 \notin H_{\mathfrak{P}}^2$. But for $2 \leqq i \leqq s$ the S-unit ε_i is a square in H, hence in $H_{\mathfrak{P}}$, hence $\varepsilon_1\varepsilon_i \notin H_{\mathfrak{P}}^2$, hence $\varepsilon_1\varepsilon_i \notin F_{\mathfrak{p}}^2$. So ε_1 and $\varepsilon_1\varepsilon_i$ are non-square units at the non-dyadic spot $\mathfrak{p}$; this implies that ε_i is a square at $\mathfrak{p}$ for $2 \leqq i \leqq s$.

q. e. d.

65:18. *Let $\varepsilon_1, \ldots, \varepsilon_s$ be a set of generators of* $\mathfrak{u} \bmod \mathfrak{u}^2$. *Then there is a set of spots $\mathfrak{p}_1, \ldots, \mathfrak{p}_s \in S$ with the following property: each ε_i is a non-square at $\mathfrak{p}_i$ and a square at the remaining $\mathfrak{p}_j$.*

Proof. This is an immediate consequence of Proposition 65:17.

q. e. d.

65:18a. *Let $\alpha \in F$ be a square at all spots in $\Omega - S$ and a unit at all spots in $S - (\mathfrak{p}_1 \cup \cdots \cup \mathfrak{p}_s)$. Then α is a square in F.*

Proof. Write $S' = S - (\mathfrak{p}_1 \cup \cdots \cup \mathfrak{p}_s)$. Suppose if possible that there is a non-square θ in F which is a square at all $\mathfrak{p} \in \Omega - S$ and a unit at all $\mathfrak{p} \in S'$. Put $E = F(\sqrt{\theta})$. These suppositions will lead us to the absurd conclusion that $J_F \subseteq P_F N_{E/F} J_E$; this conclusion is absurd since $(J_F : P_F N_{E/F} J_E) \geqq 2$ by Proposition 65:14. In virtue of the general assumption that $J_F = P_F J_F^S$ it will be enough if we show that $J_F^S \subseteq P_F N_{E/F} J_E$.

So consider a typical idèle $\mathfrak{i}$ in J_F^S. Take numbers $c_1, \ldots, c_s$ with $c_i = 0$ when $\mathfrak{i}$ is a square at $\mathfrak{p}_i$, and $c_i = 1$ when $\mathfrak{i}$ is a non-square at $\mathfrak{p}_i$. Let ε be the S-unit

$$\varepsilon = \varepsilon_1^{c_1} \ldots \varepsilon_s^{c_s}.$$

Then ε is a non-square at exactly those $\mathfrak{p}_i$ where $\mathfrak{i}$ is a non-square. But each $\mathfrak{p}_i$ is non-dyadic. Hence the idèle $(\varepsilon)\mathfrak{i}$ is a square at each $\mathfrak{p}_i$. If we apply Example 65:4 to this idèle and the set of spots S' we find that $(\varepsilon)\mathfrak{i}$ is in $N_{E/F} J_E$. Hence $\mathfrak{i} \in P_F N_{E/F} J_E$. q. e. d.

65:19. *Every element of a global field is a square at an infinite number of spots.*

Proof. Let F be the global field, ε the element. We can assume that ε is a non-square in F. Then ε is an S-unit for a sufficiently small set of spots S. Suppose $\Omega - S$ contains at least two spots. Regard ε as one of the generators in a set of generators of $\mathfrak{u} \bmod \mathfrak{u}^2$ and apply Proposition 65:17.

q. e. d.

65:20. *Let ε be an S-unit in F. Then there are distinct spots $\mathfrak{q}$ and $\mathfrak{q}'$ in S such that* (i) *ε is a square at $\mathfrak{q}$ and $\mathfrak{q}'$,* (ii) *any S-unit δ which is a square at $\mathfrak{q}$ is a square at $\mathfrak{q}'$, and conversely.*

Proof. We can assume that ε is a non-square in F, else we just replace it by one. If $s=1$ we let $\mathfrak{q}$ and $\mathfrak{q}'$ be any two spots of S at which ε is a square. So assume that $s>1$. Take generators $\varepsilon, \varepsilon_2, \ldots, \varepsilon_s$ of $\mathfrak{u}$ mod $\mathfrak{u}^2$. By Proposition 65:17 we can pick $\mathfrak{q} \in S$ in such a way that $\varepsilon, \varepsilon_2, \ldots, \varepsilon_{s-1}$ are squares at $\mathfrak{q}$ with ε_s a non-square at $\mathfrak{q}$. Pick a second spot $\mathfrak{q}'$ with the same properties as $\mathfrak{q}$. We have to show that any $\delta \in \mathfrak{u}$ which is a square at $\mathfrak{q}$ is also a square at $\mathfrak{q}'$. Write

$$\delta = \varepsilon^{\nu_1}\varepsilon_2^{\nu_2} \ldots \varepsilon_{s-1}^{\nu_{s-1}}\varepsilon_s^{\nu_s}\,\gamma$$

with $\nu_i = 0,1$ and γ in $\mathfrak{u}^2$. Then ε_s is the only quantity in the above expression which is not a square at $\mathfrak{q}$, hence $\nu_s = 0$. All the remaining quantities are squares at $\mathfrak{q}'$, hence δ is a square at $\mathfrak{q}'$. **q. e. d.**

§ 65D. Norms

65:21. $(J_F : P_F N_{E/F} J_E) = 2.$

Proof. 1) In virtue of Proposition 65:14 it is enough to prove that $(J_F : P_F N_{E/F} J_E) \leq 2$. Take a set of spots S on F which satisfies the general assumptions of this paragraph and is also so small that θ is a unit at each of its spots. We know from Proposition 65:12 that $(J_F : P_F J_F^{S,2}) = 2^s$ and from Example 65:4 that $P_F J_F^{S,2} \subseteq P_F N_{E/F} J_E$. The idea of the proof is now simply stated: put $\Phi_1 = P_F J_F^{S,2}$ and try to find a strictly ascending tower

$$\Phi_1 \subset \Phi_2 \subset \cdots \subset \Phi_s$$

of subgroups of $P_F N_{E/F} J_E$. If this can be achieved we shall have

$$\begin{aligned}(J_F : P_F N_{E/F} J_E) &= (J_F : P_F J_F^{S,2}) \div (P_F N_{E/F} J_E : P_F J_F^{S,2})\\ &\leq 2^s \div (\Phi_s : \Phi_1)\\ &\leq 2^s \div 2^{s-1} = 2\,.\end{aligned}$$

2) So we have to find the tower. Since θ is a non-square in $\mathfrak{u}$ we have a set of generators

$$\theta, \varepsilon_2, \ldots, \varepsilon_s$$

of $\mathfrak{u}$ mod $\mathfrak{u}^2$. Select a set of spots $\mathfrak{p}_1, \ldots, \mathfrak{p}_s$ as in Proposition 65:18 (with $\varepsilon_1 = \theta$). Then θ is a square at $\mathfrak{p}_2, \ldots, \mathfrak{p}_s$ and so $n_{\mathfrak{p}_j} = 1$ for $2 \leq j \leq s$. Hence $I_F^{\mathfrak{p}_j} \subseteq N_{E/F} J_E$ for $2 \leq j \leq s$ by Example 65:2. So if we define $\Phi_1 = P_F J_F^{S,2}$ and $\Phi_j = I_F^{\mathfrak{p}_j} \Phi_{j-1}$ for $2 \leq j \leq s$ we obtain a tower

$$\Phi_1 \subseteq \Phi_2 \subseteq \cdots \subseteq \Phi_s$$

of subgroups of $P_F N_{E/F} J_E$. It remains for us to prove that this tower is strictly increasing. If not we would have $I_F^{\mathfrak{p}_j} \subseteq \Phi_{j-1}$ for some $j\,(2 \leq j \leq s)$. Take an idèle $\mathfrak{i}$ in $I_F^{\mathfrak{p}_j}$ which is a prime element at $\mathfrak{p}_j$. Then $\mathfrak{i}$ is in Φ_{j-1}, hence it is in

$$I_F^{\mathfrak{p}_1} \ldots I_F^{\mathfrak{p}_{j-1}} (P_F J_F^{S,2})\,.$$

So we have an idèle $\mathfrak{k}$ which is a square at all $\mathfrak{p} \in \Omega - S$ and a unit at all $\mathfrak{p} \in S - (\mathfrak{p}_1 \cup \cdots \cup \mathfrak{p}_{j-1})$, and also a field element α, such that

$$\mathfrak{i} = (\alpha)\,\mathfrak{k}\,.$$

Comparing coordinates shows that α is a square at all $\mathfrak{p} \in \Omega - S$ and a unit at all $\mathfrak{p} \in S - (\mathfrak{p}_1 \cup \cdots \cup \mathfrak{p}_s)$. But then α is a square in F by Corollary 65:18a. And this is absurd since the equation $\mathfrak{i} = (\alpha)\,\mathfrak{k}$ also shows that α is a prime element at $\mathfrak{p}_j$. So we do indeed have $\Phi_{j-1} \subset \Phi_j$ as required. **q. e. d.**

65:22. *Suppose that θ is a unit at all spots in S, and that $J_E = P_E J_E^S$. Then $P_F^S \cap N_{E/F} J_E^S = N_{E/F} P_E^S$.*

Proof. This follows immediately from Propositions 65:13 and 65:21. **q. e. d.**

65:23. **Hasse Norm Theorem.** *Let E be a quadratic extension of the global field F. An element α of F is a norm in the extension E/F if and only if it is a local norm at all spots on F.*

Proof. We need only prove the sufficiency. Consider a non-zero element α of F which is a local norm at all spots on F. Take a set of spots S satisfying the general assumptions of this paragraph and so small that θ and α are units at all $\mathfrak{p} \in S$ with $J_E = P_E J_E^S$. Then the idèle (α) is in P_F^S; it is also in $N_{E/F} J_E^S$ by Example 65:3; so by Proposition 65:22

$$(\alpha) \in P_F^S \cap N_{E/F} J_E^S = N_{E/F} P_E^S\,.$$

Hence $\alpha \in N_{E/F}\,\mathfrak{U} \subseteq N_{E/F} E$. **q. e. d.**

§ 66. Quadratic forms over global fields

Now we can describe quadratic forms over global fields[1]. We consider a regular n-ary quadratic space V over a global field F. The corresponding bilinear and quadratic forms will be written B and Q. Completions $F_\mathfrak{p}$ are taken and fixed at each spot $\mathfrak{p}$ on F. We use $V_\mathfrak{p}$ for the $F_\mathfrak{p}$-ification $F_\mathfrak{p} V$ of V (as a vector space or as a quadratic space) and we call $V_\mathfrak{p}$ a $\mathfrak{p}$-ification or localization of V at $\mathfrak{p}$. Here $\mathfrak{p}$ can be archimedean or discrete. We say that V is isotropic at $\mathfrak{p}$ if its localization $V_\mathfrak{p}$ is isotropic; similarly we say that U and V are isometric at $\mathfrak{p}$ if their localizations are isometric. The Hasse symbol $S_\mathfrak{p} V_\mathfrak{p}$ will be written $S_\mathfrak{p} V$; its value can be computed directly from an orthogonal splitting $V \cong \langle a_1\rangle \perp \cdots \perp \langle a_n\rangle$ for V, say by the formula

$$S_\mathfrak{p} V = \prod_{1 \le i \le j \le n} \left(\frac{a_i, a_j}{\mathfrak{p}}\right),$$

where $\left(\frac{\alpha, \beta}{\mathfrak{p}}\right)$ denotes the Hilbert symbol over $F_\mathfrak{p}$.

[1] See N. Jacobson, *Bull. Am. Math. Soc.* (1940), pp. 264—268, for a reduction of the theory of hermitian forms to that of quadratic forms.

66:1. **Theorem.** *A regular quadratic space over a global field is isotropic if and only if it is isotropic at all spots on F.*

Proof. 1) We need only do the sufficiency. So consider the regular n-ary quadratic space V over the global field F. We are supposing that $n \geqq 2$ and that $V_{\mathfrak{p}}$ is isotropic at all spots $\mathfrak{p}$ on F. Fix an orthogonal basis $x_1, \ldots, x_n$ in which V has the splitting

$$V \cong \langle a_1 \rangle \perp \cdots \perp \langle a_n \rangle \quad (a_i \in F) .$$

2) The binary case. Here $d V_{\mathfrak{p}} = -1$, so $-a_1 a_2$ is a square at all $\mathfrak{p}$, hence $-a_1 a_2$ is a square in F by the Global Square Theorem, hence $d V = -1$, hence V is a hyperbolic plane, so it is isotropic.

3) The ternary case. By scaling we can assume that V has a splitting $V = L \perp P$ in which

$$L \cong \langle -\alpha \rangle , \quad P \cong \langle 1 \rangle \perp \langle -\theta \rangle .$$

We can suppose that θ is a non-square in F. Take localizations $L_{\mathfrak{p}}$, $P_{\mathfrak{p}}$ inside $V_{\mathfrak{p}}$ in the natural way. Since $V_{\mathfrak{p}}$ is isotropic we know from Proposition 42:11 that $P_{\mathfrak{p}}$ represents α at each $\mathfrak{p}$ on F. Hence the equation

$$\alpha = \xi_{\mathfrak{p}}^2 - \theta \eta_{\mathfrak{p}}^2 \quad (\xi_{\mathfrak{p}}, \eta_{\mathfrak{p}} \in F_{\mathfrak{p}})$$

has a solution at each $\mathfrak{p}$ on F. This implies that α is a local norm of the extension $E = F(\sqrt{\theta})$ over F at each spot $\mathfrak{p}$ on F. Hence by the Hasse Norm Theorem α is a norm in the extension E/F. So we have a solution

$$\alpha = \xi^2 - \theta \eta^2 \quad (\xi, \eta \in F) .$$

Hence V is isotropic[1].

4) The quaternary case. First suppose the discriminant $d V$ is 1. Take a regular ternary subspace U of V. Then $d V_{\mathfrak{p}}$ is 1 and $V_{\mathfrak{p}}$ is isotropic at all $\mathfrak{p}$ on F, hence $U_{\mathfrak{p}}$ is isotropic at all $\mathfrak{p}$ by Proposition 42:12. Hence U is isotropic by step 3). So V is isotropic and the case $d V = 1$ is settled.

Let us reduce the general quaternary case to the above. So assume V is quaternary with $d V \neq 1$. Form the quadratic extension $E = F(\sqrt{a_1 a_2 a_3 a_4})$ of F and consider the E-ification $E V$. For each spot $\mathfrak{P}$ on E the localization

$$(E V)_{\mathfrak{P}} = E_{\mathfrak{P}} x_1 \perp \cdots \perp E_{\mathfrak{P}} x_4$$

contains

$$F_{\mathfrak{P}} x_1 \perp \cdots \perp F_{\mathfrak{P}} x_4 ,$$

[1] Thus the binary and ternary cases are essentially interpretations of results of global class field theory. An examination of the proofs of § 66 will show that the rest of the global theory of quadratic forms is derived from these two cases by simple algebraic and arithmetic methods. Needless to say it would be of great interest and importance to have a direct proof of the entire theory. In the classical situation of the field of rational numbers $\mathbf{Q}$ the binary and ternary cases can be obtained by elementary methods, but then one runs into difficulty at $n = 4$, and this difficulty is usually overcome by using Dirichlet's theorem on primes in an arithmetic progression together with Hilbert's reciprocity law. For other approaches to the theory over $\mathbf{Q}$ we refer the reader to C. L. Siegel, *Am. J. Math.* (1941), pp. 658—680, and to J. W. S. Cassels, *Proc. Cam. Phil. Soc.* (1959), pp. 267—270.

and this latter space is easily seen to be isotropic since

$$V_{\mathfrak{p}} = F_{\mathfrak{p}} x_1 \perp \cdots \perp F_{\mathfrak{p}} x_4$$

is isotropic at the spot $\mathfrak{p}$ induced by $\mathfrak{P}$ on F. Hence $(EV)_{\mathfrak{P}}$ is isotropic for all $\mathfrak{P}$ on E. But $d(EV) = 1$. Hence EV is isotropic. Hence V is isotropic by Proposition 58:7.

5) Higher dimensional case. Here the proof is by induction to n. Assume $n \geqq 5$. Put

$$U = F x_1 \perp F x_2 , \quad W = F x_3 \perp \cdots \perp F x_n ,$$

so that $V = U \perp W$. Take the localizations

$$U_{\mathfrak{p}} = F_{\mathfrak{p}} x_1 \perp F_{\mathfrak{p}} x_2 , \quad W_{\mathfrak{p}} = F_{\mathfrak{p}} x_3 \perp \cdots \perp F_{\mathfrak{p}} x_n$$

inside $V_{\mathfrak{p}}$. Thus $V_{\mathfrak{p}} = U_{\mathfrak{p}} \perp W_{\mathfrak{p}}$. Put

$$T = \{\mathfrak{p} \in \Omega_F \mid W_{\mathfrak{p}} \text{ is anisotropic}\} .$$

The set T must be finite by Example 63:14. If T is empty we have W isotropic by the inductive hypothesis and we are through. So we consider a finite non-empty set T. There is a $\mu_{\mathfrak{p}}$ in $\dot{F}_{\mathfrak{p}}$ at each $\mathfrak{p}$ in T such that

$$\mu_{\mathfrak{p}} \in Q(U_{\mathfrak{p}}) , \quad -\mu_{\mathfrak{p}} \in Q(W_{\mathfrak{p}}) :$$

if $U_{\mathfrak{p}}$ is anisotropic this a consequence of the isotropy of $V_{\mathfrak{p}}$, otherwise it is a consequence of the universality of hyperbolic planes. So we have $\xi_{\mathfrak{p}}, \eta_{\mathfrak{p}} \in F_{\mathfrak{p}}$ at each $\mathfrak{p} \in T$ such that

$$Q(\xi_{\mathfrak{p}} x_1 + \eta_{\mathfrak{p}} x_2) = \xi_{\mathfrak{p}}^2 a_1 + \eta_{\mathfrak{p}}^2 a_2 = \mu_{\mathfrak{p}} .$$

Use the Weak Approximation Theorem to find ξ, η in F with ξ close to $\xi_{\mathfrak{p}}$ and η close to $\eta_{\mathfrak{p}}$ at each $\mathfrak{p} \in T$. Put $Q(\xi x_1 + \eta x_2) = \mu$. By taking good enough approximations we can make μ arbitrarily close to $\mu_{\mathfrak{p}}$, hence we can make $|\mu \mu_{\mathfrak{p}}^{-1} - 1|_{\mathfrak{p}}$ arbitrarily small at all $\mathfrak{p}$ in T. Since $\dot{F}_{\mathfrak{p}}^2$ is an open subset of $F_{\mathfrak{p}}$ we are sure that we can obtain

$$\mu \in \mu_{\mathfrak{p}} \dot{F}_{\mathfrak{p}}^2 \quad \forall\, \mathfrak{p} \in T .$$

Since $\xi x_1 + \eta x_2$ is in U we have

$$V \cong \langle \mu' \rangle \perp \langle \mu \rangle \perp W .$$

Then $\langle \mu \rangle \perp W$ is isotropic at all $\mathfrak{p}$ in T since then

$$-\mu \in -\mu_{\mathfrak{p}} \dot{F}_{\mathfrak{p}}^2 \subseteq Q(W_{\mathfrak{p}}) ,$$

and it is isotropic at all the remaining spots on F by choice of T. Hence $\langle \mu \rangle \perp W$ is isotropic. Hence V is isotropic. **q. e. d.**

66:2. **Theorem.** *An n-ary quadratic space over a function field is isotropic whenever $n \geqq 5$.*

Proof. The quadratic space in question is isotropic at all spots on the function field by Proposition 63:19, hence it is isotropic by Theorem 66:1.

q. e. d.

66:3. **Theorem.** *U and V are regular quadratic spaces over the global field F. Then U is represented by V if and only if $U_\mathfrak{p}$ is represented by $V_\mathfrak{p}$ for all $\mathfrak{p}$.*

Proof. Suppose V represents an $\alpha \in \dot{F}$ at all spots $\mathfrak{p}$. Then $<-\alpha> \perp V$ is isotropic at all $\mathfrak{p}$, hence it is isotropic by Theorem 66:1, hence V represents α. Therefore V represents α whenever it does so all spots on F. This proves the theorem in the case where $\dim U = 1$. We proceed by induction on $\dim U$. Take a non-zero α in $Q(U)$. Then $\alpha \in Q(U_\mathfrak{p}) \subseteq Q(V_\mathfrak{p})$, hence α is represented by V at all spots, hence it is represented by V. So we have splittings

$$U = L \perp U', \quad V = K \perp V'$$

in which $L \cong <\alpha>$ and $K \cong <\alpha>$. It follows easily from Witt's theorem and the representation $U_\mathfrak{p} \rightarrowtail V_\mathfrak{p}$ that $U'_\mathfrak{p} \rightarrowtail V'_\mathfrak{p}$ for all $\mathfrak{p}$. Hence $U' \rightarrowtail V'$ by the induction. Hence $U \rightarrowtail V$. **q. e. d.**

66:4. **Hasse-Minkowski Theorem.** *U and V are regular quadratic spaces over the global field F. Then U is isometric to V if and only if $U_\mathfrak{p}$ is isometric to $V_\mathfrak{p}$ for all $\mathfrak{p}$.*

Proof. By Theorem 66:3 there exists a representation $U \rightarrowtail V$. Since U is regular this representation is an isometry. **q. e. d.**

66:5. **Remark.** We have just shown that an arbitrary quadratic space V over a global field F is completely described by its local behavior at the spots of F. Using our earlier descriptions at the discrete and archimedean completions we find the following complete set of invariants for V:

(1) the dimension $\dim V$,
(2) the discriminant dV,
(3) the Hasse symbols $S_\mathfrak{p} V$ at all discrete $\mathfrak{p}$,
(4) the positive indices $\text{ind}_\mathfrak{p}^+ V$ at all real $\mathfrak{p}$.

Here $\text{ind}_\mathfrak{p}^+ V$ is used for the positive index of the localization $V_\mathfrak{p}$ of V at $\mathfrak{p}$. Of course the fourth invariant can be omitted over function fields.

66:6. **Example.** Let V be expressed in the form $V \cong <a_1> \perp \cdots \perp <a_n>$. By the Product Formula we can find a finite set of spots T on F which contains all archimedean and dyadic spots and is such that each a_i is a unit at all $\mathfrak{p} \in \Omega - T$. What is the Hasse symbol $S_\mathfrak{p} V$ at a spot $\mathfrak{p}$ in $\Omega - T$? By Example 63:12 we have $\left(\frac{a_i, a_j}{\mathfrak{p}}\right) = 1$, hence $S_\mathfrak{p} V = 1$, at each $\mathfrak{p}$ in $\Omega - T$. So in practice one has to check the Hasse invariant at only a finite number of spots. This example also shows that $V_\mathfrak{p}$ is isotropic at almost all $\mathfrak{p}$ when $n \geqq 3$.

Chapter VII

Hilbert's Reciprocity Law

The Hilbert Reciprocity Law states that

$$\prod_{\mathfrak{p}} \left(\frac{\alpha, \beta}{\mathfrak{p}}\right) = 1 .$$

The major portion of this chapter is devoted to the proof of this formula for algebraic number fields. The formula is actually true over any global field, but we shall not go into the function theoretic case here. The Hilbert Reciprocity Law gives a reciprocity law for Hasse symbols, namely

$$\prod_{\mathfrak{p}} S_{\mathfrak{p}} V = 1 ,$$

and this can be regarded as a dependence relation among the invariants of the quadratic space V. We shall investigate the full extent of this dependence in § 72.

§ 71. Proof of the reciprocity law

Our sole purpose in this paragraph is to prove Hilbert's reciprocity law. Here, as in § 65, we proceed by specializing the methods and results of global class field theory[1], which concerns itself with abelian extensions E/F, to the case of quadratic extensions E/F. Throughout § 71 we shall assume that

(i) F is an algebraic number field, Ω is the set of non-trivial spots on F, $|\ |_{\mathfrak{p}}$ is the normalized valuation at $\mathfrak{p}$,

(ii) θ is a given non-square in F,

(iii) E is the quadratic extension $E = F(\sqrt{\theta})$ of F.

Thus the assumptions about E/F are those of § 65, and in addition F is now an algebraic number field and not just an arbitrary global field. The idèle notation is the same as before.

§ 71A. Cyclotomic preliminaries

Let us recall some facts about roots of unity. Here we limit ourselves to the special case of the algebraic number field F under discussion; so the group of roots of unity is a finite cyclic subgroup of $\dot{F}$. We know that ζ is called an m-th root of unity if $\zeta^m = 1$. We call ζ a primitive m-th root of unity if it has period m. In this event

$$1, \zeta, \ldots, \zeta^{m-1}$$

[1] For a discussion of the general theory and of Artin's general law of reciprocity we refer the reader to H. FLANDERS, *Unification of class field theory*, (University of Chicago thesis, 1949). Also see E. ARTIN and J. TATE, *Class field theory* (Harvard University, 1960).

is a cyclic group of m distinct m-th roots of unity; but the number of distinct m-th roots of unity is at most m since $x^m - 1 = 0$ has at most m solutions in a field. Hence a primitive m-th root of unity generates all the m-th roots of unity in a field. If F contains m distinct m-th roots of unity it contains a primitive m-th root of unity.

71:1. **Definition.** We call an extension K/F an absolute m-cyclotomic extension if $K = F(\zeta)$ for some primitive m-th root of unity ζ.

71:2. **Definition.** We call an extension K/F an m-cyclotomic extension if it is a subextension of an absolute m-cyclotomic extension.

71:3. **Example.** The extension K/F is an absolute m-cyclotomic extension if and only if K is a splitting field of the polynomial $x^m - 1$ over F. In particular, F has an absolute m-cyclotomic extension for any natural number m.

71:4. **Example.** Let H be an extension field of F which contains a primitive m-th root of unity ζ_m and a primitive n-th root of unity ζ_n, with m relatively prime to n. Then $\zeta_m \zeta_n$ is a primitive mn-th root of unity. And $F(\zeta_m, \zeta_n) = F(\zeta_m \zeta_n)$.

71:5. **Example.** A compositum of an absolute m-cyclotomic extension with an absolute n-cyclotomic extension is an absolute mn-cyclotomic extension when m and n are relatively prime.

71:6. **Example.** Let C be an algebraically closed field containing F. Then C contains exactly one absolute m-cyclotomic extension of F, namely the splitting field H in C of $x^m - 1$ over F. If K/F is any m-cyclotomic subextension of C/F, then $F \subseteq K \subseteq H$. So a compositum of cyclotomic extensions is always cyclotomic. And if L is any subfield of C, then KL/FL is m-cyclotomic.

The absolute m-cyclotomic extension $K = F(\zeta)$ with ζ a primitive m-th root is a splitting field of the polynomial $x^m - 1$ over F, hence K/F is a galois extension. A typical element ϱ of the galois group $\mathfrak{G}(K/F)$ is completely determined by its action on ζ. But $\varrho\zeta$ is an m-th root of unity, hence $\varrho\zeta = \zeta^r$ for some r. Similarly $\tau\zeta = \zeta^t$. But then

$$\varrho\tau\zeta = \zeta^{tr} = \tau\varrho\zeta .$$

Hence $\varrho\tau = \tau\varrho$. Hence the galois group of an absolute cyclotomic extension is abelian. Hence every cyclotomic extension is galois and abelian.

Consider an arbitrary finite extension H/F of F with discrete spots $\mathfrak{P}|\mathfrak{p}$. We say that H/F is unramified at $\mathfrak{P}$ (or at $\mathfrak{P}|\mathfrak{p}$) if $H_\mathfrak{P}/F_\mathfrak{p}$ is an unramified extension of local fields at $\mathfrak{P}|\mathfrak{p}$, i. e. if

$$e(\mathfrak{P}|\mathfrak{p}) = 1 , \quad f(\mathfrak{P}|\mathfrak{p}) = n(\mathfrak{P}|\mathfrak{p}) .$$

(Of course $e(\mathfrak{P}|\mathfrak{p}) f(\mathfrak{P}|\mathfrak{p}) = n(\mathfrak{P}|\mathfrak{p})$ holds whether or not H/F is unramified at $\mathfrak{P}$.) We say that H/F is unramified at $\mathfrak{p}$ if it is unramified at all $\mathfrak{P}$ dividing $\mathfrak{p}$. If H/F is abelian, then by § 15C the local degree $n(\mathfrak{P}|\mathfrak{p})$, the ramification index $e(\mathfrak{P}|\mathfrak{p})$, and the degree of inertia $f(\mathfrak{P}|\mathfrak{p})$, depend only on $\mathfrak{p}$ and are written $n_\mathfrak{p}$, $e_\mathfrak{p}$ and $f_\mathfrak{p}$ respectively. So an abelian extension H/F is unramified at $\mathfrak{p}$ if and only if

$$e_\mathfrak{p} = 1\,, \quad f_\mathfrak{p} = n_\mathfrak{p}\,,$$

i. e. if and only if it is unramified at some $\mathfrak{P}$ dividing $\mathfrak{p}$. (Of course $e_\mathfrak{p} f_\mathfrak{p} = n_\mathfrak{p}$ holds whether or not H/F is unramified at $\mathfrak{p}$.) If H/F is quadratic we recover the special definitions given in § 65A. All this applies to cyclotomic extensions since they are abelian. For example we see by Corollary 32:6a that an m-cyclotomic extension is unramified at all discrete spots at which m is a unit.

71:7. *Let H be a finite extension of an algebraic number field F. Then H/F is unramified at almost all discrete spots $\mathfrak{P}$ on H.*

Proof. Take $H = F(\delta)$ and let $f(x) = \operatorname{irr}(x, \delta, F)$. Then by the Product Formula we know that $f(x)$ is integral and $f'(\delta)$ is a unit at almost all discrete spots $\mathfrak{P}$ on H. For any one of these $\mathfrak{P}$ the extension $H_\mathfrak{P}/F_\mathfrak{P}$ is unramified by Proposition 32:6. **q. e. d.**

71:8. *The absolute p-cyclotomic extensions of F are of degree $p - 1$ over F for almost all prime numbers p.*

Proof. Let Q denote the prime field of F. We shall consider the set of discrete spots $\mathfrak{p}$ on F for which $F_\mathfrak{p}/Q_\mathfrak{p}$ is unramified; this set consists of almost all the discrete spots on F by Proposition 71:7. Each of these spots $\mathfrak{p}$ induces a p-adic spot on Q, and almost all p-adic spots turn up in this way.

Consider any one such p-adic spot. Let it be induced by $\mathfrak{p}$ on F. We claim that $K = F(\zeta)$ with ζ a primitive p-th root of unity has degree $p - 1$ over F. To prove this we take a spot $\mathfrak{P}$ on K which induces the spot $\mathfrak{p}$ on F. We form $Q_\mathfrak{P} \subseteq F_\mathfrak{P} \subseteq K_\mathfrak{P}$. Then $\zeta - 1$ is a root in $K_\mathfrak{P}$ of the polynomial

$$\frac{(y+1)^p - 1}{y}$$

over $Q_\mathfrak{P}$; and this is easily seen to be an Eisenstein polynomial in the sense of Proposition 32:15; hence $Q_\mathfrak{P}(\zeta - 1)$ has ramification index $p - 1$ over $Q_\mathfrak{P}$. So the ramification index of $K_\mathfrak{P}/Q_\mathfrak{P}$ is at least $p - 1$. But $F_\mathfrak{P}/Q_\mathfrak{P}$ is unramified by choice of $\mathfrak{p}$. Hence the ramification index of $K_\mathfrak{P}/F_\mathfrak{P}$ is at least $p - 1$. So

$$[K:F] \geqq [K_\mathfrak{P}:F_\mathfrak{P}] \geqq p - 1\,.$$

Hence $[K:F] = p - 1$. **q. e. d.**

71:9. *H is a finite extension of F, and C is an algebraic closure of H. Then for almost all prime numbers p, every p-cyclotomic subextension*

K/F of C/F satisfies

$$[KH:H] = [K:F] .$$

Proof. The absolute p-cyclotomic extensions of H are of degree $p-1$ over H for almost all prime numbers p. We claim that a p-cyclotomic extension K corresponding to any such p satisfies the given equation. Take a primitive p-th root of unity ζ in C. Then $K \subseteq F(\zeta) \subseteq C$. We have $[H(\zeta):H] = p-1$ by choice of p. Hence $[F(\zeta):F] = p-1$. Then

$$\begin{aligned} p-1 &= [H(\zeta):H] \\ &= [H(\zeta):HK]\,[HK:H] \\ &\leqq [F(\zeta):K]\,[HK:H] \\ &\leqq [F(\zeta):K]\,[K:F] \\ &= p-1 . \end{aligned}$$

Hence we must have $[KH:H] = [K:F]$, as required. **q. e. d.**

§ 71B. Two cyclotomic special cases

Here we prove the Hilbert Reciprocity Law for two special cases (Propositions 71:12 and 71:13) that will be used in the general proof. We consider an absolute m-cyclotomic extension $K = F(\zeta)$ with ζ a primitive m-th root of unity. Assume $m > 1$. Let S denote the set of all discrete spots on F at which m is a unit. So K/F is unramified at all $\mathfrak{p}$ in S. We wish to define a Frobenius automorphism $\sigma_{\mathfrak{p}}$ in $\mathfrak{G}(K/F)$ at each $\mathfrak{p}$ in S. We take a spot $\mathfrak{P}$ on K which divides $\mathfrak{p}$. Then $K_{\mathfrak{P}}/F_{\mathfrak{P}}$ is unramified since $\mathfrak{p}$ is in S, hence by Proposition 32:12 and its corollary there is an automorphism $\sigma_{\mathfrak{p}} \in \mathfrak{G}(K_{\mathfrak{P}}/F_{\mathfrak{P}})$ called the Frobenius automorphism such that $\sigma_{\mathfrak{p}}\zeta = \zeta^{N\mathfrak{p}}$. If we now use $\sigma_{\mathfrak{p}}$ to denote the restriction of $\sigma_{\mathfrak{p}}$ to K we see that at each $\mathfrak{p}$ in S we have found an element $\sigma_{\mathfrak{p}} \in \mathfrak{G}(K/F)$ with the property that $\sigma_{\mathfrak{p}}\zeta = \zeta^{N\mathfrak{p}}$. There is clearly just one such element $\sigma_{\mathfrak{p}}$. And it is independent of the choice of the primitive m-th root of unity ζ in K. We call this $\sigma_{\mathfrak{p}}$ the Frobenius automorphism of K/F at $\mathfrak{p}\,(\mathfrak{p} \in S)$.

For each $\alpha \in \dot{F}$ which is integral at all $\mathfrak{p}$ in S we define

$$\sigma_\alpha = \prod_{\mathfrak{p} \in S} \sigma_{\mathfrak{p}}^{\operatorname{ord}_{\mathfrak{p}} \alpha} .$$

This product makes sense since $\operatorname{ord}_{\mathfrak{p}} \alpha$ is almost always 0 and since $\mathfrak{G}(K/F)$ is commutative.

71:10. **Formula.** *Let β be any element of $\dot{F}$ with $\sqrt{\beta}$ in K. Then*

$$\sigma_\alpha(\sqrt{\beta}) = \sqrt{\beta} \prod_{\mathfrak{p} \in S} \left(\frac{\alpha, \beta}{\mathfrak{p}}\right) .$$

Proof. It is enough to prove

$$\sigma_{\mathfrak{p}}^{\mathrm{ord}_{\mathfrak{p}}\alpha}(\sqrt{\beta}) = \left(\frac{\alpha, \beta}{\mathfrak{p}}\right)\sqrt{\beta}$$

for each $\mathfrak{p}$ in S. Take a $\mathfrak{P}$ on K dividing the $\mathfrak{p}$ under consideration and consider the Frobenius automorphism $\sigma_{\mathfrak{p}}$ on $K_{\mathfrak{P}}/F_{\mathfrak{P}}$. The restriction of $\sigma_{\mathfrak{p}}$ to K gives the Frobenius automorphism $\sigma_{\mathfrak{p}}$ on K/F. If β is a square at $\mathfrak{p}$ we have $\sqrt{\beta} \in F_{\mathfrak{P}}$ and so

$$\sigma_{\mathfrak{p}}^{\mathrm{ord}_{\mathfrak{p}}\alpha}(\sqrt{\beta}) = \sqrt{\beta} = \left(\frac{\alpha, \beta}{\mathfrak{p}}\right)\sqrt{\beta}$$

as required. So let us assume that β is not a square at $\mathfrak{p}$. Then $\sqrt{\beta}$ is not in the fixed field $F_{\mathfrak{P}}$ of $\sigma_{\mathfrak{p}}$, hence $\sigma_{\mathfrak{p}}(\sqrt{\beta}) = -\sqrt{\beta}$. Hence

$$\sigma_{\mathfrak{p}}^{\mathrm{ord}_{\mathfrak{p}}\alpha}(\sqrt{\beta}) = \varepsilon\sqrt{\beta}$$

with $\varepsilon = \pm 1$, the sign depending upon whether $\mathrm{ord}_{\mathfrak{p}}\alpha$ is even or odd. Furthermore for this same number ε we have $\left(\frac{\alpha, \beta}{\mathfrak{p}}\right) = \varepsilon$ by Examples 63:10 and 63:16 since $F_{\mathfrak{P}}(\sqrt{\beta})/F_{\mathfrak{P}}$ is unramified. Hence our formula is true. **q. e. d.**

71:11. **Formula.** *Every α in $\dot{F}$ which is integral at all spots in S and is sufficiently close to 1 at all discrete spots in $\Omega - S$ will satisfy*

$$\sigma_\alpha \zeta = \zeta^{(-1)^r}$$

where r is the number of real spots at which α is negative.

Proof. Let Q denote the prime field of F, let Z denote the rational integers of Q, let T_0 denote the set of discrete spots p on Q at which $|m|_p < 1$, let T denote the set of discrete spots in $\Omega - S$. Thus $T \,||\, T_0$. By Proposition 15:4 and its corollary we know that any α in F which is sufficiently close to 1 at all $\mathfrak{p}$ in T will be a unit at these spots and will satisfy

$$|N_{F/Q}\alpha - 1|_p < |m|_p \quad \text{for all } p \in T_0 .$$

So consider any α which is this close to 1 at all $\mathfrak{p}$ in T, and suppose that α is also an integer at all $\mathfrak{p}$ in S. Then $N_{F/Q}\alpha \in Z$ by Example 15:6. And so

$$N_{F/Q}\alpha \equiv 1 \bmod mZ .$$

Hence by Proposition 15:5, $(-1)^r N_{F/Q}\alpha$ is a natural number in Z which satisfies

$$(-1)^r N_{F/Q}\alpha \equiv (-1)^r \bmod mZ .$$

Hence

$$|N_{F/Q}\alpha|_\infty \equiv (-1)^r \bmod m\mathbf{Z} ,$$

where $|\ |_\infty$ is the ordinary absolute value on Q. Then

$$\sigma_\alpha \zeta = \left(\prod_{\mathfrak{p} \in S} \sigma_{\mathfrak{p}}^{\mathrm{ord}_{\mathfrak{p}}\alpha}\right)(\zeta) = \zeta^i$$

where

$$\begin{aligned} i &= \prod_{\mathfrak{p}\in S} (N\mathfrak{p})^{\mathrm{ord}_{\mathfrak{p}}\alpha} \\ &= \prod_{\mathfrak{p}\in S} 1/|\alpha|_{\mathfrak{p}} \\ &= \prod_{\text{arch}} |\alpha|_{\mathfrak{p}} \\ &= \prod_{\text{arch}} |N_{\mathfrak{p}|\infty}\alpha|_{\infty} \\ &= |N_{F/Q}\alpha|_{\infty} \\ &\equiv (-1)^r \bmod m\mathbf{Z}\,. \end{aligned}$$

So $\sigma_\alpha \zeta = \zeta^{(-1)^r}$ since ζ is an m-th root of unity. **q. e. d.**

71:12. *Suppose that the quadratic extension E/F is actually m-cyclotomic. Let $\mathfrak{q}$ be a discrete spot at which m is a unit. Then*

$$I_F^{\mathfrak{q}} \subseteq P_F N_{E/F} J_E$$

if and only if E/F has local degree 1 at $\mathfrak{q}$.

Proof. 1) If E/F has local degree 1 at $\mathfrak{q}$ we apply Example 65:2 and find $I_F^{\mathfrak{q}} \subseteq P_F N_{E/F} J_E$. Conversely we must deduce from this inclusion relation that E/F has local degree 1 at $\mathfrak{q}$. We suppose not, i. e. we suppose that the local degree at $\mathfrak{q}$ is equal to 2, and we use this to get a contradiction. We can take $F \subseteq E \subseteq K$ where $K = F(\zeta)$ with ζ a primitive m-th root of unity. We let S denote the set of discrete spots on F at which m is a unit. The preceding discussion now applies to this situation. Note that $\mathfrak{q} \in S$.

Since the local degree is 2 at $\mathfrak{q}$ we have $I_F^{\mathfrak{q}} \cap N_{E/F} J_E \subset I_F^{\mathfrak{q}}$ by § 65A, hence there is an idèle $\mathfrak{i}$ in $I_F^{\mathfrak{q}}$ which is not a local norm at $\mathfrak{q}$. But we are assuming that $I_F^{\mathfrak{q}} \subseteq P_F N_{E/F} J_E$, hence there is an α in $\dot{F}$ which is a local norm at all $\mathfrak{p} \in \Omega - \mathfrak{q}$, but not a local norm at $\mathfrak{q}$.

2) So by § 65A again we have an α in F such that

$$\left(\frac{\alpha, \theta}{\mathfrak{p}}\right) = \begin{cases} +1 & \text{if } \mathfrak{p} \in \Omega - \mathfrak{q} \\ -1 & \text{if } \mathfrak{p} = \mathfrak{q}. \end{cases}$$

We must refine this α. At each $\mathfrak{p} \in \Omega - S$ we can write

$$\alpha = a_{\mathfrak{p}}^2 - \theta b_{\mathfrak{p}}^2 \quad (a_{\mathfrak{p}}, b_{\mathfrak{p}} \in F_{\mathfrak{p}})\,.$$

Take approximations $a, b \in F$ to the $a_{\mathfrak{p}}, b_{\mathfrak{p}} \in F_{\mathfrak{p}}$ at all $\mathfrak{p}$ in $\Omega - S$. If these approximations are good enough we obtain an element $\alpha_0 = a^2 - \theta b^2$ of F which is arbitrarily close to α at all $\mathfrak{p}$ in $\Omega - S$. Hence we can make $\alpha_0 \alpha^{-1}$ arbitrarily close to 1 at all $\mathfrak{p} \in \Omega - S$. And

$$\left(\frac{\alpha_0\alpha^{-1}, \theta}{\mathfrak{p}}\right) = \left(\frac{\alpha_0, \theta}{\mathfrak{p}}\right)\left(\frac{\alpha, \theta}{\mathfrak{p}}\right) = \left(\frac{\alpha, \theta}{\mathfrak{p}}\right)$$

for all $\mathfrak{p} \in \Omega$. So if we replace α by $\alpha_0 \alpha^{-1}$, our new α will have the same property as the original one, and furthermore the new α will be arbitrarily close to 1 at all discrete $\mathfrak{p}$ in $\Omega - S$, and will be a square at all real $\mathfrak{p}$ in $\Omega - S$.

We claim that actually α can be chosen with the additional property that it is integral at all $\mathfrak{p}$ in S. For by the Strong Approximation Theorem we can find a y in F which is arbitrarily close to 1 at all discrete $\mathfrak{p}$ in $\Omega - S$, which is integral at all $\mathfrak{p}$ in S, and which makes αy integral at all $\mathfrak{p}$ in S. Then αy^2 satisfies all the properties of the last α, and it is also integral at all $\mathfrak{p}$ in S. We therefore assume that our α has this additional property.

3) Formula 71:10 applies to this situation with $\beta = \theta$. Hence $\sigma_\alpha(\sqrt{\theta}) = -\sqrt{\theta}$. And Formula 71:11 also applies if α is chosen sufficiently close to 1 at all discrete spots in $\Omega - S$. Now our present α is a square at all real spots, hence the r of the formula is 0. So $\sigma_\alpha \zeta = \zeta$. But this means that σ_α is the identity. This contradicts the fact that $\sigma_\alpha(\sqrt{\theta}) = -\sqrt{\theta}$. Hence E/F has to have local degree 1 at $\mathfrak{q}$. **q. e. d.**

71:13. *Let α be an element of the number field F such that $\left(\frac{\alpha, -1}{\mathfrak{p}}\right) = 1$ for all discrete $\mathfrak{p}$ on F. Then*

$$\prod_{\mathfrak{p} \in \Omega} \left(\frac{\alpha, -1}{\mathfrak{p}}\right) = 1 .$$

Proof. Let $K = F(\zeta)$ where ζ is a primitive 4-th root of unity, and let S be the set of all discrete non-dyadic spots on F. This gives us the situation discussed at the beginning of § 71B with $m = 4$.

An argument similar to the argument used in step 2) of Proposition 71:12 will give a new α which is an integer at all $\mathfrak{p}$ in S and arbitrarily close to 1 at all dyadic $\mathfrak{p}$, and such that the new Hilbert symbols $\left(\frac{\alpha, -1}{\mathfrak{p}}\right)$ are equal to the original ones at all $\mathfrak{p}$ in Ω. Now $\zeta^2 = -1$, so by Formula 71:10 we have $\sigma_\alpha \zeta = \zeta$. By Formula 71:11 we have $\sigma_\alpha \zeta = \zeta^{(-1)^r}$ where r is the number of real spots on F at which α is negative. Here r is even since $\zeta = \sigma_\alpha \zeta = \zeta^{(-1)^r}$. On the other hand α is negative at a real spot on F if and only if $\left(\frac{\alpha, -1}{\mathfrak{p}}\right) = -1$. Hence

$$\prod_{\mathfrak{p} \in \Omega} \left(\frac{\alpha, -1}{\mathfrak{p}}\right) = \prod_{\text{real}\, \mathfrak{p}} \left(\frac{\alpha, -1}{\mathfrak{p}}\right) = (-1)^r = 1 .$$

q. e. d.

§ 71C. The ten-field construction

Now we have reciprocity in certain special cases involving cyclotomic extensions. The next step in the proof consists in surrounding the quadratic extension E/F with cyclotomic extensions and then carrying the reciprocity laws associated with these extensions back to F.

71:14. Lemma. *Let $a \geqq 2$ be a natural number. Then there are infinitely many prime numbers p such that a has even period* $\bmod p$, *i. e. such that*

$$a^n \equiv 1 \bmod p \Rightarrow n \text{ even}.$$

Proof. Define a sequence $A_1, A_2, \ldots, A_\nu, \ldots$ by putting $A_1 = a^2$, $A_{\nu+1} = A_\nu^2$. Suppose if possible that $A_\nu + 1 \in 4\mathbf{Z}$. Then A_ν is odd, hence a is odd, hence

$$A_1 - 1 = a^2 - 1 \in 4\mathbf{Z}.$$

So $A_1 = 1 + 4\lambda$ with λ in $\mathbf{Z}$. But A_ν is a power of A_1. Hence by the binomial theorem we obtain $A_\nu - 1 \in 4\mathbf{Z}$. But we are supposing that $A_\nu + 1 \in 4\mathbf{Z}$. This is therefore impossible. Hence $A_\nu + 1 \notin 4\mathbf{Z}$ for all $\nu \geqq 1$.

Now $A_\nu + 1 > 2$; and we have just shown that $A_\nu + 1$ is not divisible by 4; hence there is a prime number $p_\nu > 2$ dividing $A_\nu + 1$. This p_ν of course cannot divide $A_\nu - 1$; but it does divide

$$A_{\nu+1} - 1 = A_\nu^2 - 1 = (A_\nu - 1)(A_\nu + 1).$$

Hence we can find prime numbers

$$p_1, p_2, \ldots, p_\nu, \ldots > 2$$

such that

$$A_{\nu+1} \equiv 1 \bmod p_\nu, \quad A_\nu \not\equiv 1 \bmod p_\nu.$$

Suppose we had $A_\mu \equiv 1 \bmod p_\nu$ for some $\mu \leqq \nu$. Then using the binomial theorem with the fact that A_ν is a power of A_μ would give us $A_\nu \equiv 1 \bmod p_\nu$ and this is false. Hence

$$A_\mu \not\equiv 1 \bmod p_\nu \quad \text{if} \quad \mu \leqq \nu.$$

This shows first that $p_\mu \neq p_\nu$ if $\mu < \nu$, for otherwise we would have $A_{\mu+1} \equiv 1 \bmod p_\mu$ with $\mu + 1 \leqq \nu$ and $p_\mu = p_\nu$. In other words the prime numbers $p_1, p_2, \ldots$ constructed above are distinct and therefore infinite in number. Secondly, it shows that

$$A_1 \not\equiv 1 \bmod p_\nu, \quad A_{\nu+1} \equiv 1 \bmod p_\nu,$$

in other words that

$$a^2 \not\equiv 1 \bmod p_\nu, \quad a^{2^{\nu+1}} \equiv 1 \bmod p_\nu.$$

Thus the period of $a \bmod p_\nu$ is a certain power of 2. **q. e. d.**

71:15. *Let $\mathfrak{p}$ be a discrete spot on F. Then for an infinite number of prime numbers p there is a p-cyclotomic extension K/F which is unramified of local degree 2 at $\mathfrak{p}$.*

Proof. By Lemma 71:14 there are infinitely many prime numbers p such that $N\mathfrak{p}$ has even order $\bmod p$. We shall show that any p which has this property and is also a unit at $\mathfrak{p}$ will work. Take $L = F(\zeta)$ with ζ a primitive p-th root of unity. Then L/F is unramified at $\mathfrak{p}$ since p is a unit there. Let f denote the degree of inertia of the extension at $\mathfrak{p}$. Then ζ is a root of unity of period prime to $N\mathfrak{p}$, hence it is an $((N\mathfrak{p})^f - 1)^{th}$ root of unity by Proposition 32:8. But the period of ζ is p. Hence p divides $((N\mathfrak{p})^f - 1)$, in other words

$$(N\mathfrak{p})^f \equiv 1 \bmod p\,.$$

Hence f is even by choice of p. So L/F is unramified of even local degree at $\mathfrak{p}$. Then by Corollary 15:10a there is a field K such that $F \subseteq K \subseteq L$ with K/F unramified of local degree 2 at $\mathfrak{p}$. Of course, K/F is p-cyclotomic. So K is the field we are after. **q. e. d.**

71:16. *Let $\mathfrak{p}_1$ and $\mathfrak{p}_2$ be discrete spots at which the quadratic extension E/F is unramified of local degree 2. Consider idèles*

$$\mathfrak{i}_1 \in I_F^{\mathfrak{p}_1} - N_{E/F} J_E\,, \quad \mathfrak{i}_2 \in I_F^{\mathfrak{p}_2} - N_{E/F} J_E\,.$$

Then

$$\mathfrak{i}_1 \mathfrak{i}_2 \in P_F N_{E/F} J_E\,.$$

Proof. 1) Let C be an algebraic closure of E. By Propositions 71:9 and 71:15 there is a prime number p_1 which is a unit at $\mathfrak{p}_1$, and a p_1-cyclotomic extension K_1/F which is unramified of local degree 2 at $\mathfrak{p}_1$, which is contained in C, and which satisfies

$$[K_1 E : E] = [K_1 : F]\,.$$

In the same way we can find a prime number p_2 which is a unit at $\mathfrak{p}_2$, and a p_2-cyclotomic extension K_2/F which is unramified of local degree 2 at $\mathfrak{p}_2$, which is contained in C, and which satisfies

$$[K_2 E K_1 : E K_1] = [K_2 : F]\,.$$

We are actually going to work inside $K_1 E K_2$ and we can now forget about C. We have

$$[K_1 E K_2 : F] = [K_1 : F]\,[E : F]\,[K_2 : F]\,.$$

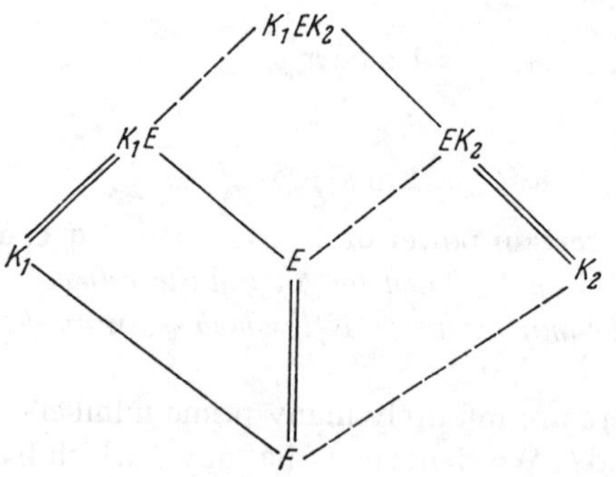

Fig. 1

Using this equation in conjunction with the fact that the degree of an extension is not increased by a field translation, we can easily check that the sides with the same ruling in Figure 1 give field extensions of equal degree.

The extension $K_1 E K_2/F$ is galois since K_1/F, E/F, K_2/F are all galois. Now the action of any automorphism of K_1EK_2/F is completely described by its action on K_1/F, E/F, K_2/F; but each of these extensions is abelian; hence any two automorphisms of K_1EK_2/F commute; hence K_1EK_2/F is abelian. So any intermediate extension is galois and abelian.

2) We fix a spot $\mathfrak{P}_1$ on K_1E which divides the given spot $\mathfrak{p}_1$ on F. Now $K_{1\mathfrak{P}_1}/F_{\mathfrak{P}_1}$ and $E_{\mathfrak{P}_1}/F_{\mathfrak{P}_1}$ are unramified extensions of degree 2, and they are subextensions of a common extension $(K_1E)_{\mathfrak{P}_1}/F_{\mathfrak{P}_1}$, hence they are equal by Example 32:11. By Proposition 11:19 we have

$$(K_1E)_{\mathfrak{P}_1} = (K_1E)\, E_{\mathfrak{P}_1} = E_{\mathfrak{P}_1}\,.$$

Hence K_1E/F is unramified of local degree 2 at $\mathfrak{p}_1$. Let F_1 denote the decomposition field of this extension at $\mathfrak{p}_1$. Then K_1E/F_1 is quadratic by Proposition 15:10, and it is unramified of local degree 2 at $\mathfrak{P}_1$. We have $E_{\mathfrak{P}_1} \supset F_{\mathfrak{P}_1} = F_{1\mathfrak{P}_1}$ so that E is not contained in F_1. Hence $K_1E = F_1E$. Similarly $K_1E = K_1F_1$.

Repeat all this with $\mathfrak{p}_2$ to obtain a spot $\mathfrak{P}_2$ on EK_2 and a decomposition field F_2 of EK_2/F at $\mathfrak{p}_2$. This parallels the situation at $\mathfrak{p}_1$ and analogous equations are obtained. In particular $EK_2 = EF_2$ and $EK_2 = F_2K_2$. We shall be interested in the compositum F_1F_2. We have

$$F_1EF_2 = K_1EF_2 = K_1EK_2$$

so that K_1EK_2/F_1F_2 is at most quadratic. Using this fact together with Figure 1 and an easy degree argument, we find that the sides with the same ruling in Figure 2 correspond to extensions of equal degree. In particular, K_1EK_2/F_1F_2 is quadratic.

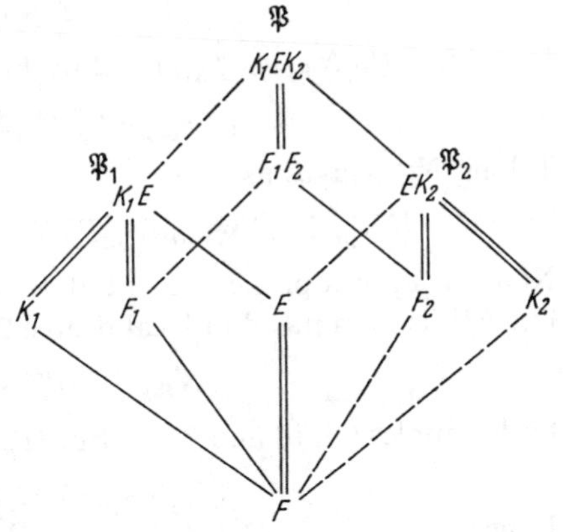

Fig. 2

3) Since K_1EK_2/F_1F_2 is quadratic it follows from the Global Square Theorem and Proposition 71:7 that there is a discrete spot $\mathfrak{P}$ on K_1EK_2 such that K_1EK_2/F_1F_2 is of local degree 2 at $\mathfrak{P}$, such that p_1 and p_2 are units at $\mathfrak{P}$, and such that K_1EK_2/F is unramified at $\mathfrak{P}$. The significance of this choice of $\mathfrak{P}$ lies in the fact that F_1F_2/F_1 and F_1F_2/F_2 have odd local degree at $\mathfrak{P}$. (We shall use the same letter $\mathfrak{P}$ for the spot induced by $\mathfrak{P}$ on a subfield of K_1EK_2.) That this is actually true follows thus: if F_1F_2/F_1 had even local degree at $\mathfrak{P}$, then we would have $E_{\mathfrak{P}} \subsetneq (F_1F_2)_{\mathfrak{P}}$ by Example 32:11, and this would imply that $K_1EK_2 = F_1EF_2$ had

local degree 1 over F_1F_2 at $\mathfrak{P}$. This is contrary to the choice of $\mathfrak{P}$. So F_1F_2/F_1 does indeed have odd local degree at $\mathfrak{P}$. Similarly with F_1F_2/F_2.

4) Take an idèle $\mathfrak{j}_1 \in I_{F_1}^{\mathfrak{P}_1}$ whose $\mathfrak{P}_1$-component is a prime element of F at $\mathfrak{p}_1$; then the $\mathfrak{P}_1$-component of $\mathfrak{j}_1$ is actually a prime element of F_1 at $\mathfrak{P}_1$ since F_1/F is unramified at $\mathfrak{P}_1$. Similarly take $\mathfrak{j}_2 \in I_{F_2}^{\mathfrak{P}_2}$. Similarly $\mathfrak{j} \in I_{F_1F_2}^{\mathfrak{P}}$.

Then $N_{F_1F_2/F_1}\mathfrak{j}$ is an element of $I_{F_1}^{\mathfrak{P}}$ and its $\mathfrak{P}$-coordinate is an odd power of a prime element of F_1 at $\mathfrak{P}$ since F_1F_2/F_1 has odd local degree at $\mathfrak{P}$. But $K_1E = F_1E$ is a quadratic extension of F_1 which is unramified of local degree 2 at $\mathfrak{P}$, by choice of $\mathfrak{P}$. Hence

$$N_{F_1F_2/F_1}\mathfrak{j} \in I_{F_1}^{\mathfrak{P}} - N_{K_1E/F_1}J_{K_1E}$$

by Examples 63:16 and 65:2. Now $K_1E = K_1F_1$ is a p_1-cyclotomic quadratic extension of F_1, and p_1 is a unit at $\mathfrak{P}$, and the local degree of the extension at $\mathfrak{P}$ is equal to 2, hence we can apply Proposition 71:12 and we find that

$$I_{F_1}^{\mathfrak{P}} \nsubseteq P_{F_1}N_{K_1E/F_1}J_{K_1E}\,.$$

We can therefore conclude from the equation

$$(I_{F_1}^{\mathfrak{P}} : I_{F_1}^{\mathfrak{P}} \cap N_{K_1E/F_1}J_{K_1E}) = 2$$

that

$$N_{F_1F_2/F_1}\mathfrak{j} \notin P_{F_1}N_{K_1E/F_1}J_{K_1E}\,.$$

The same sort of reasoning will show that

$$\mathfrak{j}_1 \notin P_{F_1}N_{K_1E/F_1}J_{K_1E}\,.$$

Now $(J_{F_1} : P_{F_1}N_{K_1E/F_1}J_{K_1E}) = 2$ by Proposition 65:21. Hence

$$\mathfrak{j}_1 N_{F_1F_2/F_1}\mathfrak{j} \in P_{F_1}N_{K_1E/F_1}J_{K_1E}\,.$$

Taking $N_{F_1/F}$ gives us

$$(N_{F_1/F}\,\mathfrak{j}_1)\,(N_{F_1F_2/F}\,\mathfrak{j}) \in P_F N_{K_1E/F}J_{K_1E} \subseteq P_F N_{E/F}J_E\,.$$

Now $N_{F_1/F}\mathfrak{j}_1$ is a prime element at $\mathfrak{p}_1$ since F_1/F has local degree 1 at $\mathfrak{p}_1$; but E/F is unramified of local degree 2 at $\mathfrak{p}_1$; hence

$$N_{F_1/F}\mathfrak{j}_1 \in I_F^{\mathfrak{p}_1} - N_{E/F}J_E$$

by Examples 63:16 and 65:2. But $(I_F^{\mathfrak{p}_1} : I_F^{\mathfrak{p}_1} \cap N_{E/F}J_E) = 2$. Hence

$$\mathfrak{i}_1\,(N_{F_1/F}\,\mathfrak{j}_1)^{-1} \in N_{E/F}J_E\,.$$

Hence

$$\mathfrak{i}_1\,(N_{F_1F_2/F}\,\mathfrak{j}) \in P_F N_{E/F}J_E\,.$$

On grounds of symmetry we have

$$\mathfrak{i}_2\,(N_{F_1F_2/F}\,\mathfrak{j}) \in P_F N_{E/F}J_E\,.$$

Hence

$$\mathfrak{i}_1\mathfrak{i}_2 \in P_F N_{E/F}J_E\,.$$

q. e. d.

71:17. *Let $\mathfrak{q}$ be a discrete spot at which the quadratic extension E/F is unramified. Then*

$$I_F^{\mathfrak{q}} \subseteq P_F N_{E/F} J_E$$

if and only if E/F has local degree 1 *at $\mathfrak{q}$.*

Proof. We suppose, if possible, that we have $I_F^{\mathfrak{q}} \subseteq P_F N_{E/F} J_E$ with E/F of local degree 2 at $\mathfrak{q}$. We shall use this to arrive at a contradiction. Let S denote the set of all discrete spots at which E/F is unramified. So S consists of almost all spots in Ω. If $\mathfrak{p}$ is any spot in S at which E/F has local degree 1, then $I_F^{\mathfrak{p}} \subseteq N_{E/F} J_E$ and so $I_F^{\mathfrak{p}} \subseteq P_F N_{E/F} J_E$. If E/F has local degree 2 at the spot $\mathfrak{p}$ in S then we can reach the same conclusion by applying Proposition 71:16 with $\mathfrak{p}_1 = \mathfrak{q}$ and $\mathfrak{p}_2 = \mathfrak{p}$. In other words, our assumption leads to the relation

$$I_F^{\mathfrak{p}} \subseteq P_F N_{E/F} J_E \quad \forall\, \mathfrak{p} \in S\,.$$

Let us use this to prove something which by Proposition 65:21 must be absurd, namely that $J_F \subseteq P_F N_{E/F} J_E$.

So consider a typical $\mathfrak{i}$ in J_F. Use the Weak Approximation Theorem to find an α in $\dot{F}$ such that $|\alpha\, \mathfrak{i}_{\mathfrak{p}}^{-1} - 1|_{\mathfrak{p}}$ is small for all $\mathfrak{p}$ in $\Omega - S$. Since $\dot{F}_{\mathfrak{p}}^2$ is an open subset of $F_{\mathfrak{p}}$ we have

$$\alpha \in \mathfrak{i}_{\mathfrak{p}} \dot{F}_{\mathfrak{p}}^2 \quad \forall\, \mathfrak{p} \in \Omega - S\,.$$

Replacing $\mathfrak{i}$ by $(\alpha)\mathfrak{i}$ allows us to assume that $\mathfrak{i}$ is a square at all $\mathfrak{p}$ in $\Omega - S$. We can also assume that $\mathfrak{i}$ is in J_F^S, because of the assumption that $I_F^{\mathfrak{p}} \subseteq P_F N_{E/F} J_E$ for all $\mathfrak{p}$ in S. But this refined idèle $\mathfrak{i}$ is a local norm at all $\mathfrak{p}$ in Ω in virtue of Example 63:16; hence it is in $N_{E/F} J_E$ by Example 65:2. **q. e. d.**

§71D. Conclusion of the proof

71:18. **Theorem.** *Let α and β be non-zero elements of an algebraic number field F. Then their Hilbert symbol is* 1 *for almost all $\mathfrak{p}$, and*

$$\prod_{\mathfrak{p} \in \Omega} \left(\frac{\alpha, \beta}{\mathfrak{p}}\right) = 1\,.$$

Proof. The first assertion follows from the Product Formula and Example 63:12.

1) We start by proving the following: given any three distinct spots $\mathfrak{p}_1, \mathfrak{p}_2, \mathfrak{p}_3$ at which β is not a square, there is a γ in F such that $\left(\frac{\gamma, \beta}{\mathfrak{p}}\right)$ is -1 at two of these spots and $+1$ at all remaining spots on F. To see this we construct the quadratic extension $E = F(\sqrt{\beta})$ of F. Since E/F has local degree 2 at $\mathfrak{p}_1$ we can pick

$$\mathfrak{i}_1 \in I_F^{\mathfrak{p}_1} - N_{E/F} J_E\,.$$

Similarly pick $\mathfrak{i}_2$, $\mathfrak{i}_3$ corresponding to $\mathfrak{p}_2$, $\mathfrak{p}_3$. Then one of the idèles $\mathfrak{i}_1\mathfrak{i}_2$, $\mathfrak{i}_2\mathfrak{i}_3$, $\mathfrak{i}_3\mathfrak{i}_1$ must be in $P_F N_{E/F} J_E$ since this group has index 2 in J_F by Proposition 65:21. Let us say

$$\mathfrak{i}_1\mathfrak{i}_2 \in P_F N_{E/F} J_E .$$

This means that there is a γ in F which is a local norm in E/F at all spots except $\mathfrak{p}_1$ and $\mathfrak{p}_2$ where it is not. Hence

$$\left(\frac{\gamma, \beta}{\mathfrak{p}}\right) = \begin{cases} -1 & \text{if } \mathfrak{p} = \mathfrak{p}_1, \mathfrak{p}_2 \\ 1 & \text{if } \mathfrak{p} \neq \mathfrak{p}_1, \mathfrak{p}_2 . \end{cases}$$

2) If the reciprocity law did not hold for α, β we would have

$$\prod_{\mathfrak{p} \in \Omega} \left(\frac{\alpha, \beta}{\mathfrak{p}}\right) = -1 .$$

In this event we could use step 1) to find a new α and a single spot $\mathfrak{q}$ such that

$$\left(\frac{\alpha, \beta}{\mathfrak{p}}\right) = \begin{cases} -1 & \text{if } \mathfrak{p} = \mathfrak{q} \\ 1 & \text{if } \mathfrak{p} \in \Omega - \mathfrak{q} . \end{cases}$$

We wish to make a slight alteration to β. We take $\beta' = -1$ when $\mathfrak{q}$ is a real archimedean spot ($\mathfrak{q}$ cannot be complex). If $\mathfrak{q}$ is discrete we let β' be an element of F which is a unit with quadratic defect $4\mathfrak{o}_\mathfrak{q}$ at $\mathfrak{q}$. Then the quadratic space

$$V \cong \langle\alpha\rangle \perp \langle\beta\rangle \perp \langle-\alpha\beta\rangle$$

is isotropic at all spots except $\mathfrak{q}$ where it is not. Clearly $V \perp \langle-\beta'\rangle$ is isotropic at $\mathfrak{q}$ where $\mathfrak{q}$ is real; by Proposition 63:17 it is also isotropic for a discrete $\mathfrak{q}$ since the discriminant β' is then a non-square at $\mathfrak{q}$. Hence $V \perp \langle-\beta'\rangle$ is isotropic by Theorem 66:1. So V represents β'. So

$$V \cong \langle\alpha'\rangle \perp \langle\beta'\rangle \perp \langle-\alpha'\beta'\rangle$$

is isotropic at all spots except $\mathfrak{q}$ where it is not. Hence

$$\left(\frac{\alpha', \beta'}{\mathfrak{p}}\right) = \begin{cases} -1 & \text{if } \mathfrak{p} = \mathfrak{q} \\ 1 & \text{if } \mathfrak{p} \in \Omega - \mathfrak{q} . \end{cases}$$

This is impossible for a real $\mathfrak{q}$ by Proposition 71:13. Hence we may suppose that $\mathfrak{q}$ is discrete. In this case we consider the quadratic extension $E = F(\sqrt{\beta'})$ of F. The information about the Hilbert symbol says that α' is a local norm of E/F at all spots except $\mathfrak{q}$ where it is not. We claim that $I_F^\mathfrak{q} \subseteq P_F N_{E/F} J_E$. Consider a typical idèle

$$\mathfrak{i} \in I_F^\mathfrak{q} - N_{E/F} J_E .$$

Then $\mathfrak{i}$ is also a local norm at all spots except $\mathfrak{q}$; hence $(\alpha')\mathfrak{i}$ is a local norm at all spots on F; so

$$\mathfrak{i} \in (\alpha')^{-1} N_{E/F} J_E \subseteq P_F N_{E/F} J_E$$

and we have established our claim that

$$I_F^{\mathfrak{q}} \subseteq P_F N_{E/F} J_E .$$

On the other hand E/F is unramified of local degree 2 at $\mathfrak{q}$ since β' is a unit of quadratic defect $4\mathfrak{o}_{\mathfrak{q}}$ at $\mathfrak{q}$. This is impossible by Proposition 71:17. **q. e. d.**

71:19. **Theorem.** *Let T be a set consisting of an even number of discrete or real spots on an algebraic number field F. Then there are α, β in $\dot{F}$ such that*

$$\left(\frac{\alpha, \beta}{\mathfrak{p}}\right) = \begin{cases} -1 & \text{if } \mathfrak{p} \in T \\ 1 & \text{if } \mathfrak{p} \in \Omega - T. \end{cases}$$

Proof. β can be any element of F which is a non-square at all spots in T. Such an element always exists: for instance the Weak Approximation Theorem provides a β which is a prime element at the discrete spots in T and a negative number at the real spots in T. Put $E = F(\sqrt{\beta})$ and define a group homomorphism $\varphi: J_F \to (\pm 1)$ by the formula

$$\varphi(\mathfrak{i}) = \prod_{\mathfrak{p} \in \Omega} \left(\frac{\mathfrak{i}_{\mathfrak{p}}, \beta}{\mathfrak{p}}\right)$$

where $\mathfrak{i} = (\mathfrak{i}_{\mathfrak{p}})_{\mathfrak{p}\in\Omega}$ denotes a typical idèle in J_F. Then $N_{E/F} J_E$ is in the kernel of φ by § 65A; and P_F is in the kernel of φ by the Hilbert Reciprocity Law; hence $P_F N_{E/F} J_E$ is in the kernel of φ; now φ is surjective by Proposition 63:13, and $(J_F : P_F N_{E/F} J_E) = 2$ by Proposition 65:21, hence $P_F N_{E/F} J_E$ is precisely the kernel of φ. Take an idèle $\mathfrak{j} \in J_F$ which is a local norm at all $\mathfrak{p} \in \Omega - T$ and is not a local norm at any $\mathfrak{p} \in T$. Then $\varphi(\mathfrak{j}) = 1$ since T contains an even number of elements, hence $\mathfrak{j} \in P_F N_{E/F} J_E$. But this relation can also be read as follows: there is an α in $\dot{F}$ which is a local norm at all $\mathfrak{p}$ in $\Omega - T$, and at no $\mathfrak{p}$ in T. This α gives the desired values to the Hilbert symbol. **q. e. d.**

71:19a. **Corollary.** *β can be any element of $\dot{F}$ which is a non-square at all spots in T.*

§ 72. Existence of forms with prescribed local behavior

72:1. **Theorem.** *A regular n-ary space $U_{\mathfrak{p}}$ is given over each completion $F_{\mathfrak{p}}$ of an algebraic number field F. In order that there exist an n-ary space V over F such that $V_{\mathfrak{p}} \cong U_{\mathfrak{p}}$ for all $\mathfrak{p}$, it is necessary and sufficient that*

(1) *there be a d_0 in F with $d U_{\mathfrak{p}} = d_0$ for all $\mathfrak{p}$,*

(2) *$S_{\mathfrak{p}} U_{\mathfrak{p}} = 1$ for almost all $\mathfrak{p}$,*

(3) $\prod_{\mathfrak{p} \in \Omega} S_{\mathfrak{p}} U_{\mathfrak{p}} = 1.$

Proof. 1) Necessity. To obtain the first condition take $d_0 = dV$. To obtain the second and third conditions consider a splitting

$$V \cong \langle\alpha_1\rangle \perp \cdots \perp \langle\alpha_n\rangle .$$

Then $S_\mathfrak{p} U_\mathfrak{p} = S_\mathfrak{p} V_\mathfrak{p}$ is a product of Hilbert symbols of the form $\left(\frac{\alpha_i, \alpha_j}{\mathfrak{p}}\right)$. Each of these is equal to 1 for almost all $\mathfrak{p}$, hence $S_\mathfrak{p} U_\mathfrak{p}$ is. Now apply the Hilbert Reciprocity Law.

2) We must prove the sufficiency. If $n = 1$ we take $V \cong \langle d_0\rangle$. So we assume that $n \geqq 2$. Let T be a finite set of spots on F which contains all archimedean spots and also all spots $\mathfrak{p}$ at which $S_\mathfrak{p} U_\mathfrak{p} = -1$. Write

$$U_\mathfrak{p} \cong \langle\alpha_1^\mathfrak{p}\rangle \perp \cdots \perp \langle\alpha_n^\mathfrak{p}\rangle \quad (\alpha_i^\mathfrak{p} \in \dot{F}_\mathfrak{p})$$

at each $\mathfrak{p}$ in T. Use the Weak Approximation Theorem to find an α_i in $\dot{F}$ such that $|\alpha_i - \alpha_i^\mathfrak{p}|_\mathfrak{p}$ is small for all $\mathfrak{p}$ in T. Since $\dot{F}_\mathfrak{p}^2$ is an open subset of $F_\mathfrak{p}$ we can obtain

$$\alpha_i \in \alpha_i^\mathfrak{p} \dot{F}_\mathfrak{p}^2 \quad \forall\, \mathfrak{p} \in T ,$$

provided our approximations are good enough. Do all this for $1 \leqq i \leqq n-1$. Then take a quadratic space W over F with

$$W \cong \langle\alpha_1\rangle \perp \cdots \perp \langle\alpha_{n-1}\rangle \perp \langle\alpha_1 \ldots \alpha_{n-1} d_0\rangle .$$

Clearly

$$U_\mathfrak{p} \cong W_\mathfrak{p} \quad \forall\, \mathfrak{p} \in T .$$

Hence $S_\mathfrak{p} W_\mathfrak{p} = S_\mathfrak{p} U_\mathfrak{p}$ for all $\mathfrak{p}$ in T. Let R denote the set of spots at which $S_\mathfrak{p} W_\mathfrak{p}$ and $S_\mathfrak{p} U_\mathfrak{p}$ are different. Of course R is a finite subset of $\Omega - T$. And $W_\mathfrak{p} \cong U_\mathfrak{p}$ for all $\mathfrak{p}$ in $\Omega - R$ by Theorem 63:20. If R is empty we are through. Otherwise R consists of those spots in $\Omega - T$ at which $S_\mathfrak{p} W_\mathfrak{p} = -1$; now

$$\prod_R S_\mathfrak{p} W_\mathfrak{p} = \prod_R S_\mathfrak{p} W_\mathfrak{p} \cdot \prod_{\Omega - R} S_\mathfrak{p} U_\mathfrak{p} = \prod_\Omega S_\mathfrak{p} W_\mathfrak{p} = 1 .$$

Hence R consists of an even number of spots in $\Omega - T$.

3) We claim that there are quadratic planes P and P' with the same discriminant over F, with $P_\mathfrak{p} \cong P'_\mathfrak{p}$ for all $\mathfrak{p}$ in $\Omega - R$, and with $P_\mathfrak{p} \not\cong P'_\mathfrak{p}$ for all $\mathfrak{p}$ in R. In fact, take any β in F which is a square at all archimedean spots and a non-square at all spots in R. By Theorem 71:19 and its corollary we can find an α in F with

$$\left(\frac{\alpha, \beta}{\mathfrak{p}}\right) = \begin{cases} -1 & \text{if } \mathfrak{p} \in R \\ 1 & \text{if } \mathfrak{p} \in \Omega - R . \end{cases}$$

Define

$$P \cong \langle 1\rangle \perp \langle -\beta\rangle , \quad P' \cong \langle\alpha\rangle \perp \langle -\alpha\beta\rangle .$$

Then $P_\mathfrak{p}$ and $P'_\mathfrak{p}$ are isometric at all archimedean spots by choice of β; applying Theorem 63:20 at the discrete spots shows that $P_\mathfrak{p} \cong P'_\mathfrak{p}$ for all $\mathfrak{p}$ in $\Omega - R$, and $P_\mathfrak{p} \not\cong P'_\mathfrak{p}$ for all $\mathfrak{p}$ in R. So we have established our claim.

4) Consider $W_\mathfrak{p}$ at any $\mathfrak{p}$ in R. This $W_\mathfrak{p}$ cannot be a hyperbolic plane since $W_\mathfrak{p}$ and $U_\mathfrak{p}$ are non-isometric spaces with the same discriminant. We are also assuming that $\dim W_\mathfrak{p} \geqq 2$.
Hence

$$P'_\mathfrak{p} \rightarrowtail (P \perp W)_\mathfrak{p} \quad \forall\, \mathfrak{p} \in R$$

by Theorem 63:21. At any $\mathfrak{p}$ in $\Omega - R$ we also have such a representation since then $P'_\mathfrak{p} \cong P_\mathfrak{p}$ by step 3). Hence there is a representation $P' \rightarrowtail (P \perp W)$ by Theorem 66:3. Hence there is a quadratic space V over F with

$$P' \perp V \cong P \perp W\,.$$

This V has discriminant $d_0 = dW$ since $dP' = dP$. For each $\mathfrak{p}$ in $\Omega - R$ we have $P_\mathfrak{p} \cong P'_\mathfrak{p}$ by choice of P, P', hence by Witt's theorem

$$V_\mathfrak{p} \cong W_\mathfrak{p} \cong U_\mathfrak{p} \quad \forall\, \mathfrak{p} \in \Omega - R\,.$$

If $\mathfrak{p}$ is in R we have $V_\mathfrak{p} \not\cong W_\mathfrak{p}$ since $P_\mathfrak{p} \not\cong P'_\mathfrak{p}$; but $U_\mathfrak{p} \not\cong W_\mathfrak{p}$ by definition of R; and

$$dV_\mathfrak{p} = dW_\mathfrak{p} = dU_\mathfrak{p} = d_0\,;$$

hence by Theorem 63:20 we must have

$$S_\mathfrak{p} V_\mathfrak{p} = -SW_\mathfrak{p} = SU_\mathfrak{p}\,;$$

so $V_\mathfrak{p} \cong U_\mathfrak{p}$ for all $\mathfrak{p}$ in R. Therefore $V_\mathfrak{p} \cong U_\mathfrak{p}$ for all $\mathfrak{p}$ and the space V has the desired property. **q. e. d.**

§ 73. The quadratic reciprocity law

We conclude this chapter by finding an expression for the Hilbert symbol in terms of the Legendre symbol over the field of rational numbers $\mathbf{Q}$. Recall the definition of the Legendre symbol: if p is an odd prime number, and if a is any rational integer that is not divisible by p, then the Legendre symbol $\left(\frac{a}{p}\right)$ is defined to be

$$\left(\frac{a}{p}\right) = +1 \quad \text{or} \quad -1$$

according as a is or is not congruent to the square of a rational integer modulo p. In other words, $\left(\frac{a}{p}\right)$ is 1 if the natural image of a in the finite field $\mathbf{Z}/p\,\mathbf{Z}$ is a square, otherwise $\left(\frac{a}{p}\right)$ is -1. Now for any finite field K of characteristic not 2 we have $(\dot{K}:\dot{K}^2) = 2$ by § 62. Hence

$$\left(\frac{a}{p}\right)\left(\frac{b}{p}\right) = \left(\frac{ab}{p}\right).$$

Note that the Local Square Theorem tells us that $\left(\frac{a}{p}\right) = 1$ if and only if a is a square in the field of p-adic numbers $\mathbb{Q}_p$.

We shall use $\left(\frac{a, b}{p}\right)$ for the Hilbert symbol over $\mathbb{Q}_p$. We know from the formulas of § 63B that the Hilbert symbol is completely determined once its values are known, first for all rational integers a and b that are prime to p, and secondly for all rational integers a that are prime to p with $b = p$. We shall therefore restrict ourselves to these special cases.

73:1. *Let p be an odd prime number, and let a and b be rational integers prime to p. Then*

$$\left(\frac{a, b}{p}\right) = 1\,, \quad \left(\frac{a, p}{p}\right) = \left(\frac{a}{p}\right).$$

Proof. The prime spot p is non-dyadic since the prime number p is odd. Apply Example 63:12, using the fact that the p-adic unit a is a square in $\mathbb{Q}_p$ if and only if $\left(\frac{a}{p}\right) = 1$. **q. e. d.**

73:2. *Let a and b be rational integers prime to 2. Then*

$$\left(\frac{a, b}{2}\right) = (-1)^{\frac{a-1}{2}\cdot\frac{b-1}{2}}, \qquad \left(\frac{a, 2}{2}\right) = (-1)^{\frac{1}{8}(a^2-1)}.$$

Proof. 1) Every odd rational integer is clearly congruent to one of the numbers

$$1, 3, 5, 7$$

modulo 8. Hence by the Local Square Theorem we can assume without loss of generality that a is one of these four numbers. The same with b.

Now if $\mathfrak{u}$ denotes the group of units of the ring of 2-adic integers $\mathbb{Z}_2$, then $(\mathfrak{u}:\mathfrak{u}^2) = 4$ by Proposition 63:9. But every element of $\mathfrak{u}$ is a square times 1, 3, 5 or 7 by the Local Square Theorem and the power series expansion of Example 31:5. Hence the numbers 1, 3, 5, 7 fall in the four distinct cosets of $\mathfrak{u}$ modulo $\mathfrak{u}^2$. In particular 5 is a non-square, hence it is a unit of quadratic defect $4\mathbb{Z}_2$. So by applying Corollary 63:11a we find that our proposition holds whenever a or b is 5. Of course the proposition holds whenever a or b is 1. We therefore restrict ourselves to $a = 3$ or 7, $b = 3$ or 7.

2) We have $7 + 2(3)^2 = 25$, hence

$$\left(\frac{7, 2}{2}\right) = 1 = (-1)^{\frac{1}{8}(7^2-1)}$$

Then

$$\left(\frac{3,2}{2}\right) = \left(\frac{7,2}{2}\right)\left(\frac{5,2}{2}\right) = -1 = (-1)^{\frac{1}{8}(3^2-1)} .$$

Hence the second formula of the proposition holds for all a.

3) It is easily seen that

$$\left(\frac{3,3}{2}\right) = \left(\frac{3,7}{2}\right) = \left(\frac{7,7}{2}\right) .$$

We will be through if we can prove that these three quantities are -1. Now by Proposition 63:13 there is a 2-adic number c such that $\left(\frac{7,c}{2}\right) = -1$ since 7 is a non-square in $\mathbf{Q}_2$. But $\left(\frac{7,2}{2}\right) = 1$ by step 2), hence we can assume that c is a 2-adic unit, hence that c is 1, 3, 5 or 7. But c cannot be 1 or 5, hence c is 3 or 7. In either event we have our result. **q. e. d.**

We cannot resist giving a proof of the famous Quadratic Reciprocity Law. This is obtained instantly from the Hilbert Reciprocity Law and the above formulas. Here then is the Quadratic Reciprocity Law with its first and second supplements.

73:3. **Theorem.** *Let p and q be distinct odd prime numbers. Then*

$$\left(\frac{p}{q}\right) \cdot \left(\frac{q}{p}\right) = (-1)^{\frac{p-1}{2} \cdot \frac{q-1}{2}} ,$$

$$\left(\frac{-1}{p}\right) = (-1)^{\frac{p-1}{2}} , \quad \left(\frac{2}{p}\right) = (-1)^{\frac{1}{8}(p^2-1)}$$

Proof. By the Hilbert Reciprocity Law we have

$$\prod_l \left(\frac{p,q}{l}\right) = 1$$

where l runs through all prime spots including ∞. But $\left(\frac{p,q}{\infty}\right) = 1$ since p and q are positive real numbers. And if l is any prime number distinct from p, q, 2 we have $\left(\frac{p,q}{l}\right) = 1$ by Proposition 73:1. Hence

$$\left(\frac{p,q}{p}\right) \cdot \left(\frac{p,q}{q}\right) = \left(\frac{p,q}{2}\right) .$$

Apply Propositions 73:1 and 73:2. This proves the reciprocity law. The first supplement is obtained in the same way from the equation

$$\prod_l \left(\frac{-1,p}{l}\right) = 1 ,$$

the second from the equation

$$\prod_l \left(\frac{2,p}{l}\right) = 1 .$$

q. e. d.

Part Four

Arithmetic Theory of Quadratic Forms over Rings

Chapter VIII

Quadratic Forms over Dedekind Domains

The rest of the book is devoted to a study of the equivalence of quadratic forms over the integers of local and global fields. Our first purpose in the present chapter is to state the nature of this problem in modern terminology and in the general setting of an arbitrary Dedekind domain. Our second purpose is to develop some technique in this general situation. The more interesting results must wait until we specialize to the fields of number theory.

We carry over the notation of Chapter II. F is a field, $\mathfrak{o} = \mathfrak{o}(S)$ is a Dedekind domain defined by a Dedekind set of spots S on F, $I = I(S)$ is the resulting group of fractional ideals, $\mathfrak{u} = \mathfrak{u}(S)$ is the group of units of F at S. So here $\mathfrak{o}$ is the ring of integers of our theory. As in § 22C we shall allow the same letter $\mathfrak{p}$ to stand for a spot in S and also for the prime ideal of $\mathfrak{o}$ that is determined by this spot.

V will denote an n-dimensional vector space over F. In the second half of the chapter we will make V into a quadratic space by providing it with a symmetric bilinear form $B: V \times V \to F$.

The general assumption that the characteristic of the underlying field F is not 2 will not be used in the first paragraph of this chapter.

§ 81. Abstract lattices

§ 81A. Definition of a lattice

Consider a subset M of V which is an $\mathfrak{o}$-module under the laws induced by the vector space structure of V over F. We define

$$FM = \{\alpha x \mid \alpha \in F, \quad x \in M\}.$$

Since M is an $\mathfrak{o}$-module, and since F is the quotient field of $\mathfrak{o}$, we have

$$FM = \{\alpha^{-1} x \mid \alpha \in \mathfrak{o}, \quad \alpha \neq 0, \quad x \in M\}.$$

From this it follows that FM is a subspace of V, in fact the subspace of V spanned by M. Given $\alpha \in F$ and $\mathfrak{a} \in I$ we put

$$\alpha M = \{\alpha x \mid x \in M\}, \quad \mathfrak{a} M = \Big\{\sum_{\text{fin}} \beta x \mid \beta \in \mathfrak{a}, \ x \in M\Big\}.$$

These are again $\mathfrak{o}$-modules and the following laws are easily seen to hold:

$$\alpha(M \cap N) = (\alpha M) \cap (\alpha N)$$

$$(\alpha\mathfrak{o})\, M = \alpha M\,, \quad (\alpha\mathfrak{a})\, M = \alpha(\mathfrak{a} M)$$

$$(\mathfrak{a} + \mathfrak{b})\, M = \mathfrak{a} M + \mathfrak{b} M\,, \quad (\mathfrak{a}\mathfrak{b})\, M = \mathfrak{a}(\mathfrak{b} M)$$

$$\mathfrak{a}(M + N) = \mathfrak{a} M + \mathfrak{a} N$$

$$F(M + N) = FM + FN\,.$$

We call the above $\mathfrak{o}$-module M a lattice in V (with respect to $\mathfrak{o}$, or with respect to the defining set of spots S) if there is a base $x_1, \ldots, x_n$ for V such that

$$M \subseteq \mathfrak{o} x_1 + \cdots + \mathfrak{o} x_n\,;$$

we say that M is a lattice on V if, in addition to the above property, we have $FM = V$. In particular, $\mathfrak{o} x_1 + \cdots + \mathfrak{o} x_n$ is a lattice on V. The single point 0 will always be regarded as a lattice.

81:1. *Let L be a lattice on the vector space V over F. Then the $\mathfrak{o}$-module M in V is a lattice in V if and only if there is a non-zero α in $\mathfrak{o}$ such that $\alpha M \subseteq L$.*

Proof. 1) First suppose that M is a lattice in V. So there is a base $x_1, \ldots, x_n$ for V such that

$$M \subseteq \mathfrak{o} x_1 + \cdots + \mathfrak{o} x_n\,.$$

Since L is on V we can find n independent elements $y_1, \ldots, y_n \in L$. Write

$$x_j = \sum_{i=1}^{n} a_{ij} y_i \quad (a_{ij} \in F)\,.$$

These a_{ij} generate a certain fractional ideal, hence there is a non-zero α in $\mathfrak{o}$ such that $\alpha a_{ij} \in \mathfrak{o}$ for all i, j. Hence

$$\alpha x_j \in \mathfrak{o} y_1 + \cdots + \mathfrak{o} y_n \subseteq L\,,$$

hence $\alpha M \subseteq L$.

2) Now the converse. We have a non-zero α in $\mathfrak{o}$ such that $\alpha M \subseteq L$. Since L is a lattice there is a base $z_1, \ldots, z_n$ for V such that $L \subseteq \mathfrak{o} z_1 + \cdots + \mathfrak{o} z_n$. Then

$$M \subseteq \alpha^{-1} L \subseteq \mathfrak{o}\left(\frac{z_1}{\alpha}\right) + \cdots + \mathfrak{o}\left(\frac{z_n}{\alpha}\right).$$

Hence M is a lattice in V. **q. e. d.**

81:1a. *Let U be a subspace of V with $M \subseteq U \subseteq V$. Then M is a lattice in V if and only if it is a lattice in U.*

Proof. Take a base $x_1, \ldots, x_r$ for U and extend it to a base $x_1, \ldots, x_n$ for V. Put $L' = \mathfrak{o}x_1 + \cdots + \mathfrak{o}x_r$ and $L = \mathfrak{o}x_1 + \cdots + \mathfrak{o}x_n$. If M is a lattice in U, then $\alpha M \subseteq L' \subseteq L$ for some non-zero α in $\mathfrak{o}$, and so M is a lattice in V. If M is a lattice in V, then $\alpha M \subseteq L$ for some non-zero α in $\mathfrak{o}$, hence $\alpha M \subseteq L \cap U = L'$, and so M is a lattice in U. **q. e. d.**

It follows immediately from the definition that every submodule of a lattice is a lattice. In particular $L \cap K$ is a lattice whenever L and K are lattices in V. And Proposition 81:1 shows that αL, $\mathfrak{a}L$, $L + K$ are lattices for any $\alpha \in F$, $\mathfrak{a} \in I$. Clearly $\mathfrak{o}x$ is a lattice for any x in V, and $\mathfrak{a}x$ is also a lattice in V. Hence $\mathfrak{a}_1 z_1 + \cdots + \mathfrak{a}_r z_r$ is a lattice for any $\mathfrak{a}_i \in I$, $z_i \in V$. In particular, every finitely generated $\mathfrak{o}$-module in V is a lattice.

§ 81B. Bases

Consider the lattice L in V. For any non-zero vector x in FL we define the coefficient of x in L to be the set

$$\mathfrak{a}_x = \{\alpha \in F \mid \alpha x \in L\}.$$

This is clearly an $\mathfrak{o}$-module in F, and it follows from the fact that L spans FL that it is not zero. Now

$$\mathfrak{a}_x x = L \cap Fx,$$

hence $\mathfrak{a}_x x$ is a lattice in Fx, hence $\alpha(\mathfrak{a}_x x) \subseteq \mathfrak{o}x$ for some non-zero α in $\mathfrak{o}$. Therefore $\alpha\mathfrak{a}_x \subseteq \mathfrak{o}$, so that $\mathfrak{a}_x$ is actually a fractional ideal in F. Note that

$$\alpha\mathfrak{a}_{\alpha x} = \mathfrak{a}_x \qquad \forall\, \alpha \in \dot{F}.$$

It is clear that

$$\mathfrak{a}_x \supseteq \mathfrak{o} \Leftrightarrow x \in L.$$

We say that x is a maximal vector of L if $\mathfrak{a}_x = \mathfrak{o}$. So x is a maximal vector of L if and only if

$$L \cap Fx = \mathfrak{o}x.$$

Every line in FL contains a maximal vector of L when the class number of F at S is equal to 1, i. e. when every fractional ideal is principal. For consider the line Fy in FL. Put $\mathfrak{a}_y = \alpha\mathfrak{o}$ with $\alpha \in \dot{F}$, then put $x = \alpha y$; we have $\mathfrak{a}_x = \mathfrak{o}$, hence x is a maximal vector of L that falls in the line Fy.

81:2. *Given a lattice L on V, a hyperplane U in V, and a vector x_0 in $V - U$. Then among all vectors in $x_0 + U$ there is at least one whose coefficient with respect to L is absolutely largest. Let this coefficient be $\mathfrak{a}$. Then for any vector $x_0 + u_0$ $(u_0 \in U)$ with coefficient $\mathfrak{a}$ we have*

$$L = \mathfrak{a}(x_0 + u_0) + (L \cap U).$$

Proof. 1) We claim that the set

$$\mathfrak{a} = \{\alpha \in F \mid \alpha x_0 \in L + U\}$$

is a fractional ideal in F. It is clearly a non-zero $\mathfrak{o}$-module in F. And by Proposition 81:1 there is a non-zero β in $\mathfrak{o}$ such that $\beta L \subseteq \mathfrak{o} x_0 + U$. Hence

$$(\beta \mathfrak{a})\, x_0 \subseteq \beta L + U \subseteq \mathfrak{o} x_0 + U .$$

So $\beta \mathfrak{a} \subseteq \mathfrak{o}$. Hence $\mathfrak{a}$ is indeed a fractional ideal as claimed.

Now the coefficient of any vector in $x_0 + U$ is contained in $\mathfrak{a}$ by definition of $\mathfrak{a}$. Hence the first part of the proposition will be proved if we can find a vector u in U such that $\mathfrak{a}(x_0 + u) \subseteq L$. Since $\mathfrak{a}\mathfrak{a}^{-1} = \mathfrak{o}$ we can find an expression

$$\alpha_1 \beta_1 + \cdots + \alpha_r \beta_r = 1 \qquad (\alpha_i \in \mathfrak{a},\ \beta_i \in \mathfrak{a}^{-1}) .$$

Now each α_i provides an expression

$$\alpha_i x_0 = l_i + u_i \qquad (l_i \in L, \quad u_i \in U)$$

by definition of $\mathfrak{a}$. Then

$$x_0 = \sum_1^r \beta_i l_i + \sum_1^r \beta_i u_i .$$

But $\beta_i \mathfrak{a} \subseteq \mathfrak{o}$ for $1 \leq i \leq r$. Hence

$$\mathfrak{a}\left(x_0 - \sum_1^r \beta_i u_i\right) \subseteq L .$$

So we have found $u \in U$ such that $\mathfrak{a}(x_0 + u) \subseteq L$ and the first part of the proposition is proved.

2) We are given that $\mathfrak{a}$ is the coefficient of $x_0 + u_0$, hence

$$\mathfrak{a}(x_0 + u_0) + (L \cap U) \subseteq L .$$

We must reverse this inclusion relation. So consider a typical vector in L which has the form $\alpha(x_0 + u)$ with $\alpha \in F, u \in U$. Then $\alpha x_0 \in L + U$ and so $\alpha \in \mathfrak{a}$ by definition of $\mathfrak{a}$. Hence

$$\alpha(u - u_0) = \alpha(x_0 + u) - \alpha(x_0 + u_0) \in L .$$

Therefore

$$\alpha(x_0 + u) = \alpha(x_0 + u_0) + \alpha(u - u_0) \in \mathfrak{a}(x_0 + u_0) + (L \cap U) .$$

q. e. d.

81:3. **Theorem.** *L is a lattice on the vector space V and $x_1, \ldots, x_n$ is a base for V. Then there is a base $y_1, \ldots, y_n$ with*

$$y_j \in F x_1 + \cdots + F x_j \quad (1 \leq j \leq n) ,$$

and there are fractional ideals $\mathfrak{a}_1, \ldots, \mathfrak{a}_n$, such that

$$L = \mathfrak{a}_1 y_1 + \cdots + \mathfrak{a}_n y_n .$$

Proof. Let U be the hyperplane $U = F x_1 + \cdots + F x_{n-1}$. Then by Proposition 81:2

$$L = (L \cap U) + \mathfrak{a}_n y_n$$

for some fractional ideal $\mathfrak{a}_n$ and some $y_n \in V - U$. Proceed by induction on $n = \dim V$.

q. e. d.

81:4. **Example.** Let $x_1, \ldots, x_n$ be a base for V and let $L = \mathfrak{a}_1 x_1 + \ldots + \mathfrak{a}_n x_n$ with $\mathfrak{a}_i \in I$. Then the coefficient with respect to L of any vector of the form

$$\alpha_1 x_1 + \cdots + \alpha_r x_r \qquad (\alpha_i \in \dot{F})$$

is equal to

$$(\mathfrak{a}_1 \alpha_1^{-1}) \cap \cdots \cap (\mathfrak{a}_r \alpha_r^{-1}) \, .$$

In particular, the coefficient of x_i is equal to $\mathfrak{a}_i$.

81:5. *Let L be a lattice on the vector space V. Then there is a fractional ideal $\mathfrak{a}$ and a base $z_1, \ldots, z_n$ for V such that*

$$L = \mathfrak{a} z_1 + \mathfrak{o} z_2 + \cdots + \mathfrak{o} z_n \, .$$

Proof. Let us write $L = \mathfrak{a}_1 y_1 + \cdots + \mathfrak{a}_n y_n$ in the manner of Theorem 81:3. If $n = 1$ we have $L = \mathfrak{a}_1 y_1$ and we are through. The case of a general $n \geqq 3$ follows by successive applications of the case $n = 2$. So let us assume that $n = 2$. By Proposition 22:5 we can find $\alpha_1, \alpha_2 \in \dot{F}$ such that

$$\alpha_1 \mathfrak{a}_1^{-1} + \alpha_2 \mathfrak{a}_2^{-1} = \mathfrak{o} \, .$$

Put $x = \alpha_1 y_1 + \alpha_2 y_2$. Then the coefficient of x in L is equal to

$$(\mathfrak{a}_1 \alpha_1^{-1}) \cap (\mathfrak{a}_2 \alpha_2^{-1}) = (\alpha_1 \mathfrak{a}_1^{-1} + \alpha_2 \mathfrak{a}_2^{-1})^{-1} = \mathfrak{o}$$

by Examples 22:4 and 81:4. Hence $L = \mathfrak{o} x + \mathfrak{b} y$ by Theorem 81:3.

q. e. d.

81:6. **Example.** An $\mathfrak{o}$-module in V is a lattice if and only if it is finitely generated.

We say that the base $z_1, \ldots, z_n$ for V is adapted to the lattice L if there are fractional ideals $\mathfrak{a}_1, \ldots, \mathfrak{a}_n$ such that

$$L = \mathfrak{a}_1 z_1 + \cdots + \mathfrak{a}_n z_n \, .$$

Theorem 81:3 asserts that there is a base for FL that is adapted to L, where L is any lattice in V.

Consider a lattice L in V. It follows immediately from the fact that F is the quotient field of $\mathfrak{o}$ that a set of vectors in L is independent over $\mathfrak{o}$ if and only if it is independent over F. Hence a set of vectors of L is maximal independent over $\mathfrak{o}$ if and only if it is a base for FL. In particular, any two such sets must contain the same number of elements. This number is called the rank of L and is written rank L. Thus

$$\operatorname{rank} L = \dim FL \, .$$

A set of vectors is called a base for L if it is a base in the sense of $\mathfrak{o}$-modules, i. e. if it is independent and spans L over $\mathfrak{o}$. So $x_1, \ldots, x_r$ is a base for L if and only if it is a base for FL with

$$L = \mathfrak{o}x_1 + \cdots + \mathfrak{o}x_r .$$

A lattice which has a base is called free. Any two bases of a free lattice L contain the same number of elements; this number is called the dimension of L and is written $\dim L$; we have

$$\dim L = \dim FL = \operatorname{rank} L .$$

Every lattice L is almost free in the sense that it can be expressed in the form

$$L = \mathfrak{a}x_1 + \mathfrak{o}x_2 + \cdots + \mathfrak{o}x_r$$

with $\mathfrak{a}$ a fractional ideal and $x_1, \ldots, x_r$ a base for FL. Clearly every lattice is free when the class number is 1, i. e. when every fractional ideal is principal.

§ 81 C. Change of base

Consider two lattices L and K on the same vector space V and let $x_1, \ldots, x_n$ and $y_1, \ldots, y_n$ be bases in which

$$L = \mathfrak{a}_1 x_1 + \cdots + \mathfrak{a}_n x_n \qquad (\mathfrak{a}_i \in I)$$

$$K = \mathfrak{b}_1 y_1 + \cdots + \mathfrak{b}_n y_n \qquad (\mathfrak{b}_i \in I) .$$

Let

$$y_j = \sum_i a_{ij} x_i , \quad x_j = \sum_i b_{ij} y_i$$

be the equations relating these two bases. So (a_{ij}) is the inverse of the matrix (b_{ij}).

81:7. *$K \subseteq L$ if and only if $a_{ij}\mathfrak{b}_j \subseteq \mathfrak{a}_i$ for all i, j.*

Proof. We have $K \subseteq L$ if and only if $\mathfrak{b}_j y_j \subseteq L$, i. e. if and only if

$$\mathfrak{b}_j (a_{1j} x_1 + \cdots + a_{ij} x_i + \cdots) \subseteq \mathfrak{a}_1 x_1 + \cdots + \mathfrak{a}_i x_i + \cdots$$

for $1 \leqq j \leqq n$. This is true if and only if $a_{ij}\mathfrak{b}_j \subseteq \mathfrak{a}_i$ for $1 \leqq i \leqq n$, $1 \leqq j \leqq n$.

q. e. d.

81:8. *Suppose $K \subseteq L$. Then $K = L$ if and only if*

$$\mathfrak{a}_1 \ldots \mathfrak{a}_n = \mathfrak{b}_1 \ldots \mathfrak{b}_n \cdot \det(a_{ij}) .$$

Proof. First consider $K = L$. Then by Proposition 81:7 we have $a_{ij}\mathfrak{b}_j \subseteq \mathfrak{a}_i$ for all i, j, hence

$$\begin{aligned} \det(a_{ij}) &= \Sigma \pm a_{1\alpha} \ldots a_{n\omega} \\ &\in \Sigma (\mathfrak{a}_1 \mathfrak{b}_\alpha^{-1}) \ldots (\mathfrak{a}_n \mathfrak{b}_\omega^{-1}) \\ &= (\mathfrak{a}_1 \ldots \mathfrak{a}_n)(\mathfrak{b}_1 \ldots \mathfrak{b}_n)^{-1} . \end{aligned}$$

Hence

$$(\mathfrak{b}_1 \ldots \mathfrak{b}_n) \det(a_{ij}) \subseteq (\mathfrak{a}_1 \ldots \mathfrak{a}_n) .$$

Similarly

$$(\mathfrak{a}_1 \dots \mathfrak{a}_n) \det(b_{ij}) \subseteq (\mathfrak{b}_1 \dots \mathfrak{b}_n) .$$

Now $\det(a_{ij})$ is the inverse of $\det(b_{ij})$. Hence the result follows.

Now the converse. Since $K \subseteq L$ we have $a_{ij} \in \mathfrak{a}_i \mathfrak{b}_j^{-1}$ for all relevant i, j. The cofactor A_{ij} of a_{ij} is equal to

$$A_{ij} = \Sigma \pm a_{1\alpha} \dots a_{n\omega}$$

in which the first index avoids i and the second j. Hence

$$A_{ij} \mathfrak{a}_i \mathfrak{b}_j^{-1} \subseteq (\mathfrak{a}_1 \dots \mathfrak{a}_n)(\mathfrak{b}_1 \dots \mathfrak{b}_n)^{-1} = \mathfrak{o} \cdot \det(a_{ij}) .$$

Therefore

$$b_{ji} \mathfrak{a}_i = \frac{A_{ij}}{\det(a_{ij})} \mathfrak{a}_i \subseteq \mathfrak{b}_j .$$

This is true for all i, j. Hence $L \subseteq K$ by Proposition 81:7. Hence $L = K$.

q. e. d.

Recall that the elements of $\mathfrak{o}$ are the integers of our theory. Accordingly we say that an $n \times n$ matrix (a_{ij}) with entries in F is integral (with respect to $\mathfrak{o}$, or with respect to the defining set of spots S) if each of its entries is in $\mathfrak{o}$. We shall call (a_{ij}) unimodular if it is integral with $\det(a_{ij})$ a unit of F at S. By looking at cofactors one sees that the inverse of a unimodular matrix is integral, and hence unimodular. The defining equation of the inverse of a matrix shows that if an integral matrix has an inverse, and if this inverse is integral, then both the matrix and its inverse are unimodular. In other words, an integral matrix is unimodular if and only if it is invertible with an integral inverse.

81:9. *Suppose L is a free lattice with base $x_1, \dots, x_n$ and consider vectors $y_1, \dots, y_n$ determined by*

$$y_j = \sum_i a_{ij} x_i \qquad (a_{ij} \in F) .$$

Then these vectors form a base for L if and only if the matrix (a_{ij}) is unimodular.

Proof. This is an easy application of Proposition 81:8. **q. e. d.**

81:10. **Example.** Let $v_1, \dots, v_r$ be vectors in V, let ε be a unit in $\mathfrak{o}$, and let $\alpha_2, \dots, \alpha_r$ be elements of $\mathfrak{o}$. Then

$$\mathfrak{o} v_1 + \mathfrak{o} v_2 + \dots + \mathfrak{o} v_r = \mathfrak{o} \bar{v}_1 + \mathfrak{o} v_2 + \dots + \mathfrak{o} v_r$$

where $\bar{v}_1 = \varepsilon v_1 + \alpha_2 v_2 + \dots + \alpha_r v_r$.

§ 81D. Invariant factors

81:11. **Theorem**[1]. *Given lattices L and K on the non-zero vector space V.*

---

[1] This theorem can be used to derive structure theorems for finitely generated modules over the Dedekind domain $\mathfrak{o}$. These structure theorems reduce to the Fundamental Theorem of Abelian Groups when $\mathfrak{o}$ is the ring of rational integers $\mathbf{Z}$.

Then there is a base $x_1, \ldots, x_n$ for V in which

$$\begin{cases} L = \mathfrak{a}_1 x_1 + \cdots + \mathfrak{a}_n x_n \\ K = \mathfrak{a}_1 \mathfrak{r}_1 x_1 + \cdots + \mathfrak{a}_n \mathfrak{r}_n x_n \end{cases}$$

where $\mathfrak{a}_i$ and $\mathfrak{r}_i$ are fractional ideals with

$$\mathfrak{r}_1 \supseteq \mathfrak{r}_2 \supseteq \cdots \supseteq \mathfrak{r}_n .$$

The $\mathfrak{r}_i$ determined in this way are unique.

Proof. 1) We can suppose that $K \subseteq L$ (if necessary we can replace K by αK where α is a suitable non-zero element of $\mathfrak{o}$). For any x in $\dot{V}$ we let $\mathfrak{a}_x$ denote the coefficient of x in L and $\mathfrak{b}_x$ the coefficient of x in K. Then put $\mathfrak{r}_x = \mathfrak{b}_x / \mathfrak{a}_x$. Since we are taking $K \subseteq L$ we have $\mathfrak{b}_x \subseteq \mathfrak{a}_x$ and so $\mathfrak{r}_x \subseteq \mathfrak{o}$.

2) We can therefore take a $v \in \dot{V}$ for which $\mathfrak{r}_v$ is maximal (though we are not yet sure it will be absolutely largest). By Theorem 81:3 there is a hyperplane U such that

$$L = \mathfrak{a}_v v + (L \cap U) .$$

We claim that $\mathfrak{b}_{v+u} \subseteq \mathfrak{b}_v$ for any $u \in U$. If not, then by Proposition 81:2 we can find a $u \in U$ for which $\mathfrak{b}_{v+u} \supset \mathfrak{b}_v$. But

$$\mathfrak{a}_{v+u}(v+u) \subseteq L = \mathfrak{a}_v v + (L \cap U) ,$$

so that $\mathfrak{a}_{v+u} \subseteq \mathfrak{a}_v$. Hence

$$\mathfrak{r}_{v+u} = \frac{\mathfrak{b}_{v+u}}{\mathfrak{a}_{v+u}} \supset \frac{\mathfrak{b}_v}{\mathfrak{a}_v} = \mathfrak{r}_v$$

and this contradicts the choice of v. So we do indeed have $\mathfrak{b}_{v+u} \subseteq \mathfrak{b}_v$ for all u in U. Now apply Proposition 81:2 again. We obtain

$$K = \mathfrak{b}_v v + (K \cap U) .$$

3) An inductive argument gives us expressions

$$\begin{cases} L = \mathfrak{a}_v v + (\mathfrak{a}_w w + \cdots + \mathfrak{a}_z z) \\ K = \mathfrak{a}_v \mathfrak{r}_v v + (\mathfrak{a}_w \mathfrak{r}_w w + \cdots + \mathfrak{a}_z \mathfrak{r}_z z) \end{cases}$$

with $\mathfrak{r}_w \supseteq \cdots \supseteq \mathfrak{r}_z$. We shall therefore be through with the first part of the theorem if we can prove that $\mathfrak{r}_v \supseteq \mathfrak{r}_w$. By Example 22:9 we can pick $\alpha, \beta \in \dot{F}$ in such a way that

$$\begin{cases} \alpha(\mathfrak{a}_v^{-1}\mathfrak{r}_v^{-1}) + \beta(\mathfrak{a}_w^{-1}\mathfrak{r}_w^{-1}) = \mathfrak{o} \\ \alpha \mathfrak{a}_v^{-1} + \beta \mathfrak{a}_w^{-1} = \mathfrak{r}_v + \mathfrak{r}_w . \end{cases}$$

Put $x = \alpha v + \beta w$. By Examples 22:4 and 81:4 we obtain

$$\mathfrak{a}_x = (\mathfrak{a}_v \alpha^{-1}) \cap (\mathfrak{a}_w \beta^{-1}) = (\alpha \mathfrak{a}_v^{-1} + \beta \mathfrak{a}_w^{-1})^{-1} = (\mathfrak{r}_v + \mathfrak{r}_w)^{-1} .$$

And similarly $\mathfrak{b}_x = \mathfrak{o}$. Hence $\mathfrak{r}_x = \mathfrak{r}_v + \mathfrak{r}_w$. So $\mathfrak{r}_w \subseteq \mathfrak{r}_v$ by choice of v, and this is what we required.

4) Now the uniqueness. Let us call $\mathfrak{r}_i$ the i-th invariant factor of K in L (a formal definition will be made once the theorem is proved). Our purpose is to prove that the invariant factors are indeed invariant, i.e. that they are independent of the base used in defining them. We make a start by remarking that the product of the invariant factors is invariant: the reader may easily verify this by using Proposition 81:8.

Consider the invariant factors $\mathfrak{r}_1, \ldots, \mathfrak{r}_n$ in the given base and let $\mathfrak{r}'_1, \ldots, \mathfrak{r}'_n$ be the invariant factors with respect to some other base. Suppose if possible that the second base gives rise to a different set of invariant factors. Take the first i $(1 \leqq i \leqq n)$ for which $\mathfrak{r}_i \neq \mathfrak{r}'_i$. We can suppose that we actually have $\mathfrak{r}_i + \mathfrak{r}'_i \supset \mathfrak{r}'_i$. We put

$$J = K + (\mathfrak{r}'_i L) .$$

Consider the invariant factors of J in L. For $1 \leqq \lambda \leqq i-1$ the λ-th invariant factor is $\mathfrak{r}_\lambda$ in either base; for $i \leqq \lambda \leqq n$ it is $\mathfrak{r}_\lambda + \mathfrak{r}'_i$ in the first base and $\mathfrak{r}'_i$ in the second. But this means that the product of all invariant factors in the first base is strictly larger than the product in the second base. We have already remarked that these products must be equal. So we have a contradiction. Hence $\mathfrak{r}_\lambda = \mathfrak{r}'_\lambda$ for $1 \leqq \lambda \leqq n$. **q. e. d.**

81:12. **Definition.** The invariants $\mathfrak{r}_1 \supseteqq \cdots \supseteqq \mathfrak{r}_n$ of the last theorem are called the invariant factors of K in L. And $\mathfrak{r}_i$ is called the i-th invariant factor of K in L (for $1 \leqq i \leqq n$).

Suppose we have lattices K and L on V with $K \subseteqq L$. In this event the invariant factors of K in L are all integral ideals. Referring to Theorem 81:11 and § 22D we find

$$(L:K) = \prod_{1 \leqq i \leqq n} (\mathfrak{a}_i : \mathfrak{a}_i \mathfrak{r}_i) = \prod_{1 \leqq i \leqq n} (\mathfrak{o} : \mathfrak{r}_i) .$$

In the important situations (e.g. local fields and global fields) the indices $(\mathfrak{o} : \mathfrak{r}_i)$ are all finite; hence $(L:K) < \infty$; and the number of lattices between K and L is finite.

§ 81E. Localization

Let us give some attention to the completions $F_\mathfrak{p}$ of F at the spots $\mathfrak{p}$ in S. We let $|\ |_\mathfrak{p}$ stand for any fixed valuation on $F_\mathfrak{p}$ at $\mathfrak{p}$. Of course the Dedekind theory of ideals applies to $F_\mathfrak{p}$ at $\mathfrak{p}$, and the theory of lattices applies to vector spaces over $F_\mathfrak{p}$. We let $I_\mathfrak{p}$ denote the group of fractional ideals of $F_\mathfrak{p}$ at $\mathfrak{p}$. Recall that $\mathfrak{o}_\mathfrak{p}$, $\mathfrak{u}_\mathfrak{p}$, $\mathfrak{m}_\mathfrak{p}$ are used instead of $\mathfrak{o}(\mathfrak{p})$, $\mathfrak{u}(\mathfrak{p})$, $\mathfrak{m}(\mathfrak{p})$ on $F_\mathfrak{p}$ at $\mathfrak{p}$. Here the same symbol $\mathfrak{p}$ denotes two different spots, the one on F and the other on $F_\mathfrak{p}$. And $\mathfrak{p}$ also denotes two different prime ideals, the prime ideal $\mathfrak{o} \cap \mathfrak{m}_\mathfrak{p}$ of $\mathfrak{o}$ and the prime ideal $\mathfrak{m}_\mathfrak{p}$ of $\mathfrak{o}_\mathfrak{p}$.

A typical fractional ideal of F at S has the form

$$\prod_{\mathfrak{p} \in S} \mathfrak{p}^{\nu_\mathfrak{p}}$$

with all $\nu_{\mathfrak{p}}$ in $\mathbf{Z}$ and almost all of them 0; and a typical fractional ideal of $F_{\mathfrak{p}}$ at $\mathfrak{p}$ has the form

$$\mathfrak{p}^{\mu_{\mathfrak{p}}} \qquad (\mu_{\mathfrak{p}} \in \mathbf{Z}) .$$

We introduce the surjective homomorphism

$$I(S) \to I_{\mathfrak{p}}$$

defined by the mapping

$$\prod_{\mathfrak{p} \in S} \mathfrak{p}^{\nu_{\mathfrak{p}}} \rightarrowtail \mathfrak{p}^{\nu_{\mathfrak{p}}} .$$

The image of $\mathfrak{a} \in I(S)$ under this mapping will be written $\mathfrak{a}_{\mathfrak{p}}$. We call $\mathfrak{a}_{\mathfrak{p}}$ the $\mathfrak{p}$-ification or localization at $\mathfrak{p}$ of the ideal $\mathfrak{a}$.

The following laws follow easily from the results of §§ 22B and 22C:

$$(\alpha\mathfrak{a})_{\mathfrak{p}} = \alpha\mathfrak{a}_{\mathfrak{p}} , \quad (\mathfrak{a}\mathfrak{b})_{\mathfrak{p}} = \mathfrak{a}_{\mathfrak{p}}\mathfrak{b}_{\mathfrak{p}} ,$$

$$(\mathfrak{a} \cap \mathfrak{b})_{\mathfrak{p}} = \mathfrak{a}_{\mathfrak{p}} \cap \mathfrak{b}_{\mathfrak{p}} , \quad (\mathfrak{a} + \mathfrak{b})_{\mathfrak{p}} = \mathfrak{a}_{\mathfrak{p}} + \mathfrak{b}_{\mathfrak{p}} ,$$

$$|\mathfrak{a}_{\mathfrak{p}}|_{\mathfrak{p}} = |\mathfrak{a}|_{\mathfrak{p}} .$$

This last equation shows that $\mathfrak{a}$ is always contained in $\mathfrak{a}_{\mathfrak{p}}$, and in fact

$$\mathfrak{a} = \bigcap_{\mathfrak{p} \in S} (F \cap \mathfrak{a}_{\mathfrak{p}}) .$$

We have

$$\mathfrak{a} \subseteq \mathfrak{b} \Leftrightarrow \mathfrak{a}_{\mathfrak{p}} \subseteq \mathfrak{b}_{\mathfrak{p}} \quad \forall \mathfrak{p} \in S ,$$

and in particular

$$\mathfrak{a} = \mathfrak{b} \Leftrightarrow \mathfrak{a}_{\mathfrak{p}} = \mathfrak{b}_{\mathfrak{p}} \quad \forall \mathfrak{p} \in S .$$

It is easily seen, again using the equation $|\mathfrak{a}_{\mathfrak{p}}|_{\mathfrak{p}} = |\mathfrak{a}|_{\mathfrak{p}}$, that $\mathfrak{a}_{\mathfrak{p}}$ is the $\mathfrak{o}_{\mathfrak{p}}$-ideal generated by $\mathfrak{a}$ in $F_{\mathfrak{p}}$.

81:13. $\mathfrak{a}_{\mathfrak{p}}$ *is the closure of* $\mathfrak{a}$ *in* $F_{\mathfrak{p}}$.

Proof. We know that

$$\mathfrak{a}_{\mathfrak{p}} = \{\alpha \in F_{\mathfrak{p}} \mid |\alpha|_{\mathfrak{p}} \leqq |\mathfrak{a}|_{\mathfrak{p}}\} ;$$

but the map $\alpha \rightarrowtail |\alpha|_{\mathfrak{p}}$ of $F_{\mathfrak{p}}$ into $\mathbf{R}$ is continuous; hence $\mathfrak{a}_{\mathfrak{p}}$ is closed in $F_{\mathfrak{p}}$. Why is $\mathfrak{a}_{\mathfrak{p}}$ the closure of $\mathfrak{a}$? We must consider a typical $\alpha \in \mathfrak{a}_{\mathfrak{p}}$ and an $\varepsilon > 0$, and we must find an $a \in \mathfrak{a}$ such that $|a - \alpha|_{\mathfrak{p}} < \varepsilon$. We can assume that $0 < \varepsilon < |\alpha|_{\mathfrak{p}} \leqq |\mathfrak{a}|_{\mathfrak{p}}$. By the Strong Approximation Theorem we can find an a in F such that

$$\begin{cases} |a - \alpha|_{\mathfrak{p}} < \varepsilon \\ |a|_{\mathfrak{q}} \leqq |\mathfrak{a}|_{\mathfrak{q}} \qquad \forall \mathfrak{q} \in S - \mathfrak{p} . \end{cases}$$

The Principle of Domination then insists that $|a|_{\mathfrak{p}} = |\alpha|_{\mathfrak{p}}$. Hence $|a|_{\mathfrak{q}} \leqq |\mathfrak{a}|_{\mathfrak{q}}$ for all $\mathfrak{q}$ in S. So a is in $\mathfrak{a}$. **q. e. d.**

Now consider an n-dimensional vector space V over F. As in § 66 we use $V_{\mathfrak{p}}$ for the localization $F_{\mathfrak{p}}V$ of V at $\mathfrak{p}$. Let L be a lattice in V (with respect to the set of spots S). By the $\mathfrak{p}$-ification or localization $L_{\mathfrak{p}}$

of L in $V_\mathfrak{p}$ at the spot $\mathfrak{p}$ in S we mean the $\mathfrak{o}_\mathfrak{p}$-module generated by L in $V_\mathfrak{p}$. By the definition of a lattice there is a base $x_1, \ldots, x_n$ for V such that

$$L \subseteqq \mathfrak{o}x_1 + \cdots + \mathfrak{o}x_n ,$$

hence

$$L_\mathfrak{p} \subseteqq \mathfrak{o}_\mathfrak{p}x_1 + \cdots + \mathfrak{o}_\mathfrak{p}x_n ,$$

so $L_\mathfrak{p}$ is a lattice in $V_\mathfrak{p}$ (with respect to $\mathfrak{o}_\mathfrak{p}$, or with respect to the spot $\mathfrak{p}$ on $F_\mathfrak{p}$).

If L and K are lattices in V, then it is easily seen that

$$(L + K)_\mathfrak{p} = L_\mathfrak{p} + K_\mathfrak{p} \qquad \forall \mathfrak{p} \in S .$$

Now $(\mathfrak{a}x)_\mathfrak{p} = \mathfrak{a}_\mathfrak{p}x$ for any $\mathfrak{a} \in I(S)$ and any x in V, hence

$$(\mathfrak{a}_1 z_1 + \cdots + \mathfrak{a}_r z_r)_\mathfrak{p} = \mathfrak{a}_{1\mathfrak{p}} z_1 + \cdots + \mathfrak{a}_{r\mathfrak{p}} z_r$$

for any $\mathfrak{a}_i \in I(S)$ and any $z_i \in V$. If we take a base $x_1, \ldots, x_n$ for V such that

$$L = \mathfrak{a}_1 x_1 + \cdots + \mathfrak{a}_r x_r ,$$

then

$$L_\mathfrak{p} = \mathfrak{a}_{1\mathfrak{p}} x_1 + \cdots + \mathfrak{a}_{r\mathfrak{p}} x_r .$$

Hence

$$V \cap L_\mathfrak{p} = (F \cap \mathfrak{a}_{1\mathfrak{p}}) x_1 + \cdots + (F \cap \mathfrak{a}_{r\mathfrak{p}}) x_r ,$$

so

$$\bigcap_{\mathfrak{p} \in S} (V \cap L_\mathfrak{p}) = L .$$

We therefore have

$$L \subseteqq K \iff L_\mathfrak{p} \subseteqq K_\mathfrak{p} \qquad \forall \mathfrak{p} \in S ,$$

and in particular,

$$L = K \iff L_\mathfrak{p} = K_\mathfrak{p} \qquad \forall \mathfrak{p} \in S .$$

If $FL = FK$, then the Invariant Factor Theorem of § 81D shows that $L_\mathfrak{p} = K_\mathfrak{p}$ for almost all $\mathfrak{p}$ in S.

81:14. *An $\mathfrak{o}_\mathfrak{p}$-lattice $J_{(\mathfrak{p})}$ is given on $V_\mathfrak{p}$ at each $\mathfrak{p}$ in S. Suppose there is an $\mathfrak{o}$-lattice L on V with $L_\mathfrak{p} = J_{(\mathfrak{p})}$ for almost all $\mathfrak{p}$. Then there is an $\mathfrak{o}$-lattice K on V with $K_\mathfrak{p} = J_{(\mathfrak{p})}$ for all $\mathfrak{p}$ in S.*

Proof. 1) First we prove the following contention: given a spot $\mathfrak{p}$ in S, there is a base $y_1, \ldots, y_n$ for V such that

$$J_{(\mathfrak{p})} = \mathfrak{o}_\mathfrak{p} y_1 + \cdots + \mathfrak{o}_\mathfrak{p} y_n .$$

To prove this we fix a base $x_1, \ldots, x_n$ for V; multiplying each of these basis vectors by a scalar which is sufficiently small at $\mathfrak{p}$ allows us to assume that each x_i is in $J_{(\mathfrak{p})}$. Take a base for $J_{(\mathfrak{p})}$,

$$J_{(\mathfrak{p})} = \mathfrak{o}_\mathfrak{p} \eta_1 + \cdots + \mathfrak{o}_\mathfrak{p} \eta_n \qquad (\eta_i \in V_\mathfrak{p}) ,$$

and write each

$$\eta_j = \alpha_{1j} x_1 + \cdots + \alpha_{nj} x_n \qquad (\alpha_{ij} \in F_\mathfrak{p}) .$$

Now $F_\mathfrak{p}$ is the closure of F at $\mathfrak{p}$, hence we can find, for each i and j, an element $a_{ij} \in F$ such that

$$|a_{ij} - \alpha_{ij}|_\mathfrak{p} < 1$$

is as small as we wish. If the approximations are good enough we will obtain, in virtue of the continuity of multiplication and addition in the topological field $F_\mathfrak{p}$, the inequality

$$|\det(a_{ij}) - \det(\alpha_{ij})|_\mathfrak{p} < |\det(\alpha_{ij})|_\mathfrak{p} .$$

Hence $|\det(a_{ij})|_\mathfrak{p} = |\det(\alpha_{ij})|_\mathfrak{p}$ by the Principle of Domination. Put

$$y_j = a_{1j}x_1 + \cdots + a_{nj}x_n$$

for $1 \leqq j \leqq n$. Then

$$y_j - \eta_j \in \mathfrak{o}_\mathfrak{p} x_1 + \cdots + \mathfrak{o}_\mathfrak{p} x_n \subseteq J_{(\mathfrak{p})}$$

by choice of the a_{ij}, hence

$$\mathfrak{o}_\mathfrak{p} y_1 + \cdots + \mathfrak{o}_\mathfrak{p} y_n \subseteq J_{(\mathfrak{p})} .$$

If we write $y_j = \sum_i \gamma_{ij}\eta_i$ with all $\gamma_{ij} \in F_\mathfrak{p}$ we have

$$(\gamma_{ij}) = (\alpha_{ij})^{-1}(a_{ij}) ,$$

hence $\det(\gamma_{ij})$ is a unit in $\mathfrak{o}_\mathfrak{p}$, hence

$$\mathfrak{o}_\mathfrak{p} y_1 + \cdots + \mathfrak{o}_\mathfrak{p} y_n = J_{(\mathfrak{p})}$$

by Proposition 81:8. This proves our contention.

2) It is enough to prove the following: given a single spot $\mathfrak{p} \in S$ there is an $\mathfrak{o}$-lattice K on V such that

$$K_\mathfrak{q} = \begin{cases} L_\mathfrak{q} & \text{if } \mathfrak{q} \in S - \mathfrak{p} \\ J_{(\mathfrak{p})} & \text{if } \mathfrak{q} = \mathfrak{p} . \end{cases}$$

The final result will follow by successive applications of this special case.

In virtue of step 1) we can find an $\mathfrak{o}$-lattice J on V with $J_\mathfrak{p} = J_{(\mathfrak{p})}$. Use the Invariant Factor Theorem of § 81D to find a base $z_1, \ldots, z_n$ for V such that

$$\begin{cases} L = \mathfrak{a}_1 z_1 + \cdots + \mathfrak{a}_n z_n \\ J = \mathfrak{b}_1 z_1 + \cdots + \mathfrak{b}_n z_n \end{cases}$$

Construct fractional ideals $\mathfrak{c}_i\,(1 \leqq i \leqq n)$ with

$$\mathfrak{c}_{i\mathfrak{q}} = \begin{cases} \mathfrak{a}_{i\mathfrak{q}} & \text{if } \mathfrak{q} \in S - \mathfrak{p} \\ \mathfrak{b}_{i\mathfrak{p}} & \text{if } \mathfrak{q} = \mathfrak{p} . \end{cases}$$

Then

$$K = \mathfrak{c}_1 z_1 + \cdots + \mathfrak{c}_n z_n$$

has the required property. **q. e. d.**

§ 82. Lattices in quadratic spaces

An additional structure is imposed on the situation under discussion: the vector space V is made into a quadratic space by giving it a symmetric bilinear form B with associated quadratic form Q. We shall call a lattice L in V binary, ternary, quaternary, quinary, . . ., n-ary, according as its rank is 2, 3, 4, 5, . . ., n.

§ 82A. Statement of the problem

Consider a lattice L in V. Let U be some other quadratic space over F and consider a lattice K in U. We say that K is represented by L, and write $K \rightarrowtail L$, if there is a representation $\sigma: FK \rightarrowtail FL$ such that $\sigma K \subseteq L$. We say that there is an isometry of K into L, and write $K \succ\!\!- L$, if there is an isometry $\sigma: FK \succ\!\!- FL$ such that $\sigma K \subseteq L$. We say that K and L are isometric, and write

$$K \rightarrowtail\!\!\!\!\rightarrow L\,, \quad \text{or} \quad K \cong L\,,$$

if there is an isometry $\sigma: FK \rightarrowtail\!\!\!\!\rightarrow FL$ such that $\sigma K = L$.

We pose the following fundamental question: can we determine, say by means of invariants, whether or not two given lattices L and K are isometric. This should be regarded as the integral analogue of the earlier work on quadratic spaces. And just as the question for spaces is a geometric interpretation of the classical problem on the fractional equivalence of quadratic forms, so the question for lattices can be regarded as a geometric interpretation of a classical question on the integral equivalence of quadratic forms. In fact it will be shown in § 82B that finding the lattices isometric to a given free lattice is the same as finding the integral equivalence class of a quadratic form.

This poses the question. What about its solution? It is certainly too much to expect an answer over an arbitrary Dedekind domain at the present time. However, a complete solution is known when $\mathfrak{o}$ is the ring of integers of a local field, and considerable work has been done over the integers of a global field. These theories will be presented in the remaining chapters of this book.

If a solution to the space problem is known over the field F (for instance if F is either a local field or a global field) then there is no loss of generality in assuming that FK and FL are identical. Under these circumstances we can ask our question in the following form: given lattices L and K on the quadratic space V, does there exist an element σ of the orthogonal group $O(V)$ such that $\sigma K = L$?

§ 82B. The free case

Let M be an $m \times m$ symmetric matrix and N an $n \times n$ symmetric matrix, both over the field F. We write

$$M \rightarrowtail N \qquad (\text{over } \mathfrak{o})$$

and say that M is integrally represented by N if there is an $n \times m$ matrix T with coefficients in $\mathfrak{o}$ such that

$$M = {}^tTNT\,,$$

where ${}^tT$ stands for the transpose of T. If this can be done with a unimodular matrix T we say that M is integrally equivalent to N and write

$$M \cong N \qquad (\text{over } \mathfrak{o})\,.$$

Integral equivalence of matrices is clearly an equivalence relation on the set of all $n \times n$ symmetric matrices over F. This equivalence relation depends, of course, on the set of spots S used in defining the ring $\mathfrak{o} = \mathfrak{o}(S)$. We let

$$\mathrm{cls}_S N \quad \text{or} \quad \mathrm{cls} N$$

denote the set of $n \times n$ matrices which are integrally equivalent to the $n \times n$ symmetric matrix N over F, and we call this set the class of N at S. These classes partition the set of $n \times n$ symmetric matrices over F at S.

If L is a free lattice on V there is a base $x_1, \ldots, x_n$ for V such that

$$L = \mathfrak{o}x_1 + \cdots + \mathfrak{o}x_n\,.$$

By the matrix of L in the base $x_1, \ldots, x_n$ we mean the matrix $N = (B(x_i, x_j))$, i.e. the matrix of V in $x_1, \ldots, x_n$; we write

$$L \cong N \quad \text{in} \quad x_1, \ldots, x_n\,.$$

If there is at least one base $x_1, \ldots, x_n$ for which this holds, then we say that L has the matrix N and we write

$$L \cong N\,.$$

If $x'_1, \ldots, x'_n$ is another base for L with

$$x'_j = \sum_\lambda t_{\lambda j} x_\lambda \qquad (t_{\lambda j} \in \mathfrak{o})$$

then $T = (t_{\lambda j})$ is unimodular by Proposition 81:9, and the matrix N' of L in $x'_1, \ldots, x'_n$ is equal to

$$N' = {}^tTNT\,.$$

In other words a change of base leads from N to a matrix N' in $\mathrm{cls}_S N$, and every matrix of $\mathrm{cls}_S N$ can be obtained in this way. We can therefore associate one entire class of integrally equivalent matrices with a free lattice on a given quadratic space. And every class of integrally equivalent symmetric matrices can be obtained from a suitable free lattice on a suitable quadratic space.

Given any symmetric matrix N we have agreed to let $\langle N \rangle$ stand for a quadratic space having the matrix N. We shall also use the symbol $\langle N \rangle$ to denote a free lattice with the matrix N (in a suitable quadratic space).

82:1. *Let K and L be free lattices with matrices M and N on the quadratic spaces U and V, respectively. Then*

(1) $K \rightarrowtail L$ *if and only if* $M \rightarrowtail N$ *(over* $\mathfrak{o}$*)*

(2) $K \cong L$ *if and only if* $M \cong N$ *(over* $\mathfrak{o}$*).*

Proof. The proof parallels the proof of Proposition 41:2. **q. e. d.**

Consider the discriminant $d_B(x_1, \ldots, x_n)$ of a base $x_1, \ldots, x_n$ of the free lattice L on V. If we take another base $x'_1, \ldots, x'_n$ for L the equation $N' = {}^tTNT$ shows us that

$$d_B(x'_1, \ldots, x'_n) = \varepsilon^2 d_B(x_1, \ldots, x_n)$$

for some unit ε in $\mathfrak{o}$. Hence the canonical image of $d_B(x_1, \ldots, x_n)$ in $0 \cup (\dot{F}/\mathfrak{u}^2)$ is independent of the base chosen for L, it is called the discriminant of L, and it is written

$$d_B L \quad \text{or} \quad dL\,.$$

When L consists of the single point 0 we take

$$dL = 1\,.$$

We shall often write $dL = \alpha$ with α in F; this will really mean that dL is the canonical image of α in $0 \cup (\dot{F}/\mathfrak{u}^2)$. It is equivalent to saying that L has a base $L = \mathfrak{o}x_1 + \cdots + \mathfrak{o}x_n$ in which

$$d_B(x_1, \ldots, x_n) = \alpha\,.$$

§ 82C. The class of a lattice

Consider a regular non-zero quadratic space V over F, and lattices $K, L, \ldots$ on V. We say that K and L are in the same class if

$$K = \sigma L \quad \text{for some} \quad \sigma \in O(V)\,.$$

This is clearly an equivalence relation on the set of all lattices on V, and we accordingly obtain a partition of this set into equivalence classes. We use

$$\text{cls}\, L$$

to denote the class of L. The fundamental question of § 82A can now be regarded as a question of characterizing the class $\text{cls}\, L$. We define the proper class

$$\text{cls}^+ L$$

to be the set of all lattices K on V such that

$$K = \sigma L \quad \text{for some} \quad \sigma \in O^+(V)\,.$$

The proper classes also put a partition on the set of all lattices on V, and this partition is finer than the partition into classes. In fact it is easily verified, using the fact that $O^+(V)$ has index 2 in $O(V)$, that each class contains either one or two proper classes. Of course the class and the proper class depend on several factors, such as the underlying set of spots S, the supporting vector space V, and the underlying bilinear form B.

We define the group of units of L to be the subgroup

$$O(L) = \{\sigma \in O(V) \mid \sigma L = L\}$$

of $O(V)$. We put

$$O^+(L) = O(L) \cap O^+(V) .$$

The determinant map

$$\det : O(L) \rightarrowtail (\pm 1)$$

has kernel $O^+(L)$, hence $O^+(L)$ is a normal subgroup of $O(L)$ with

$$1 \leqq (O(L) : O^+(L)) \leqq 2 .$$

We shall see later that it is possible for this index to be either 1 or 2. It is 2 when there is at least one reflexion on V which is a unit of L; otherwise it is 1. We define the set

$$O^-(L) = O(L) \cap O^-(V) .$$

Then

$$O(L) = O^+(L) \cup O^-(L), \quad O^+(L) \cap O^-(L) = \emptyset .$$

82:2. **Example.** $\sigma O(L)\sigma^{-1} = O(\sigma L)$ for any σ in $O(V)$.

82:3. **Example.** $\operatorname{cls} L = \operatorname{cls}^+ L$ if and only if $(O(L) : O^+(L)) = 2$.

82:4. **Example.** $\operatorname{cls} L = \operatorname{cls}^+ L$ if $\dim V$ is odd, since -1_V is in $O(L)$ but not in $O^+(L)$.

82:5. **Example.** Suppose L is free and let $x_1, \ldots, x_n$ be a base for V with

$$L = \mathfrak{o} x_1 + \cdots + \mathfrak{o} x_n .$$

Let M denote the matrix of V in this base. According to § 43A there is a group isomorphism of $O(V)$ onto the group of automorphs of M, obtained by sending an isometry σ onto its matrix T in the base $x_1, \ldots, x_n$. This isomorphism carries rotations to proper automorphs. What does it do to $O(L)$? It carries the units of L to the integral automorphs, i.e. to the automorphs with integral entries. And to $O^+(L)$? These elements are carried to the proper integral automorphs of M. Thus $(O(L) : O^+(L)) = 2$ if and only if M has an integral automorph of determinant -1. So $\operatorname{cls} L = \operatorname{cls}^+ L$ if and only if there is an improper integral automorph of M.

82:6. **Example.** Let K and L be free lattices on the same quadratic space V with matrices M and N respectively. Then $\operatorname{cls} K = \operatorname{cls} L$ if and only if $\operatorname{cls} M = \operatorname{cls} N$.

§ 82D. Orthogonal splittings

Consider the quadratic space V provided with its symmetric bilinear form B and its associated quadratic form Q. Let L be any lattice in V.

We say that L is a direct sum of sublattices $L_1, \ldots, L_r$ if it is their direct sum as $\mathfrak{o}$-modules, i.e. if every element $x \in L$ can be expressed in one and only one way in the form

$$x = x_1 + \cdots + x_r \qquad (x_i \in L) .$$

We write

$$L = L_1 \oplus \cdots \oplus L_r$$

for the direct sum. We know, for example, that $L = L_1 \oplus L_2$ if and only if

$$L = L_1 + L_2 \quad \text{with} \quad L_1 \cap L_2 = 0 .$$

Suppose L is the direct sum $L = L_1 \oplus \cdots \oplus L_r$ with

$$B(L_i, L_j) = 0 \quad \text{for} \quad 1 \leqq i < j \leqq r .$$

We then say that L is the orthogonal sum of the sublattices $L_1, \ldots, L_r$, or that L has the orthogonal splitting

$$L = L_1 \perp \cdots \perp L_r .$$

We call the L_i the components of the splitting. We also use the notation

$$\mathop{\perp}_{1}^{r} L_i \quad \text{and} \quad \mathop{\oplus}_{1}^{r} L_i$$

for orthogonal sums and direct sums respectively. We formally define

$$\mathop{\perp}_{\emptyset} L_i = 0 \quad \text{and} \quad \mathop{\oplus}_{\emptyset} L_i = 0 .$$

We say that a sublattice K splits L, or that K is a component of L, if there is a sublattice J of L with

$$L = K \perp J .$$

If $X_1, \ldots, X_r$ are sublattices of different quadratic spaces, not necessarily contained in V, we write

$$L \cong X_1 \perp \cdots \perp X_r$$

to signify that L has a splitting

$$L = L_1 \perp \cdots \perp L_r$$

in which each sublattice L_i is isometric to X_i.

Suppose we are given quadratic spaces $V_i (1 \leqq i \leqq r)$ over F and lattices L_i in V_i. Then we know that there exists a quadratic space V over F such that

$$V \cong V_1 \perp \cdots \perp V_r .$$

Hence there always exists a quadratic space V which contains a lattice L such that

$$L \cong L_1 \perp \cdots \perp L_r .$$

There is a slight modification of the preceding construction which we shall call adjunction. Consider quadratic spaces U and V over F. Then there is a quadratic space W which contains the quadratic space V and is such that

$$W = U' \perp V \quad \text{with} \quad U' \cong U .$$

We say that W is constructed from V by adjunction of U. Now let K and L be lattices in U and V respectively. Then there is a lattice J in W such that

$$J = K' \perp L \quad \text{with} \quad K' \cong K .$$

We say that J has been constructed from L by an adjunction of K.

As an example of the notation consider the equation

$$L \cong \langle M \rangle \perp \langle N \rangle ,$$

with M and N symmetric matrices over F. This means that L is a free lattice in a quadratic space with

$$L \cong \left(\begin{array}{c|c} M & 0 \\ \hline 0 & N \end{array}\right) .$$

Similarly

$$L \cong \langle \alpha_1 \rangle \perp \cdots \perp \langle \alpha_n \rangle$$

with all α_i in F means that L has a base $x_1, \ldots, x_n$ in which $Q(x_i) = \alpha_i$ for $1 \leqq i \leqq n$ and $B(x_i, x_j) = 0$ for $1 \leqq i < j \leqq n$. A base with this property is called an orthogonal base for L.

It is easily seen that the direct sum of lattices

$$L = L_1 \oplus \cdots \oplus L_r$$

implies the direct sum of spaces

$$FL = FL_1 \oplus \cdots \oplus FL_r .$$

Similarly with orthogonal sums. These equations show that

$$\operatorname{rank} L = \operatorname{rank} L_1 + \cdots + \operatorname{rank} L_r .$$

We define the radical of a lattice L in a quadratic space V to be the sublattice

$$\operatorname{rad} L = \{x \in L \mid B(x, L) = 0\} .$$

We call L regular if $\operatorname{rad} L = 0$. It is easily seen that

$$\operatorname{rad} FL = F \operatorname{rad} L$$

and

$$\operatorname{rad} L = L \cap \operatorname{rad} FL .$$

In particular, L is regular if and only if FL is regular. We have

$$\operatorname{rad}(L \perp K) = (\operatorname{rad} L) \perp (\operatorname{rad} K) .$$

In particular, $L \perp K$ is regular if and only if L and K are regular. Suppose $J = L \perp K$ with L regular; then K is uniquely determined by J and L, and in fact

$$K = \{x \in J \mid B(x, L) = 0\} .$$

Let us show that L has a radical splitting, i.e. that there is a lattice K such that $L = K \perp \operatorname{rad} L$. If L is regular just take $K = L$. Now assume that L is not regular and take a base $x_1, \ldots, x_n$ for FL in which $x_1, \ldots, x_r$ span the radical of FL. By Theorem 81:3 there is a base $y_1, \ldots, y_n$ for FL in which

$$L = \mathfrak{a}_1 y_1 + \cdots + \mathfrak{a}_r y_r + \cdots + \mathfrak{a}_n y_n$$

and

$$\operatorname{rad} FL = F y_1 + \cdots + F y_r .$$

Then

$$\operatorname{rad} L = L \cap \operatorname{rad} FL = \mathfrak{a}_1 y_1 + \cdots + \mathfrak{a}_r y_r .$$

So taking $K = \mathfrak{a}_{r+1} y_{r+1} + \cdots + \mathfrak{a}_n y_n$ gives us the desired radical splitting $L = K \perp \operatorname{rad} L$. Incidentally, the lattice K in any radical splitting $L = K \perp \operatorname{rad} L$ is always regular since the equations

$$\operatorname{rad} L = \operatorname{rad} K \perp \operatorname{rad}(\operatorname{rad} L) = \operatorname{rad} K \perp \operatorname{rad} L$$

imply that $\operatorname{rad} K$ is 0.

If $L = K \perp \operatorname{rad} L$ and $L_1 = K_1 \perp \operatorname{rad} L_1$ are two radical splittings, then it can be shown (using a proof somewhat similar to the proof of Proposition 42:8) that L is isometric to L_1 if and only if

$$K \cong K_1 \quad \text{with} \quad \operatorname{rad} L \cong \operatorname{rad} L_1 .$$

82:7. *Let L be a lattice in the quadratic space V and let K be a regular sublattice of L. Suppose K has a splitting*

$$K = K_1 \perp \cdots \perp K_r .$$

Then K is a component of L if and only if each K_i is.

Proof. If K splits L, then it is clear that each K_i does too. Conversely suppose L is split by each K_i. Since FK is regular we have a splitting

$$FL = FK \perp U$$

in which U is the orthogonal complement of FK in FL, by Proposition 42:4. We claim that

$$L = K \perp (U \cap L) .$$

In order to prove this it is enough to show that $L = K + (U \cap L)$. So consider a typical x in L. For $1 \leqq \lambda \leqq r$ we can write $x = x_\lambda + y_\lambda$ with $x_\lambda \in K_\lambda$ and $B(y_\lambda, K_\lambda) = 0$, since K_λ splits L. Then

$$B\left(x - \sum_\lambda x_\lambda, \quad K_i\right) = B(x - x_i, K_i) = B(y_i, K_i) = 0$$

for $1 \leqq i \leqq r$. Hence

$$B\left(x - \sum_{\lambda} x_\lambda, FK\right) = 0 .$$

So

$$x - \sum_{\lambda} x_\lambda \in U \cap L .$$

Hence

$$x = \left(\sum_{\lambda} x_\lambda\right) + \left(x - \sum_{\lambda} x_\lambda\right) \in K + (U \cap L) .$$

q. e. d.

§ 82E. Scale, norm and volume

Consider a lattice L in the quadratic space V. By the scale

$$\mathfrak{s}L$$

of L we mean the $\mathfrak{o}$-module generated by the subset $B(L, L)$ of F. Clearly

$$\mathfrak{s}L = \left\{\sum_{\text{fin}} B(x, y) \mid x, y \in L\right\} .$$

Since L is finitely generated we can write $L = \mathfrak{o}z_1 + \cdots + \mathfrak{o}z_r$, hence

$$\mathfrak{s}L \subseteq \sum_{i,j} B(z_i, z_j)\mathfrak{o} ,$$

so $\mathfrak{s}L$ is either a fractional ideal or 0. We define the norm

$$\mathfrak{n}L$$

to be the $\mathfrak{o}$-module generated by the subset $Q(L)$ of F. Since $Q(L) \subseteq B(L,L)$ it follows that $\mathfrak{n}L$ is also either a fractional ideal or 0. Now for any x, y in L we have

$$2B(x, y) = Q(x + y) - Q(x) - Q(y) \in \mathfrak{n}L .$$

Hence

$$2\mathfrak{s}L \subseteq \mathfrak{n}L \subseteq \mathfrak{s}L .$$

We also have

$$Q(FL) = F^2 Q(L) .$$

So all the sets

$$\mathfrak{s}L, \quad \mathfrak{n}L, \quad B(FL, FL), \quad Q(FL)$$

are 0 if one of them is. If L has the splitting $L = J \perp K$, then it is easily verified that

$$\mathfrak{s}L = \mathfrak{s}J + \mathfrak{s}K, \quad \mathfrak{n}L = \mathfrak{n}J + \mathfrak{n}K .$$

82:8. *Let the lattice L in the quadratic space V have the form*

$$L = \mathfrak{a}_1 z_1 + \cdots + \mathfrak{a}_r z_r$$

with the $\mathfrak{a}_i$ in I and the z_i in V. Then

(1) $\mathfrak{s}L = \sum_{i\,j} \mathfrak{a}_i \mathfrak{a}_j B(z_i, z_j)$, (2) $\mathfrak{n}L = \sum_{i} \mathfrak{a}_i^2 Q(z_i) + 2\mathfrak{s}L$.

Proof. 1) First we do the case $r = 1$. Clearly

$$\mathfrak{n}(\mathfrak{a}_1 z_1) \subseteq \mathfrak{s}(\mathfrak{a}_1 z_1) \subseteq \mathfrak{a}_1^2 Q(z_1) .$$

Let us prove that $\mathfrak{a}_1^2 Q(z_1) \subseteq \mathfrak{n}(\mathfrak{a}_1 z_1)$. Write $\mathfrak{a}_1 = \alpha\mathfrak{o} + \beta\mathfrak{o}$ with α, β in $\mathfrak{a}_1$. Then

$$\alpha^2 Q(z_1) = Q(\alpha z_1) \in Q(\mathfrak{a}_1 z_1) \subseteq \mathfrak{n}(\mathfrak{a}_1 z_1) ,$$

and similarly $\beta^2 Q(z_1) \in \mathfrak{n}(\mathfrak{a}_1 z_1)$. Hence by Example 22:3

$$\mathfrak{a}_1^2 Q(z_1) = (\alpha^2\mathfrak{o} + \beta^2\mathfrak{o})\, Q(z_1) \subseteq \mathfrak{n}(\mathfrak{a}_1 z_1) .$$

So

$$\mathfrak{n}(\mathfrak{a}_1 z_1) = \mathfrak{s}(\mathfrak{a}_1 z_1) = \mathfrak{a}_1^2 Q(z_1) .$$

This proves the proposition in the case $r = 1$.

2) Next we find the scale for any $r > 1$. Clearly

$$B(x, y) \in \sum_{i,j} \mathfrak{a}_i \mathfrak{a}_j B(z_i, z_j) ,$$

for any x, y in L, hence

$$\mathfrak{s}L \subseteq \sum_{i,j} \mathfrak{a}_i \mathfrak{a}_j B(z_i, z_j) .$$

Conversely, for any $\alpha_i \in \mathfrak{a}_i$ and any $\beta_j \in \mathfrak{a}_j$ we have

$$\alpha_i \beta_j B(z_i, z_j) = B(\alpha_i z_i, \beta_j z_j) \in \mathfrak{s}L ,$$

hence

$$\mathfrak{a}_i \mathfrak{a}_j B(z_i, z_j) \subseteq \mathfrak{s}L ,$$

hence

$$\sum_{i,j} \mathfrak{a}_i \mathfrak{a}_j B(z_i, z_j) = \mathfrak{s}L .$$

3) Finally the norm equation. For a typical vector

$$x = \alpha_1 z_1 + \cdots + \alpha_r z_r \qquad (\alpha_i \in \mathfrak{a}_i)$$

in L we have

$$Q(x) = \sum_i \alpha_i^2 Q(z_i) + 2 \sum_{i<j} \alpha_i \alpha_j B(z_i, z_j) ,$$

hence

$$\mathfrak{n}L \subseteq \sum_i \mathfrak{a}_i^2 Q(z_i) + 2\mathfrak{s}L .$$

On the other hand we have $2\mathfrak{s}L \subseteq \mathfrak{n}L$, and by step 1)

$$\mathfrak{a}_i^2 Q(z_i) = \mathfrak{n}(\mathfrak{a}_i z_i) \subseteq \mathfrak{n}L ,$$

hence

$$\mathfrak{n}L = \sum_i \mathfrak{a}_i^2 Q(z_i) + 2\mathfrak{s}L .$$

q. e. d.

82:8a. *For any fractional ideal* $\mathfrak{a}$ *we have* $\mathfrak{s}(\mathfrak{a}L) = \mathfrak{a}^2(\mathfrak{s}L)$ *and* $\mathfrak{n}(\mathfrak{a}L) = \mathfrak{a}^2(\mathfrak{n}L)$.

82:9. *Let K be a regular non-zero lattice in the quadratic space V. Then there is a lattice L on V which is split by K and has the same scale and norm as K.*

Proof. Since K is regular we have a splitting $V = (FK) \perp U$. Take any lattice J on U. Now $\mathfrak{s}K$ and $\mathfrak{n}K$ are non-zero, hence they are fractional ideals, hence there is a non-zero α in $\mathfrak{o}$ such that

$$\mathfrak{s}(\alpha J) = \alpha^2(\mathfrak{s}J) \subseteq \mathfrak{s}K\,,$$

and

$$\mathfrak{n}(\alpha J) = \alpha^2(\mathfrak{n}J) \subseteq \mathfrak{n}K\,.$$

Then $L = K \perp (\alpha J)$ is a lattice on V with the desired properties. **q. e. d.**

If L is not 0 there is a base $x_1, \ldots, x_r$ for FL such that

$$L = \mathfrak{a}_1 x_1 + \cdots + \mathfrak{a}_r x_r$$

with the $\mathfrak{a}_i$ in I. We define the volume to be

$$\mathfrak{v}L = \mathfrak{a}_1^2 \ldots \mathfrak{a}_r^2 \cdot d(x_1, \ldots, x_r)\,.$$

This quantity is 0 when L is not regular, it is a fractional ideal in F when L is regular. Is it independent of the choice of base for FL? Consider another adaptation

$$L = \mathfrak{b}_1 y_1 + \cdots + \mathfrak{b}_r y_r$$

with

$$y_j = \sum_i a_{ij} x_i\,.$$

Then by § 41B we have

$$d(y_1, \ldots, y_r) = (\det a_{ij})^2 \cdot d(x_1, \ldots, x_r)\,,$$

hence by Proposition 81:8 we obtain

$$\mathfrak{b}_1^2 \ldots \mathfrak{b}_r^2 \cdot d(y_1, \ldots, y_r) = \mathfrak{a}_1^2 \ldots \mathfrak{a}_r^2 \cdot d(x_1, \ldots, x_r)\,.$$

So $\mathfrak{v}L$ is indeed well-defined. If L is the lattice 0 we define $\mathfrak{v}L = \mathfrak{o}$. We note that

$$\mathfrak{v}L = \mathfrak{v}J \cdot \mathfrak{v}K$$

when L has a splitting $L = J \perp K$.

82:10. *Let L be a non-zero lattice in the quadratic space V. Then*

$$\mathfrak{v}L \subseteq (\mathfrak{s}L)^r$$

where r denotes the rank of L.

Proof. Take an adaptation

$$L = \mathfrak{a}_1 x_1 + \cdots + \mathfrak{a}_r x_r$$

in the usual way. Put $g_{ij} = B(x_i, x_j)$. So $\mathfrak{a}_i \mathfrak{a}_j g_{ij} \subseteq \mathfrak{s}L$ by Proposition 82:8.

By definition of the determinant we obtain

$$\begin{aligned} \mathfrak{v}L &= (\mathfrak{a}_1^2 \ldots \mathfrak{a}_r^2) \cdot (\sum \pm g_{1\alpha} \cdots g_{r\omega}) \\ &\subseteq \sum (\mathfrak{a}_1 \mathfrak{a}_\alpha g_{1\alpha}) \ldots (\mathfrak{a}_r \mathfrak{a}_\omega g_{r\omega}) \\ &\subseteq (\mathfrak{s}L)^r . \end{aligned}$$

q. e. d.

82:11. *Let K and L be non-zero lattices on the regular quadratic space V with $K \subseteq L$. Let $\mathfrak{a}$ be the product of the invariant factors of K in L. Then $\mathfrak{a}$ is an integral ideal and*

$$\mathfrak{v}K = \mathfrak{a}^2(\mathfrak{v}L), \quad \mathfrak{a}L \subseteq K .$$

Proof. Let $\mathfrak{r}_1, \ldots, \mathfrak{r}_n$ be the invariant factors of K in L. Then there is a base $x_1, \ldots, x_n$ for V and there are fractional ideals $\mathfrak{a}_1, \ldots, \mathfrak{a}_n$ such that

$$\begin{cases} L = \mathfrak{a}_1 x_1 + \cdots + \mathfrak{a}_n x_n \\ K = \mathfrak{a}_1 \mathfrak{r}_1 x_1 + \cdots + \mathfrak{a}_n \mathfrak{r}_n x_n . \end{cases}$$

Here we have all $\mathfrak{r}_i \subseteq \mathfrak{o}$ since $K \subseteq L$, hence $\mathfrak{a} \subseteq \mathfrak{o}$. If we compute volumes with respect to the above base we find $\mathfrak{v}K = \mathfrak{a}^2(\mathfrak{v}L)$. Finally, we have $\mathfrak{a} \subseteq \mathfrak{r}_i$ for $1 \leq i \leq n$, hence $\mathfrak{a}L \subseteq K$. q. e. d.

82:11a. $\mathfrak{v}K \subset \mathfrak{v}L$ *if* $K \subset L$.

82:12. **Example.** Consider the lattice L on the regular quadratic space V and let σ be an element of $O(V)$ such that $\sigma L \subseteq L$. Then σ, being an isometry, will preserve volume. Hence $\sigma L = L$, i.e. σ is actually in $O(L)$.

§ 82F. The dual of a lattice

Consider a lattice L in the quadratic space V. Suppose that L is regular. We define the dual of L to be

$$L^\# = \{x \in FL \mid B(x, L) \subseteq \mathfrak{o}\} .$$

If L is the trivial lattice 0, then $L^\# = 0$. Suppose L is not 0. Then we have a base $x_1, \ldots, x_r$ for FL and fractional ideals $\mathfrak{a}_1, \ldots, \mathfrak{a}_r$ such that

$$L = \mathfrak{a}_1 x_1 + \cdots + \mathfrak{a}_r x_r .$$

We claim that

$$L^\# = \mathfrak{a}_1^{-1} y_1 + \cdots + \mathfrak{a}_r^{-1} y_r$$

where $y_1, \ldots, y_r$ is the dual base of $x_1, \ldots, x_r$ on FL. Now $B(\mathfrak{a}_i x_i, \mathfrak{a}_j^{-1} y_j) = 0$ if $i \neq j$, and $B(\mathfrak{a}_i x_i, \mathfrak{a}_i^{-1} y_i) \subseteq \mathfrak{o}$ otherwise. Hence

$$\mathfrak{a}_1^{-1} y_1 + \cdots + \mathfrak{a}_r^{-1} y_r \subseteq L^\# .$$

On the other hand, if we take a typical vector $z = \beta_1 y_1 + \cdots + \beta_r y_r$ in $L^\#$ we must have

$$\beta_i \mathfrak{a}_i = B(\beta_i y_i, \mathfrak{a}_i x_i) = B(z, \mathfrak{a}_i x_i) \subseteq B(z, L) \subseteq \mathfrak{o} ,$$

and so $\beta_i \in \mathfrak{a}_i^{-1}$. Hence we have established our claim. And incidentally this also shows that $L^\#$ is a lattice. As immediate consequences we have

$$FL^\# = FL\,, \quad L^{\#\#} = L\,,$$

and

$$(\mathfrak{a}L)^\# = \mathfrak{a}^{-1}L^\#$$

for any fractional ideal $\mathfrak{a}$. In virtue of the fact that

$$d(x_1, \ldots, x_r)\, d(y_1, \ldots, y_r) = 1$$

(see Example 42:5) we must have

$$\mathfrak{v}L^\# = (\mathfrak{v}L)^{-1}.$$

If L has a splitting $L = J \perp K$, then

$$L^\# = J^\# \perp K^\#\,.$$

Finally, it is easily seen that if L, J, K are all on the same space FK, then

$$L \supseteq K \;\Rightarrow\; L^\# \subseteq K^\#$$

and

$$(J + K)^\# = J^\# \cap K^\#\,.$$

§ 82G. Modular lattices

Consider a lattice L of rank r in the quadratic space V. Suppose that the scale of L is the fractional ideal $\mathfrak{a}$. Then we know from Proposition 82:10 that $\mathfrak{s}L = \mathfrak{a}$ and $\mathfrak{v}L \subseteq \mathfrak{a}^r$. Suppose that L actually satisfies

$$\mathfrak{s}L = \mathfrak{a} \quad \text{and} \quad \mathfrak{v}L = \mathfrak{a}^r\,.$$

In this event we call L $\mathfrak{a}$-modular, or just modular. For any α in $\dot{F}$ we call L α-modular if it is $\mathfrak{a}$-modular with $\mathfrak{a} = \alpha\mathfrak{o}$. We call L unimodular if it is $\mathfrak{a}$-modular with $\mathfrak{a} = \mathfrak{o}$. It is clear from Proposition 82:10 that L is $\mathfrak{a}$-modular if

$$\mathfrak{s}L \subseteq \mathfrak{a} \quad \text{and} \quad \mathfrak{v}L = \mathfrak{a}^r\,.$$

An $\mathfrak{a}$-modular lattice is regular, non-zero, of scale $\mathfrak{a}$, and of volume

$$\mathfrak{v}L = (\mathfrak{s}L)^r\,.$$

If L is $\mathfrak{a}$-modular, then αL is $\alpha^2\mathfrak{a}$-modular for any α in $\dot{F}$, and $\mathfrak{b}L$ is $\mathfrak{b}^2\mathfrak{a}$-modular for any $\mathfrak{b}$ in I. If we take a non-trivial splitting $L = J \perp K$, then it is easily seen that L is $\mathfrak{a}$-modular if and only if both J and K are. The lattice $\mathfrak{o}x$ with x an anisotropic vector of V is $Q(x)$-modular.

82:13. *The free lattice L in the quadratic space V has the matrix M. Then L is a unimodular lattice if and only if M is a unimodular matrix.*

Proof. Take a base $x_1, \ldots, x_r$ for L in which L has the matrix M. So $L = \mathfrak{o}x_1 + \cdots + \mathfrak{o}x_r$ and $\mathfrak{v}L = (\det M)\mathfrak{o}$. If L is unimodular, then all $B(x_i, x_j)$ are in $\mathfrak{o}$ since $\mathfrak{s}L = \mathfrak{o}$, hence M is an integral matrix. Furthermore $(\det M)\mathfrak{o} = \mathfrak{o}$ and so $\det M$ is a unit. Hence M is unimodular. Now let M be unimodular. Then M is integral and so

$$\mathfrak{s}L = \sum_{i,j} B(x_i, x_j)\mathfrak{o} \subseteq \mathfrak{o} .$$

And $\det M$ is a unit so that $\mathfrak{v}L = \mathfrak{o}$. Hence L is unimodular. **q. e. d.**

82:14. *L is a non-zero regular lattice in a quadratic space V. Then L is $\mathfrak{a}$-modular if and only if $\mathfrak{a}L^{\#} = L$.*

Proof. First suppose that $\mathfrak{a}L^{\#} = L$. Then

$$B(L, L) = B(L, \mathfrak{a}L^{\#}) \subseteq \mathfrak{a} ,$$

and so $\mathfrak{s}L \subseteq \mathfrak{a}$. Furthermore

$$\mathfrak{v}L = \mathfrak{v}(\mathfrak{a}L^{\#}) = \mathfrak{a}^{2r}(\mathfrak{v}L^{\#}) = \mathfrak{a}^{2r}(\mathfrak{v}L)^{-1},$$

hence $\mathfrak{v}L = \mathfrak{a}^r$. Therefore L is $\mathfrak{a}$-modular.

Now assume that L is $\mathfrak{a}$-modular. Then $B(L, \mathfrak{a}^{-1}L) \subseteq \mathfrak{o}$ so that $\mathfrak{a}^{-1}L \subseteq L^{\#}$. On the other hand we have

$$\mathfrak{v}(\mathfrak{a}^{-1}L) = \mathfrak{a}^{-2r}(\mathfrak{v}L) = (\mathfrak{v}L)^{-1} = \mathfrak{v}L^{\#} .$$

Hence $\mathfrak{a}^{-1}L = L^{\#}$ by Corollary 82:11a. **q. e. d.**

82:14a. *Suppose L is $\mathfrak{a}$-modular. Then*

$$L = \{x \in FL \mid B(x, L) \subseteq \mathfrak{a}\} .$$

Proof. Since L is $\mathfrak{a}$-modular we have $B(L, L) \subseteq \mathfrak{s}L = \mathfrak{a}$ and so

$$L \subseteq \{x \in FL \mid B(x, L) \subseteq \mathfrak{a}\} .$$

Conversely consider an x in FL with $B(x, L) \subseteq \mathfrak{a}$. Then

$$B(x, L^{\#}) = B(x, \mathfrak{a}^{-1}L) \subseteq \mathfrak{o} ,$$

hence $x \in L^{\#\#} = L$. **q. e. d.**

82:14b. *L is unimodular if and only if $L^{\#} = L$.*

82:15. *L is a lattice in a quadratic space V and J is an $\mathfrak{a}$-modular sublattice of L. Then J splits L if and only if $B(J, L) \subseteq \mathfrak{a}$.*

Proof. If J splits L we have $L = J \perp K$ and then

$$B(J, L) = B(J, J) \subseteq \mathfrak{a} .$$

Conversely suppose J satisfies the condition $B(J, L) \subseteq \mathfrak{a}$. Since J is modular it is regular, hence by Proposition 42:4 we have a splitting $FL = (FJ) \perp U$. We claim that

$$L = J \perp (L \cap U) .$$

It is enough to show that a typical x in L is also in $J + (L \cap U)$. Write

$$x = y + z \qquad (y \in FJ, z \in U)$$

Then

$$B(y, J) = B(x, J) \subseteq B(L, J) \subseteq \mathfrak{a} .$$

But y is in FJ. Hence y is in J by Corollary 82:14a. So z is in L and we are through. **q. e. d.**

82:15a. *If J is an $\mathfrak{a}$-modular sublattice of L with $\mathfrak{s}L = \mathfrak{a}$, then J splits L.*

82:16. *Let L be an $\mathfrak{a}$-modular lattice and let x be an isotropic vector in L. Then there is a binary lattice J which splits L and contains x.*

Proof. By Theorem 81:3 there is a base for FL which is adapted to L and includes x among its members, say

$$L = \mathfrak{b}x + \cdots.$$

Then by § 82F we have

$$L^{\#} = \mathfrak{b}^{-1}y + \cdots$$

where y is a vector in FL with $B(x, y) = 1$. Now $\mathfrak{a}L^{\#} = L$ since L is $\mathfrak{a}$-modular. Hence

$$J = \mathfrak{b}x + \mathfrak{b}^{-1}\mathfrak{a}y$$

is a sublattice of L which contains x. But $\mathfrak{d}J$ is easily computed and found to be $\mathfrak{a}^2$. And $\mathfrak{s}J \subseteq \mathfrak{s}L \subseteq \mathfrak{a}$. Hence J is $\mathfrak{a}$-modular and therefore splits L. **q. e. d.**

82:17. *Suppose $\mathfrak{o}$ is a principal ideal domain. Then a non-zero lattice L in a quadratic space V is $\mathfrak{a}$-modular if and only if $B(x, L) = \mathfrak{a}$ for every maximal vector x in L.*

Proof. First suppose L is $\mathfrak{a}$-modular. Then by Theorem 81:3 we can find a base $x, \ldots$ for L in which

$$L = \mathfrak{o}x + \cdots.$$

So

$$L^{\#} = \mathfrak{o}y + \cdots$$

where y is a vector with $B(x, y) = 1$. Now $\mathfrak{a}L^{\#} = L$ since L is $\mathfrak{a}$-modular, hence

$$\mathfrak{a} \supseteq B(x, L) \supseteq B(x, \mathfrak{a}y) \supseteq \mathfrak{a},$$

so $B(x, L) = \mathfrak{a}$.

Conversely suppose $B(x, L) = \mathfrak{a}$ holds for every maximal vector in L. Every vector y in L falls in $\mathfrak{o}x$ where x is a maximal vector of L in the line Fy, hence L is regular and $\mathfrak{s}L = \mathfrak{a}$. So $B(\mathfrak{a}^{-1}L, L) \subseteq \mathfrak{o}$. This proves that $\mathfrak{a}^{-1}L \subseteq L^{\#}$ and hence that $L \subseteq \mathfrak{a}L^{\#}$. On the other hand $B(\mathfrak{a}L^{\#}, L) \subseteq \mathfrak{a}$. If y is a vector of $FL - L$, then $y = \alpha x$ with x maximal in L and α not in $\mathfrak{o}$, hence

$$B(y, L) = \alpha B(x, L) = \alpha\mathfrak{a}$$

and this is not contained in $\mathfrak{a}$; so no vector y of $FL - L$ can fall in $\mathfrak{a}L^{\#}$. Hence $\mathfrak{a}L^{\#} \subseteq L$. So $\mathfrak{a}L^{\#} = L$ and L is $\mathfrak{a}$-modular. **q. e. d.**

§ 82H. Maximal lattices

Let L be a non-zero lattice on the regular quadratic space V, and let $\mathfrak{a}$ be a fractional ideal. We say that L is $\mathfrak{a}$-maximal on V if $\mathfrak{n}L \subseteq \mathfrak{a}$ and if for every lattice K on V which contains L we have

$$\mathfrak{n}K \subseteq \mathfrak{a} \Rightarrow K = L.$$

We say that L is maximal on V if it is $\mathfrak{a}$-maximal for some $\mathfrak{a}$. If L is an $\mathfrak{a}$-maximal lattice on V, then it is easily seen that αL is $\alpha^2\mathfrak{a}$-maximal for every α in $\dot{F}$, and $\mathfrak{b}L$ is $\mathfrak{b}^2\mathfrak{a}$-maximal for every $\mathfrak{b}$ in I; if $L = J \perp K$ is a non-trivial splitting of the $\mathfrak{a}$-maximal lattice L, then J and K are $\mathfrak{a}$-maximal on FJ and FK respectively.

82:18. *Let L be a lattice on a non-zero regular quadratic space V and suppose $\mathfrak{n}L \subseteq \mathfrak{a}$ where $\mathfrak{a}$ is a fractional ideal. Then there is an $\mathfrak{a}$-maximal lattice K on V with $K \supseteq L$.*

Proof. Ler r denote the dimension of V. First observe that for any L on V with $\mathfrak{n}L \subseteq \mathfrak{a}$ we have

$$\mathfrak{v}L \subseteq (\mathfrak{s}L)^r \subseteq \left(\frac{1}{2}\mathfrak{a}\right)^r.$$

The Unique Factorization Theorem shows that the number of fractional ideals between $\mathfrak{v}L$ and $\left(\frac{1}{2}\mathfrak{a}\right)^r$ is some finite number ν. If L is not $\mathfrak{a}$-maximal we have a lattice L_1 on V with $L \subset L_1$ and $\mathfrak{n}L_1 \subseteq \mathfrak{a}$. If L_1 is $\mathfrak{a}$-maximal we are through. Otherwise repeat the preceding step. Continue in this way to obtain a chain

$$L \subset L_1 \subset \cdots \subset L_t$$

with each $\mathfrak{n}L_i \subseteq \mathfrak{a}$. This gives rise to the chain of fractional ideals

$$\mathfrak{v}L \subset \mathfrak{v}L_1 \subset \cdots \subset \mathfrak{v}L_t \subseteq \left(\frac{1}{2}\mathfrak{a}\right)^r.$$

So $t \leq \nu$. Thus the process must terminate before ν steps. In other words we obtain an $\mathfrak{a}$-maximal lattice L_t before ν steps. This L_t is our K.

q. e. d.

82:18a. *Every non-zero regular quadratic space contains an $\mathfrak{a}$-maximal lattice for every fractional ideal $\mathfrak{a}$.*

82:19. *Let L be a lattice on the r-dimensional regular quadratic space V and let $\mathfrak{a}$ be a fractional ideal such that $\mathfrak{n}L \subseteq \mathfrak{a}$. Then the ideal*

$$2^r(\mathfrak{v}L) \div (\mathfrak{a})^r$$

is integral. If this ideal has no integral square factors, then L is $\mathfrak{a}$-maximal.

Proof. The ideal $2^r(\mathfrak{v}L) \div (\mathfrak{a})^r$ is integral since $(\mathfrak{v}L) \subseteq (\mathfrak{s}L)^r \subseteq \left(\frac{1}{2}\mathfrak{a}\right)^r$. Suppose this ideal has no integral square factors. If L is not $\mathfrak{a}$-maximal we can find a lattice K on V with $L \subset K$ and $\mathfrak{n}K \subseteq \mathfrak{a}$. Then there is a proper integral ideal $\mathfrak{b}$ with $\mathfrak{v}L = \mathfrak{b}^2(\mathfrak{v}K)$ by Proposition 82:11. And

$2^r(\mathfrak{v}K)/\mathfrak{a}^r$ is integral since $\mathfrak{n}K \subseteq \mathfrak{a}$. This implies that

$$2^r(\mathfrak{v}L)/\mathfrak{a}^r = \mathfrak{b}^2 2^r(\mathfrak{v}K)/\mathfrak{a}^r$$

has an integral square factor, and this is contrary to hypothesis. **q. e. d.**

82:20. *Let L be an $\mathfrak{a}$-maximal lattice and let x be an isotropic vector in L. Then there is a binary lattice J which splits L and contains x.*

Proof. Let V denote the regular quadratic space on which L is $\mathfrak{a}$-maximal. Let $\mathfrak{b}$ denote the coefficient of x in L. We have

$$\mathfrak{s}L \subseteq \frac{1}{2}(\mathfrak{n}L) \subseteq \frac{1}{2}\mathfrak{a}\,.$$

Hence

$$B(2\mathfrak{a}^{-1}\mathfrak{b}x, L) \subseteq B(2\mathfrak{a}^{-1}L, L) \subseteq \mathfrak{o}\,,$$

hence $(2\mathfrak{a}^{-1}\mathfrak{b})\,x \subseteq L^\#$, so $2\mathfrak{a}^{-1}\mathfrak{b} \subseteq \mathfrak{c}$ where $\mathfrak{c}$ denotes the coefficient of x in $L^\#$. We claim that $2\mathfrak{a}^{-1}\mathfrak{b} = \mathfrak{c}$. Suppose not. Then $2\mathfrak{a}^{-1}\mathfrak{b} \subset \mathfrak{c}$, so that $\frac{1}{2}\mathfrak{a}\mathfrak{c}x + L$ is a lattice in V which properly contains L. Now an easy computation gives

$$Q\left(\frac{1}{2}\mathfrak{a}\mathfrak{c}x + L\right) \subseteq \mathfrak{a}\,.$$

So $\frac{1}{2}\mathfrak{a}\mathfrak{c}x + L$ is a lattice which properly contains L and has its norm contained in $\mathfrak{a}$. This denies the maximality of L. So we do indeed have $2\mathfrak{a}^{-1}\mathfrak{b} = \mathfrak{c}$. By Theorem 81:3 we have a base for V which includes the vector x and such that

$$L^\# = 2\mathfrak{a}^{-1}\mathfrak{b}x + \cdots.$$

Then by § 82F there is a vector y with $B(x, y) = 1$ such that

$$L = L^{\#\#} = \frac{1}{2}\mathfrak{a}\mathfrak{b}^{-1}y + \cdots.$$

So we have a sublattice

$$J = \mathfrak{b}x + \frac{1}{2}\mathfrak{a}\mathfrak{b}^{-1}y$$

of L with $\mathfrak{s}J \subseteq \mathfrak{s}L \subseteq \frac{1}{2}\mathfrak{a}$ and $\mathfrak{v}J = \frac{1}{4}\mathfrak{a}^2$; therefore J is $\frac{1}{2}\mathfrak{a}$-modular. By Proposition 82:15 J will split L. **q. e. d.**

82:21. *Let L be a lattice on the hyperbolic plane V. Then L is $2\mathfrak{a}$-maximal if and only if L is $\mathfrak{a}$-modular with $\mathfrak{n}L \subseteq 2\mathfrak{a}$.*

Proof. Take x, y in V with $Q(x) = Q(y) = 0$ and $B(x, y) = 1$. First suppose that L is $2\mathfrak{a}$-maximal on V. By Theorem 81:3 we can write

$$L = \mathfrak{b}x + \mathfrak{c}(\alpha x + y)$$

for some α in F and some $\mathfrak{b}$, $\mathfrak{c}$ in I. Now $\mathfrak{b}\mathfrak{c} \subseteq \mathfrak{s}L \subseteq \mathfrak{a}$ by Proposition 82:8. If we had $\mathfrak{b}\mathfrak{c} \subset \mathfrak{a}$, then

$$\mathfrak{a}\mathfrak{c}^{-1}x + \mathfrak{c}(\alpha x + y)$$

would be a lattice of norm $2\mathfrak{a}$ which strictly contained L and this would deny the maximality of L. Hence $\mathfrak{b}\mathfrak{c} = \mathfrak{a}$ and

$$L = \mathfrak{b}x + \mathfrak{a}\mathfrak{b}^{-1}(\alpha x + y) .$$

So $\mathfrak{s}L \subseteq \mathfrak{a}$ and $\mathfrak{v}L = \mathfrak{a}^2$. Hence L is $\mathfrak{a}$-modular. And $\mathfrak{n}L \subseteq 2\mathfrak{a}$ since L is $2\mathfrak{a}$-maximal.

Conversely, suppose that L is $\mathfrak{a}$-modular with $\mathfrak{n}L \subseteq 2\mathfrak{a}$. Then

$$2^2(\mathfrak{v}L) \div (2\mathfrak{a})^2 = \mathfrak{o} ,$$

hence L is $2\mathfrak{a}$-maximal by Proposition 82:19. **q. e. d.**

82:21a. *Suppose either of the equivalent conditions of the proposition is satisfied. Let Fx and Fy be the isotropic lines of V. Then the base x, y for V is adapted to L.*

Proof. We can assume that $B(x, y) = 1$. In the proof of the proposition we obtained

$$L = \mathfrak{b}x + \mathfrak{a}\mathfrak{b}^{-1}(\alpha x + y) .$$

Now by Proposition 82:8 $\mathfrak{a}^2\mathfrak{b}^{-2}(2\alpha) \subseteq \mathfrak{n}L \subseteq 2\mathfrak{a}$, hence $\mathfrak{a}\mathfrak{b}^{-1}\alpha \subseteq \mathfrak{b}$, hence $L = \mathfrak{b}x + \mathfrak{a}\mathfrak{b}^{-1}y$. **q. e. d.**

82:22. **Example.** Suppose every fractional ideal is principal, and let α be an element of $\dot{F}$. Consider a lattice L on the hyperbolic plane V. Then L is $2\alpha\mathfrak{o}$-maximal if and only if

$$L \cong \begin{pmatrix} 0 & \alpha \\ \alpha & 0 \end{pmatrix} .$$

Hence in this situation any two $\mathfrak{a}$-maximal lattices on V are isometric.

82:23. *Let V be an isotropic regular quadratic space and let K and L be maximal lattices on V. Then there is a splitting $V = U \perp W$ in which U is a hyperbolic plane and*

$$L = (L \cap U) \perp (L \cap W) , \quad K = (K \cap U) \perp (K \cap W) .$$

Proof. Let L be $\mathfrak{a}$-maximal, let K be $\mathfrak{b}$-maximal. For each non-zero vector x in V we let $\mathfrak{a}_x$ denote the coefficient of x with respect to L and we let $\mathfrak{b}_x$ denote the coefficient of x with respect to K. We put

$$\mathfrak{r} = \mathfrak{b}/\mathfrak{a} , \quad \mathfrak{r}_x = \mathfrak{b}_x/\mathfrak{a}_x .$$

Now $\alpha K \subseteq L$ for some non-zero α in F, hence $\mathfrak{r}_x \subseteq \alpha^{-1}\mathfrak{o}$; hence we can pick an isotropic vector x in V for which $\mathfrak{r}_x$ is maximal (among all the isotropic vectors of V). By Proposition 82:20 and Corollary 82:21a there is an isotropic vector y such that $B(x, y) = 1$ and

$$L = (\mathfrak{a}_x x + \mathfrak{a}_y y) \perp \cdots .$$

Since $\mathfrak{a}_x x + \mathfrak{a}_y y$ is $\mathfrak{a}$-maximal its scale $\mathfrak{a}_x\mathfrak{a}_y$ must be equal to $\frac{1}{2}\mathfrak{a}$ by Proposition 82:21. Now $\mathfrak{b}_x x + \mathfrak{b}_y y$ is contained in K, hence $\mathfrak{b}_x\mathfrak{b}_y \subseteq \frac{1}{2}\mathfrak{b}$.

Therefore

$$\mathfrak{r}_x\mathfrak{r}_y \subseteq \mathfrak{r} .$$

Using Proposition 82:20 and Corollary 82:21a again, we find an isotropic vector z such that

$$K = (\mathfrak{b}_z z + \mathfrak{b}_y y) \perp \cdots .$$

This time we obtain $\mathfrak{b}_z\mathfrak{b}_y = \frac{1}{2}\mathfrak{b}$ and $\mathfrak{a}_z\mathfrak{a}_y \subseteq \frac{1}{2}\mathfrak{a}$, hence $\mathfrak{r}_z\mathfrak{r}_y \supseteq \mathfrak{r}$. Therefore

$$\mathfrak{r}_x\mathfrak{r}_y \subseteq \mathfrak{r} \subseteq \mathfrak{r}_z\mathfrak{r}_y .$$

But $\mathfrak{r}_x$ was chosen maximal. Hence $\mathfrak{r}_x\mathfrak{r}_y = \mathfrak{r}$ and so

$$\mathfrak{b}_x\mathfrak{b}_y = (\mathfrak{a}_x\mathfrak{a}_y)\,(\mathfrak{r}_x\mathfrak{r}_y) = \frac{1}{2}\,\mathfrak{a}\,\mathfrak{b}/\mathfrak{a} = \frac{1}{2}\,\mathfrak{b} .$$

Therefore $\mathfrak{b}_x x + \mathfrak{b}_y y$ has scale $\frac{1}{2}\mathfrak{b}$ and volume $\frac{1}{4}\mathfrak{b}^2$; so it is $\frac{1}{2}\mathfrak{b}$-modular. Hence

$$\begin{cases} K = (\mathfrak{b}_x x + \mathfrak{b}_y y) \perp \cdots \\ L = (\mathfrak{a}_x x + \mathfrak{a}_y y) \perp \cdots \end{cases}$$

Then $U = Fx + Fy$ gives the desired splitting of V. **q. e. d.**

§ 82I. The lattice $L^{\mathfrak{a}}$

Consider a lattice L in the quadratic space V, and a fractional ideal $\mathfrak{a}$ in I. Suppose that L is regular. We define $L^{\mathfrak{a}}$ as the sublattice

$$L^{\mathfrak{a}} = \{x \in L \mid B(x, L) \subseteq \mathfrak{a}\}$$

of L. For the trivial lattice we have $L^{\mathfrak{a}} = 0$, so let us assume that L is not 0. We immediately have

$$B(L^{\mathfrak{a}}, L) \subseteq \mathfrak{a}\,, \quad \mathfrak{s}L^{\mathfrak{a}} \subseteq \mathfrak{a}\,,$$

and

$$L^{\mathfrak{a}} = L \quad \Leftrightarrow \quad \mathfrak{s}L \subseteq \mathfrak{a} .$$

It is easily seen that

$$L^{\mathfrak{a}} = \mathfrak{a}L^{\#} \cap L .$$

In particular this shows that $FL^{\mathfrak{a}} = FL$. For any splitting $L = J \perp K$ we have

$$(J \perp K)^{\mathfrak{a}} = J^{\mathfrak{a}} \perp K^{\mathfrak{a}} .$$

Now consider another fractional ideal $\mathfrak{b}$. Then

$$\mathfrak{a}\mathfrak{b}^{-1}L^{\mathfrak{b}} \subseteq L^{\mathfrak{a}} \subseteq L^{\mathfrak{b}} \quad \text{if} \quad \mathfrak{a} \subseteq \mathfrak{b}$$

follows directly from the definitions. If L is $\mathfrak{b}$-modular, then

$$L^{\mathfrak{a}} = \begin{cases} L & \text{if} \quad \mathfrak{b} \subseteq \mathfrak{a} \\ \mathfrak{a}\mathfrak{b}^{-1}L & \text{if} \quad \mathfrak{b} \supseteq \mathfrak{a}\,, \end{cases}$$

since

$$L^{\mathfrak{a}} = \mathfrak{a}L^{\#} \cap L = \mathfrak{a}\mathfrak{b}^{-1}\,L \cap L .$$

§ 82J. Scaling

Consider the lattice L in the quadratic space V. Let α be a non-zero scalar. Recall that V^α denotes the vector space V provided with a new bilinear form $B^\alpha(x, y) = \alpha B(x, y)$. We shall use L^α to denote the lattice L when it is regarded as a lattice in V^α. Thus the properties of L^α as a lattice are identical to those of L; the superscript α merely emphasizes that our interest has shifted from the properties of B on L to those of B^α on L. We easily see that

$$\mathfrak{s}L^\alpha = \alpha(\mathfrak{s}L)\,, \quad \mathfrak{n}L^\alpha = \alpha(\mathfrak{n}L)\,, \quad \mathfrak{v}L^\alpha = \alpha^r(\mathfrak{v}L)$$

where r denotes the rank of L. If L is $\mathfrak{a}$-modular in V, then L^α is $\alpha\mathfrak{a}$-modular in V^α. If L is $\mathfrak{a}$-maximal on V, then L^α is $\alpha\mathfrak{a}$-maximal on V^α. If $\varphi: L \rightarrowtail K$ is a representation, then so is $\varphi: L^\alpha \rightarrowtail K^\alpha$. So for a lattice L on the regular non-zero space V we have

$$O(L) = O(L^\alpha)\,, \quad O^+(L) = O^+(L^\alpha)\,,$$

and

$$\operatorname{cls} L = \operatorname{cls} L^\alpha, \quad \operatorname{cls}^+ L = \operatorname{cls}^+ L^\alpha.$$

If L is a free lattice with matrix M, then L^α is free with matrix αM; so here

$$dL^\alpha = \alpha^r(dL)\,.$$

Suppose L is regular and let $\mathfrak{a}$ be a fractional ideal. Then the notation L^α should not be confused with the notation $L^\mathfrak{a}$. For instance, $L^{\alpha\mathfrak{o}}$ denotes a certain sublattice of L in the quadratic space V, while L^α denotes the original lattice L in the quadratic space V^α. It is easily verified that

$$(L^\alpha)^\mathfrak{a} = (L^{\alpha^{-1}\mathfrak{a}})^\alpha.$$

§ 82K. Localization

Consider the lattice L in the quadratic space V. Let $\mathfrak{p}$ be any spot in the Dedekind set of spots S on which our ideal theory is based. Consider the quadratic space $V_\mathfrak{p}$ (i. e. the $F_\mathfrak{p}$-ification $F_\mathfrak{p} V$ of the quadratic space V) and the localization $L_\mathfrak{p}$ of L in $V_\mathfrak{p}$. Take a base $x_1, \ldots, x_r$ for FL in which L has the form

$$L = \mathfrak{a}_1 x_1 + \cdots + \mathfrak{a}_r x_r\,.$$

Then by § 81E

$$L_\mathfrak{p} = \mathfrak{a}_{1\mathfrak{p}} x_1 + \cdots + \mathfrak{a}_{r\mathfrak{p}} x_r\,,$$

so by Proposition 82:8 we have

$$\mathfrak{s}L_\mathfrak{p} = \sum_{i,j} \mathfrak{a}_{i\mathfrak{p}} \mathfrak{a}_{j\mathfrak{p}} B(x_i, x_j) = \Big(\sum_{i,j} \mathfrak{a}_i \mathfrak{a}_j B(x_i, x_j)\Big)_\mathfrak{p} = (\mathfrak{s}L)_\mathfrak{p}\,.$$

Similarly with norm and volume. In other words,

$$\mathfrak{s}L_\mathfrak{p} = (\mathfrak{s}L)_\mathfrak{p}\,, \quad \mathfrak{n}L_\mathfrak{p} = (\mathfrak{n}L)_\mathfrak{p}\,, \quad \mathfrak{v}L_\mathfrak{p} = (\mathfrak{v}L)_\mathfrak{p}$$

holds for all $\mathfrak{p}$ in S. In particular two lattices L and K have the same scale (or norm or volume) if and only if $L_\mathfrak{p}$ and $K_\mathfrak{p}$ have the same scale (or norm or volume) at all $\mathfrak{p}$ in S.

From this it follows, using the definition of a modular lattice, that L is $\mathfrak{a}$-modular if and only if $L_\mathfrak{p}$ is $\mathfrak{a}_\mathfrak{p}$-modular for all $\mathfrak{p}$ in S.

It is also true that L is $\mathfrak{a}$-maximal on V if and only if $L_\mathfrak{p}$ is $\mathfrak{a}_\mathfrak{p}$-maximal on $V_\mathfrak{p}$ for all $\mathfrak{p}$ in S. First suppose that $L_\mathfrak{p}$ is $\mathfrak{a}_\mathfrak{p}$-maximal for all $\mathfrak{p}$ in S. If L is not $\mathfrak{a}$-maximal, there is a lattice K on V such that $K \supset L$ and $\mathfrak{n}K \subseteq \mathfrak{a}$. Then by § 81E there is a $\mathfrak{q}$ in S at which $K_\mathfrak{q} \supset L_\mathfrak{q}$. But $\mathfrak{n}K_\mathfrak{q} \subseteq \mathfrak{a}_\mathfrak{q}$. This denies the maximality of $L_\mathfrak{q}$. Conversely let us be given an $\mathfrak{a}$-maximal lattice L on V, and let us suppose that $L_\mathfrak{q}$ is not $\mathfrak{a}_\mathfrak{q}$-maximal at some spot $\mathfrak{q}$ in S. Then there is a lattice $J_{(\mathfrak{q})}$ on $V_\mathfrak{q}$ which strictly contains $L_\mathfrak{q}$ and has $\mathfrak{n}J_{(\mathfrak{q})} \subseteq \mathfrak{a}_\mathfrak{q}$. By Proposition 81:14 there is a lattice K on V with

$$K_\mathfrak{p} = \begin{cases} L_\mathfrak{p} & \text{if } \mathfrak{p} \in S - \mathfrak{q} \\ J_{(\mathfrak{q})} & \text{if } \mathfrak{p} = \mathfrak{q}\,. \end{cases}$$

Here we have $K \supset L$. But $\mathfrak{n}K_\mathfrak{p} \subseteq \mathfrak{a}_\mathfrak{p}$ for all $\mathfrak{p}$ in S. Hence $\mathfrak{n}K \subseteq \mathfrak{a}$. This denies the maximality of L.

Chapter IX

Integral Theory of Quadratic Forms over Local Fields

This chapter classifies quadratic forms under integral equivalence over local fields[1].

We continue using the notation of the last chapter, except that the field F is now taken to be a local field, S is the single spot $\mathfrak{p}$, $\mathfrak{o}$ is the ring of integers $\mathfrak{o}(\mathfrak{p})$, $\mathfrak{p}$ is the maximal ideal $\mathfrak{m}(\mathfrak{p})$, and $\mathfrak{u}$ is the group of units $\mathfrak{u}(\mathfrak{p})$. We let $|\ | = |\ |_\mathfrak{p}$ be a valuation, say the normalized valuation, in the spot $\mathfrak{p}$. The letter π denotes a prime element of F at $\mathfrak{p}$. Ideal theory and lattice theory are with respect to the single spot $\mathfrak{p}$.

The quadratic space V provided with its symmetric bilinear form B and its associated quadratic form Q is assumed to be regular and non-zero. We shall consider non-zero regular lattices $L, K, J, \ldots$ in V. Note that every line Fx in FL contains a maximal vector of L since the class number of F at $\mathfrak{p}$ is 1. And every non-zero lattice is free. In particular the discriminant dL of § 82B is available for use.

§ 91. Generalities

Here F can be dyadic or non-dyadic. Note that L has the following property: if we take any two vectors x, y in L for which $|B(x, y)|$ is

[1] For the local integral theory of hermitian forms see R. Jacobowitz, *Am. J. Math.* (1962), pp. 441—465.

largest, then

$$B(L, L) = B(x, y)\,\mathfrak{o} = \mathfrak{s}L\,.$$

Similarly if we take any vector $x \in L$ for which $|Q(x)|$ is largest, then

$$Q(x)\,\mathfrak{o} = \mathfrak{n}L\,.$$

§ 91 A. Maximal lattices

91:1. **Theorem.** *V is an anisotropic quadratic space over a local field, and L is an $\mathfrak{a}$-maximal lattice on V. Then*

$$L = \{x \in V \mid Q(x) \in \mathfrak{a}\}\,.$$

In particular, all $\mathfrak{a}$-maximal lattices on V are equal.

Proof. First let us show that the set

$$X = \{x \in V \mid Q(x) \in \mathfrak{a}\}$$

is closed under addition. Consider typical vectors x, y in X. Thus $Q(x)$, $Q(y) \in \mathfrak{a}$ and we must show that $Q(x+y) \in \mathfrak{a}$. Suppose not. Then

$$2B(x, y) = Q(x+y) - Q(x) - Q(y) \notin \mathfrak{a}\,,$$

hence $|2B(x, y)| > |\mathfrak{a}|$ and so

$$\left|\frac{Q(x)\,Q(y)}{B(x, y)^2}\right| < |4|\,.$$

Now the discriminant $d_B(x, y)$ is equal to

$$-B(x, y)^2\left\{1 - \frac{Q(x)\,Q(y)}{B(x, y)^2}\right\},$$

so $-d_B(x, y)$ is a square in $\dot{F}$ by the Local Square Theorem. So $Fx + Fy$ is a hyperbolic plane. This is impossible since V is given anisotropic.

Hence X is indeed closed under addition. Hence it is an $\mathfrak{o}$-module. Suppose X is not equal to L. Then $L \subset X$ by definition of X, so we can pick $z \in X - L$. Then $L + \mathfrak{o}z$ is a lattice contained in the $\mathfrak{o}$-module X, so

$$Q(L + \mathfrak{o}z) \subseteq Q(X) \subseteq \mathfrak{a}\,,$$

hence $\mathfrak{n}(L + \mathfrak{o}z) \subseteq \mathfrak{a}$. This contradicts the maximality of L. **q. e. d.**

91:2. **Theorem.** *Let K and L be $\mathfrak{a}$-maximal lattices on the regular quadratic space V over the local field F. Then* $\operatorname{cls} K = \operatorname{cls} L$.

Proof. We can suppose that V is isotropic by Theorem 91:1. By Proposition 82:23 we have a splitting

$$V = H_1 \perp \cdots \perp H_r \perp H_0$$

in which $H_1, \ldots, H_r$ are hyperbolic planes and H_0 is either 0 or an anisotropic space, with

$$\begin{cases} L = (L \cap H_1) \perp \cdots \perp (L \cap H_r) \perp (L \cap H_0)\,, \\ K = (K \cap H_1) \perp \cdots \perp (K \cap H_r) \perp (K \cap H_0)\,. \end{cases}$$

Now $L \cap H_0$ and $K \cap H_0$ are either both 0 or both $\mathfrak{a}$-maximal on H_0, hence they are equal by Theorem 91:1. And $L \cap H_i$ and $K \cap H_i$ are isometric for $1 \leqq i \leqq r$ by Example 82:22. So for $0 \leqq i \leqq r$ we have $\sigma_i \in O(H_i)$ with $\sigma_i(L \cap H_i) = K \cap H_i$. Then

$$\sigma = \sigma_1 \perp \cdots \perp \sigma_r \perp \sigma_0$$

is an element of $O(V)$ with $\sigma L = K$. Hence $K \in \operatorname{cls} L$. Hence $\operatorname{cls} K = \operatorname{cls} L$.

q. e. d.

91:3. **Example.** If L is an $\mathfrak{a}$-maximal lattice on a regular quadratic space V over a local field F, then $Q(V) \cap \mathfrak{a} = Q(L)$. For clearly $Q(V) \cap \mathfrak{a} \supseteq Q(L)$. On the other hand if α is a non-zero element of $Q(V) \cap \mathfrak{a}$, there is a vector x in V with $Q(x) = \alpha \in \mathfrak{a}$. This vector x is contained in an $\mathfrak{a}$-maximal lattice M on V by Propositions 82:9 and 82:18. But $\operatorname{cls} L = \operatorname{cls} M$ by Theorem 91:2. Hence $\alpha \in Q(L)$ as required. In particular if $\dim V \geqq 4$ we must have $Q(L) = \mathfrak{a}$ since every regular quadratic space V with $\dim V \geqq 4$ over a local field is universal by Remark 63:18. If $\dim V = 3$ we can use Remark 63:18 to show that V represents either all units or all prime elements, hence for any β in $\mathfrak{a}$ we have either $Q(L) \supseteq \beta\,\mathfrak{u}$ or $Q(L) \supseteq \pi\beta\,\mathfrak{u}$.

§ 91B. The group of units of a lattice

Consider the lattice L on the regular, non-zero quadratic space V over the local field F. Let u be a maximal anisotropic vector of L. We claim that the symmetry τ_u of V is in $O(L)$ if and only if

$$\frac{2B(u, L)}{Q(u)} \subseteq \mathfrak{o}\,.$$

First suppose τ_u is a unit of L. Then $\tau_u L = L$ and so

$$\frac{2B(u, x)}{Q(u)}\, u = x - \tau_u x \in L$$

for all x in L. But the coefficient of u is $\mathfrak{o}$ since u is maximal in L, hence $2B(u, x)/Q(u) \in \mathfrak{o}$ as required. Conversely if this condition is satisfied it is clear that $\tau_u L \subseteq L$ and so $\tau_u L = L$ by Example 82:12, i. e. $\tau_u \in O(L)$ as required.

91:4. *Let L be a lattice on the regular non-zero quadratic space V over the local field F. Then $O(L)$ contains a symmetry of V.*

Proof. Take $u \in L$ with $Q(u)\,\mathfrak{o} = \mathfrak{n}L$. Then u is clearly a maximal vector in L; and

$$2B(u, L) \subseteq 2\mathfrak{s}L \subseteq \mathfrak{n}L = Q(u)\,\mathfrak{o}\,,$$

so that $2B(u, L)/Q(u) \subseteq \mathfrak{o}$; hence τ_u is a unit of L. **q. e. d.**

91:4a. $(O(L) : O^+(L)) = 2$ *and so* $\operatorname{cls} L = \operatorname{cls}^+ L$.

91:5. *Let L be a maximal lattice on an anisotropic quadratic space V over a local field. Then $O(L) = O(V)$.*

Proof. We must show that a typical $\sigma \in O(V)$ is in $O(L)$. Now by Theorem 91:1 there is a fractional ideal $\mathfrak{a}$ such that

$$L = \{x \in V \mid Q(x) \in \mathfrak{a}\}.$$

For any x in L we have $Q(\sigma x) = Q(x) \in \mathfrak{a}$ since σ is an isometry, hence $\sigma L \subseteq L$, hence $\sigma \in O(L)$. **q. e. d.**

91:6. *Let V be a regular quadratic space over a local field with* $\dim V \geqq 3$. *Then $\theta(O^+(V)) = \dot{F}$.*

Proof. If $\dim V \geqq 4$ the space is universal by Remark 63:18, hence there is a symmetry with any preassigned spinor norm, hence there is a rotation with any preassigned spinor norm. We are left with the ternary case. By Remark 63:18 the space V will represent all α in $\dot{F}$ for which

$$V \perp \langle -\alpha \rangle$$

has discriminant not equal to 1. So if the discriminant of V is a unit, V will represent all prime elements and at least one unit; now every element of $\dot{F}$ is a product of exactly two such elements times a square in $\dot{F}$; hence $\theta(O^+(V)) = \dot{F}$. A similar argument applies when the discriminant of V is a prime element. **q. e. d.**

91:7. *Let V be a regular quadratic space over a local field with* $\dim V \geqq 3$. *Then there is a lattice L on V with $\theta(O^+(L)) = \dot{F}$.*

Proof. If V is anisotropic we take any maximal lattice L on V. Then by Propositions 91:5 and 91:6 we have

$$\theta(O^+(L)) = \theta(O^+(V)) = \dot{F}.$$

Hence we may assume that V is isotropic. Take a base $x, y, z, \ldots$ for V in which

$$V \cong \left\langle \begin{matrix} 0 & 1 \\ 1 & 0 \end{matrix} \right\rangle \perp \langle \alpha \rangle \perp \cdots.$$

By suitably scaling the bilinear form defining V we may assume that $\alpha = 2\pi$ where π is a prime element. Let L be the lattice

$$L = (\mathfrak{o} x + \mathfrak{o} y) \perp (\mathfrak{o} z) \perp \cdots.$$

Then for any unit ε we have

$$\frac{2B(x + \varepsilon y, L)}{Q(x + \varepsilon y)} = \frac{2\mathfrak{o}}{2\varepsilon} \subseteq \mathfrak{o},$$

hence the symmetry $\tau_{x+\varepsilon y}$ is a unit of L. Now this symmetry has spinor norm $2\varepsilon \dot{F}^2$. Hence $\theta(O^+(L)) \supseteq \mathfrak{u}\dot{F}^2$. Using the same argument with z instead of $x + \varepsilon y$ we obtain a symmetry τ_z which is a unit of L and has spinor norm $2\pi\dot{F}^2$; then $\theta(\tau_z\tau_{x+\varepsilon y}) = \varepsilon\pi\dot{F}^2$; hence $\theta(O^+(L)) \supseteq \pi\mathfrak{u}\dot{F}^2$ and we are through. **q. e. d.**

91:8. *L is a maximal lattice on a regular quadratic space V over a local field with* $\dim V \geqq 3$. *Then* $\theta(O^+(L)) \supseteq \mathfrak{u}$.

Proof. By scaling V we can assume that L is $2\mathfrak{o}$-maximal. If V is anisotropic, then $O^+(L) = O^+(V)$ by Proposition 91:5 and so $\theta(O^+(L)) = \dot{F} \supseteq \mathfrak{u}$ by Proposition 91:6. Now suppose that V is isotropic. By the results of § 82H there is a splitting $L = U \perp \cdots$ in which U has a base $U = \mathfrak{o}x + \mathfrak{o}y$ with matrix $\begin{pmatrix} 0 & 1 \\ 1 & 0 \end{pmatrix}$. Consider any ε in $\mathfrak{u}$. Then

$$\frac{2B(x+\varepsilon y, L)}{Q(x+\varepsilon y)} = \frac{2\mathfrak{o}}{2\varepsilon} \subseteq \mathfrak{o}\,,$$

hence the symmetry $\tau_{x+\varepsilon y}$ is a unit of L. Now this symmetry has spinor norm $2\varepsilon\dot{F}^2$. Hence $\theta(O^+(L)) \supseteq \mathfrak{u}\dot{F}^2$. **q. e. d.**

§ 91 C. Jordan splittings

Consider a non-zero regular lattice L in the quadratic space V. We claim that L splits into 1- and 2-dimensional modular lattices. If there is an x in L with $Q(x)\,\mathfrak{o} = \mathfrak{s}L$, then $J = \mathfrak{o}x$ is an $\mathfrak{s}L$-modular sublattice of L. Otherwise $Q(x)\,\mathfrak{o} \subset \mathfrak{s}L$ for all x in L; in this event we can find a binary $\mathfrak{s}L$-modular sublattice J of L: we pick x, y in L with $B(x,y)\mathfrak{o} = \mathfrak{s}L$; then the vectors x, y have discriminant

$$d_B(x, y) = Q(x)\,Q(y) - B(x, y)^2$$

and this is not zero by the Principle of Domination; hence $J = \mathfrak{o}x + \mathfrak{o}y$ is a binary lattice; a direct computation shows that

$$\mathfrak{s}J = B(x, y)\,\mathfrak{o} = \mathfrak{s}L\,, \quad \mathfrak{v}J = B(x, y)^2\mathfrak{o} = (\mathfrak{s}L)^2,$$

so that J is actually a binary $\mathfrak{s}L$-modular sublattice of L. Hence L always contains a 1- or 2-dimensional $\mathfrak{s}L$-modular sublattice J. Then L has a splitting $L = J \perp K$ by Corollary 82:15a. Now repeat the argument on K, etc. Ultimately we obtain a splitting of L into 1- or 2-dimensional modular components. This establishes our claim.

If we group the modular components of the above splitting in a suitable way we find that L has a splitting

$$L = L_1 \perp \cdots \perp L_t$$

in which each component is modular and

$$\mathfrak{s}L_1 \supset \cdots \supset \mathfrak{s}L_t.$$

Any such splitting is called a Jordan splitting of L. We have therefore proved that *every non-zero regular lattice L in a quadratic space V over a local field F has at least one Jordan splitting*. The rest of this chapter is really a study of the extent of the uniqueness of the Jordan splittings of L.

We shall need the lattice $L^{\mathfrak{a}}$ of § 82I. First suppose that L is $\mathfrak{b}$-modular. If $\mathfrak{b}=\mathfrak{a}$, then $L^{\mathfrak{a}}=L$ and so $L^{\mathfrak{a}}$ is $\mathfrak{a}$-modular. If $\mathfrak{b}\subset\mathfrak{a}$, then $L^{\mathfrak{a}}=L$ is $\mathfrak{b}$-modular with $\mathfrak{b}\subset\mathfrak{a}$. And if $\mathfrak{b}\supset\mathfrak{a}$, then $L^{\mathfrak{a}}=\mathfrak{a}\mathfrak{b}^{-1}L$ is $\mathfrak{a}^2\mathfrak{b}^{-1}$-modular with $\mathfrak{a}^2\mathfrak{b}^{-1}\subset\mathfrak{a}$. So for a $\mathfrak{b}$-modular lattice L we have the following result: $L^{\mathfrak{a}}$ is $\mathfrak{a}$-modular if and only if $\mathfrak{a}=\mathfrak{b}$, otherwise it is $\mathfrak{c}$-modular with $\mathfrak{c}\subset\mathfrak{a}$.

What does this mean in general? Let us consider any non-zero regular lattice L in the Jordan splitting

$$L=L_1\perp\cdots\perp L_t.$$

We know that $\mathfrak{s}L^{\mathfrak{a}}\subseteq\mathfrak{a}$. And we have a splitting

$$L^{\mathfrak{a}}=L_1^{\mathfrak{a}}\perp\cdots\perp L_t^{\mathfrak{a}}$$

into modular components. If L_i is not $\mathfrak{a}$-modular, then $L_i^{\mathfrak{a}}$ will be $\mathfrak{c}$-modular with $\mathfrak{c}\subset\mathfrak{a}$. And if L_i is $\mathfrak{a}$-modular, then $L_i^{\mathfrak{a}}$ will be equal to L_i. Hence we find that $\mathfrak{s}L^{\mathfrak{a}}=\mathfrak{a}$ *if and only if there is an* $\mathfrak{a}$*-modular component in the given Jordan splitting. Otherwise* $\mathfrak{s}L^{\mathfrak{a}}\subset\mathfrak{a}$. Now consider L_j in the given splitting, and suppose it is $\mathfrak{a}$-modular. If we group the components of the splitting $L_1^{\mathfrak{a}}\perp\cdots\perp L_t^{\mathfrak{a}}$ we obtain a Jordan splitting of $L^{\mathfrak{a}}$ in which the first component is L_j. So we have proved that an $\mathfrak{a}$-modular component in a given Jordan splitting of the lattice L is the first component in some Jordan splitting of the lattice $L^{\mathfrak{a}}$.

91:9. **Theorem.** *Let L be a lattice in the quadratic space V over the local field F, and let*

$$L=L_1\perp\cdots\perp L_t\,,\quad L=K_1\perp\cdots\perp K_T$$

be two Jordan splittings of L. Then $t=T$. And for $1\leqq\lambda\leqq t$ we have

(1) $\mathfrak{s}L_\lambda=\mathfrak{s}K_\lambda$, $\dim L_\lambda=\dim K_\lambda$,

(2) $\mathfrak{n}L_\lambda=\mathfrak{s}L_\lambda$ *if and only if* $\mathfrak{n}K_\lambda=\mathfrak{s}K_\lambda$.

Proof. We shall use the results of the preceding discussion in the proof.

1) Suppose there is an $\mathfrak{a}$-modular component in the first Jordan splitting. Then $\mathfrak{s}L^{\mathfrak{a}}=\mathfrak{a}$. Hence there is an $\mathfrak{a}$-modular component in the second splitting. Hence, on grounds of symmetry, we have $t=T$ and $\mathfrak{s}L_\lambda=\mathfrak{s}K_\lambda$ for $1\leqq\lambda\leqq t$.

Consider a typical λ with $1\leqq\lambda\leqq t$. We must prove that $\dim L_\lambda=\dim K_\lambda$ and that $\mathfrak{n}L_\lambda=\mathfrak{s}L_\lambda$ if and only if $\mathfrak{n}K_\lambda=\mathfrak{s}K_\lambda$. Now L_λ and K_λ are first components in Jordan splittings of $L^{\mathfrak{a}}$ where $\mathfrak{a}=\mathfrak{s}L_\lambda=\mathfrak{s}K_\lambda$. Hence we can assume that $\lambda=1$. And by suitably scaling the bilinear form B on V we can assume that $\mathfrak{s}L=\mathfrak{o}$. On grounds of symmetry we are therefore reduced to proving the following: given $\mathfrak{s}L_1=\mathfrak{s}K_1=\mathfrak{s}L=\mathfrak{o}$, prove that $\dim L_1\leqq\dim K_1$, and also that $\mathfrak{n}L_1=\mathfrak{o}$ implies $\mathfrak{n}K_1=\mathfrak{o}$. This we now do.

2) We shall need the projection map $\varphi\colon FL_1\rightarrowtail FK_1$. This is defined as follows: each x in FL_1 is an element of FL and hence has a unique

expression $x = y + z$ with $y \in FK_1$ and $z \in F(K_2 \perp \cdots \perp K_t)$. Put $\varphi x = y$. Then φx is a well-defined element of FK_1, and it is easily seen that the map $\varphi: FL_1 \rightarrow FK_1$ determined by $x \rightarrow \varphi x$ is F-linear.

Now when x in L_1 is expressed as $x = y + z$ in the above way we have $y \in K_1$ and $z \in K_2 \perp \cdots \perp K_t$. Hence $\varphi L_1 \subseteq K_1$. And for all x, x' in L_1 we have

$$B(\varphi x, \varphi x') = B(x - z, x' - z') \equiv B(x, x') \bmod \mathfrak{p}$$

since

$$B(K_2 \perp \cdots \perp K_t, L) \subseteq \mathfrak{p}\,.$$

In other words

$$B(\varphi x, \varphi x') \equiv B(x, x')\,, \quad Q(\varphi x) \equiv Q(x)$$

modulo $\mathfrak{p}$ for all x, x' in L_1.

Suppose we had $\varphi x = 0$ for a non-zero vector x of FL_1. Then we would have $\varphi x = 0$ for a maximal vector x of the lattice L_1. Since L_1 is unimodular there is a vector y in L_1 with $B(x, y) = 1$ by Proposition 82:17. Then

$$1 \equiv B(x, y) \equiv B(\varphi x, \varphi y) \equiv 0 \bmod \mathfrak{p}\,,$$

and this is absurd. Hence φ is an isomorphism of FL_1 into FK_1. Hence

$$\dim L_1 = \dim FL_1 \leqq \dim FK_1 = \dim K_1\,,$$

as required.

3) Finally we have to prove that $\mathfrak{n}K_1 = \mathfrak{o}$ if $\mathfrak{n}L_1 = \mathfrak{o}$. Since $\mathfrak{n}L_1 = \mathfrak{o}$ we can find a vector x in L_1 with $Q(x) = \varepsilon$ where ε is in $\mathfrak{u}$. Then

$$Q(\varphi x) \equiv Q(x) \equiv \varepsilon \bmod \mathfrak{p}\,.$$

Hence $Q(\varphi x) \in \mathfrak{u}$. Hence $\mathfrak{n}K_1 \supseteq \mathfrak{o}$. Hence $\mathfrak{n}K_1 = \mathfrak{o}$. **q. e. d.**

Consider non-zero regular lattices L and K in quadratic spaces V and U over the same field F. Let

$$L = L_1 \perp \cdots \perp L_t\,, \quad K = K_1 \perp \cdots \perp K_T$$

be Jordan splittings of L and K. We say that these Jordan splittings are of the same type if $t = T$ and, whenever $1 \leqq \lambda \leqq t$, we have

$$\mathfrak{s}L_\lambda = \mathfrak{s}K_\lambda\,, \quad \dim L_\lambda = \dim K_\lambda$$

and

$$\mathfrak{n}L_\lambda = \mathfrak{s}L_\lambda \quad \text{if and only if} \quad \mathfrak{n}K_\lambda = \mathfrak{s}K_\lambda\,.$$

We know from Theorem 91:9 that any two Jordan splittings of L are of the same type. And the same with K. We say that the lattices L and K are of the same Jordan type if their Jordan splittings are of the same type. Isometric lattices are of the same Jordan type.

91:10. **Notation.** Given a lattice L and a Jordan splitting $L = L_1 \perp \cdots \perp L_t$, we put

$$L_{(i)} = L_1 \perp \cdots \perp L_i$$

and we call

$$L_{(1)} \subset L_{(2)} \subset \cdots \subset L_{(t)}$$

the Jordan chain associated with the given splitting. We put

$$L^*_{(i)} = L_i \perp \cdots \perp L_t$$

and we call

$$L^*_{(1)} \supset L^*_{(2)} \supset \cdots \supset L^*_{(t)}$$

the inverse Jordan chain associated with the given splitting. Clearly

$$L = L_{(i)} \perp L^*_{(i+1)} .$$

A Jordan chain is determined by one and only one Jordan splitting of L.

91:11. **Example.** Let $L_{(1)} \subset \cdots \subset L_{(t)}$ and $K_{(1)} \subset \cdots \subset K_{(t)}$ be Jordan chains of lattices L and K of the same Jordan type. Then $L_{(i)}$ and $K_{(i)}$ are lattices of the same Jordan type. Also $\mathfrak{v} L_{(i)} = \mathfrak{v} K_{(i)}$. Hence there is a unit ε such that

$$d L_{(i)} / d K_{(i)} = \varepsilon .$$

The same applies to inverse Jordan chains of L and K.

§ 92. Classification of lattices over non-dyadic fields

Throughout this paragraph we assume that the local field under discussion is non-dyadic. We consider a non-zero regular lattice L in the quadratic space V. We know from § 82E that $2\mathfrak{s}L \subseteq \mathfrak{n}L \subseteq \mathfrak{s}L$. But 2 is a unit in $\mathfrak{o}$ since F is non-dyadic. Hence $\mathfrak{n}L = \mathfrak{s}L$, i. e. norm and scale are equal over non-dyadic fields. We can therefore pick $x \in L$ with

$$Q(x)\,\mathfrak{o} = \mathfrak{n}L = \mathfrak{s}L .$$

Then $J = \mathfrak{o}x$ is an $\mathfrak{s}L$-modular sublattice of L, hence L has a splitting $L = J \perp K$. If we repeat on K, etc. we ultimately find a splitting

$$L = \mathfrak{o}x_1 \perp \cdots \perp \mathfrak{o}x_r .$$

In other words, in the non-dyadic case every non-zero regular lattice has an orthogonal base.

92:1. *Let L be a unimodular lattice on the quadratic space V over the non-dyadic local field F. Then there is a unit ε such that*

$$L \cong \langle 1\rangle \perp \cdots \perp \langle 1\rangle \perp \langle\varepsilon\rangle .$$

Proof. Since L has an orthogonal base we can write

$$L \cong \langle\varepsilon_1\rangle \perp \cdots \perp \langle\varepsilon_n\rangle \qquad (\varepsilon_i \in \mathfrak{u}) .$$

Put $\varepsilon = \varepsilon_1 \ldots \varepsilon_n$. Then

$$FL \cong \langle 1\rangle \perp \cdots \perp \langle 1\rangle \perp \langle\varepsilon\rangle$$

by the criterion of Theorem 63:20 in virtue of the fact that the Hilbert symbol $\left(\frac{\delta, \delta'}{\mathfrak{p}}\right)$ is 1 whenever δ, δ' are units in a non-dyadic local field.

So there is a lattice K over V with

$$K \cong \langle 1\rangle \perp \cdots \perp \langle 1\rangle \perp \langle \varepsilon\rangle .$$

Now K and L are $\mathfrak{o}$-maximal on V by Proposition 82:19, hence they are isometric by Theorem 91:2. **q. e. d.**

92:1a. *There are essentially two unimodular lattices of given dimension over a given non-dyadic local field.*

92:1b. $Q(L) \supseteq \mathfrak{u}$ *if* $\dim L = 2$. *And* $Q(L) = \mathfrak{o}$ *if* $\dim L \geqq 3$.

Proof. If $\dim L = 2$ we apply the proposition twice to obtain

$$L \cong \langle 1\rangle \perp \langle \varepsilon\rangle \cong \langle \delta\rangle \perp \langle \delta\varepsilon\rangle$$

for any δ in $\mathfrak{u}$. Hence $\delta \in Q(L)$ as required. If $\dim L \geqq 3$ we must prove that $\alpha \in Q(L)$ for any α in $\mathfrak{p}$. But by the case $\dim L = 2$ we have

$$L \cong \langle -1\rangle \perp \langle 1+\alpha\rangle \perp \cdots .$$

Hence $\alpha \in Q(L)$. **q. e. d.**

92:2. **Theorem.**[1] *Let L and K be lattices of the same Jordan type on the regular quadratic space V over the non-dyadic local field F. Consider Jordan splittings*

$$L = L_1 \perp \cdots \perp L_t\,, \quad K = K_1 \perp \cdots \perp K_t\,.$$

Then $\mathrm{cls}\,L = \mathrm{cls}\,K$ *if and only if*

$$dL_i = dK_i \quad \textit{for} \quad 1 \leqq i \leqq t\,.$$

Proof. 1) First suppose that $\mathrm{cls}\,L = \mathrm{cls}\,K$. Then $K = \sigma L$ for some σ in $O(V)$, so we can actually suppose that $K = L$. Consider L_i and K_i. By suitably scaling the bilinear form B on V we can assume that L_i and K_i are unimodular. Then L_i and K_i are the first components of Jordan splittings of L°. We may therefore assume that $i = 1$.

We saw in step 2) of the proof of Theorem 91:9 that there is an F-linear isomorphism $\varphi: FL_1 \rightarrowtail FK_1$ with $\varphi L_1 \subseteq K_1$ and such that

$$B(\varphi x, \varphi y) \equiv B(x, y) \mod \mathfrak{p}$$

for all x, y in L_1. Take a base

$$L_1 = \mathfrak{o}x_1 + \cdots + \mathfrak{o}x_r\,.$$

Then φL_1 is a sublattice of K_1 which is on FK_1 and has a base

$$\varphi L_1 = \mathfrak{o}(\varphi x_1) + \cdots + \mathfrak{o}(\varphi x_r)\,.$$

Write $d_B(x_1, \ldots, x_r) = \varepsilon$. Then ε is a unit since L_1 is unimodular. Now

$$d_B(\varphi x_1, \ldots, \varphi x_r) \equiv \varepsilon \mod \mathfrak{p}.$$

Therefore $d_B(\varphi x_1, \ldots, \varphi x_r)$ is a unit. Hence $\mathfrak{v}(\varphi L_1) = \mathfrak{o} = \mathfrak{v}K_1$. So $\varphi L_1 = K_1$ by Corollary 82:11a. So K_1 has the base

$$K_1 = \mathfrak{o}(\varphi x_1) + \cdots + \mathfrak{o}(\varphi x_r)\,.$$

[1] A similar theorem holds for representations in the non-dyadic case. See O. T. O'Meara, *Am. J. Math.* (1958), p. 850.

Hence $dK_1 = \varepsilon(1+\alpha)$ with α in $\mathfrak{p}$. Hence $dK_1 = \varepsilon$ by the Local Square Theorem. Hence $dK_1 = dL_1$.

2) Conversely suppose that $dL_i = dK_i$ for $1 \leq i \leq t$. Then it follows easily from Proposition 92:1 that there is an isometry σ_i of FL_i onto FK_i such that $\sigma_i L_i = K_i$. Put

$$\sigma = \sigma_1 \perp \cdots \perp \sigma_t .$$

Then σ is an element of $O(V)$ such that $\sigma L = K$. Hence $K \in \operatorname{cls} L$. Hence $\operatorname{cls} K = \operatorname{cls} L$. **q. e. d.**

92:2a. **Corollary.** *$\operatorname{cls} L = \operatorname{cls} K$ if and only if $FL_i \cong FK_i$ for $1 \leq i \leq t$.*

92:2b. **Corollary.** *Let $L_{(1)} \subset \cdots \subset L_{(t)}$ and $K_{(1)} \subset \cdots \subset K_{(t)}$ be the Jordan chains associated with the given splittings of L and K. Then $\operatorname{cls} L = \operatorname{cls} K$ if and only $FL_{(i)} \cong FK_{(i)}$ for $1 \leq i \leq t$.*

92:3. **Theorem.** *Let L and K be isometric lattices on the regular quadratic spaces V and U over the non-dyadic local field F. Suppose there are splittings $L = L' \perp L''$ and $K = K' \perp K''$ with L' isometric to K'. Then L'' is isometric to K''.*

Proof. This is an easy application of Theorem 92:2 and is left as an exercise to the reader. **q. e. d.**

92:4. **Theorem.** *Let L be a lattice on a regular n-ary quadratic space over a non-dyadic local field. Then every element of $O(L)$ is a product of at most $2n-1$ symmetries in $O(L)$.*

Proof. Let V be the quadratic space in question. The proof is by induction to n. The case $n = 1$ is trivial[1], so we assume that $n > 1$. By suitably scaling V we can assume that $\mathfrak{s}L = \mathfrak{o}$. Consider a typical σ in $O(L)$. This σ must be expressed as a product of symmetries in $O(L)$. Fix y in L with $Q(y)$ equal to some unit ε. Then

$$Q(y - \sigma y) + Q(y + \sigma y) = 4\varepsilon$$

and this is a unit since the field is non-dyadic. Hence either $Q(y-\sigma y)$ or $Q(y+\sigma y)$ is a unit. In the first instance the criterion of § 91B shows that the symmetry $\tau_{y-\sigma y}$ is a unit of L, and we have

$$\tau_{y-\sigma y} y = \sigma y .$$

In the second instance τ_y and $\tau_{y+\sigma y}$ are elements of $O(L)$, and we have

$$\tau_{y+\sigma y} \tau_y y = \sigma y .$$

So in either case there is an element ϱ which is a product of one or two symmetries in $O(L)$ such that $\varrho y = \sigma y$. By Corollary 82:15a we have a splitting $L = \mathfrak{o}y \perp K$. Then $\varrho^{-1}\sigma$ induces an isometry on FK; this isometry is a unit of K and hence by the inductive hypothesis it is a

---

[1] Starting the induction at $n = 2$ will show that at most $2n - 2$ symmetries are needed when $n \geq 2$.

product

$$\tau_1 \tau_2 \cdots$$

of at most $2n-3$ symmetries in $O(K)$. Then

$$\varrho^{-1}\sigma = (1 \perp \tau_1)(1 \perp \tau_2) \ldots .$$

Hence

$$\sigma = \varrho(1 \perp \tau_1)(1 \perp \tau_2) \ldots$$

is a product of at most $2n-1$ symmetries in $O(L)$. **q. e. d.**

92:5. *Let L be a modular lattice on a quadratic space over a non-dyadic local field with* $\dim L \geqq 2$. *Then* $\theta(O^+(L)) = \mathfrak{u}\dot{F}^2$.

Proof. By suitably scaling the space we can assume without loss of generality that L is unimodular. Take a typical symmetry in $O(L)$. This symmetry has the form τ_y with y a maximal anisotropic vector in L. Since τ_y is a unit of L we have $B(y, L) \subseteq Q(y)\mathfrak{o}$ by the criterion of § 91B. But $B(y, L) = \mathfrak{o}$ since L is unimodular. Hence $Q(y) \in \mathfrak{u}$. So $\theta(\tau_y) \subseteq \mathfrak{u}\dot{F}^2$. Therefore

$$\theta(O^+(L)) \subseteq \mathfrak{u}\dot{F}^2 .$$

We then obtain equality here by applying Corollary 92:1b to find a symmetry τ in $O(L)$ with $\theta(\tau) = \varepsilon\dot{F}^2$ for any ε in $\mathfrak{u}$. **q. e. d.**

92:6. **Example.** Let L be a lattice on the regular quadratic space V over the non-dyadic local field F. Consider a maximal anisotropic vector y of L. The criterion of § 91B says that the symmetry τ_y is a unit of L if and only if $B(\mathfrak{o}y, L) \subseteq Q(y)\mathfrak{o}$. By Proposition 82:15 we know that this condition is equivalent to saying that $\mathfrak{o}y$ splits L since $\mathfrak{o}y$ is $Q(y)$-modular. Hence for any maximal anisotropic vector y in L we have

$$\tau_y \in O(L) \Leftrightarrow \mathfrak{o}y \text{ splits } L.$$

92:7. **Example.** Suppose V is a regular quadratic space over a non-dyadic local field with invariants $dV = 1$ and $S_\mathfrak{p} V = 1$. We claim that there is a lattice L on V with

$$\theta(O(L)) = \dot{F}^2 .$$

By Theorem 63:20 we have

$$V \cong \langle 1 \rangle \perp \cdots \perp \langle 1 \rangle .$$

Hence there is a lattice L on V with

$$L \cong \langle 1 \rangle \perp \langle \pi^2 \rangle \perp \cdots \perp \langle \pi^{2(n-1)} \rangle .$$

By Theorem 92:2 we know that if $\mathfrak{o}y$ splits L, then $Q(y) \in \pi^{2i}\mathfrak{u}^2$ for some i $(0 \leqq i \leqq n-1)$. Hence by Example 92:6 all symmetries in $O(L)$ have spinor norm $\dot{F}^2$. Hence by Theorem 92:4 all elements of $O(L)$ have spinor norm $\dot{F}^2$.

92:8. **Example.** Suppose V satisfies the conditions of Example 92:7, and suppose further that $\dim V$ is even. Let ε be any unit in $\mathfrak{o}$. Then there is a lattice L on V with

$$\theta(O^-(L)) = \varepsilon \dot{F}^2 .$$

92:9. **Example.** Let L be a lattice on a regular n-ary quadratic space over a non-dyadic local field, with

$$\mathfrak{s}L \subseteq \mathfrak{o} \quad \text{and} \quad \mathfrak{v}L \supset \mathfrak{p}^{\frac{1}{2}n(n-1)}$$

We claim that

$$\theta(O^+(L)) \supseteq \mathfrak{u} .$$

This follows immediately if we can prove that the number t in the Jordan splitting

$$L = L_1 \perp \cdots \perp L_t$$

is less than n: for then we will have $\dim L_i \geq 2$ for at least one $i\,(1 \leq i \leq t)$ and we can apply Proposition 92:5 to this L_i. Suppose if possible that $t = n$. Then

$$\mathfrak{v}L_i = \mathfrak{s}L_i \subseteq \mathfrak{p}^{i-1} ,$$

hence

$$\mathfrak{v}L \subseteq \mathfrak{p}^{0+1+\cdots+(n-1)} = \mathfrak{p}^{\frac{1}{2}n(n-1)}$$

and this is contrary to hypothesis.

§ 93. Classification of lattices over dyadic fields

Throughout this paragraph F is a dyadic local field. Thus F has characteristic 0[1] and the residue class field of F at $\mathfrak{p}$ is a finite field of characteristic 2. So the residue class field is perfect, and the congruence

$$\varepsilon' \equiv \varepsilon\,\delta^2 \bmod \pi$$

has a solution δ for any given units ε and ε'. For any given α, β in $\dot{F}$ we shall write

$$\alpha \cong \beta$$

if α/β is an element of $\mathfrak{u}^2$. This defines an equivalence relation. And $\alpha \cong \beta$ if and only if α/β is a unit with quadratic defect $\mathfrak{d}(\alpha/\beta) = 0$. For any fractional ideal $\mathfrak{a}$ we write

$$\alpha \cong \beta \bmod \mathfrak{a}$$

if α/β is a unit and $\alpha \equiv \beta\varepsilon^2 \bmod \mathfrak{a}$ for some unit ε. This also defines an equivalence relation. And $\alpha \cong \beta \bmod \mathfrak{a}$ if and only if α/β is a unit with

[1] See C. H. Sah, *Am. J. Math.* (1960), pp. 812—830, for the integral theory over local fields of characteristic 2.

$\mathfrak{d}(\alpha/\beta) \subseteq \mathfrak{a}/\beta$. In virtue of the perfectness of the residue class field we have $\alpha \cong \beta \bmod \alpha\mathfrak{p}$ when $\alpha/\beta \in \mathfrak{u}$.

The letter Δ will denote a fixed unit of quadratic defect $4\mathfrak{o}$. It is assumed that Δ has the form $\Delta = 1 + 4\varrho$ for some fixed unit ϱ in F. Of course $\Delta \cong 1 \bmod 4\mathfrak{o}$.

V will be a regular n-ary quadratic space over F, L will be a non-zero regular lattice in V. As usual

$$2(\mathfrak{s}L) \subseteq \mathfrak{n}L \subseteq \mathfrak{s}L\,.$$

But we now have

$$2(\mathfrak{s}L) \subset \mathfrak{s}L\,,$$

and it is this strict inclusion that makes the dyadic theory of quadratic forms distinctly different from the non-dyadic theory of the last paragraph.

93:1. **Notation.** We let $\mathfrak{o}^2$ denote the set of squares of elements of $\mathfrak{o}$. The symbol $\mathfrak{o}^2$ already has a meaning in the sense of ideal theory, namely $\mathfrak{o}^2$ is the product of the fractional ideal $\mathfrak{o}$ with itself. However this product is equal to $\mathfrak{o}$ so there is never any need to use the symbol $\mathfrak{o}^2$ in this sense. For us then $\mathfrak{o}^2$ will be the set

$$\mathfrak{o}^2 = \{\alpha^2 \mid \alpha \in \mathfrak{o}\}\,.$$

93:2. **Notation.** Given a non-zero scalar α and a fractional ideal $\mathfrak{a}$ we shall write

$$dL \cong \alpha \bmod \mathfrak{a}$$

if $dL = \beta$ for some $\beta \in \dot{F}$ which satisfies $\beta \cong \alpha \bmod \mathfrak{a}$. This is the same as saying that L has a base $x_1, \ldots, x_n$ in which

$$d_B(x_1, \ldots, x_n) \cong \alpha \bmod \mathfrak{a}\,.$$

Given two lattices L and K we write

$$dL/dK \cong \alpha \bmod \mathfrak{a}$$

if there are non-zero scalars β, γ such that $dL = \beta$, $dK = \gamma$, and $\beta/\gamma \cong \alpha \bmod \mathfrak{a}$. This is equivalent to saying that there are bases $x_1, \ldots, x_n$ and $y_1, \ldots, y_m$ for L and K such that

$$d_B(x_1, \ldots, x_n)/d_B(y_1, \ldots, y_m) \cong \alpha \bmod \mathfrak{a}\,.$$

§ 93A. The norm group $\mathfrak{g}L$ and the weight $\mathfrak{w}L$

It is easily seen that the set $Q(L) + 2(\mathfrak{s}L)$ is an additive subgroup of F. We shall call this subgroup the norm group of L and we shall write it

$$\mathfrak{g}L = Q(L) + 2(\mathfrak{s}L)\,.$$

The norm group $\mathfrak{g}L$ is a finer object than the norm ideal $\mathfrak{n}L$ which it generates. We have

$$2(\mathfrak{s}L) \subseteq \mathfrak{g}L \subseteq \mathfrak{n}L\,,$$

and $\mathfrak{n}L = a\mathfrak{o}$ holds for any element a of $\mathfrak{g}L$ with largest value. Given two regular non-zero lattices K and L in V, then

$$K \subseteq L \Rightarrow \mathfrak{g}K \subseteq \mathfrak{g}L,$$

and

$$\mathfrak{g}(K \perp L) = \mathfrak{g}K + \mathfrak{g}L.$$

The sets $Q(L)$ and $\mathfrak{g}L$ don't have to be equal. For instance if $L = \mathfrak{o}x$, then $Q(L) = \mathfrak{o}^2 Q(x)$ contains no fractional ideals and so it cannot be all of $\mathfrak{g}L$. On the other hand, *if L is any lattice in V with $\mathfrak{s}L = \mathfrak{o}$ and if L contains a sublattice H of the form $H \cong \begin{pmatrix} 0 & 1 \\ 1 & 0 \end{pmatrix}$ then we do indeed have*

$$\mathfrak{g}L = Q(L).$$

In order to prove this we must show that a typical element

$$Q(x) + 2\alpha \qquad (x \in L, \alpha \in \mathfrak{o})$$

of $\mathfrak{g}L$ is also in $Q(L)$. Now we have a splitting $L = H \perp K$ since H is unimodular and $\mathfrak{s}L = \mathfrak{o}$, hence we can write $x = h + k$ with $h \in H$, $k \in K$. Since $Q(H) = 2\mathfrak{o}$ we can find $h' \in H$ with $Q(h') = Q(h) + 2\alpha$. Then

$$Q(h' + k) = Q(h') + Q(k) = Q(h) + 2\alpha + Q(k) = Q(x) + 2\alpha.$$

Hence $Q(x) + 2\alpha$ is in $Q(L)$ as required. So $Q(L)$ and $\mathfrak{g}L$ are sometimes equal. As a matter of fact the result that we have just proved will enable us to arrange $\mathfrak{g}L = Q(L)$ whenever we please.

93:3. *Let L be a lattice of scale $\mathfrak{o}$ on a hyperbolic space V over a dyadic local field. Then there is a unimodular lattice K on V with $L \subseteq K \subseteq V$ such that $\mathfrak{g}K = \mathfrak{g}L$.*

Proof. As we ascend a tower of lattices on V we obtain an ascending tower of volumes in F. Hence we can find a lattice K on V with

$$L \subseteq K, \quad \mathfrak{s}K = \mathfrak{o}, \quad \mathfrak{g}K = \mathfrak{g}L,$$

and such that there is no lattice K' on V which has these properties and strictly contains K. This K will be the lattice required by the proposition.

We assert that every isotropic vector x in V which satisfies $B(x, K) \subseteq \mathfrak{o}$ is actually in K. For consider $K' = \mathfrak{o}x + K$. Then $\mathfrak{s}K' = \mathfrak{o}$ and $Q(K') \subseteq Q(K) + 2\mathfrak{o} = \mathfrak{g}K$, so that

$$L \subseteq K', \quad \mathfrak{s}K' = \mathfrak{o}, \quad \mathfrak{g}K' = \mathfrak{g}L.$$

Hence $K' = K$ by choice of K. Therefore x is in K as asserted.

Hence every maximal isotropic vector x of K satisfies the equation $B(x, K) = \mathfrak{o}$. Pick $y \in K$ with $B(x, y) = 1$. Then $\mathfrak{o}x + \mathfrak{o}y$ is a unimodular sublattice of K, so we have a splitting $K = (\mathfrak{o}x + \mathfrak{o}y) \perp J$. But FJ is isotropic since FK is hyperbolic. Hence J has a splitting $J = (\mathfrak{o}x' + \mathfrak{o}y') \perp J'$ with x' isotropic and $\mathfrak{o}x' + \mathfrak{o}y'$ unimodular. Repeat. Ultimately we obtain a splitting of K into binary unimodular lattices. So K is unimodular. **q. e. d.**

We let $\mathfrak{m}L$ denote the largest fractional ideal contained in the norm group $\mathfrak{g}L$. Thus

$$2(\mathfrak{s}L) \subseteq \mathfrak{m}L \subseteq \mathfrak{g}L \subseteq \mathfrak{n}L\,.$$

Let us show that

$$\text{ord}_{\mathfrak{p}}\,\mathfrak{n}L + \text{ord}_{\mathfrak{p}}\,\mathfrak{m}L \quad \text{is even.}$$

Suppose not. Pick $a \in \mathfrak{g}L$ with $a\mathfrak{o} = \mathfrak{n}L$ and write $\mathfrak{m}L = a\mathfrak{p}^{2r+1}$ with an $r \geqq 0$. We claim that $a\mathfrak{p}^{2r} \subseteq \mathfrak{g}L$. We have to show that any element of the form $a\,\varepsilon\,\pi^{2r}$ with $\varepsilon \in \mathfrak{u}$ is in $\mathfrak{g}L$. By the perfectness of the residue class field we can solve the congruence

$$a\,\varepsilon\,\pi^{2r} \equiv a\,\delta^2\,\pi^{2r} \bmod a\mathfrak{p}^{2r+1}$$

for some unit δ. But $a\mathfrak{p}^{2r+1} \subseteq \mathfrak{g}L$ by definition of $\mathfrak{m}L$. And $a(\delta\pi^r)^2$ is in $\mathfrak{g}L$. Hence $a\,\varepsilon\,\pi^{2r}$ is in $\mathfrak{g}L$. So $\text{ord}_{\mathfrak{p}}\,\mathfrak{n}L + \text{ord}_{\mathfrak{p}}\,\mathfrak{m}L$ is even.

We define the weight $\mathfrak{w}L$ by the equation

$$\mathfrak{w}L = \mathfrak{p}(\mathfrak{m}L) + 2(\mathfrak{s}L)\,.$$

So $\mathfrak{m}L$ depends only on $\mathfrak{g}L$, while $\mathfrak{w}L$ depends on L. We have

$$\mathfrak{p}(\mathfrak{m}L) \subseteq \mathfrak{w}L \subseteq \mathfrak{m}L\,, \quad 2(\mathfrak{s}L) \subseteq \mathfrak{w}L \subseteq \mathfrak{n}L\,.$$

Also

$$\text{ord}_{\mathfrak{p}}\,\mathfrak{n}L + \text{ord}_{\mathfrak{p}}\,\mathfrak{w}L \quad \text{is odd} \quad \Leftrightarrow \quad \mathfrak{w}L = \mathfrak{p}(\mathfrak{m}L)$$

and

$$\text{ord}_{\mathfrak{p}}\,\mathfrak{n}L + \text{ord}_{\mathfrak{p}}\,\mathfrak{w}L \quad \text{is even} \quad \Rightarrow \quad \mathfrak{w}L = \mathfrak{m}L = 2(\mathfrak{s}L)\,.$$

Hence

$$\mathfrak{w}L = \mathfrak{n}L \quad \Leftrightarrow \quad 2(\mathfrak{s}L) = \mathfrak{n}L\,.$$

It is easily seen that

$$K \subseteq L \quad \Rightarrow \quad \mathfrak{w}K \subseteq \mathfrak{w}L$$

and that

$$\mathfrak{g}K = \mathfrak{g}L \quad \text{with} \quad \mathfrak{s}K = \mathfrak{s}L \quad \Rightarrow \quad \mathfrak{w}K = \mathfrak{w}L\,.$$

We call the scalar a a norm generator of L if it is a scalar of largest value in $\mathfrak{g}L$. Thus a is a norm generator of L if and only if $a \in \mathfrak{g}L \subseteq a\mathfrak{o}$, i. e. if and only if

$$a \in \mathfrak{g}L \quad \text{with} \quad a\mathfrak{o} = \mathfrak{n}L\,.$$

We call the scalar b a weight generator of L if it is a scalar of largest value in $\mathfrak{w}L$. Thus b is a weight generator of L if and only if

$$b\mathfrak{o} = \mathfrak{w}L\,.$$

If a and b are norm and weight generators of L, then

$$|2(\mathfrak{s}L)| \leqq |b| < |\mathfrak{s}L|\,, \quad |b| \leqq |a|\,,$$

$$\text{ord}_{\mathfrak{p}}\,a + \text{ord}_{\mathfrak{p}}\,b \quad \text{is even} \quad \Rightarrow \quad |b| = |2(\mathfrak{s}L)|$$

$$|b| = |a| \quad \Leftrightarrow \quad |a| = |2(\mathfrak{s}L)|\,.$$

Let us prove that *every element of* $\mathfrak{g}L$ *whose order has opposite parity to the order of* $\mathfrak{n}L$ *is in* $\mathfrak{w}L$. In other words, if a is a norm generator of L, and if $\beta \in \mathfrak{g}L$ with $\mathrm{ord}_{\mathfrak{p}}\, a + \mathrm{ord}_{\mathfrak{p}}\, \beta$ odd, then $\beta \in \mathfrak{w}L$. It is enough to prove that $\beta\mathfrak{o} \subseteqq \mathfrak{g}L$, for then $\beta\mathfrak{o} \subseteqq \mathfrak{m}L$ by definition of $\mathfrak{m}L$, hence $\beta\mathfrak{o} \subseteqq \mathfrak{p}(\mathfrak{m}L) \subseteqq \mathfrak{w}L$. Consider a typical non-zero γ in $\beta\mathfrak{o}$. We have to show that $\gamma \in \mathfrak{g}L$. Now by Proposition 63:11 we have $\xi, \eta \in F$ such that

$$\gamma \equiv a\xi^2 + \beta\,\eta^2 \bmod 4\,\gamma\,.$$

By the Principle of Domination both ξ and η are in $\mathfrak{o}$, since $\gamma \in \beta\,\mathfrak{o} \subseteqq a\mathfrak{o}$. But $\mathfrak{g}L$ stands multiplication by elements of $\mathfrak{o}^2$, by definition of $\mathfrak{g}L$. And $4\,\gamma\,\mathfrak{o} \subseteqq 4(\mathfrak{n}L) \subseteqq 2(\mathfrak{s}L)$. Hence $\gamma \in \mathfrak{g}L$, as asserted.

93:4. *Let L be a non-zero regular lattice over a dyadic local field with norm and weight generators a and b. Then*

$$\mathfrak{g}L = a\mathfrak{o}^2 + b\mathfrak{o}\,.$$

Proof. The set $\mathfrak{g}L$ stands multiplication by elements of $\mathfrak{o}^2$, hence $a\mathfrak{o}^2 \subseteqq \mathfrak{g}L$, hence $a\mathfrak{o}^2 + \mathfrak{w}L$ is contained in $\mathfrak{g}L$. Conversely consider a typical element α of $\mathfrak{g}L$. We wish to express α as an element of $a\mathfrak{o}^2 + \mathfrak{w}L$. By Proposition 63:11 we have scalars ξ and η such that

$$\alpha \equiv a\xi^2 + a\,\pi\,\eta^2 \bmod 4\,\alpha\,.$$

By the Principle of Domination we see that $\xi \in \mathfrak{o}$, since $\alpha \in a\mathfrak{o}$. Then α and $a\xi^2$ are in $\mathfrak{g}L$, and $4\alpha\,\mathfrak{o} \subseteqq 2(\mathfrak{s}L) \subseteqq \mathfrak{g}L$, hence $a\,\pi\,\eta^2$ in in $\mathfrak{g}L$, hence it is in $\mathfrak{w}L$ since its order has opposite parity to the order of $\mathfrak{n}L$. **q. e. d.**

93:5. **Example.** Let us give a general method for finding norm and weight generators for L (computable methods will be given later in § 94). To find a norm generator simply take any element a of largest value in $Q(L)$. To find a weight generator first take $b_0 \in Q(L)$ of largest value such that

$$\mathrm{ord}_{\mathfrak{p}}\, a + \mathrm{ord}_{\mathfrak{p}}\, b_0 \text{ is odd.}$$

If $b_0\mathfrak{o} \supseteqq 2(\mathfrak{s}L)$, then $\mathfrak{w}L = b_0\mathfrak{o}$ and b_0 is a weight generator of L. If $b_0\mathfrak{o} \subseteqq 2(\mathfrak{s}L)$, or if b_0 does not exist, then $\mathfrak{w}L = 2(\mathfrak{s}L)$ and we can take any b for which $b\mathfrak{o} = 2(\mathfrak{s}L)$.

93:6. **Example.** Let a be a norm generator of L, and let a' be some other scalar. Then a' is a norm generator of L if and only if $a \cong a' \bmod \mathfrak{w}L$.

93:7. **Example.** Consider a non-zero scalar α. Then $\mathfrak{g}(\alpha L) = \alpha^2(\mathfrak{g}L)$, $\mathfrak{m}(\alpha L) = \alpha^2(\mathfrak{m}L)$, and $\mathfrak{w}(\alpha L) = \alpha^2(\mathfrak{w}L)$. If a is a norm generator of L, then $\alpha^2 a$ is a norm generator of αL.

93:8. **Example.** What happens when we scale the quadratic space by a non-zero scalar α? We obtain $\mathfrak{g}(L^\alpha) = \alpha(\mathfrak{g}L)$, $\mathfrak{m}(L^\alpha) = \alpha(\mathfrak{m}L)$, and $\mathfrak{w}(L^\alpha) = \alpha(\mathfrak{w}L)$. If a is a norm generator of L, then αa is a norm generator of L^α. If $\alpha = -1$ we obtain

$$\mathfrak{g}L^{-1} = \mathfrak{g}L\,,\quad \mathfrak{m}L^{-1} = \mathfrak{m}L\,,\quad \mathfrak{w}L^{-1} = \mathfrak{w}L\,,$$

and a is a norm generator for L^{-1} as well as for L.

§ 93B. The matrix $A(\alpha, \beta)$

We shall use the symbol $A(\alpha, \beta)$ to denote the 2×2 matrix

$$A(\alpha, \beta) = \begin{pmatrix} \alpha & 1 \\ 1 & \beta \end{pmatrix}$$

whenever α, β are scalars which satisfy the conditions

$$\alpha, \beta \in \mathfrak{o}, \qquad -1 + \alpha\beta \in \mathfrak{u}.$$

These conditions simply guarantee that the matrix $A(\alpha, \beta)$ is unimodular. Whenever the symbol $A(\alpha, \beta)$ appears it will be understood that α, β satisfy the above conditions, even if this is not explicitly stated at the time. We use $\xi A(\alpha, \beta)$ to denote ordinary multiplication of the matrix $A(\alpha, \beta)$ by the scalar ξ. Thus

$$\xi A(\alpha, \beta) = \begin{pmatrix} \xi\alpha & \xi \\ \xi & \xi\beta \end{pmatrix}.$$

93:9. **Example.** If $L \cong A(\alpha, 0)$, then $L \cong A(\alpha+2\beta, 0)$ for any β in $\mathfrak{o}$. For if we take a base $L = \mathfrak{o}x + \mathfrak{o}y$ in which L has the matrix $A(\alpha, 0)$, then $L = \mathfrak{o}(x + \beta y) + \mathfrak{o}y$ also gives a base for L and the matrix of L in this base is $A(\alpha+2\beta, 0)$.

93:10. **Example.** Let L be a binary unimodular lattice and let a be a norm generator of L that is also in $Q(L)$. We claim that (i) $L \cong A(a, \beta)$ for some $\beta \in \mathfrak{w}L$, (ii) if $\mathfrak{w}L \supset 2\mathfrak{o}$ then the β in the above matrix for L is a weight generator of L. To prove this we pick any $x \in L$ with $Q(x) = a$. Then x is a maximal vector in L, so there is a vector y in L with $L = \mathfrak{o}x + \mathfrak{o}y$. Now a is a norm generator of L, hence by Proposition 93:4 there is a ξ in $\mathfrak{o}$ and an η in $\mathfrak{w}L$ with $Q(y) = a\xi^2 + \eta$. Then $L = \mathfrak{o}x + \mathfrak{o}(y + \xi x)$ and

$$Q(y + \xi x) = (a\xi^2 + \eta) + (\xi^2 a) + 2\xi\, B(y, x) \;\in\; \mathfrak{w}L\,.$$

But $\mathfrak{w}L \subset \mathfrak{s}L = \mathfrak{o}$, hence $B(x, y + \xi x)$ is a unit since L is unimodular. Put $z = (y + \xi x)/B(x, y + \xi x)$. Then L has a matrix of the desired type in the base $L = \mathfrak{o}x + \mathfrak{o}z$. This proves the first part of our contention. Now the second part. Here we are given $\mathfrak{w}L \supset 2\mathfrak{o}$. We recall from Example 93:5 that there is a number $b \in Q(L)$ such that $b\mathfrak{o} = \mathfrak{w}L$ with $\mathrm{ord}_{\mathfrak{p}} a + \mathrm{ord}_{\mathfrak{p}} b$ odd. Then $|\beta| \leqq |b|$, and $|2| < |b|$, and

$$b = a\xi^2 + 2\xi\,\eta + \beta\eta^2$$

with $\xi, \eta \in \mathfrak{o}$. If we had $|\beta| < |b|$ we would have $|b| = |a\xi^2|$ by the Principle of Domination, and this is absurd since $\mathrm{ord}_{\mathfrak{p}} a + \mathrm{ord}_{\mathfrak{p}} b$ is odd. Hence $|\beta| = |b|$, so $\mathfrak{w}L = \beta\mathfrak{o}$ and β is a weight generator of L.

93:11. **Example.** Let L be a binary unimodular lattice with $\mathfrak{n}L \subsetneqq 2\mathfrak{o}$. We say that

$$L \cong A(0, 0) \quad \text{or} \quad L \cong A(2, 2\varrho)\,.$$

We know that L is $2\mathfrak{o}$-maximal by Proposition 82:19. So if FL is isotropic we will have $L \cong A(0, 0)$ by Example 82:22. Now suppose that

FL is not isotropic. Then $L \cong A(2\alpha, 2\beta)$ by Example 93:10, and both α and β will have to be units since otherwise we would have $d(FL) = -1$ by the Local Square Theorem. Construct a lattice K on a quadratic space FK with $K \cong A(2, 2\varrho)$. Then a direct calculation of Hasse symbols shows that $S_\mathfrak{p}(FL) = S_\mathfrak{p}(FK)$, hence FL and FK are isometric by Theorem 63:20, hence L and K are isometric by Theorem 91:2, hence $L \cong A(2, 2\varrho)$.

93:12. *Let L be a lattice on a regular quadratic space V over a dyadic local field. Suppose that L has a splitting* $L = J \perp K$ *and that J has a base* $J = \mathfrak{o}x + \mathfrak{o}y$ *in which* $J \cong A(\alpha, 0)$ *with* $\alpha \in \mathfrak{o}$. *Put*

$$J' = \mathfrak{o}(x + z) + \mathfrak{o}y \quad \textit{with} \quad z \in K^\mathfrak{o}.$$

Then there is a splitting $L = J' \perp K'$. *And K' is isometric to K.*

Proof. J' is unimodular and $B(J', L) \subseteq \mathfrak{o}$ since $z \in K^\mathfrak{o}$. Hence we have a splitting $L = J' \perp K'$ by Proposition 82:15.

For any $u \in FK$ we define

$$\varphi u = u - B(u, z)\, y\,.$$

Then $Q(\varphi u) = Q(u)$ since $Q(y) = 0$ and $B(u, y) = 0$, hence $\varphi: FK \to V$ is a representation; but FK is regular; hence $\varphi: FK \rightarrowtail V$ is an isometry by Proposition 42:7. Now

$$B(\varphi u, x + z) = B(u - B(u, z)\, y, x + z) = B(u, z) - B(u, z)\, B(y, x) = 0.$$

And similarly $B(\varphi u, y) = 0$. Hence $B(\varphi(FK), FJ') = 0$ and we have an isometry $\varphi: FK \rightarrowtail FK'$.

Now $B(u, z) \in \mathfrak{o}$ whenever $u \in K$ since $z \in K^\mathfrak{o}$. Hence $\varphi K \subseteq L$. Hence $\varphi K \subseteq FK' \cap L = K'$. But φ preserves volumes since it is an isometry. Hence $\varphi K = K'$. So we have found an isometry of K onto K'. **q. e. d.**

93:13. **Example.** Suppose the lattice L on the regular quadratic space V has a splitting

$$L \cong \langle \xi A(\alpha, 0)\rangle \perp K$$

with $\xi \in \dot{F}$ and $\alpha \in \mathfrak{o}$. Let β be any element of $\mathfrak{g}K^{\xi\mathfrak{o}}$. Then

$$L \cong \langle \xi A(\alpha + \xi^{-1}\beta, 0)\rangle \perp K.$$

This is obtained from Proposition 93:12 by scaling. In particular, if L has a splitting

$$L \cong \langle A(\alpha, 0)\rangle \perp K$$

with $\alpha \in \mathfrak{o}$ and $\mathfrak{s}K \subseteq \mathfrak{o}$, then

$$L \cong \langle A(\alpha + \beta, 0)\rangle \perp K$$

for any β in $Q(K)$.

§ 93C. Two cancellation laws

93:14. **Theorem.** *A lattice L on a regular quadratic space over a dyadic local field has splittings* $L = J \perp K$ *and* $L = J' \perp K'$ *with J isometric to J' and* $J \cong A(0, 0)$. *Then K is isometric to K'.*

Proof. 1) Take bases $J = \mathfrak{o}x + \mathfrak{o}y$ and $J' = \mathfrak{o}x' + \mathfrak{o}y'$ in which $J \cong A(0, 0)$ and $J' \cong A(0, 0)$. Note that J and J' are contained in $L^{\mathfrak{o}}$. Hence $J + J' \subseteq L^{\mathfrak{o}}$ and $B(J, J') \subseteq \mathfrak{o}$.

2) First we do the special case where $x = x'$. We can express y' in the form

$$y' = \alpha x + y + z \quad (\alpha \in \mathfrak{o}, z \in K)$$

since y' is in $L = J \perp K$ and $B(x, y') = 1$. Then $z = y' - \alpha x - y$ is in $(J + J') \cap K \subseteq L^{\mathfrak{o}} \cap K$, hence z is in $K^{\mathfrak{o}}$. We can therefore apply Proposition 93:12 to the sublattices

$$J = \mathfrak{o}x + \mathfrak{o}(y + \alpha x)$$

$$J' = \mathfrak{o}x + \mathfrak{o}(y + \alpha x + z)$$

of L. This gives us an isometry of K' onto K.

3) Next we do the case where $B(J, J') = \mathfrak{o}$. We can suppose that $B(x, y')$, say, is a unit. Making a slight change to x', y' allows us to assume that $B(x, y') = 1$. Put $J'' = \mathfrak{o}x + \mathfrak{o}y'$. Then $J'' \cong A(0, 0)$ is a unimodular sublattice of $J + J' \subseteq L^{\mathfrak{o}}$, hence there is a splitting $L = J'' \perp K''$. But we can apply the special result of step 2) to J and J'', and also to J'' and J'. Hence $K \cong K''$ and $K'' \cong K'$. Hence $K \cong K'$.

4) Finally we consider $B(J, J') \subseteq \mathfrak{p}$. Here we put $J'' = \mathfrak{o}x + \mathfrak{o}(y + y')$. Then J'' is a unimodular sublattice of $L^{\mathfrak{o}}$ with $\mathfrak{n}J'' \subseteq 2\mathfrak{o}$. Hence $J'' \cong A(0, 0)$ by Example 93:11. And we have a splitting $L = J'' \perp K''$. But here $B(J, J'') = \mathfrak{o}$ and $B(J'', J') = \mathfrak{o}$. Hence $K \cong K''$ and $K'' \cong K'$ by step 3). So $K \cong K'$. **q. e. d.**

By a hyperbolic adjunction to a lattice L on a quadratic space we mean the adjunction of a lattice J of the form

$$J \cong \langle \xi_1 A(0, 0) \rangle \perp \cdots \perp \langle \xi_r A(0, 0) \rangle \quad (\xi_i \in \dot{F}) .$$

If $\xi_i \mathfrak{o} = \mathfrak{a}$ for each i, then J is $\mathfrak{a}$-modular and we call the adjunction of J an $\mathfrak{a}$-modular hyperbolic adjunction.

93:14a. **Corollary.** *A lattice L on a regular quadratic space over a dyadic local field has splittings $L = J \perp K$ and $L = J_1 \perp K_1$ with J isometric to J_1. Suppose that J is $\mathfrak{a}$-modular with*

$$\mathfrak{g}J \subseteq \mathfrak{g}K^{\mathfrak{a}} \quad \text{and} \quad \mathfrak{g}J_1 \subseteq \mathfrak{g}K_1^{\mathfrak{a}} .$$

Then K is isometric to K_1.

Proof. By scaling we can assume that $\mathfrak{a} = \mathfrak{o}$. Adjoining J^{-1} to L shows that we may assume, without loss of generality, that J and J_1 are isometric unimodular lattices on hyperbolic spaces with $\mathfrak{g}J \subseteq \mathfrak{g}K^{\mathfrak{o}}$ and $\mathfrak{g}J_1 \subseteq \mathfrak{g}K_1^{\mathfrak{o}}$. Now by Proposition 82:17 a unimodular lattice on a hyperbolic space has the form

$$\langle A(\alpha_1, 0) \rangle \perp \cdots \perp \langle A(\alpha_r, 0) \rangle ,$$

hence we have a splitting

$$L \cong \langle A(\alpha_1, 0)\rangle \perp \cdots \perp \langle A(\alpha_r, 0)\rangle \perp K$$

in which the α_i are in $\mathfrak{g}J$ and hence in $\mathfrak{g}K^{\mathfrak{o}}$. Successive application of Example 93:13 now gives

$$L \cong \langle A(0, 0)\rangle \perp \cdots \perp \langle A(0, 0)\rangle \perp K\,.$$

Similarly

$$L \cong \langle A(0, 0)\rangle \perp \cdots \perp \langle A(0, 0)\rangle \perp K_1\,.$$

Then $K \cong K_1$ by Theorem 93:14. **q. e. d.**

The general cancellation law for lattices on quadratic spaces over non-dyadic fields (Theorem 92:3) does not extend to the dyadic case. For instance there is an isometry

$$\langle A(0,0)\rangle \perp \langle 1\rangle \cong \langle A(1,0)\rangle \perp \langle 1\rangle$$

by Example 93:13; but $\langle A(0,0)\rangle$ and $\langle A(1,0)\rangle$ are not isometric since their norms are not equal.

§ 93D. Unimodular lattices

93:15. *A unimodular lattice L in a quadratic space over a dyadic local field has an orthogonal base if and only if $\mathfrak{n}L = \mathfrak{s}L$. If $\mathfrak{n}L \subset \mathfrak{s}L$, then L is an orthogonal sum of binary sublattices.*

Proof. If L has an orthogonal base it contains a 1-dimensional unimodular lattice; any such lattice has norm $\mathfrak{o}$; hence $\mathfrak{n}L = \mathfrak{o} = \mathfrak{s}L$.

Conversely let us suppose that $\mathfrak{n}L = \mathfrak{s}L = \mathfrak{o}$. Then there is an x in L with $Q(x)$ in $\mathfrak{u}$. The lattice $\mathfrak{o}x$ is a unimodular sublattice of L and therefore splits L. Hence we have a splitting

$$L \cong \langle \varepsilon_1\rangle \perp \cdots \perp \langle \varepsilon_r\rangle \perp \langle A(\alpha_1, \beta_1)\rangle \perp \cdots \perp \langle A(\alpha_t, \beta_t)\rangle$$

in which $r \geqq 1$ and $t \geqq 0$, by § 91C and Example 93:10. Consider the 3-dimensional sublattice $K \cong \langle \varepsilon\rangle \perp \langle A(\alpha, \beta)\rangle$. If neither α nor β is a unit, then $A(\alpha + \varepsilon, \beta)$ is unimodular and we obtain a new splitting

$$K \cong \langle \varepsilon'\rangle \perp \langle A(\alpha + \varepsilon, \beta)\rangle$$

in which $\alpha + \varepsilon$ is a unit. We may therefore assume that α, say, is a unit. Then

$$K \cong \langle \varepsilon\rangle \perp \langle \alpha\rangle \perp \langle \delta\rangle\,.$$

By successively applying the 3-dimensional case to L we ultimately obtain

$$L \cong \langle \varepsilon_1\rangle \perp \cdots \perp \langle \varepsilon_n\rangle$$

as required.

To prove the last part we take a splitting of L into 1- and 2-dimensional components. If a 1-dimensional component appeared in this splitting it would be unimodular and hence would have norm $\mathfrak{o}$. This is impossible since $\mathfrak{n}L \subset \mathfrak{s}L = \mathfrak{o}$. Hence L splits into binary components.

q. e. d.

93:16. **Theorem.** *Let L and K be unimodular lattices on the same quadratic space over a dyadic local field. Then* $\operatorname{cls} L = \operatorname{cls} K$ *if and only if* $\mathfrak{g}L = \mathfrak{g}K$.[1]

Proof. By making the same unimodular hyperbolic adjunction to L and K we can assume, in virtue of Theorem 93:14 and § 93A, that $Q(L) = \mathfrak{g}L = \mathfrak{g}K = Q(K)$. Now adjoin the lattice $L \perp L^{-1}$ to each of L and K, and let the resulting lattices be denoted L' and K' respectively. So

$$L' \cong L \perp L^{-1} \perp L\,, \quad K' \cong L \perp L^{-1} \perp K\,.$$

Now $\mathfrak{g}(L \perp L^{-1}) = \mathfrak{g}L = \mathfrak{g}K$. Hence by Corollary 93:14a it will be enough to prove that L' is isometric to K'. But the component $L^{-1} \perp K$ of K' is a unimodular lattice on a hyperbolic space, hence K' has a splitting

$$K' \cong L \perp \langle A(\alpha_1, 0)\rangle \perp \cdots \perp \langle A(\alpha_r, 0)\rangle$$

with all α_i in $\mathfrak{g}L = Q(L)$. Then

$$K' \cong L \perp \langle A(0,0)\rangle \perp \cdots \perp \langle A(0,0)\rangle$$

by Example 93:13. Similarly we find

$$L' \cong L \perp \langle A(0,0)\rangle \perp \cdots \perp \langle A(0,0)\rangle\,.$$

Hence $K' \cong L'$. Hence $K \cong L$. In other words, $\operatorname{cls} K = \operatorname{cls} L$. **q. e. d.**

93:17. **Example.** Let L be a binary unimodular lattice on a quadratic space over a dyadic local field, let a be a norm generator of L that is in $Q(L)$, let b be a weight generator of L, let the discriminant dL be written in the form $dL = -(1+\alpha)$ with $\mathfrak{d}(1+\alpha) = \alpha\mathfrak{o}$. We claim that

$$\alpha a^{-1} \in b\mathfrak{o}$$

and

$$L \cong A(a, -\alpha a^{-1}).$$

By Example 93:10 we have $L \cong A(a, b\lambda)$ with $\lambda \in \mathfrak{o}$, hence $-(1 - ab\lambda) = -\varepsilon^2(1+\alpha)$ with $\varepsilon \in \mathfrak{u}$, hence

$$\alpha\mathfrak{o} = \mathfrak{d}(1+\alpha) = \mathfrak{d}(1 - ab\lambda) \subseteq ab\lambda\mathfrak{o}\,.$$

So $\alpha a^{-1} \in b\mathfrak{o}$. We must prove that $L \cong K$ where K is a lattice with $K \cong A(a, -\alpha a^{-1})$. Now

$$FL \cong \langle a\rangle \perp \langle -a(1+\alpha)\rangle \cong FK\,,$$

and a is a norm generator of both L and K, hence by Theorem 93:16 it suffices to prove that $\mathfrak{w}L = \mathfrak{w}K$. If $\mathfrak{w}L = 2\mathfrak{o}$, then $\alpha a^{-1} \in 2\mathfrak{o} \subseteq \mathfrak{w}K$ and so $\mathfrak{w}K = 2\mathfrak{o}$ by Example 93:10. If $\mathfrak{w}L \supset 2\mathfrak{o}$, then λ is a unit by Example 93:10. We have $\mathfrak{d}(1 - ab\lambda) = ab\lambda\mathfrak{o}$ since here $\operatorname{ord}_{\mathfrak{p}} a + \operatorname{ord}_{\mathfrak{p}} b$ is odd with $|a|_{\mathfrak{p}} > |b|_{\mathfrak{p}} > |2|_{\mathfrak{p}}$, and so $\alpha a^{-1}\mathfrak{o} = b\mathfrak{o}$. Hence $\mathfrak{w}K \supseteq \alpha a^{-1}\mathfrak{o} \supset 2\mathfrak{o}$. Hence $\mathfrak{w}K = \alpha a^{-1}\mathfrak{o}$ by Example 93:10. So $\mathfrak{w}K = \mathfrak{w}L$ as required.

[1] Hence $\operatorname{cls} L = \operatorname{cls} K$ if and only if $Q(L) = Q(K)$, i. e. if and only if L and K represent the same numbers.

93:18. **Example.** (i) Let us describe the unimodular lattices of dimension $\geqq 3$. We consider a unimodular lattice L on a quadratic space V over a dyadic local field. Let a be any norm generator, let b be any weight generator, let d be the discriminant of L. Regard d as an element of $\mathfrak{u}$. Note that $b\mathfrak{o} = 2\mathfrak{o}$ when $\mathrm{ord}_\mathfrak{p} a + \mathrm{ord}_\mathfrak{p} b$ is even, by § 93A.

(ii) First we dispose of all cases with $\dim L \geqq 3$ and $\mathrm{ord}_\mathfrak{p} a + \mathrm{ord}_\mathfrak{p} b$ even. We claim that

$$L \cong \langle A(0,0)\rangle \perp \cdots.$$

To see this we take a splitting $L = J \perp K$ in which J is binary and $\mathfrak{n}K = \mathfrak{n}L$. Let a_1 be an element of $Q(K)$ such that $a_1\mathfrak{o} = \mathfrak{n}K$. Then a_1 is a norm generator of L and so $\mathfrak{g}L = a_1\mathfrak{o}^2 + 2\mathfrak{o}$. Hence by Example 93:10 we have a splitting

$$L \cong \langle A(a_1\xi^2 + 2\eta, 2\zeta)\rangle \perp K \cong \langle A(2\eta', 2\zeta)\rangle \perp K'$$

with ξ, η, ζ, η' in $\mathfrak{o}$. But K' represents an element a_2 such that $a_2\mathfrak{o} = \mathfrak{n}L$, hence it represents an element of the form 2ε with ε a unit, hence by the perfectness of the residue class field we can write

$$L \cong \langle A(2\pi\eta'', 2\zeta)\rangle \perp K''$$

with η'' in $\mathfrak{o}$. Hence by Example 93:11,

$$L \cong \langle A(0,0)\rangle \perp K''.$$

(iii) Next we consider $\dim L = 4$ with $\mathrm{ord}_\mathfrak{p} a + \mathrm{ord}_\mathfrak{p} b$ odd. Here we suppose that d has been expressed in the form $d = 1 + \alpha$ with $\mathfrak{d}(d) = \alpha\mathfrak{o}$. It follows from Example 93:10 that we must have $\alpha \in ab\,\mathfrak{o}$. Take lattices J, K on quadratic spaces over F with

$$J \cong \langle A(a, -\alpha a^{-1})\rangle \perp \langle A(b, 0)\rangle$$
$$K \cong \langle A(a, -(\alpha - 4\varrho)\, a^{-1})\rangle \perp \langle A(b, 4\varrho\, b^{-1})\rangle.$$

Let us prove that L is isometric to J or to K, but not to both. First we note that every number represented by J is in $a\mathfrak{o}^2 + b\mathfrak{o} = \mathfrak{g}L$, hence $\mathfrak{g}J \subseteq \mathfrak{g}L$; but a is a norm generator of J and J represents the number b with $\mathrm{ord}_\mathfrak{p} a + \mathrm{ord}_\mathfrak{p} b$ odd, so $b \in \mathfrak{w}J$, hence

$$\mathfrak{g}J = a\mathfrak{o}^2 + \mathfrak{w}J \supseteq \mathfrak{g}L.$$

Similarly with K. Hence we have proved that

$$\mathfrak{g}J = \mathfrak{g}L = \mathfrak{g}K.$$

It is easily checked that

$$dJ = dL = dK.$$

If we compute Hasse symbols using the rules of § 63B we obtain

$$S_\mathfrak{p}(FJ) = -S_\mathfrak{p}(FK).$$

This shows first of all that L cannot be isometric to both J and K.

Secondly, it shows that FL is isometric to one of FJ or FK, by Theorem 63:20, hence L is isometric to J or K by Theorem 93:16.

(iv) Now let $\dim L = 3$ with $\mathrm{ord}_{\mathfrak{p}} a + \mathrm{ord}_{\mathfrak{p}} b$ odd. We claim that

$$L \cong \langle A(b, 0)\rangle \perp \langle -d\rangle$$

or

$$L \cong \langle A(b, 4\varrho b^{-1})\rangle \perp \langle -d(1-4\varrho)\rangle,$$

but not both. This can be derived from the 4-dimensional case. Take a unit ε in $Q(L)$ and adjoin $\langle\varepsilon\rangle$ to L to obtain a lattice K with $K \cong \langle\varepsilon\rangle \perp L$. Then $\mathfrak{g}L = \mathfrak{g}K$ and $\mathfrak{w}L = \mathfrak{w}K$ so that

$$K \cong \langle A(\varepsilon, \ldots)\rangle \perp \langle A(b, 0)\rangle$$

or

$$K \cong \langle A(\varepsilon, \ldots)\rangle \perp \langle A(b, 4\varrho b^{-1})\rangle$$

Therefore

$$\langle\varepsilon\rangle \perp L \cong \langle\varepsilon\rangle \perp \langle\ldots\rangle \perp \langle A(b, 0)\rangle$$

or

$$\langle\varepsilon\rangle \perp L \cong \langle\varepsilon\rangle \perp \langle\ldots\rangle \perp \langle A(b, 4\varrho b^{-1})\rangle$$

By Corollary 93:14a we can cancel $\langle\varepsilon\rangle$, so L has the desired form. A computation with Hasse symbols shows that L cannot have both the given forms.

(v) If $\dim L \geqq 5$ we claim that

$$L \cong \langle A(0, 0)\rangle \perp \ldots,$$

and so $\mathfrak{g}L = Q(L)$. We can assume that $\mathrm{ord}_{\mathfrak{p}} a + \mathrm{ord}_{\mathfrak{p}} b$ is odd. First we take a quaternary unimodular sublattice of L which has norm $a\mathfrak{o}$. If this quaternary lattice has weight $b\mathfrak{p}^{2r+1}$ $(r \geqq 0)$ it will have the form $\langle A(0, 0)\rangle \perp \ldots$ by (ii) and we will be through. Otherwise it will have weight $b\mathfrak{p}^{2r}$ $(r \geqq 0)$ in which case we obtain

$$L \cong \langle A(b\pi^{2r}, 2\alpha)\rangle \perp K$$

for some α in $\mathfrak{o}$ by (iii). Here $\mathfrak{n}K$ is still $a\mathfrak{o}$. If we repeat the above procedure on a suitable ternary or quaternary sublattice of K we either obtain $K \cong \langle A(0, 0)\rangle \perp \cdots$, or else

$$L \cong \langle A(b\pi^{2r}, 2\alpha)\rangle \perp \langle A(b\pi^{2t}, 2\beta)\rangle \perp \cdots$$

for some $t \geqq 0$ and some β in $\mathfrak{o}$. In the first event we are through. Otherwise we can assume that $r \geqq t$, say. Then

$$\begin{aligned} L &\cong \langle A(2b\pi^{2r}, 2\alpha)\rangle \perp \cdots \\ &\cong \langle A(0, 0)\rangle \perp \cdots. \end{aligned}$$

(vi) As a special example let us consider the case $\dim L = 4$ with $dL = 1$. Then FL is isotropic if $\mathrm{ord}_{\mathfrak{p}} a + \mathrm{ord}_{\mathfrak{p}} b$ is even, and we actually have

$$L \cong \langle A(a, 0)\rangle \perp \langle A(0, 0)\rangle$$

by (ii) and Example 93:9. If $\operatorname{ord}_{\mathfrak{p}} a + \operatorname{ord}_{\mathfrak{p}} b$ is odd, then

$$L \cong \langle A(a, 0)\rangle \perp \langle A(b, 0)\rangle$$

or

$$L \cong \langle A(a, 4a^{-1}\varrho)\rangle \perp \langle A(b, 4b^{-1}\varrho)\rangle .$$

The first of these is clearly isotropic, and the second is not by Example 63:15 (i). So we have the following result: if $\dim L = 4$ and $dL = 1$, then

$$L \cong \langle A(a, 4a^{-1}\lambda)\rangle \perp \langle A(b, 4b^{-1}\lambda)\rangle$$

with $\lambda = 0$ when FL is isotropic (in particular when $\operatorname{ord}_{\mathfrak{p}} a + \operatorname{ord}_{\mathfrak{p}} b$ is even) and with $\lambda = \varrho$ when FL is not isotropic.

93:19. *Let L be a modular lattice in a quadratic space over a dyadic local field with $\mathfrak{s}L \subseteq \mathfrak{p}$ and $\dim L \geqq 3$. Consider any $\lambda \in \mathfrak{g}L$. Then*

$$\langle A(\alpha, \beta)\rangle \perp L \cong \langle A(\alpha + \lambda, \beta)\rangle \perp K$$

with $\mathfrak{g}L \subseteq \mathfrak{g}K$.

Proof. Put $\mathfrak{s}L = \xi\mathfrak{o}$ with $\xi \in \dot{F}$. Then by Examples 93:13 and 93:18 (v) we have

$$\begin{aligned} \langle A(\alpha, \beta)\rangle &\perp \langle \xi A(0, 0)\rangle \perp L \\ &\cong \langle A(\alpha, \beta)\rangle \perp \langle \xi A(\xi^{-1}\lambda, 0)\rangle \perp L \\ &\cong \langle A(\alpha + \lambda, \beta)\rangle \perp \langle \xi A(\ldots, \ldots)\rangle \perp L \\ &\cong \langle A(\alpha + \lambda, \beta)\rangle \perp \langle \xi A(0, 0)\rangle \perp K . \end{aligned}$$

We have $\mathfrak{g}L \subseteq 2\xi\mathfrak{o} + \mathfrak{g}K = \mathfrak{g}K$ and, by Theorem 93:14,

$$\langle A(\alpha, \beta)\rangle \perp L \cong \langle A(\alpha + \lambda, \beta)\rangle \perp K .$$

q. e. d.

93:20. **Example.** Let L be a modular lattice on the quadratic space V over the dyadic local field F, and suppose that $\dim V \geqq 3$. We claim that

$$\theta(O^+(L)) \supseteq \mathfrak{u} .$$

By scaling V we can assume that L is unimodular. Let b denote any weight generator of L. By examining the different cases in Example 93:18 we see that there is always a maximal vector y in L with $Q(y) = b$. This vector satisfies

$$\frac{2B(y, L)}{Q(y)} = \frac{2\mathfrak{o}}{b} \subseteq \mathfrak{o} ,$$

hence the symmetry τ_y is a unit of L by § 91B. So there is always a symmetry in $O(L)$ with spinor norm $b\dot{F}^2$. But εb is also a weight generator of L, for any ε in $\mathfrak{u}$. Hence $\varepsilon \in \theta(O^+(L))$. Hence $\theta(O^+(L)) \supseteq \mathfrak{u}$.

§ 93E. The fundamental invariants

Consider the non-zero regular lattice L in the quadratic space V over a dyadic local field. Let L have the Jordan splitting

$$L = L_1 \perp \cdots \perp L_t .$$

Put

$$\mathfrak{s}_i = \mathfrak{s} L_i$$

for $1 \leqq i \leqq t$. Thus $\mathfrak{s} L = \mathfrak{s}_1$. And L_i is $\mathfrak{s}_i$-modular with

$$\mathfrak{s}_1 \supset \mathfrak{s}_2 \supset \cdots \supset \mathfrak{s}_t .$$

Note that $\mathfrak{s} L^{\mathfrak{s}_i} = \mathfrak{s}_i$ by § 91C. We define

$$\mathfrak{g}_i = \mathfrak{g} L^{\mathfrak{s}_i} , \quad \mathfrak{w}_i = \mathfrak{w} L^{\mathfrak{s}_i} .$$

Thus

$$\mathfrak{g}_1 \supseteqq \mathfrak{g}_2 \supseteqq \cdots \supseteqq \mathfrak{g}_t$$

$$\mathfrak{w}_1 \supseteqq \mathfrak{w}_2 \supseteqq \cdots \supseteqq \mathfrak{w}_t$$

since $L^{\mathfrak{s}_i} \supseteqq L^{\mathfrak{s}_j}$ when $\mathfrak{s}_i \supseteqq \mathfrak{s}_j$. We take a norm generator a_i for $L^{\mathfrak{s}_i}$ and we fix it. Thus we have

$$a_1 \mathfrak{o} \supseteqq \cdots \supseteqq a_t \mathfrak{o}$$

and also

$$\mathfrak{g}_i = a_i \mathfrak{o}^2 + \mathfrak{w}_i .$$

Other relations among a_i, $\mathfrak{s}_i$, $\mathfrak{w}_i$ can be deduced from § 93A. The invariants

$$\boxed{t, \dim L_i, \mathfrak{s}_i, \mathfrak{w}_i, a_i}$$

($1 \leqq i \leqq t$) will be called the fundamental invariants of the lattice L. The number t is, of course, the number of modular components of any Jordan splitting of L. We shall call the $\dim L_i$ the fundamental dimensions, the $\mathfrak{s}_i$ the fundamental scales, etc., of the lattice L. The norm group of L is equal to the first fundamental norm group $\mathfrak{g}_1$ of L. All the fundamental invariants other than the a_i are unique for a given L. By Example 93:6, scalars $a'_1, \ldots, a'_t$ will be fundamental norm generators for L if and only if

$$a'_i \cong a_i \bmod \mathfrak{w}_i \quad \text{for} \quad 1 \leqq i \leqq t .$$

Now consider another lattice L' in a quadratic space V' over the same field. Let $L' = L'_1 \perp \cdots \perp L'_{t'}$, denote a Jordan splitting of L', and let

$$t', \dim L'_i, \mathfrak{s}'_i, \mathfrak{w}'_i, a'_i$$

be a set of fundamental invariants of L'. Let $\mathfrak{g}'_i$ be the fundamental norm groups of L'. We say that L and L' are of the same fundamental type if

$$\boxed{\begin{gathered} t = t', \dim L_i = \dim L'_i, \mathfrak{s}_i = \mathfrak{s}'_i \\ \mathfrak{g}_i = \mathfrak{g}'_i \end{gathered}}$$

for $1 \leqq i \leqq t$. This is equivalent to

$$\boxed{\begin{array}{c} t = t',\ \dim L_i = \dim L'_i,\ \mathfrak{s}_i = \mathfrak{s}'_i \\ \mathfrak{w}_i = \mathfrak{w}'_i\,,\quad a_i \cong a'_i \bmod \mathfrak{w}_i \end{array}}$$

for $1 \leqq i \leqq t$. It is clear that isometric lattices are of the same fundamental type and that an isometry preserves the fundamental invariants.

Suppose L has the same fundamental type as L'. Then the fundamental norm generators satisfy $a_i \cong a'_i \bmod \mathfrak{w}'_i$ and so $\{a_1, \ldots, a_t\}$ can be regarded as a set of fundamental norm generators of L'. When this is done we say that we have chosen the same set of fundamental invariants for L and L', or simply that L and L' have the same set of fundamental invariants.

The lattice L_i is the first component in some Jordan splitting of $L^{\mathfrak{s}_i}$. Hence $\mathfrak{n}L_i = \mathfrak{s}_i$ if and only if $\mathfrak{n}L^{\mathfrak{s}_i} = \mathfrak{s}_i$, i. e. if and only if $a_i\mathfrak{o} = \mathfrak{s}_i$. So lattices of the same fundamental type are also of the same Jordan type.

Let us introduce some additional notation. We put

$$s_i = \operatorname{ord}_{\mathfrak{p}} \mathfrak{s}_i, \quad u_i = \operatorname{ord}_{\mathfrak{p}} a_i$$

for $1 \leqq i \leqq t$. These quantities clearly depend just on the fundamental type. We define fractional ideals $\mathfrak{f}_i$ for $1 \leqq i \leqq t-1$ by the following equations: we put

$$\mathfrak{s}_i^2 \mathfrak{f}_i = \sum \mathfrak{d}(\alpha\,\beta)$$

with $\alpha \in \mathfrak{g}_i$, $\beta \in \mathfrak{g}_{i+1}$ when $u_i + u_{i+1}$ is odd, and we put

$$\mathfrak{s}_i^2 \mathfrak{f}_i = \sum \mathfrak{d}(\alpha\,\beta) + 2\mathfrak{p}^{\frac{1}{2}(u_i+u_{i+1})+s_i}$$

with $\alpha \in \mathfrak{g}_i$, $\beta \in \mathfrak{g}_{i+1}$ when $u_i + u_{i+1}$ is even. These ideals $\mathfrak{f}_i$ also depend just on the fundamental type. It is clear that in all cases

$$\mathfrak{f}_i \subseteq \mathfrak{p}\,.$$

The modular lattice L_i is the first component in some Jordan splitting

$$L^{\mathfrak{s}_i} = L_i \perp \cdots$$

of $L^{\mathfrak{s}_i}$. Hence we have

$$\mathfrak{g}L_i \subseteq \mathfrak{g}L^{\mathfrak{s}_i} = \mathfrak{g}_i\,.$$

We shall call the given Jordan splitting $L = L_1 \perp \cdots \perp L_t$ saturated if

$$\mathfrak{g}L_i = \mathfrak{g}_i \quad \text{for} \quad 1 \leqq i \leqq t\,.$$

If $L = L_1 \perp \cdots \perp L_t$ is saturated, then a_i is a norm generator of L_i and

$$\mathfrak{w}_i = \mathfrak{w}L_i\,,$$

since L_i and $L^{\mathfrak{s}_i}$ have the same norm group and the same scale.

93:21. *Suppose L has a Jordan splitting $L = L_1 \perp \cdots \perp L_t$ in which $\dim L_i \geqq 3$ for $1 \leqq i \leqq t$. Then L has a saturated Jordan splitting.*

Proof. 1) First let us prove the result in the case where all $\dim L_i \geqq 7$. Take a set of fundamental norm generators $a_1, \ldots, a_t$ and a set of fundamental weight generators $b_1, \ldots, b_t$. So

$$\mathfrak{g}_i = a_i \mathfrak{o}^2 + b_i \mathfrak{o} \quad \text{for} \quad 1 \leqq i \leqq t\,.$$

Consider a single index i and put $\mathfrak{s}_i = \xi \mathfrak{o}$ with ξ in $\dot{F}$. Then by Example 93:18 L_i has a splitting

$$L_i \cong \langle \xi A(0,0) \rangle \perp \langle \xi A(0,0) \rangle \perp \cdots$$

and this induces a splitting

$$L \cong \langle \xi A(0,0) \rangle \perp \langle \xi A(0,0) \rangle \perp J\,.$$

Now

$$\mathfrak{g} J^{\xi \mathfrak{o}} = 2\xi \mathfrak{o} + \mathfrak{g} J^{\xi \mathfrak{o}} = \mathfrak{g} L^{\xi \mathfrak{o}} = \mathfrak{g}_i\,,$$

hence a_i, b_i are in $\mathfrak{g} J^{\xi \mathfrak{o}}$. If we now apply Example 93:13 twice we find

$$L \cong \langle \xi A(\xi^{-1} a_i, 0) \rangle \perp \langle \xi A(\xi^{-1} b_i, 0) \rangle \perp J\,.$$

Hence we have the following: if we start with a Jordan splitting $L = L_1 \perp \cdots \perp L_t$ there is another Jordan splitting $L = K_1 \perp \cdots \perp K_t$ in which $a_i, b_i \in \mathfrak{g} K_i$ and $K_j \cong L_j$ for $j \neq i$. It then follows from the properties of the weight (§ 93A) that $b_i \mathfrak{o} \subseteq \mathfrak{g} K_i$. Hence

$$\mathfrak{g}_i = a_i \mathfrak{o}^2 + b_i \mathfrak{o} \subseteq \mathfrak{g} K_i \subseteq \mathfrak{g}_i\,,$$

and so $\mathfrak{g}_i = \mathfrak{g} K_i$. Do all this for $i = 1$, then for $i = 2$, etc. Ultimately we obtain a saturated Jordan splitting for L.

2) Now assume that $\dim L_i \geqq 3$ for $1 \leqq i \leqq t$. Make a 4-dimensional $\mathfrak{s}_1$-modular hyperbolic adjunction of a lattice H_1 to L. Then repeat with $\mathfrak{s}_2$, etc. We obtain a lattice $\bar{L}$ with the same fundamental norm groups as L and a Jordan splitting

$$\bar{L} = (H_1 \perp L_1) \perp \cdots \perp (H_t \perp L_t)$$

with $\dim(H_i \perp L_i) \geqq 7$ for $1 \leqq i \leqq t$. But $\bar{L}$ has a saturated Jordan splitting

$$\bar{L} = K_1 \perp \cdots \perp K_t$$

by step 1). Now $K_i \cong H_i \perp J_i$ by Example 93:18 since $\dim H_i = 4$ and $\dim K_i \geqq 7$. Hence we have a splitting

$$\bar{L} \cong (H_1 \perp J_1) \perp \cdots \perp (H_t \perp J_t)$$

with $\mathfrak{g} J_i = \mathfrak{g} K_i = \mathfrak{g}_i$. By Theorem 93:14

$$L \cong J_1 \perp \cdots \perp J_t\,.$$

Hence L has a saturated splitting. q. e. d.

93:22. **Example.** Suppose the Jordan splitting $L = L_1 \perp \cdots \perp L_t$ has $\dim L_i \geqq 3$ for $1 \leqq i \leqq t$. Then it follows from the definition of a saturated splitting and from Proposition 93:21 that the given Jordan splitting is saturated if and only if it has the following property: if $L = L'_1 \perp \cdots \perp L'_t$ is any Jordan splitting of L, then $\mathfrak{g}L_i \supseteq \mathfrak{g}L'_i$ for $1 \leqq i \leqq t$. So in this case a saturated splitting is one that maximizes all the groups $\mathfrak{g}L_i$.

93:23. **Example.** What happens to the fundamental invariants under scaling? If L has the Jordan splitting $L = L_1 \perp \cdots \perp L_t$, then L^α has the Jordan splitting $L^\alpha = L_1^\alpha \perp \cdots L_t^\alpha$. Hence the new fundamental scales are

$$\alpha\,\mathfrak{s}_1 \supset \cdots \supset \alpha\,\mathfrak{s}_t\,.$$

We have

$$\mathfrak{g}\,(L^\alpha)^{\alpha\,\mathfrak{s}_i} = \mathfrak{g}\,(L^{\mathfrak{s}_i})^\alpha = \alpha\,\mathfrak{g}L^{\mathfrak{s}_i} = \alpha\,\mathfrak{g}_i\,,$$

hence the new fundamental norm groups are

$$\alpha\,\mathfrak{g}_1 \supseteq \cdots \supseteq \alpha\,\mathfrak{g}_t\,.$$

Similarly with weights. So we obtain a new set of fundamental invariants

$$t,\ \dim L_i,\ \alpha\,\mathfrak{s}_i,\ \alpha\,\mathfrak{w}_i,\ \alpha\,a_i$$

for L^α. It follows from the definition that the ideals $\mathfrak{f}_1, \ldots, \mathfrak{f}_{t-1}$ are the same for L^α as for L. A Jordan splitting is saturated for L if and only if it is saturated for L^α.

93:24. **Example.** What happens to the fundamental invariants as we pass from L to $L^\#$? Consider a Jordan splitting $L = L_1 \perp \cdots \perp L_t$. Then by § 82F and Proposition 82:14 we get a Jordan splitting

$$L^\# = L_t^\# \perp \cdots \perp L_1^\# = \pi^{-s_t} L_t \perp \cdots \perp \pi^{-s_1} L_1$$

in which the fundamental scales of $L^\#$ are

$$\mathfrak{s}_t^{-1} \supset \cdots \supset \mathfrak{s}_1^{-1}\,.$$

The fundamental norm groups and weights of L and $L^\#$ are not altered by making $\mathfrak{s}_i$-modular hyperbolic adjunctions to L. Let us therefore assume that $\dim L_i \geqq 3$ for $1 \leqq i \leqq t$. Suppose the given Jordan splitting of L is saturated; then it follows easily from Example 93:22 and the fact that $L^{\#\#} = L$ that the Jordan splitting

$$L^\# = \pi^{-s_t} L_t \perp \cdots \perp \pi^{-s_1} L_1$$

is saturated. Hence by Example 93:7 we have

$$\mathfrak{g}_i^\# = \pi^{-2s_{t-i+1}} \mathfrak{g}_{t-i+1}\,, \quad \mathfrak{w}_i^\# = \pi^{-2s_{t-i+1}} \mathfrak{w}_{t-i+1}\,,$$

$$a_i^\# = \pi^{-2s_{t-i+1}} a_{t-i+1}$$

where the $\#$ symbol refers to the invariants of $L^\#$. Hence for $1 \leqq i \leqq t-1$ we have

$$\mathfrak{f}_i^\# = \mathfrak{f}_{t-i}\,.$$

93:25. **Example.** It follows from Example 93:24 that the fundamental norm groups and weights satisfy

$$\pi^{-2s_t}\mathfrak{g}_t \supseteq \cdots \supseteq \pi^{-2s_1}\mathfrak{g}_1,$$

$$\pi^{-2s_t}\mathfrak{w}_t \supseteq \cdots \supseteq \pi^{-2s_1}\mathfrak{w}_1.$$

Hence

$$\mathfrak{s}_t^{-2}a_t \supseteq \cdots \supseteq \mathfrak{s}_1^{-2}\,a_1\,.$$

93:26. **Example.** The ideals $\mathfrak{f}_i$ can be expressed in terms of the fundamental invariants. Consider the lattice L with fundamental invariants

$$t,\ \dim L_i,\ \mathfrak{s}_i,\ \mathfrak{w}_i,\ a_i\,.$$

Consider any i with $1 \leqq i \leqq t-1$. Then it is clear by inspection that

$$\mathfrak{s}_i^2\mathfrak{f}_i = a_i\,a_{i+1}\,\mathfrak{o}$$

when $u_i + u_{i+1}$ is odd. And a straightforward computation using the fact that $a_i\,\mathfrak{s}_i^{-2} \subseteq a_{i+1}\,\mathfrak{s}_{i+1}^{-2}$ will show that

$$\mathfrak{s}_i^2\mathfrak{f}_i = \mathfrak{d}\,(a_i a_{i+1}) + a_i\mathfrak{w}_{i+1} + a_{i+1}\mathfrak{w}_i + 2\mathfrak{p}^{\frac{1}{2}(u_i+u_{i+1})+s_i}$$

when $u_i + u_{i+1}$ is even.

93:27. **Example.** Let us continue with the preceding example and compute $\mathfrak{s}_i^2\mathfrak{f}_i$ in certain special cases that will be needed later. We know that $u_{i+1} \geqq u_i$ since $\mathfrak{g}_{i+1} \subseteq \mathfrak{g}_i$. We claim that

$$\mathfrak{s}_i^2\mathfrak{f}_i = \begin{cases} a_{i+1}\,\mathfrak{w}_i & \text{if} \quad u_{i+1} = u_i \\ a_i a_{i+1}\,\mathfrak{o} & \text{if} \quad u_{i+1} = u_i + 1 \\ a_i\,\mathfrak{w}_{i+1} & \text{if} \quad u_{i+1} = u_i + 2\,, \quad s_{i+1} = s_i + 1\,. \end{cases}$$

The first two computations are direct. Let us do the third. We know from Example 93:25 that $\pi^{-2s_{i+1}}\mathfrak{g}_{i+1} \supseteq \pi^{-2s_i}\,\mathfrak{g}_i$, so in the present situation we have

$$\pi^2\,\mathfrak{g}_i \subseteq \mathfrak{g}_{i+1} \subseteq \mathfrak{g}_i\,,$$

hence $\pi^2 a_i$ may be taken as an $(i+1)^{th}$ fundamental norm generator, in other words we may assume that $a_{i+1} = \pi^2 a_i$. We also have

$$\pi^2\,\mathfrak{w}_i \subseteq \mathfrak{w}_{i+1} \subseteq \mathfrak{w}_i\,.$$

Now substitute these values in the second formula of Example 93:26. We find that $\mathfrak{s}_i^2\,\mathfrak{f}_i = a_i\,\mathfrak{w}_{i+1}$ as required.

§ 93F. Determination of the class

93:28. **Theorem**[1]. *Let L and K be lattices on a regular quadratic space over a dyadic local field F, suppose that L and K have the same fundamental*

[1] A different classification involving the so-called "Gauss sums" can be found in O. T. O'Meara, *Am. J. Math.* (1957), pp. 687—709.

invariants, and let

$$L_{(1)} \subset \cdots \subset L_{(t)} \text{ and } K_{(1)} \subset \cdots \subset K_{(t)}$$

be Jordan chains for L and K. Then $\operatorname{cls} L = \operatorname{cls} K$ *if and only if the following conditions hold for* $1 \leqq i \leqq t-1$:

(i) $dL_{(i)}/dK_{(i)} \cong 1 \bmod \mathfrak{f}_i$

(ii) $FL_{(i)} \rightarrow FK_{(i)} \perp \langle a_{i+1} \rangle$ *when* $\mathfrak{f}_i \subset 4a_{i+1}\mathfrak{w}_{i+1}^{-1}$

(iii) $FL_{(i)} \rightarrow FK_{(i)} \perp \langle a_i \rangle$ *when* $\mathfrak{f}_i \subset 4a_i\mathfrak{w}_i^{-1}$.

Proof. Let V denote the quadratic space in question. Let

$$L = L_1 \perp \cdots \perp L_t\,, \quad K = K_1 \perp \cdots \perp K_t$$

denote Jordan splittings which determine the given Jordan chains. The fundamental invariants

$$t, \dim L_i, \mathfrak{s}_i, \mathfrak{w}_i, a_i$$

refer to both L and K since it is given that L and K have the same fundamental invariants. Recall the auxiliarly notation $s_i = \operatorname{ord}_\mathfrak{p} \mathfrak{s}_i$ and $u_i = \operatorname{ord}_\mathfrak{p} a_i$. Let $b_1, \ldots, b_t$ denote a set of fundamental weight generators.

Proof of the necessity. Here we are given that $\operatorname{cls} L = \operatorname{cls} K$. Hence there is an isometry $\sigma \in O(V)$ such that $K = \sigma L$. Now σ preserves the fundamental invariants, and if (i), (ii), (iii) are established for K and σL they can be carried back to K and L by σ^{-1}, hence we can assume that $K = L$. The above Jordan chains and splittings now refer to the same lattice L.

1) The first step is to prove (i) and (iii) for the following special case: $i = 1$, FL_1 4-dimensional hyperbolic, L_1 unimodular, $\mathfrak{g}L_1 = \mathfrak{g}K_1 = \mathfrak{g}_1$. If $u_1 = u_2$ we have $\mathfrak{f}_1 = a_1\mathfrak{w}_1$ by Example 93:27; but $dL_1 = 1$ and $dK_1 \cong 1 \bmod a_1\mathfrak{w}_1$ by Example 93:10; hence $dL_1/dK_1 \cong 1 \bmod \mathfrak{f}_1$ and we have (i); condition (iii) is vacuous since $\mathfrak{w}_1 \supseteq 2\mathfrak{s}_1 = 2\mathfrak{o}$ implies that

$$\mathfrak{f}_1 = a_1\mathfrak{w}_1 \supseteq 4a_1\mathfrak{w}_1^{-1}.$$

Hence the special case is proved when $u_1 = u_2$. We therefore assume that $u_1 < u_2$.

We now choose a base

$$L_1 = (\mathfrak{o}x_1 + \mathfrak{o}x_2) \perp (\mathfrak{o}x_3 + \mathfrak{o}x_4)$$

in which

$$L_1 \cong \langle A(a_1', 0) \rangle \perp \langle A(a_1'', 0) \rangle$$

where a_1' and a_1'' are norm generators of $L^{\mathfrak{s}_1}$; the existence of such a base follows easily from the fact that FL_1 is hyperbolic and that L_1 has norm $a_1\mathfrak{o}$. Now each of the vectors x_i $(1 \leqq i \leqq 4)$ is in $L = K_1 \perp K_{(2)}^*$, hence we can find vectors

$$y_i \in K_1\,, \quad Y_i \in K_{(2)}^*$$

for $1 \leqq i \leqq 4$ such that

$$x_i = y_i + Y_i .$$

Then $Y_i \in L^{\mathfrak{s}_2}$ and so

$$\begin{cases} B(x_i, x_j) \equiv B(y_i, y_j) \bmod \mathfrak{s}_2 \\ Q(x_i) \equiv Q(y_i) \quad \bmod \mathfrak{g}_2 \end{cases}$$

for $1 \leqq i \leqq 4$ and $1 \leqq j \leqq 4$. In particular we have

$$B(x_i, x_j) \equiv B(y_i, y_j) \bmod \mathfrak{p} ,$$

so $d_B(y_1, \ldots, y_4)$ is a unit, hence $\mathfrak{o} y_1 + \cdots + \mathfrak{o} y_4$ is a 4-dimensional sublattice of K_1 with volume equal to $\mathfrak{o}$, hence

$$K_1 = \mathfrak{o} y_1 + \cdots + \mathfrak{o} y_4 .$$

Now solve the pair of equations

$$\begin{cases} B(y_1, y_3) + \xi_3 Q(y_1) + \eta_3 B(y_1, y_2) = 0 \\ B(y_2, y_3) + \xi_3 B(y_1, y_2) + \eta_3 Q(y_2) = 0 \end{cases}$$

for ξ_3, η_3. We have

$$Q(y_1) Q(y_2) - B(y_1, y_2)^2 \in \mathfrak{u} ,$$

and so the solutions ξ_3, η_3 are in $\mathfrak{s}_2$. Then

$$y_3' = y_3 + \xi_3 y_1 + \eta_3 y_2$$

is a vector of K_1 which is orthogonal to both y_1 and y_2. Similarly obtain

$$y_4' = y_4 + \xi_4 y_1 + \eta_4 y_2$$

orthogonal to y_1 and y_2 with ξ_4 and η_4 in $\mathfrak{s}_2$. Then

$$K_1 = (\mathfrak{o} y_1 + \mathfrak{o} y_2) \perp (\mathfrak{o} y_3' + \mathfrak{o} y_4') .$$

We find that we still have

$$\begin{cases} B(x_i, x_j) \equiv B(y_i', y_j') \bmod \mathfrak{s}_2 \\ Q(x_i) \equiv Q(y_i') \quad \bmod \mathfrak{g}_2 \end{cases}$$

for $3 \leqq i \leqq 4$ and $3 \leqq j \leqq 4$. By a direct calculation using the definition of $\mathfrak{f}_1$ we find that

$$d_B(y_1, y_2) \cong -1 \bmod \mathfrak{f}_1 , \quad d_B(y_3', y_4') \cong -1 \bmod \mathfrak{f}_1 ,$$

hence $d K_1 \cong 1 \bmod \mathfrak{f}_1$. Hence $d L_1 / d K_1 \cong 1 \bmod \mathfrak{f}_1$. This proves (i) in the special case under discussion at the moment.

We must prove (iii). We assume that $\mathfrak{f}_1 \subset 4\, a_1 \mathfrak{w}_1^{-1}$. We have $Q(y_1)\mathfrak{o} = a_1 \mathfrak{o} = \mathfrak{n} K_1$ since $Q(y_1) \equiv Q(x_1) \bmod \mathfrak{g}_2$, and since $\mathfrak{g}_2 \subsetneqq a_1 \mathfrak{p}$ by the assumption that $u_1 < u_2$. Similarly with y_3'. So we have just found a splitting

$$K_1 = J \perp J'$$

in which J and J' have norm $a_1\mathfrak{o}$ and

$$dJ \cong -1 \bmod \mathfrak{f}_1\,, \quad dJ' \cong -1 \bmod \mathfrak{f}_1\,.$$

Then by Example 93:17,

$$J \cong A(a_1\alpha^2 + \beta, \gamma)$$

with $\alpha \in \mathfrak{o}$, $\beta \in \mathfrak{w}_1$, $\gamma \in a_1^{-1}\mathfrak{f}_1 \subseteq 4\mathfrak{w}_1^{-1}\mathfrak{p}$. Hence

$$\begin{aligned} K_1 \perp \langle a_1\rangle &\cong \langle A(a_1\alpha^2 + \beta, \gamma)\rangle \perp J' \perp \langle a_1\rangle \\ &\cong \langle A(2a_1\alpha^2 + \beta, \gamma)\rangle \perp J' \perp \langle a_1'''\rangle\,. \end{aligned}$$

Now

$$(2a_1\alpha^2 + \beta)\,\gamma \in (\mathfrak{w}_1)\,(4\mathfrak{w}_1^{-1})\,\mathfrak{p} = 4\mathfrak{p}\,,$$

hence $\langle A(2a_1\alpha^2 + \beta, \gamma)\rangle$ is isotropic by the Local Square Theorem and Proposition 42:9, hence

$$K_1 \perp \langle a_1\rangle \cong \langle A(0, \ldots)\rangle \perp J' \perp \langle a_1'''\rangle\,.$$

But $a_1''' \cong a_1 \bmod \mathfrak{f}_1$ by a consideration of the discriminant, and $\mathfrak{f}_1 \subseteq 2\mathfrak{o}$ so that a_1''' is a norm generator of L. We may therefore repeat the last argument with $J' \perp \langle a_1'''\rangle$ instead of $J \perp \langle a_1\rangle$. This gives

$$K_1 \perp \langle a_1\rangle \cong \langle A(0, \ldots)\rangle \perp \langle A(0, \ldots)\rangle \perp \langle \cdots\rangle\,.$$

Hence

$$FK_1 \perp \langle a_1\rangle \cong \langle A(0, 0)\rangle \perp \langle A(0, 0)\rangle \perp \langle \cdots\rangle\,.$$

Hence $FL_1 \rightarrowtail FK_1 \perp \langle a_1\rangle$, as required.

2) Now we prove (i) and (ii) in general. By making suitable hyperbolic adjunctions we can increase all the fundamental dimensions without changing the remaining fundamental invariants. If we can prove conditions (i) and (ii) for the enlarged lattice, then we shall have them for the original lattice L by cancellation. In effect, this allows us to assume that all fundamental dimensions are $\geqq 3$. Next we take a saturated splitting $L = J_1 \perp \cdots \perp J_t$ and make the adjunction $J_1 \perp \cdots \perp J_t$ to L. The only fundamental invariants that are changed by this adjunction are the fundamental dimensions, and it suffices to prove conditions (i) and (ii) for the enlarged lattice. In effect this allows us to assume that the given Jordan splittings are saturated. A similar argument involving the adjunction of the lattice

$$L^{-1} = L_1^{-1} \perp \cdots \perp L_t^{-1}$$

allows us to assume that each space FL_i (but not necessarily each FK_i) is hyperbolic.

Suppose we wish to verify (i) and (ii) for a specific value of i. By suitably scaling V we can assume that L_{i+1} and K_{i+1} are unimodular. Now $FL_{(i+1)}^*$ is a hyperbolic space. Hence by Proposition 93:3 there is a unimodular lattice J on $FL^*_{(i+1)}$ with $L^*_{(i+1)} \subseteq J \subseteq FL^*_{(i+1)}$ and $\mathfrak{g}J = \mathfrak{g}L^*_{(i+1)}$. Put $L' = L + J$. By considering the Jordan splitting

$$L' = L_1 \perp \cdots \perp L_i \perp J$$

it is easily verified that the first $i+1$ fundamental invariants other than the fundamental dimensions are the same for L and L'. Obviously the first i fundamental dimensions also agree. So the above splitting for L' is saturated. If $1 \leqq \lambda \leqq i$ the lattice K_λ is $\mathfrak{s}_\lambda$-modular with $\mathfrak{s}_\lambda \supseteqq \mathfrak{o}$, and

$$B(K_\lambda, L') \subseteq B(K_\lambda, L) + B(K_\lambda, J) \subseteq B(K_\lambda, L) + B(L', J) \subseteq \mathfrak{s}_\lambda .$$

Hence K_λ splits L' by Proposition 82:15. Hence by Proposition 82:7 there is a splitting

$$L' = K_1 \perp \cdots \perp K_i \perp I .$$

It is easily seen that this splitting is saturated. If we can prove (i) and (ii) for the lattice L' at i, then we shall have these conditions for L at i. In effect this means that we can make the further assumption that $i = t - 1$ with L_t unimodular.

By Example 93:18 we have a splitting $L_t = L'_t \perp L''_t$ with $\dim L'_t = 4$ and

$$L''_t \cong \langle A(0,0)\rangle \perp \cdots \perp \langle A(0,0)\rangle .$$

Similarly with K_t. Apply Theorem 93:14 to cancel off L''_t and K''_t. This allows us to assume that $\dim L_t = \dim K_t = 4$.

So we have reduced things to the following: FL_t is 4-dimensional hyperbolic, L_t is unimodular, $\mathfrak{g}L_t = \mathfrak{g}K_t = \mathfrak{g}_t$, and we must prove (i) and (ii) for $i = t - 1$.

Form $L^\#$. Then L_t and K_t are first components in the Jordan splittings of $L^\#$ that are derived from the given Jordan splittings of L. We know that FL_t is 4-dimensional hyperbolic, that L_t is unimodular, and that $\mathfrak{g}L_t = \mathfrak{g}K_t = \mathfrak{g}_t = \mathfrak{g}^\#_1$. The special conditions of step 1) are therefore met by the above Jordan splittings of $L^\#$, hence $L^\#$ satisfies (i) and (iii). So

$$dL_t/dK_t \cong 1 \bmod \mathfrak{f}^\#_1 .$$

But $\mathfrak{f}^\#_1 = \mathfrak{f}_{t-1}$ by Example 93:24, and $dL_{(t-1)} \cdot dL_t = dK_{(t-1)} \cdot dK_t$, hence

$$dL_{(t-1)}/dK_{(t-1)} \cong 1 \bmod \mathfrak{f}_{t-1} .$$

We have therefore completely proved (i). Now (ii). We suppose that $\mathfrak{f}_{t-1} \subset 4\, a_t \mathfrak{w}_t^{-1}$. By Example 93:24 this condition translates into $\mathfrak{f}^\#_1 \subset 4\, a^\#_1 (\mathfrak{w}^\#_1)^{-1}$. So by step 1), $FL_t \rightarrow FK_t \perp \langle -a^\#_1\rangle$. Hence

$$FL_t \perp \langle \cdots \rangle \cong FK_t^{-1} \perp \langle a_t\rangle$$

since FL_t is hyperbolic. Adjoining V to both sides gives

$$FL_{(t-1)} \perp FL_t \perp FL_t \perp \langle \cdots \rangle \cong FK_{(t-1)} \perp FK_t \perp FK_t^{-1} \perp \langle a_t\rangle .$$

Hence $FL_{(t-1)} \rightarrowtail FK_{(t-1)} \perp \langle a_t \rangle$ by Witt's theorem.

3) Finally (iii). Consider any i for which $\mathfrak{f}_i \subset 4 a_i \mathfrak{w}_i^{-1}$. Then

$$\mathfrak{f}^{\#}_{t-i} \subset 4\, a^{\#}_{t-i+1}\, (\mathfrak{w}^{\#}_{t-i+1})^{-1}$$

by Example 93:24. But this is condition (ii) for the lattice $L^{\#}$ at $t-i$. And (ii) was completely established in step 2) of this proof. Hence

$$FK^{*}_{(i+1)} \rightarrowtail FL^{*}_{(i+1)} \perp \langle a^{\#}_{t-i+1} \rangle\,.$$

Adjoining V to each side gives

$$\begin{aligned} &FL_{(i)} \perp FL^{*}_{(i+1)} \perp FK^{*}_{(i+1)} \perp \langle \cdots \rangle \\ &\cong FK_{(i)} \perp FK^{*}_{(i+1)} \perp FL^{*}_{(i+1)} \perp \langle a_i \rangle\,. \end{aligned}$$

Hence $FL_{(i)} \rightarrowtail FK_{(i)} \perp \langle a_i \rangle$ by Witt's theorem. This completes the proof of the necessity.

Proof of the sufficiency. 1) We shall prove the sufficiency by induction on the quantity $s_t - s_1 \geqq 0$. We start with $s_t = s_1$. Then L and K are modular of the same scale. And $\mathfrak{g}L = \mathfrak{g}K$ since L and K have the same fundamental invariants. Hence $\operatorname{cls} L = \operatorname{cls} K$ by Theorem 93:16. So we are through with the case $s_t = s_1$. We therefore assume that $s_t - s_1 > 0$ (and so $t > 1$) and we proceed with the induction.

2) By making suitable hyperbolic adjunctions we can increase all the fundamental dimensions without altering the other fundamental invariants or the given conditions. By Theorem 93:14 it is enough to prove the isometry of the enlarged lattices. In effect this allows us to assume that all fundamental dimensions are $\geqq 3$. Now take a saturated splitting $L = J_1 \perp \cdots \perp J_t$ and make the adjunction of $J_1 \perp \cdots \perp J_t$ to L and to K. The only fundamental invariants that are changed by this adjunction are the fundamental dimensions; and if the given conditions hold for L and K they will also hold for $L \perp L$ and $L \perp K$; if we can prove that $L \perp L$ and $L \perp K$ are isometric, then we can obtain an isometry of L onto K by using Corollary 93:14a to cancel off one J_i at a time. In effect this allows us to assume that the given splittings of L and K are saturated. A similar argument involving the adjunction of the lattice

$$L^{-1} = L_1^{-1} \perp \cdots \perp L_t^{-1}$$

allows us to assume that each space FL_i (but not necessarily each FK_i) is hyperbolic. Now by Example 93:18 we know that each L_i (or K_i) is of the form

$$L_i \cong \langle \xi\, A(0,0) \rangle \perp \cdots \perp \langle \xi\, A(0,0) \rangle \perp L_i'$$

with $\dim L_i' = 4$. By making suitable cancellations to the L_i and K_i we see that we can assume that $\dim L_i = \dim K_i = 4$ for $1 \leqq i \leqq t$. Finally we scale V to make L_1 and K_1 unimodular.

So we have reduced things to the following: we are given lattices L and K with the same fundamental invariants in saturated Jordan splittings

$$L = L_1 \perp \cdots \perp L_t, \quad K = K_1 \perp \cdots \perp K_t$$

which satisfy conditions (i)—(iii), the first components L_1 and K_1 are unimodular, and all spaces FL_i are 4-dimensional hyperbolic, and we must prove, using the inductive hypothesis, that L and K are isometric.

Note that Example 93:18 (vi) asserts that L_1 has the form

$$L_1 \cong \langle A(a_1, 0)\rangle \perp \langle A(b_1, 0)\rangle .$$

3) The splitting $L = L_1 \perp \cdots \perp L_t$ will be left fixed throughout. The computational part of the proof consists in successively changing the Jordan splitting $K_1 \perp K_2$ until a new splitting is obtained in which K_1 is isometric to L_1. It is understood that each splitting employed in this procedure induces a saturated splitting on K. We shall use (without further reference) the rules of Example 93:13 and Proposition 93:19 to effect these changes to $K_1 \perp K_2$.

In connection with this matter we must mention that conditions (i)—(iii) will continue to hold if the given splitting $K_1 \perp \cdots \perp K_t$ is replaced by any other Jordan splitting $K_1' \perp \cdots \perp K_t'$. For $dL_{(i)}/dK_{(i)} \cong 1 \bmod \mathfrak{f}_i$ by hypothesis, and $dK_{(i)}/dK'_{(i)} \cong 1 \bmod \mathfrak{f}_i$ by the necessity of this theorem, hence $dL_{(i)}/dK'_{(i)} \cong 1 \bmod \mathfrak{f}_i$ and so (i) holds in the new splitting. Let us do (ii). Assume we have $\mathfrak{f}_i \subset 4\, a_{i+1}\, \mathfrak{w}_{i+1}^{-1}$. Then

$$FL_{(i)} \perp \langle a'_{i+1}\rangle \cong FK_{(i)} \perp \langle a_{i+1}\rangle$$

by hypothesis; now the discriminant shows that we can take

$$a'_{i+1} \cong a_{i+1} \bmod a_{i+1}\, \mathfrak{f}_i\,;$$

but $a_{i+1}\, \mathfrak{f}_i \subset 4\, a_{i+1}^2\, \mathfrak{w}_{i+1}^{-1} \subseteq \mathfrak{w}_{i+1}$; so a'_{i+1} is a fundamental norm generator; hence we have an isometry

$$FK_{(i)} \perp \langle \cdots \rangle \cong FK'_{(i)} \perp \langle a'_{i+1}\rangle$$

by the necessity. Then

$$FL_{(i)} \perp \langle a'_{i+1}\rangle \perp FK_{(i)} \perp \langle \cdots \rangle \cong FK_{(i)} \perp \langle a_{i+1}\rangle \perp FK'_{(i)} \perp \langle a'_{i+1}\rangle$$

and so $FL_{(i)} \rightarrowtail FK'_{(i)} \perp \langle a_{i+1}\rangle$ as required. We establish (iii) in the same way.

This shows that we will be through once we obtain a new saturated Jordan splitting $K = K_1 \perp \cdots \perp K_t$ in which $L_1 \cong K_1$. For the new splitting will still satisfy conditions (i)—(iii). And we can assume that $L_1 = K_1$ by Witt's theorem, so $L^*_{(2)}$ and $K^*_{(2)}$ become lattices on the same quadratic space $FL^*_{(2)}$. But the first ,..., $(t-1)^{th}$ fundamental invariants of $L^*_{(2)}$ and $K^*_{(2)}$ are equal to the second , ..., t^{th} fundamental invariants of L and K since the splittings $L_1 \perp \cdots \perp L_t$ and $K_1 \perp \cdots \perp K_t$ are saturated. Hence $L^*_{(2)}$ and $K^*_{(2)}$ satisfy conditions (i)—(iii), hence they are isometric by the inductive hypothesis. Of course this implies the isometry of L and K.

4) Now we start the computational part of the proof. First suppose that $u_2 = u_1$. So $a_2\mathfrak{o} = a_1\mathfrak{o}$. In this event we have $\mathfrak{f}_1 = a_2\mathfrak{w}_1$ by Example 93:27. We know that $dL_1 = 1$. Let us arrange $dK_1 = 1$. By condition (i) we have $dK_1 \cong 1 \bmod \mathfrak{f}_1$, hence $dK_1 = 1 + a_2\lambda$ for some $\lambda \in \mathfrak{w}_1$. Then

$$\begin{aligned}\langle A(0,0)\rangle \perp K_1 \perp K_2 &\cong \langle A(-\lambda, 0)\rangle \perp K_1 \perp K_2\\ &\cong \langle A(-\lambda, a_2)\rangle \perp K_1 \perp K_2'\\ &\cong \langle A(0,0)\rangle \perp K_1' \perp K_2'\,.\end{aligned}$$

Hence $K_1 \perp K_2$ has a saturated splitting $K_1' \perp K_2'$ in which $dK_1' = 1$. In other words we can assume that $dK_1 = 1$.

By Example 93:18 (vi) there is a splitting

$$K_1 \cong \langle A(a_1, 4a_1^{-1}\lambda)\rangle \perp \langle A(b_1, 4b_1^{-1}\lambda)\rangle$$

in which $\lambda = 0$ when FK_1 is isotropic and $\lambda = \varrho$ when FK_1 is not isotropic. If $\mathfrak{w}_2 \subset 4\mathfrak{w}_1^{-1}$, then $\mathfrak{f}_1 = a_2\mathfrak{w}_1 \subset 4a_2\mathfrak{w}_2^{-1}$ and condition (ii) holds. Hence in this case we have $FL_1 \cong FK_1$, so FK_1 is isotropic, therefore $\lambda = 0$. Otherwise $\mathfrak{w}_2 \supseteq 4\mathfrak{w}_1^{-1} = 4b_1^{-1}\mathfrak{o}$. Then

$$K_1 \perp K_2 \cong \langle A(a_1, 4a_1^{-1}\lambda)\rangle \perp \langle A(b_1, 0)\rangle \perp K_2'\,.$$

Now K_2' represents a number whose order is equal to the order of $4a_1^{-1}\varrho$, hence by the perfectness of the residue class field, the Local Square Theorem, and Example 93:17, there is a splitting

$$K_1 \perp K_2 \cong \langle A(a_1, 0)\rangle \perp \langle A(b_1, 0)\rangle \perp K_2''\,.$$

So we can take $\lambda = 0$ in all cases. Then $K_1 \cong L_1$. Hence $K \cong L$ by the observation at the end of step 3).

5) Next suppose $a_1\mathfrak{o} = 2\mathfrak{o}$. By step 4) we can assume that $a_2\mathfrak{o} \subseteq 2\mathfrak{p}$. Hence $\mathfrak{f}_1 \subseteq 4\mathfrak{p}$. So $dK_1 \cong 1 \bmod 4\mathfrak{p}$, hence $dK_1 = 1$ by the Local Square Theorem. Hence by Example 93:18 K_1 has the form

$$K_1 \cong \langle A(0,0)\rangle \perp \langle A(0,0)\rangle\,.$$

But L_1 has the same form. Therefore $L_1 \cong K_1$, and $L \cong K$ by the observation at the end of step 3).

6) Now let $u_2 = u_1 + 1$. We can assume that $a_1\mathfrak{o} \supset 2\mathfrak{o}$ by step 5). Then $a_2 \in \mathfrak{w}_1$ since $a_2 \in \mathfrak{g}_2 \subseteq \mathfrak{g}_1$ with $\operatorname{ord}_{\mathfrak{p}} a_1 + \operatorname{ord}_{\mathfrak{p}} a_2$ odd, hence $a_1\mathfrak{p} = a_2\mathfrak{o}$ is contained in $\mathfrak{g}_1$, hence $a_1\mathfrak{p}$ is the maximal ideal in $\mathfrak{g}_1$, hence $\mathfrak{g}_1 = a_1\mathfrak{p}$. And $b_1\mathfrak{o} = \mathfrak{w}_1 = a_2\mathfrak{o}$ since $a_1\mathfrak{o} \supset 2\mathfrak{o}$. Now $\mathfrak{f}_1 = a_1 a_2\mathfrak{o}$ by Example 93:27, so we can proceed as in step 4) to get a new splitting $K_1 \perp K_2$ in which $dK_1 = 1$. By Example 93:18 there is therefore a splitting

$$K_1 \cong \langle A(a_1, 4a_1^{-1}\lambda)\rangle \perp \langle A(b_1, 4b_1^{-1}\lambda)\rangle$$

in which $\lambda = 0$ if FK_1 is isotropic and $\lambda = \varrho$ if FK_1 is not isotropic. If $\mathfrak{w}_2 \subset 4a_1^{-1}\mathfrak{o}$ we have $\mathfrak{f}_1 = a_1 a_2 \mathfrak{o} \subset 4a_2 \mathfrak{w}_2^{-1}$ and condition (ii) holds. This implies that $FL_1 \cong FK_1$ and so $\lambda = 0$. If $\mathfrak{w}_2 \supseteq 4a_1^{-1}\mathfrak{o}$ we can proceed as in step 4) to get a new splitting $K_1 \perp K_2$ with $\lambda = 0$. So in any case we can arrange $\lambda = 0$. Then $L_1 \cong K_1$. Hence $L \cong K$. So the case $u_2 = u_1 + 1$ is proved.

7) Next we settle all cases in which $s_2 = 1$, i. e. in which the second component is $\mathfrak{p}$-modular. By steps 4), 5), 6) we can assume that $a_2\mathfrak{o} \subsetneqq a_1\mathfrak{p}^2$ and $a_1\mathfrak{o} \supset 2\mathfrak{o}$. Then in fact $a_2\mathfrak{o} = a_1\mathfrak{p}^2$ since $\pi^2\mathfrak{g}_1 \subseteqq \mathfrak{g}_2 \subseteqq \mathfrak{g}_1$. We have $\mathfrak{f}_1 = a_1\mathfrak{w}_2$ by Example 93:27, so we can arrange the splitting $K_1 \perp K_2$ to be such that $dK_1 = 1$ as in step 4). Continuing as in step 4) we obtain a new splitting

$$K_1 \perp K_2 \cong \langle A(a_1, 0)\rangle \perp \langle A(b_1, 0)\rangle \perp K_2'$$

and then $K \cong L$. (There are two variations in this part of the proof: first, one uses (iii) instead of (ii), secondly one needs the inclusion $a_2\mathfrak{o} = a_1\mathfrak{p}^2 \supseteqq 4a_1^{-1}\mathfrak{o}$.)

8) Finally we consider the following situation: $s_2 > 1$, $a_2\mathfrak{o} \subsetneqq a_1\mathfrak{p}^2$, $a_1\mathfrak{o} \supset 2\mathfrak{o}$. Perform the adjunction of the same lattice J to L and K, where

$$J \cong \langle \pi A(0, 0)\rangle .$$

Put $L' = L \perp J$ and $K' = K \perp J$. Suppose we can show that L' and K' satisfy the conditions of the theorem. Then $s_t' - s_1' = s_t - s_1$ and $s_2' = 1$, so L' and K' will be isometric by steps 2) and 7). Hence L and K will be isometric by Theorem 93:14. The rest of the proof therefore concerns itself with showing that L' and K' satisfy the conditions of the theorem.

We have Jordan splittings

$$L' = L_1 \perp L_{1^1/_2} \perp L_2 \perp \cdots \perp L_t\,, \quad K' = K_1 \perp K_{1^1/_2} \perp K_2 \perp \cdots \perp K_t$$

in which $L_{1^1/_2} = J = K_{1^1/_2}$. It is easily seen that L' and K' have the same fundamental invariants. And it is clear what happens to the invariants t, $\dim L_i$, $\mathfrak{s}_i$ as we pass from L to L'. What about the new fundamental norm groups, weights, and norm generators? We find that we can take

$$\mathfrak{g}_i' = \mathfrak{g}_i\,, \mathfrak{w}_i' = \mathfrak{w}_i\,, a_i' = a_i \quad (i = 1, 2, \ldots, t)\,,$$

hence from the definition

$$\mathfrak{f}_i' = \mathfrak{f}_i \quad (i = 2, \ldots, t)\,.$$

We have

$$(L')^{\mathfrak{p}} = \pi L_1 \perp J \perp L_2 \perp \cdots \perp L_t\,,$$

hence $\mathfrak{n}(L')^{\mathfrak{p}} = \pi^2 a_1 \mathfrak{o}$ and we can take $a_{1^1/_2}' = \pi^2 a_1$ as a fundamental norm generator. By Example 93:25,

$$\pi^2\,\mathfrak{g}_1 \subseteqq \mathfrak{g}_{1^1/_2}' \subseteqq \mathfrak{g}_1\,, \quad \pi^2\,\mathfrak{w}_1 \subseteqq \mathfrak{w}_{1^1/_2}' \subseteqq \mathfrak{w}_1\,,$$

and

$$\mathfrak{g}_2 \subseteq \mathfrak{g}'_{1^1/_2}\,, \quad \mathfrak{w}_2 \subseteq \mathfrak{w}'_{1^1/_2}\,.$$

Hence from the definition

$$\mathfrak{f}_1 \subseteq \mathfrak{f}'_1\,, \quad \mathfrak{f}_1 \subseteq \mathfrak{f}'_{1^1/_2}\,.$$

It is now obvious that L' and K' satisfy (i) for all i. Also that they satisfy (ii) and (iii) for $i = 2, \ldots, t$. Also that they satisfy (ii) for $i = 1^1/_2$. We see that (ii) is vacuous for $i = 1$ since then

$$\mathfrak{w}'_{1^1/_2}\,\mathfrak{f}'_1 = a_1\,\mathfrak{w}'_{1^1/_2}\,\mathfrak{w}'_{1^1/_2} \supseteq 4a_1\,\mathfrak{p}^2 = 4a'_{1^1/_2}\,\mathfrak{o}\,.$$

We leave the verification of (iii) for $i = 1$ and $i = 1^1/_2$ to the reader.

q. e. d.

§ 93G. The 2-adic case

The ideal $2\mathfrak{o}$ in a dyadic local field F is always contained in the maximal ideal $\mathfrak{p}$ since the residue class field has characteristic 2. Let us suppose for the rest of this subparagraph that we actually have $2\mathfrak{o} = \mathfrak{p}$, or equivalently that 2 is a prime element in F. We shall call such a field a 2-adic local field in order to signify that 2 is one of its prime elements. (The field of 2-adic numbers $\mathbf{Q}_2$ is an example of a 2-adic local field.) Otherwise let the situation be the one already under discussion in this paragraph. In particular let L be a lattice in a regular quadratic space V over F with the Jordan splitting

$$L = L_1 \perp \cdots \perp L_t\,.$$

If $\mathfrak{a}$ is any fractional ideal in F, then there are no fractional ideals properly situated between $2\mathfrak{a}$ and $\mathfrak{a}$. Hence $\mathfrak{n}L$ is either $\mathfrak{s}L$ or $2(\mathfrak{s}L)$. So by Theorem 91:9 the ideal $\mathfrak{n}L_i$ is independent of the Jordan splitting $L = L_1 \perp \cdots \perp L_t$. We define

$$\mathfrak{n}_i = \mathfrak{n}L_i \quad \text{for} \quad 1 \leq i \leq t\,.$$

(This definition applies only in the 2-adic case since otherwise the $\mathfrak{n}_i$ are not uniquely determined by the Jordan splitting.) We call the quantities

$$\boxed{t,\ \dim L_i,\ \mathfrak{s}_i,\ \mathfrak{n}_i}$$

the Jordan invariants of L. It is clear that two lattices have the same Jordan invariants if and only if they are of the same Jordan type. We put

$$u_i = \operatorname{ord}_{\mathfrak{p}} \mathfrak{n}_i\,.$$

We have

$$\mathfrak{n}L_{i+1} \subseteq \mathfrak{s}L_{i+1} \subseteq 2(\mathfrak{s}L_i) \subseteq \mathfrak{n}L_i\,,$$

and so

$$\mathfrak{n}_1 \supseteq \mathfrak{n}_2 \supseteq \cdots \supseteq \mathfrak{n}_t\,.$$

What is the connection between the Jordan invariants and the fundamental invariants of L? The maximal ideal $\mathfrak{m}L$ in the norm group

$\mathfrak{g}L$ must be all of $\mathfrak{n}L$ since $2(\mathfrak{s}L) \subsetneqq \mathfrak{m}L \subseteqq \mathfrak{n}L$ with $\mathrm{ord}_\mathfrak{p}\mathfrak{m}L + \mathrm{ord}_\mathfrak{p}\mathfrak{n}L$ even. Hence

$$\mathfrak{g}L = \mathfrak{n}L \quad \text{and} \quad \mathfrak{w}L = 2(\mathfrak{s}L)\,.$$

Then $\mathfrak{g}L^{\mathfrak{s}_i} = \mathfrak{n}L^{\mathfrak{s}_i} = \mathfrak{n}L_i$ and so

$$\mathfrak{g}_i = \mathfrak{n}_i\,, \quad \mathfrak{w}_i = 2\mathfrak{s}_i \quad \text{for} \quad 1 \leqq i \leqq t\,.$$

In particular this shows that the quantity $u_i = \mathrm{ord}_\mathfrak{p}\mathfrak{n}_i$ is equal to the quantity $u_i = \mathrm{ord}_\mathfrak{p} a_i$ defined in § 93E. We can take any element of order u_i as a fundamental norm generator, in particular we can take $a_i = 2^{u_i}$. The lattice L now has the following set of fundamental invariants:

$$t,\ \dim L_i,\ \mathfrak{s}_i,\ 2\mathfrak{s}_i,\ 2^{u_i}\,.$$

This shows that any two lattices with the same Jordan invariants (over a 2-adic local field) have the same fundamental invariants.

93:29. **Theorem.**[1] *Let L and K be lattices on a regular quadratic space over a 2-adic local field F, suppose that L and K have the same Jordan invariants, and let*

$$L_{(1)} \subset \cdots \subset L_{(t)} \quad \textit{and} \quad K_{(1)} \subset \cdots \subset K_{(t)}$$

be Jordan chains for L and K. Then $\mathrm{cls}\,L = \mathrm{cls}\,K$ *if and only if the following conditions hold for* $1 \leqq i \leqq t-1$:

(i) $dL_{(i)}/dK_{(i)} \cong 1 \bmod \mathfrak{n}_i\,\mathfrak{n}_{i+1}/\mathfrak{s}_i^2$

(ii) $FL_{(i)} \rightarrow FK_{(i)} \perp \langle 2^{u_i}\rangle$ *when* $\mathfrak{n}_{i+1} \subsetneqq 4\mathfrak{n}_i$.

Proof. 1) We obtain the necessity by suitably interpreting Theorem 93:28. Take $a_i = 2^{u_i}$ for $1 \leqq i \leqq t$. Then it follows by direct calculation from Example 93:26 that

$$\mathfrak{s}_i^2\,\mathfrak{f}_i \supseteqq \mathfrak{n}_i\,\mathfrak{n}_{i+1}$$

and that

$$\mathfrak{s}_i^2\,\mathfrak{f}_i \supset \mathfrak{n}_i\,\mathfrak{n}_{i+1} \;\Rightarrow\; \mathfrak{f}_i \subsetneqq 8\mathfrak{o}\,.$$

Hence (i) is a consequence of condition (i) of Theorem 93:28 and the Local Square Theorem. We find that $\mathfrak{n}_{i+1} \subsetneqq 4\mathfrak{n}_i$ implies that $\mathfrak{f}_i \subset 4\mathfrak{n}_i\mathfrak{w}_i^{-1}$, therefore (ii) is a consequence of condition (iii) of Theorem 93:28.

2) The sufficiency is proved in essentially the same way as the sufficiency of Theorem 93:28. Let V denote the quadratic space in question. Let

$$L = L_1 \perp \cdots \perp L_t\,, \quad K = K_1 \perp \cdots \perp K_t$$

denote Jordan splittings which determine the given Jordan chains. We shall prove the sufficiency by induction on the quantity t. If $t = 1$ we

[1] See O. T. O'Meara, *Am. J. Math.* (1958), pp. 843—878, for the solution of the representation problem over 2-adic fields. A general solution over arbitrary dyadic fields is not known at present, but much progress has been made on this difficult problem by C. Riehm, *On the integral representations of quadratic forms over local fields* (Princeton University thesis, 1961).

have $\mathfrak{g}L_1 = \mathfrak{n}_1 = \mathfrak{g}K_1$ and so $\mathrm{cls}\,L = \mathrm{cls}\,K$ by Theorem 93:16. We therefore assume that $t > 1$ and proceed with the induction.

By making suitable hyperbolic adjunctions we can assume that $\dim L_i \geqq 3$ for $1 \leqq i \leqq t$, in virtue of Theorem 93:14. By adjoining the lattice

$$L^{-1} = L_1^{-1} \perp \cdots \perp L_t^{-1}$$

we see that we can assume that each space FL_i is hyperbolic, in virtue of Corollary 93:14a. By making suitable hyperbolic cancellations we can assume that $\dim L_i = \dim K_i = 4$ for $1 \leqq i \leqq t$. By scaling V we can assume that L_1 and K_1 are unimodular.

It is enough to find a new Jordan splitting

$$K = K_1' \perp K_2' \perp K_3 \perp \cdots \perp K_t$$

in which $K_1' \cong L_1$. For then we can assume that $K_1' = L_1$ by Witt's theorem. So the lattices

$$L_2 \perp L_3 \perp \cdots \perp L_t \quad \text{and} \quad K_2' \perp K_3 \perp \cdots \perp K_t$$

are on the same quadratic space, and they clearly satisfy the conditions of the theorem. The inductive hypothesis then asserts that these lattices are isometric. Hence $L \cong K$.

First consider the case $\mathfrak{n}_2 = 2\mathfrak{o}$. Then

$$L_1 \cong \langle A(2^{u_1}, 0)\rangle \perp \langle A(2, 0)\rangle .$$

But

$$K_1 \cong \langle A(2^{u_1}, 2\alpha)\rangle \perp \langle A(2, 2\beta)\rangle$$

with $\alpha, \beta \in \mathfrak{o}$, hence by Proposition 93:19

$$K_1 \perp K_2 \cong \langle A(2^{u_1}, 0)\rangle \perp \langle A(2, 0)\rangle \perp K_2'$$

since $\mathfrak{g}K_2 = \mathfrak{n}_2 = 2\mathfrak{o}$. So $K_1 \perp K_2$ has a splitting $K_1' \perp K_2'$ with $K_1' \cong L_1$. This settles the case $\mathfrak{n}_2 = 2\mathfrak{o}$.

Next let $\mathfrak{n}_2 = 4\mathfrak{o}$. If $\mathfrak{n}_1 = 2\mathfrak{o}$, then $dL_1/dK_1 \cong 1 \bmod 8\mathfrak{o}$, hence $dK_1 = dL_1 = 1$, so

$$L_1 \cong \langle A(2, 0)\rangle \perp \langle A(2, 0)\rangle \cong K_1 .$$

We therefore assume that $\mathfrak{n}_1 = \mathfrak{o}$. Here we get $dK_1 \cong 1 \bmod 4\mathfrak{o}$, so

$$K_1 \cong \langle A(1, 4\lambda)\rangle \perp \langle A(2, 2\mu)\rangle$$

with λ equal to 0 or ϱ, and the same with μ. But $\mathfrak{n}_2 = 4\mathfrak{o} \subsetneqq 4\mathfrak{n}_1$, hence

$$FL_1 \perp \langle \varepsilon\rangle \cong FK_1 \perp \langle 1\rangle$$

where ε is the unit $(1-4\lambda)(1-4\mu)$. A computation of Hasse symbols, say, will show that $\mu = 0$. Then

$$K_1 \perp K_2 \cong \langle A(1, 0)\rangle \perp \langle A(2, 0)\rangle \perp K_2'$$

since $\mathfrak{g}K_2 = \mathfrak{n}_2 = 4\mathfrak{o}$. But

$$L_1 \cong \langle A(1, 0)\rangle \perp \langle A(2, 0)\rangle .$$

Hence $K \cong L$.

Finally let $\mathfrak{n}_2 \subsetneq 8\mathfrak{o}$. Then $dL_1/dK_1 \cong 1 \bmod 8\mathfrak{o}$, hence $dK_1 = dL_1 = 1$. But $\mathfrak{n}_2 \subsetneq 8\mathfrak{o} \subsetneq 4\mathfrak{n}_1$, so that

$$FL_1 \perp \langle 1 \rangle \cong FK_1 \perp \langle 1 \rangle ,$$

hence FK_1 is isotropic. Hence

$$L_1 \cong \langle A(2^{u_1}, 0) \rangle \perp \langle A(2, 0) \rangle \cong K_1 .$$

q. e. d.

§ 94. Effective determination of the invariants

Let us show how to compute the invariants of the local integral theory. We suppose that L is a regular lattice given in the base $L = \mathfrak{o}x_1 + \cdots + \mathfrak{o}x_n$, and that the symmetric bilinear form B is determined by specifying the values $B(x_i, x_j) = a_{ij}$. (In practice it is just the symmetric matrix (a_{ij}) that is given.) To find the scale we simply take the largest of the ideals $B(x_i, x_j)\,\mathfrak{o}$, i. e.

$$\mathfrak{s}L = \sum_{i,j} (a_{ij}\,\mathfrak{o}) .$$

To find the norm we take the largest of the ideals $Q(x_i)\,\mathfrak{o}$ and $2(\mathfrak{s}L)$, i. e.

$$\mathfrak{n}L = \sum_{i} (a_{ii}\,\mathfrak{o}) + 2(\mathfrak{s}L) .$$

These rules are justified by Proposition 82:8.

Next we show how to obtain a Jordan splitting from the given base. First the non-dyadic case. If there is no x_i with $Q(x_i)\,\mathfrak{o} = \mathfrak{s}L$, then there will be two distinct basis vectors, x_1 and x_2 say, with $B(x_1, x_2)\,\mathfrak{o} = \mathfrak{s}L$; applying the Principle of Domination to the equation

$$Q(x_1 + x_2) = Q(x_1) + Q(x_2) + 2B(x_1, x_2)$$

shows that $Q(x_1 + x_2)\,\mathfrak{o} = \mathfrak{s}L$. (Note that this will not work in the dyadic case.) Hence a minor adjustment to the given base allows us to assume that $Q(x_1)\,\mathfrak{o} = \mathfrak{s}L$. All the coefficients $B(x_i, x_1)/Q(x_1)$ are now in $\mathfrak{o}$, hence the vectors $x_1, y_2, \ldots, y_n$ with

$$y_i = x_i - \frac{B(x_i, x_1)}{Q(x_1)}\, x_1$$

form a new base for L and in fact we have a splitting

$$L = \mathfrak{o}x_1 \perp (\mathfrak{o}y_2 + \cdots + \mathfrak{o}y_n) .$$

Repeat all this on $\mathfrak{o}y_2 + \cdots + \mathfrak{o}y_n$, etc. Ultimately we arrive at an orthogonal base for L, and a Jordan splitting is formed by suitably grouping these basis vectors.

The dyadic procedure is somewhat longer. If there is a vector, say x_1, with $Q(x_1)\,\mathfrak{o} = \mathfrak{s}L$, use the non-dyadic method to split off $\mathfrak{o}x_1$. Otherwise $Q(x_i)\,\mathfrak{o} \subset \mathfrak{s}L$ for $1 \leq i \leq n$. Then

$$B(x_1, x_2)\,\mathfrak{o} = \mathfrak{s}L , \quad Q(x_1)\,\mathfrak{o} \subset \mathfrak{s}L ,$$

say. This implies that the equations

$$\begin{cases} B(x_1, x_i) + \xi_i Q(x_1) + \eta_i B(x_1, x_2) = 0 \\ B(x_2, x_i) + \xi_i B(x_1, x_2) + \eta_i Q(x_2) = 0 \end{cases}$$

have integral solutions ξ_i, η_i for $3 \leqq i \leqq n$. Hence the vectors $x_1, x_2, y_3, \ldots, y_n$ with

$$y_i = x_i + \xi_i x_1 + \eta_i x_2$$

form a base for L, and in fact there is a splitting

$$L = (\mathfrak{o} x_1 + \mathfrak{o} x_2) \perp (\mathfrak{o} y_3 + \cdots + \mathfrak{o} y_n)$$

with $\mathfrak{o} x_1 + \mathfrak{o} x_2$ modular. Repeat on $\mathfrak{o} y_3 + \cdots + \mathfrak{o} y_n$, etc. Ultimately we obtain a Jordan splitting for L.

The Jordan splitting is all that is needed in the non-dyadic theory. The Jordan invariants are also needed in the 2-adic theory, but these can be read off from the Jordan splitting. So Theorems 92:2 and 93:29 can be applied in practice.

The general dyadic theory requires a knowledge of the fundamental invariants a_i and $\mathfrak{w}_i$ (the invariants $\mathfrak{f}_i$ are obtained from the fundamental invariants using Example 93:26). First let us find a norm generator a and the weight $\mathfrak{w}$ of L. The norm generator is easy: if $\mathfrak{n}L = 2(\mathfrak{s}L)$ take any a with $a\mathfrak{o} = 2(\mathfrak{s}L)$, otherwise take $a = Q(x_i)$ for any value of i which makes $Q(x_i)\,\mathfrak{o}$ largest. The weight is then given by the formula

$$\mathfrak{w} = \sum_j a\,\mathfrak{d}(Q(x_j)/a) + 2(\mathfrak{s}L)\,.$$

We leave the verification of this formula to the reader. In order to find the fundamental invariants $a_1, \ldots, a_t$ and $\mathfrak{w}_1, \ldots, \mathfrak{w}_t$ we apply this procedure to each of the lattices $L^{\mathfrak{s}_1}, \ldots, L^{\mathfrak{s}_t}$. These lattices can be expressed in terms of the Jordan splitting $L = L_1 \perp \cdots \perp L_t$ by the formula

$$L^{\mathfrak{s}_i} = (\pi^{s_i - s_1} L_1) \perp \cdots \perp (\pi^{s_i - s_{i-1}} L_{i-1}) \perp L_i \perp \cdots \perp L_t\,.$$

§ 95. Special subgroups of $O_n(V)$

We conclude this chapter by giving the structure of the groups

$$\Omega_n \cap Z_n\,, \quad O'_n/\Omega_n\,, \quad O^+_n/O'_n$$

of a regular n-ary quadratic space V over an arbitrary local field. Recall that we first raised this question over a general field in § 56, and that we have already described these groups over complete archimedean fields and over finite fields in §§ 61 and 62.

95:1. *Let V be a regular n-ary quadratic space over a local field F. Then $O'_n = \Omega_n$ with the following exception: V is a quaternary anisotropic space over a non-dyadic field. In the exceptional case we have $(O'_4 : \Omega_4) = 2$.*

Proof. 1) If V is isotropic, in particular if $n \geqq 5$, then $O'_n = \Omega_n$ by Corollary 55:6a. If $1 \leqq n \leqq 3$ we have $O'_n = \Omega_n$ by Proposition 55:5. So the quaternary anisotropic case remains, and we must prove that $O'_4 = \Omega_4$ if F is dyadic and that $(O'_4 : \Omega_4) = 2$ if F is non-dyadic. Note that $dV = 1$ by Proposition 63:17.

If α is a non-zero scalar in $Q(V)$ we let $\tau_{\langle\alpha\rangle}$ stand for a symmetry with respect to a vector x with $Q(x) = \alpha$. This is not a precise notation since $\tau_{\langle\alpha\rangle}$ is not uniquely determined by α. However the coset $\tau_{\langle\alpha\rangle}\Omega_4$ is well-defined in virtue of Proposition 55:3. In fact we can write

$$\tau_{\langle\alpha_1\rangle} \cdots \tau_{\langle\alpha_r\rangle} \Omega_4 \quad (\alpha_i \in \dot{F})$$

without fear of ambiguity, and this coset is even independent of the order in which the α_i appear.

2) We are ready to do the dyadic case. We must prove that a typical element σ of O'_4 is actually in Ω_4. Express σ as a product of four symmetries, say

$$\sigma = \tau_{\langle\alpha_1\rangle} \cdots \tau_{\langle\alpha_4\rangle} \quad (\alpha_i \in \dot{F}) .$$

Here $\alpha_1\alpha_2\alpha_3\alpha_4$ is in $\dot{F}^2$ since σ is in O'_4 .Clearly $\alpha_1, \alpha_2, \alpha_3, \alpha_4$ fall in at most four distinct cosets of $\dot{F}$ modulo $\dot{F}^2$. On the other hand $(\dot{F}:\dot{F}^2) \geqq 8$ by Proposition 63:9. Hence there is a ξ in $\dot{F}$ such that $\xi\alpha_1, \xi\alpha_2, \xi\alpha_3, \xi\alpha_4$ are non-squares in $\dot{F}$. Now V represents $-\xi$ since V is universal, hence there is a ternary subspace U of V such that $V \cong \langle -\xi\rangle \perp U$. Here we have $dU = -\xi$ since $dV = 1$, hence the discriminant of the space $\langle -\alpha_i\rangle \perp U$ is a non-square, hence this space is isotropic, hence U represents $\alpha_1, \alpha_2, \alpha_3, \alpha_4$. We can therefore write

$$\sigma\Omega_4 = \tau_{\langle\alpha_1\rangle} \cdots \tau_{\langle\alpha_4\rangle} \Omega_4$$

in such a way that each of the symmetries appearing in this equation is with respect to a line in U. But $O'_3(U) = \Omega_3(U)$. Hence $\sigma\Omega_4 = \Omega_4$. So $\sigma \in \Omega_4$, as required.

3) From now on we can assume that F is non-dyadic. Let Δ denote a fixed non-square unit in F. Take a typical σ in O'_4 and express it in the form

$$\sigma = \tau_{\langle\alpha_1\rangle} \cdots \tau_{\langle\alpha_4\rangle} \quad (\alpha_i \in \dot{F}) .$$

Here again $\alpha_1\alpha_2\alpha_3\alpha_4 \in \dot{F}^2$. If two of the α_i are in the same coset of $\dot{F}$ modulo $\dot{F}^2$, then so are the remaining two, hence

$$\sigma\Omega_4 = \tau_{\langle\alpha_i\rangle}\,\tau_{\langle\alpha_i\rangle}\,\tau_{\langle\alpha_j\rangle}\,\tau_{\langle\alpha_j\rangle}\,\Omega_4 = \Omega_4 .$$

Otherwise we can write

$$\sigma\Omega_4 = \tau_{\langle 1\rangle}\,\tau_{\langle\Delta\rangle}\,\tau_{\langle\pi\rangle}\,\tau_{\langle\pi\Delta\rangle}\,\Omega_4 .$$

Hence O_4' consists of at most two cosets of Ω_4, namely

$$\Omega_4 \text{ and } \tau_{\langle 1\rangle}\,\tau_{\langle \Delta\rangle}\,\tau_{\langle \pi\rangle}\,\tau_{\langle \pi\Delta\rangle}\,\Omega_4\,.$$

Hence $(O_4'\colon\Omega_4) \leqq 2$.

4) Finally we must prove that $(O_4'\colon\Omega_4) = 2$ for the non-dyadic case now under discussion. In order to do this it will be enough to produce an element of O_4' which is not in Ω_4. Since V is anisotropic there is exactly one $\mathfrak{o}$-maximal lattice on V by Theorem 91:1. Let L denote this lattice. By Proposition 63:17 we can take a lattice $L_1 \perp L_2$ on V with

$$L_1 \cong \langle 1\rangle \perp \langle -\Delta\rangle \quad\text{and}\quad L_2 \cong \langle \pi\rangle \perp \langle -\pi\Delta\rangle\,.$$

Then $L_1 \perp L_2 \subseteq L$. But L cannot be unimodular, for if it were the space V would be isotropic by Example 63:14. Hence by a volume argument

$$L = L_1 \perp L_2\,.$$

We shall need the lattice $L^{\mathfrak{p}}$ defined in § 82I. Note that

$$L^{\mathfrak{p}} = \mathfrak{p}L_1 \perp L_2 \supseteqq \mathfrak{p}L\,.$$

Since $L^{\mathfrak{p}}$ is an additive subgroup of L we can form the factor group $L/L^{\mathfrak{p}}$. Let $\bar{V}$ denote this factor group and let bar denote the natural group homomorphism

$$L \to \bar{V} = L/L^{\mathfrak{p}}\,.$$

Also let bar denote the natural ring homomorphism of $\mathfrak{o}$ onto the residue class field $\bar{F}$ of F. We put a scalar multiplication on $\bar{V}$ by defining

$$\bar{\alpha}\,\bar{x} = \overline{\alpha x} \quad (\alpha \in \mathfrak{o}\,,\quad x \in L)\,.$$

This is well-defined since $\mathfrak{p}L \subseteq L^{\mathfrak{p}}$, and it makes $\bar{V}$ into a vector space over $\bar{F}$. It is easily seen that any base for the lattice L_1 becomes a base for the vector space $\bar{V}$, hence $\dim \bar{V} = 2$. Put a symmetric bilinear form on $\bar{V}$ by defining

$$B(\bar{x}, \bar{y}) = \overline{B(x, y)} \quad \forall\, x, y \in L\,.$$

This is well-defined and it makes $\bar{V}$ into a regular binary space with discriminant $-\bar{\Delta}$. Each σ in O_4 is a unit of L by Proposition 91:5; and $\sigma L^{\mathfrak{p}} = L^{\mathfrak{p}}$ by definition of $L^{\mathfrak{p}}$; so σ induces a mapping $\bar{\sigma}\colon \bar{V} \rightarrowtail \bar{V}$ by means of the equation

$$\bar{\sigma}\,\bar{x} = \overline{\sigma x} \quad \forall\, x \in L;$$

it is easily seen that $\bar{\sigma}$ is an isometry of $\bar{V}$ onto $\bar{V}$; and that $\overline{\sigma_1\sigma_2} = \bar{\sigma}_1\,\bar{\sigma}_2$ for all σ_1, σ_2 in O_4. Hence we have a natural homomorphism

$$O_4(V) \rightarrowtail O_2(\bar{V})\,.$$

By Corollary 92:1b we can find vectors $x, y \in L_1$ and $u, v \in L_2$ with

$$Q(x) = 1,\quad Q(y) = \Delta,\quad Q(u) = \pi,\quad Q(v) = \pi\Delta\,.$$

Define

$$\sigma = \tau_x \tau_y \tau_u \tau_v \in O_4' .$$

We assert that σ is not in Ω_4. Once we have proved this we shall be through. Suppose if possible that $\sigma \in \Omega_4$. Then $\bar{\sigma} \in \Omega_2(\bar{V}) = O_2'(\bar{V})$. On the other hand it is easily seen from the defining equation of a symmetry that

$$\bar{\sigma} = \tau_{\bar{x}} \tau_{\bar{y}} .$$

Hence $\theta(\bar{\sigma}) = Q(\bar{x})\, Q(\bar{y}) = \bar{\Delta}$. But $\bar{\Delta}$ is not a square in $\bar{F}$ by the Local Square Theorem. So $\bar{\sigma}$ is not in $O_2'(\bar{V})$. This is absurd. Hence $(O_4':\Omega_4) = 2$ as required. **q. e. d.**

95:1a. *In the exceptional case*

$$\tau_{\langle 1\rangle} \tau_{\langle \Delta\rangle} \tau_{\langle \pi\rangle} \tau_{\langle \pi\Delta\rangle} \Omega_4 \quad \textit{and} \quad \Omega_4$$

are the two distinct cosets of O_4' *modulo* Ω_4. *Here* Δ *can be any non-square unit of* F.

95:2. *Let* V *be a regular* n*-ary quadratic space over a local field with* $n \geqq 3$, *and let* σ *be an element of* Ω_n *of the form* $\sigma = \tau_{\langle\alpha_1\rangle} \dots \tau_{\langle\alpha_r\rangle}$ *with* $r \leqq n$ *and all* α_i *in* $\dot{F}$. *Then there is a regular ternary subspace of* V *which represents* $\alpha_1, \dots, \alpha_r$.

Proof. If V is isotropic we take any regular ternary subspace which contains an isotropic vector of V. We may therefore assume that V is a quaternary anisotropic space and that $r = 4$. In the dyadic case we can use the argument used in step 2) of the proof of Proposition 95:1 to find a regular ternary subspace of V which represents $\alpha_1, \dots, \alpha_4$. So let us suppose that F is non-dyadic. If $\alpha_1, \dots, \alpha_4$ fell in distinct cosets of $\dot{F}$ modulo $\dot{F}^2$ we could not have σ in Ω_4, by Corollary 95:1a. Hence we can assume that $\alpha_1\alpha_2 \in \dot{F}^2$ and $\alpha_3\alpha_4 \in \dot{F}^2$, say. There is clearly a ternary (in fact a binary) subspace of V that represents α_1 and α_3. This space represents $\alpha_1, \dots, \alpha_4$. **q. e. d.**

If V is any regular quadratic space over a local field other than an anisotropic quaternary space over a non-dyadic local field, then the condition for $\Omega_n \cap Z_n$ to be $\{\pm 1_V\}$ is the same as the condition for -1_V to be in O_n', namely $dV = 1$ with n even. Let us settle the exceptional case. We claim that in the exceptional case we will have $\Omega_4 \cap Z_4 = \{\pm 1_V\}$ if and only if -1 is a non-square in F. For suppose that -1 is a non-square in F. Then $-\Delta \in \dot{F}^2$. Hence

$$V \cong \langle 1\rangle \perp \langle 1\rangle \perp \langle \pi\rangle \perp \langle \pi\rangle ,$$

therefore

$$-1_V = \tau_{\langle 1\rangle} \tau_{\langle 1\rangle} \tau_{\langle \pi\rangle} \tau_{\langle \pi\rangle} \in \Omega_4 ,$$

and so $\Omega_4 \cap Z_4 = \{\pm 1_V\}$. Conversely suppose that $-1_V \in \Omega_4$. If -1 were a square we would have

$$\tau_{\langle 1\rangle}\,\tau_{\langle\Delta\rangle}\,\tau_{\langle\pi\rangle}\,\tau_{\langle\pi\Delta\rangle}\,\Omega_4 = \tau_{\langle 1\rangle}\,\tau_{\langle-\Delta\rangle}\,\tau_{\langle\pi\rangle}\,\tau_{\langle-\pi\Delta\rangle}\,\Omega_4 = -1_V\,\Omega_4 = \Omega_4$$

and this would contradict Corollary 95:1a. So -1 is a non-square.

Finally we recall from Proposition 91:6 that the group O_n^+/O_n' has the form

$$O_n^+/O_n' \cong \dot{F}/\dot{F}^2$$

for $n \geqq 3$. The same applies if $n = 2$ with V isotropic. If $n = 2$ with V anisotropic, then one can use Example 63:15 to show that O_n^+/O_n' is isomorphic to a subgroup of index 2 in $\dot{F}/\dot{F}^2$.

We have therefore fully described the groups

$$\Omega_n \cap Z_n\,,\quad O_n'/\Omega_n\,,\quad O_n^+/O_n'$$

over local fields.

Chapter X

Integral Theory of Quadratic Forms over Global Fields

We conclude this book by introducing the genus and the spinor genus of a lattice on a quadratic space over a global field, and by studying the relation between these two new objects and the class. We shall use these relations to obtain sufficient conditions under which two lattices are in the same class.

We continue with the notation of Chapter VIII, except that the field F is now a global field and S is a Dedekind set of spots *which consists of almost all spots on* F. We let $\mathfrak{o}$ be the ring of integers $\mathfrak{o}(S)$, $\mathfrak{u}$ the group of units $\mathfrak{u}(S)$. As usual we let $\mathfrak{p}$ stand either for a prime spot in S or for the prime ideal which it determines in $\mathfrak{o}$. (There will be one exception to this notation: in § 101A we shall let F denote an arbitrary valuated field.) Ω_F or Ω will stand for the set of all non-trivial spots on F, $|\ |_{\mathfrak{p}}$ will be the normalized valuation on $F_{\mathfrak{p}}$ at a spot $\mathfrak{p}$ in Ω. If $\mathfrak{p}$ is discrete we let $\mathfrak{o}_{\mathfrak{p}}$, $\mathfrak{u}_{\mathfrak{p}}$, $\mathfrak{m}_{\mathfrak{p}}$ stand for the ring of integers, the group of units, and the maximal ideal of $F_{\mathfrak{p}}$ at $\mathfrak{p}$. $\mathfrak{a}_{\mathfrak{p}}$ will be the localization at $\mathfrak{p}$ of the fractional ideal $\mathfrak{a}$ of F at S.

V will be a regular non-zero n-ary quadratic space over F with symmetric bilinear form B and associated quadratic form Q. We shall consider lattices $L, K, \ldots$ on V, always with respect to the underlying set of spots S. We let $V_{\mathfrak{p}}$ denote a fixed localization of the quadratic space V at a spot $\mathfrak{p}$ in Ω. The lattice $L_{\mathfrak{p}}$ will be the localization of L in $V_{\mathfrak{p}}$ at any spot $\mathfrak{p}$ in S. The notation $O(V)$, $O^+(V), \ldots$ for the subgroups of the orthogonal group will be carried over from Chapters IV and V.

§ 101. Elementary properties of the orthogonal group over arithmetic fields

§ 101 A. The orthogonal group over valuated fields

In this subparagraph F denotes an arbitrary valuated field, not necessarily the global field F under discussion in this chapter. Let $|\ |$ or $|\ |_{\mathfrak{p}}$ be the given valuation on F, and let $\mathfrak{p}$ be the spot which it determines. V is an n-dimensional vector space over F. A base $x_1, \ldots, x_n$ is taken and fixed for V. The norms $\|\ \|$ which we are about to define are with respect to the same base $x_1, \ldots, x_n$ unless otherwise stated. Recall our earlier notation: $L_F(V)$ denotes the algebra of linear transformations of V into itself, and $M_n(F)$ denotes the algebra of $n \times n$ matrices over F.

Practically no proofs will be given here. All assertions can be verified either by inspection or by simple direct calculation.

First we define the norm on V. Given any vector x in V, express it in the form

$$x = \alpha_1 x_1 + \cdots + \alpha_n x_n \quad (\alpha_i \in F)$$

and then define the norm of x by the equation

$$\|x\|_{\mathfrak{p}} = \max_i |\alpha_i|_{\mathfrak{p}} .$$

Use $\|\ \|$ instead of $\|\ \|_{\mathfrak{p}}$ whenever convenient. So $\|\ \|$ is a real-valued function with the following properties:

(1) $\|x\| > 0$ if $x \in \dot{V}$, and $\|0\| = 0$

(2) $\|\alpha x\| = |\alpha|\ \|x\| \qquad \forall\ \alpha \in F, \quad x \in V$

(3) $\|x + y\| \leqq \|x\| + \|y\| \quad \forall\ x, y \in V$.

In other words $\|\ \|$ is a norm in the sense of § 11G. And we can make V into a metric (topological) space by defining the distance between the vectors x and y to be $\|x - y\|$. As usual,

$$|\|x\| - \|y\|| \leqq \|x - y\| .$$

In the case of a non-archimedean field we have

$$\|x + y\| \leqq \max(\|x\|, \|y\|) \quad \forall\ x, y \in V$$

with

$$\|x + y\| = \max(\|x\|, \|y\|) \quad \text{if} \quad \|x\| \neq \|y\| .$$

In particular, in the non-archimedean case there is a neighborhood of any given point $x_0 \neq 0$ throughout which

$$\|x\| = \|x_0\| .$$

Each of the mappings

$$(x, y) \rightarrow x + y \quad \text{of} \quad V \times V \quad \text{into } V,$$
$$x \rightarrow -x \quad \text{of} \quad V \quad \text{into } V,$$
$$(\alpha, x) \rightarrow \alpha x \quad \text{of} \quad F \times V \quad \text{into } V,$$

is continuous. This means, to use the language of topological groups, that V is a topological vector space over the topological field F. The map

$$(y_1, \ldots, y_r) \rightarrow y_1 + \cdots + y_r$$

of $V \times \cdots \times V$ into V is continuous. So is the map $x \rightarrow \|x\|$ of V into $\mathbf{R}$.

Now do the same thing with $L_F(V)$. Consider a typical σ in $L_F(V)$, write

$$\sigma x_j = \sum_i \alpha_{ij} x_i \quad (\alpha_{ij} \in F)$$

for $1 \leqq j \leqq n$ and define the norm of σ by the equation

$$\|\sigma\|_{\mathfrak{p}} = \max_{i,j} |\alpha_{ij}|_{\mathfrak{p}} = \max_j \|\sigma x_j\|_{\mathfrak{p}} .$$

Use $\| \;\|$ instead of $\| \;\|_{\mathfrak{p}}$ whenever convenient. Then $\| \;\|$ makes $L_F(V)$ into a normed vector space, i. e. we have

(1) $\|\sigma\| > 0$ if $\sigma \in L_F(V)$ with $\sigma \neq 0$, and $\|0\| = 0$
(2) $\|\alpha\sigma\| = |\alpha| \, \|\sigma\| \quad \forall\, \alpha \in F, \; \sigma \in L_F(V)$
(3) $\|\sigma + \tau\| \leqq \|\sigma\| + \|\tau\| \quad \forall\, \sigma, \tau \in L_F(V)$.

And $L_F(V)$ is provided with a metric topology in which the distance between σ and τ is defined to be $\|\sigma - \tau\|$. As usual,

$$|\|\sigma\| - \|\tau\|| \leqq \|\sigma - \tau\| .$$

In the case of a non-archimedean field we have

$$\|\sigma + \tau\| \leqq \max(\|\sigma\|, \|\tau\|) \quad \forall\, \sigma, \tau \in L_F(V)$$

with

$$\|\sigma + \tau\| = \max(\|\sigma\|, \|\tau\|) \quad \text{if} \quad \|\sigma\| \neq \|\tau\| .$$

In particular, in the non-archimedean case there is a neighborhood of any given $\sigma_0 \neq 0$ throughout which

$$\|\sigma\| = \|\sigma_0\| .$$

We again have continuity of addition, of taking negatives, and of scalar multiplication, so that $L_F(V)$ is also a topological vector space over the given topological field. All this parallels the discussion for V. But we also have multiplicative laws to consider. We find that

$$\|\sigma x\| \leqq \begin{cases} n \|\sigma\| \, \|x\| & \text{in general} \\ \|\sigma\| \, \|x\| & \text{if non-archimedean,} \end{cases}$$

for all σ in $L_F(V)$ and all x in V. Similarly for σ, τ in $L_F(V)$ we have

$$\|\sigma\tau\| \leqq \begin{cases} n\,\|\sigma\|\,\|\tau\| & \text{in general} \\ \|\sigma\|\,\|\tau\| & \text{if non-archimedean.} \end{cases}$$

Hence the mapping

$$(\sigma, \tau) \rightarrow \sigma\tau \quad \text{of} \quad L_F(V) \times L_F(V) \quad \text{into} \quad L_F(V)$$

is continuous. This makes $L_F(V)$ into a topological ring as well as a topological vector space (as the name suggests, a topological ring is a ring with a topology in which addition, the taking of negatives, and multiplication, are all continuous). The mappings

$$(\sigma_1, \ldots, \sigma_r) \rightarrow \sigma_1 + \cdots + \sigma_r$$

and

$$(\sigma_1, \ldots, \sigma_r) \rightarrow \sigma_1 \ldots \sigma_r$$

are continuous. So are the mappings

$$(\sigma, x) \rightarrow \sigma x \quad \text{of} \quad L_F(V) \times V \quad \text{into} \quad V\,,$$

$$\sigma \rightarrow \det \sigma \quad \text{of} \quad L_F(V) \quad \text{into} \quad F\,.$$

The continuity of the determinant map shows that the general linear group $GL(V)$ is an open subset of $L_F(V)$. If we restrict ourselves to $GL(V)$ we find that the mapping

$$\sigma \rightarrow \sigma^{-1} \quad \text{of} \quad GL(V) \quad \text{into} \quad GL(V)$$

is continuous.

We can introduce a norm $\|\ \|_{\mathfrak{p}}$ on the algebra of $n \times n$ matrices $M_n(F)$ by defining

$$\|(a_{ij})\|_{\mathfrak{p}} = \max_{i,j} |a_{ij}|_{\mathfrak{p}}$$

for a typical matrix (a_{ij}) over F. We shall write $\|a_{ij}\|_{\mathfrak{p}}$ or $\|a_{ij}\|$ instead of $\|(a_{ij})\|_{\mathfrak{p}}$. Note that the norm $\|\sigma\|$ of a linear transformation σ is equal to the norm of its matrix in the base $x_1, \ldots, x_n$.

101:1. Example. What happens to the norm under a change of base? Take a new base $x_1', \ldots, x_n'$ for V with

$$x_j' = \sum_i a_{ij} x_i \quad \text{and} \quad x_j = \sum_i b_{ij} x_i'\,.$$

Let $\|\ \|'$ denote the norm with respect to the base $x_1', \ldots, x_n'$. Then in general we have

$$\frac{\|x\|}{n\|a_{ij}\|} \leqq \|x\|' \leqq n\,\|b_{ij}\|\,\|x\|$$

and

$$\frac{\|\sigma\|}{n^2\|a_{ij}\|\,\|b_{ij}\|} \leqq \|\sigma\|' \leqq n^2\,\|a_{ij}\|\,\|b_{ij}\|\,\|\sigma\|$$

for any x in V and any σ in $L_F(V)$. In the non-archimedean case we have

$$\frac{\|x\|}{\|a_{ij}\|} \leqq \|x\|' \leqq \|b_{ij}\| \, \|x\|$$

and

$$\frac{\|\sigma\|}{\|a_{ij}\| \, \|b_{ij}\|} \leqq \|\sigma\|' \leqq \|a_{ij}\| \, \|b_{ij}\| \, \|\sigma\| .$$

To conclude we suppose that V has been made into a regular quadratic space by providing it with the symmetric bilinear form B and the quadratic form Q. Then there is a positive constant λ such that

$$|B(x, y)| \leqq \lambda \|x\| \, \|y\| \quad \forall\, x, y \in V ,$$

so

$$|Q(x)| \leqq \lambda \|x\|^2 \quad \forall\, x \in V .$$

The mappings

$$(x, y) \rightarrowtail B(x, y) \quad \text{of} \quad V \times V \quad \text{into } F ,$$

$$x \rightarrowtail Q(x) \quad \text{of} \quad V \quad \text{into } F ,$$

are continuous.

101:2. **Example.** The continuity of the map $x \rightarrowtail Q(x)$ shows that the set of anisotropic vectors of V is an open subset of V. Let u denote an anisotropic vector of V. Then the mapping

$$u \rightarrowtail \tau_u$$

of the set of anisotropic vectors of V into $O_n(V)$ is continuous (here τ_u denotes the symmetry of V with respect to u). To prove this one considers the defining equation

$$\tau_u x = x - \frac{2B(u, x)}{Q(u)} u$$

of a symmetry. First one shows, using the continuity of the maps $u \rightarrowtail 2B(u, x)$, $u \rightarrowtail Q(u)$, etc., that the mapping $u \rightarrowtail \tau_u x$ is continuous for each fixed x in V. One then deduces the continuity of $u \rightarrowtail \tau_u$ at u_0 from the equation

$$\|\tau_u - \tau_{u_0}\| = \max_j \|\tau_u x_j - \tau_{u_0} x_j\| .$$

Hence $u \rightarrowtail \tau_u$ is continuous.

101:3. **Example.** The continuity of the determinant map tells us that $O^+(V)$ and $O^-(V)$ are closed subsets of $O(V)$. Hence $O^+(V)$ is an open and closed subgroup of $O(V)$.

101:4. **Example.** (i) Suppose that the field F under discussion is actually a local field. Then for any σ in $O(V)$ we have $\det\sigma = \pm 1$, hence $\det\sigma$ is a unit, hence $\|\sigma\| \geqq 1$. Now let M be the lattice $M = \mathfrak{o} x_1 + \cdots + \mathfrak{o} x_n$, where $x_1, \ldots, x_n$ is the base used in defining $\| \ \|$. Then $\|\sigma\| = 1$ if and only if $\|\sigma\| \leqq 1$, this is equivalent to $\sigma M \subseteq M$, and hence to $\sigma M = M$. So the elements of $O(M)$ are precisely the isometries of V with $\|\sigma\| = 1$. In particular the set of isometries σ with $\|\sigma\| = 1$ is a group.

(ii) Consider a second lattice L on V. We claim that $\sigma L = L$ holds for all σ in $O(V)$ which are sufficiently close to 1_V. Take a base $x_1', \ldots, x_n'$ for L, and let $\| \ \|'$ denote norms with respect to this new base. Then by Example 101:1 we see that all σ which are sufficiently close to 1_V satisfy $\|\sigma - 1_V\|' < 1$. Each such σ satisfies $\|\sigma\|' = 1$, hence $\sigma L = L$.

(iii) Consider a third lattice K on V, and suppose that $K = \lambda L$ for some λ in $O(V)$. We claim that $\sigma L = K$ holds for all σ in $O(V)$ which are sufficiently close to λ. By choosing σ sufficiently close to λ we can make

$$\|\lambda^{-1}\sigma - 1_V\| \leqq \|\lambda^{-1}\| \, \|\sigma - \lambda\|$$

arbitrarily small. But all $\lambda^{-1}\sigma$ which are sufficiently close to 1_V make $\lambda^{-1}\sigma L = L$. Hence all σ which are sufficiently close to λ make $\sigma L = \lambda L = K$.

§ 101B. The orthogonal group over global fields

We return to the situation described at the beginning of the chapter: quadratic forms over global fields. F is again a global field and V is a quadratic space over F. Since each $V_\mathfrak{p}$ is a vector space over the valuated field $F_\mathfrak{p}$ we can introduce norms $\| \ \|_\mathfrak{p}$ on $V_\mathfrak{p}$ and $L_{F_\mathfrak{p}}(V_\mathfrak{p})$ with respect to any given base of $V_\mathfrak{p}$, in particular with respect to any given base for V over F. We shall *always assume that all norms under discussion are with respect to a common base for* V. We let $x_1, \ldots, x_n$ denote the vectors of this base. If a new base $x_1', \ldots, x_n'$ is taken for V, and if we consider the corresponding norms $\| \ \|_\mathfrak{p}'$ at each $\mathfrak{p}$ on F, then it follows from Example 101:1 and the Product Formula of § 33B that we have

$$\| \ \|_\mathfrak{p} = \| \ \|_\mathfrak{p}' \qquad \text{for almost all } \mathfrak{p}.$$

Consider a typical linear transformation σ in $L_F(V)$. By considering the effect of σ on a base for V we see that there is a unique linear transformation $\sigma_\mathfrak{p}$ on $V_\mathfrak{p}$ which induces σ on V. We call $\sigma_\mathfrak{p}$ the $\mathfrak{p}$-ification or localization at $\mathfrak{p}$ of σ. It is easily verified, again by considering a base for V, that we have the rules

$$(\sigma + \tau)_\mathfrak{p} = \sigma_\mathfrak{p} + \tau_\mathfrak{p}\,, \quad (\sigma\tau)_\mathfrak{p} = \sigma_\mathfrak{p}\tau_\mathfrak{p}$$
$$(\alpha\sigma)_\mathfrak{p} = \alpha\sigma_\mathfrak{p}\,, \quad \det\sigma_\mathfrak{p} = \det\sigma$$

for all σ, τ in $L_F(V)$ and all α in F. In particular, the mapping $\sigma \rightarrow \sigma_\mathfrak{p}$ gives us an injective ring homomorphism

$$L_F(V) \rightarrowtail L_{F_\mathfrak{p}}(V_\mathfrak{p})\,.$$

If σ is an isometry, then so is $\sigma_\mathfrak{p}$. If σ is a rotation, then so is $\sigma_\mathfrak{p}$. If τ_u is the symmetry of V with respect to the anisotropic line Fu, then a geometric argument shows that $(\tau_u)_\mathfrak{p}$ is the symmetry of $V_\mathfrak{p}$ with respect to the line $F_\mathfrak{p}u$; we express this symbolically by the equation

$$(\tau_u)_\mathfrak{p} = \tau_u\,.$$

In keeping with functional notation we let $A_\mathfrak{p}$ denote the image of a subset A of $L_F(V)$ in $L_{F_\mathfrak{p}}(V_\mathfrak{p})$ under localization at $\mathfrak{p}$. We have

$$\begin{cases} O_n(V)_\mathfrak{p} \subseteq O_n(V_\mathfrak{p}), & O_n^+(V)_\mathfrak{p} \subseteq O_n^+(V_\mathfrak{p}), \\ \Omega_n(V)_\mathfrak{p} \subseteq \Omega_n(V_\mathfrak{p}), & O_n'(V)_\mathfrak{p} \subseteq O_n'(V_\mathfrak{p}). \end{cases}$$

101:5. **Conventions.** In some situations things become clearer if the notation is relaxed and σ is used for the localization $\sigma_\mathfrak{p}$ of σ, in other situations the strict notation $\sigma_\mathfrak{p}$ is preferable. We shall use both. We shall also use $\varphi_\mathfrak{p}$ (or $\sigma_\mathfrak{p}$) to denote a typical element of $L_{F_\mathfrak{p}}(V_\mathfrak{p})$; of course this does not necessarily mean that $\varphi_\mathfrak{p}$ is the localization of a linear transformation φ of V.

101:6. **Example.** Let S be a Dedekind set of spots on F, let L be a lattice in the vector space V over F, and let σ be an element of $L_F(V)$. We claim that

$$\sigma_\mathfrak{p} L_\mathfrak{p} = (\sigma L)_\mathfrak{p} \quad \forall\, \mathfrak{p} \in S.$$

To prove this we express L in the form

$$L = \mathfrak{a}_1 y_1 + \cdots + \mathfrak{a}_r y_r$$

where the $\mathfrak{a}_i$ are fractional ideals and the y_i are elements of V. Then

$$\begin{aligned} \sigma_\mathfrak{p} L_\mathfrak{p} &= \sigma_\mathfrak{p}(\mathfrak{a}_{1\mathfrak{p}} y_1 + \cdots + \mathfrak{a}_{r\mathfrak{p}} y_r) \\ &= \mathfrak{a}_{1\mathfrak{p}}(\sigma y_1) + \cdots + \mathfrak{a}_{r\mathfrak{p}}(\sigma y_r) \\ &= (\mathfrak{a}_1(\sigma y_1) + \cdots + \mathfrak{a}_r(\sigma y_r))_\mathfrak{p} \\ &= (\sigma L)_\mathfrak{p}. \end{aligned}$$

This proves our claim. As an immediate consequence we have

$$O_n(L)_\mathfrak{p} \subseteq O_n(L_\mathfrak{p}) \quad \text{and} \quad O_n^+(L)_\mathfrak{p} \subseteq O_n^+(L_\mathfrak{p}),$$

for all $\mathfrak{p}$ in S.

101:7. **Weak Approximation Theorem for Rotations.** *Let V be a regular quadratic space over a global field F and let T be a finite set of spots on F. Suppose $\varphi_\mathfrak{p}$ is given in $O^+(V_\mathfrak{p})$ at each $\mathfrak{p}$ in T. Then for each $\varepsilon > 0$ there is a σ in $O^+(V)$ such that*

$$\|\sigma - \varphi_\mathfrak{p}\|_\mathfrak{p} < \varepsilon \quad \forall\, \mathfrak{p} \in T.$$

Proof. Express each $\varphi_\mathfrak{p}$ as a product of symmetries,

$$\varphi_\mathfrak{p} = \tau_{u_1^\mathfrak{p}} \cdots \tau_{u_r^\mathfrak{p}},$$

where the $u_i^\mathfrak{p}$ are anisotropic vectors in $V_\mathfrak{p}$. Here the number r is even, and we can suppose that the same r applies for all $\mathfrak{p}$ by adjoining squares of symmetries wherever necessary. Using the Weak Approximation Theorem of § 11E on the coordinates of the vectors $u_1^\mathfrak{p}$ in the underlying base $x_1, \ldots, x_n$ we can obtain a vector u_1 in V such that $\|u_1 - u_1^\mathfrak{p}\|_\mathfrak{p}$ is arbitrarily small at all $\mathfrak{p}$ in T. If the approximation is good enough we obtain an anisotropic vector u_1 such that τ_{u_1} is arbitrarily close to $\tau_{u_1^{\mathfrak{p}}}$ at each $\mathfrak{p}$ in T, in virtue of the continuity of the map $u \rightarrow \tau_u$. Now do all this for $i = 1, 2, \ldots, r$ and in this way obtain anisotropic vectors $u_1, \ldots, u_r$ of V with

$$\|\tau_{u_i} - \tau_{u_i^{\mathfrak{p}}}\|_{\mathfrak{p}}$$

arbitrarily small for all $\mathfrak{p}$ in T and for $1 \leqq i \leqq r$. If good enough approximations are taken all around we can arrange to have

$$\|\tau_{u_1} \ldots \tau_{u_r} - \tau_{u_1^{\mathfrak{p}}} \ldots \tau_{u_r^{\mathfrak{p}}}\|_{\mathfrak{p}} < \varepsilon$$

at all $\mathfrak{p}$ in T, in virtue of the continuity of multiplication in $L_{F_{\mathfrak{p}}}(V_{\mathfrak{p}})$. Put $\sigma = \tau_{u_1} \ldots \tau_{u_r} \in O^+(V)$. Then $\|\sigma_{\mathfrak{p}} - \varphi_{\mathfrak{p}}\|_{\mathfrak{p}} < \varepsilon$ for all $\mathfrak{p}$ in T or, in the relaxed notation, $\|\sigma - \varphi_{\mathfrak{p}}\|_{\mathfrak{p}} < \varepsilon$. **q. e. d.**

101:8. *Let V be a regular quadratic space over a global field F with $\dim V \geqq 3$. Then $\theta(O^+(V))$ consists of the set of elements of $\dot{F}$ which are positive at all real spots $\mathfrak{p}$ at which $V_{\mathfrak{p}}$ is anisotropic.*

Proof. Let R denote the set of real spots $\mathfrak{p}$ of F at which $V_{\mathfrak{p}}$ is anisotropic. (Needless to say, R may be empty.) Every element of $\theta(O^+(V))$ is a product of elements of the form

$$Q(y_1) \ldots Q(y_{2r})$$

where the y_i are anisotropic vectors of V. Each $Q(y_i)$ is negative at those $\mathfrak{p}$ in R at which $V_{\mathfrak{p}}$ is negative definite, hence the above product is positive at these $\mathfrak{p}$; the same applies at the spots $\mathfrak{p}$ of R at which $V_{\mathfrak{p}}$ is positive definite; hence the above product is positive at all $\mathfrak{p}$ in R.

Conversely, let α be any element of $\dot{F}$ which is positive at all $\mathfrak{p}$ in R. First let $n \geqq 4$. Take any fixed non-zero element β in $Q(V)$. Then $\alpha\beta$ is represented by $V_{\mathfrak{p}}$ at all $\mathfrak{p}$ in R; and $V_{\mathfrak{p}}$ is universal at all remaining spots on F; hence $\alpha\beta$ is represented by $V_{\mathfrak{p}}$ at all $\mathfrak{p}$ on F; hence $\alpha\beta$ is represented by V, by Theorem 66:3. Hence $\alpha\beta^2 = (\alpha\beta)\beta$ is in $\theta(O^+(V))$, hence α is.

Now let $n = 3$. Scaling V by its discriminant shows that we can assume that the discriminant dV is actually equal to 1. Then $V_{\mathfrak{p}}$ is positive definite at all $\mathfrak{p}$ in R. Let T denote the set of discrete spots $\mathfrak{p}$ on F at which $V_{\mathfrak{p}}$ is anisotropic. So $R \cup T$ consists of all the spots $\mathfrak{p}$ on F at which $V_{\mathfrak{p}}$ is anisotropic. Use the Weak Approximation Theorem of § 11E to find an element β of F which is positive at all $\mathfrak{p}$ in R and which has the following property: both $-\beta$ and $-\alpha\beta$ are non-squares at each $\mathfrak{p}$ in T. Then by Proposition 63:17 we see that $\langle -\beta\rangle \perp V$ is isotropic at all $\mathfrak{p}$ in T, it is also isotropic at all $\mathfrak{p}$ in R, hence it is isotropic at all $\mathfrak{p}$, hence it is isotropic, hence V represents β. Similarly V represents $\alpha\beta$. Then $\alpha\beta^2 = (\alpha\beta)\beta$ is in $\theta(O^+(V))$, hence α is. **q. e. d.**

101:8a. *If F is a function field, then*

$$\theta(O^+(V)) = \dot{F}\,.$$

101:9. *Let V be a regular quaternary quadratic space over a global field F. Consider a set of spots T on F such that $\Omega_4(V_\mathfrak{p}) \subset O_4'(V_\mathfrak{p})$ for all $\mathfrak{p}$ in T. Then there is a σ in $O_4'(V)$ such that $\sigma_\mathfrak{p} \notin \Omega_4(V_\mathfrak{p})$ for all $\mathfrak{p}$ in T and $\sigma_\mathfrak{p} \in \Omega_4(V_\mathfrak{p})$ at all remaining $\mathfrak{p}$ on F.*

Proof. By scaling V we can assume that $1 \in Q(V)$. The set T is a subset of the set of non-dyadic spots $\mathfrak{p}$ at which $V_\mathfrak{p}$ is anisotropic. Hence T is a finite set. Let W denote the remaining set of spots $\mathfrak{p}$ (archimedean or discrete) at which $V_\mathfrak{p}$ is anisotropic. Again W is a finite set, possibly empty. Using the Weak Approximation Theorem and the fact that $\dot{F}_\mathfrak{p}^2$ is open in $F_\mathfrak{p}$ we can find an element Δ in F which is a non-square unit at all $\mathfrak{p}$ in T and a square at all $\mathfrak{p}$ in W. Similarly we can find an element π in F which is a prime element at all $\mathfrak{p}$ in T and a square at all $\mathfrak{p}$ in W. Then Δ is represented by $V_\mathfrak{p}$ at each $\mathfrak{p}$ in T since a regular quaternary space over a local field is universal; and Δ is represented by $V_\mathfrak{p}$ at each $\mathfrak{p}$ in W since Δ is a square there and 1 is in $Q(V)$; and Δ is represented by $V_\mathfrak{p}$ at all remaining spots $\mathfrak{p}$ on F since $V_\mathfrak{p}$ is then isotropic by definition of T and W. Hence $\Delta \in Q(V)$ by Theorem 66:3. Similarly $\pi \in Q(V)$. Similarly $\pi\Delta \in Q(V)$. Hence

$$1,\ \Delta,\ \pi,\ \pi\Delta \ \in\ Q(V)\,.$$

There must therefore exists a σ in $O_4(V)$ of the form

$$\sigma = \tau_{\langle 1\rangle}\,\tau_{\langle \Delta\rangle}\,\tau_{\langle \pi\rangle}\,\tau_{\langle \pi\Delta\rangle}\,.$$

(Here, as in § 95, $\tau_{\langle\alpha\rangle}$ denotes a symmetry with respect to a vector x with $Q(x) = \alpha$.) Then σ is in $O_4'(V)$ by definition of $O_4'(V)$. And $\sigma_\mathfrak{p} \notin \Omega_4(V_\mathfrak{p})$ at each $\mathfrak{p}$ in T by Corollary 95:1a. And $\sigma_\mathfrak{p} \in \Omega_4(V_\mathfrak{p})$ at each $\mathfrak{p}$ in W since we then have

$$\sigma_\mathfrak{p} = \tau_{\langle 1\rangle}\,\tau_{\langle 1\rangle}\,\tau_{\langle 1\rangle}\,\tau_{\langle 1\rangle} \in \Omega_4(V_\mathfrak{p})\,.$$

And $V_\mathfrak{p}$ is isotropic at each of the remaining spots $\mathfrak{p}$ on F, hence $\Omega_4(V_\mathfrak{p}) = O_4'(V_\mathfrak{p})$, hence $\sigma_\mathfrak{p} \in \Omega_4(V_\mathfrak{p})$. **q. e. d.**

101:10. **Example.** V is a regular n-ary space over the global field F and S is a Dedekind set consisting of almost all spots on F. Consider lattices $L, K, \ldots$ on V with respect to S. We always have $\mathrm{cls}^+L = \mathrm{cls}\,L$ when n is odd (see Example 82:4). This is also true for even n over local fields (see Corollary 91:4a). It is not true in general for global fields. In fact we claim that there is a lattice K on V with $\mathrm{cls}^+K \subset \mathrm{cls}\,K$ whenever n is even. Let us construct such a K. We start with an arbitrary lattice L on V. Write $dV = \varepsilon$ with ε in $\dot{F}$. We know from Example 66:6 that the Hasse symbol $S_\mathfrak{p}V$ is 1 at almost all $\mathfrak{p}$ on F. Let W denote a set of non-dyadic spots on F which consists of almost all spots in S, such that ε is a W-unit and $S_{\mathfrak{p}} V = 1$ at all $\mathfrak{p}$ in W, such that $J_F = P_F J_F^W$, and finally such that $L_{\mathfrak{p}}$ is unimodular at all $\mathfrak{p}$ in W. Pick spots $\mathfrak{q}$ and $\mathfrak{q}'$ in W such that ε is a square at $\mathfrak{q}$ and $\mathfrak{q}'$, and such that a W-unit is a square at $\mathfrak{q}$ if and only if it is a square at $\mathfrak{q}'$. The existence of such a pair of spots was established in Proposition 65:20. By Proposition 81:14 and Example 92:8 there is a lattice K on V with $K_{\mathfrak{p}} = L_{\mathfrak{p}}$ whenever $\mathfrak{p} \in S \dot{-} (\mathfrak{q} \cup \mathfrak{q}')$ and

$$\theta(O^-(K_{\mathfrak{q}})) = \Delta_{\mathfrak{q}} \dot{F}_{\mathfrak{q}}^2, \quad \theta(O^-(K_{\mathfrak{q}'})) = \dot{F}_{\mathfrak{q}'}^2,$$

where $\Delta_{\mathfrak{q}}$ is a non-square unit in $F_{\mathfrak{q}}$. We say that this K has the desired properties, i. e. $\text{cls}^+ K \subset \text{cls} K$, i. e. $O^+(K) = O(K)$. For suppose not. Then there is a reflexion σ of V such that $\sigma K = K$. Write $\theta(\sigma) = \alpha$ with α in $\dot{F}$. If $\mathfrak{p} \in W - (\mathfrak{q} \cup \mathfrak{q}')$, then

$$\alpha \in \theta(O(L_{\mathfrak{p}})) = \mathfrak{u}_{\mathfrak{p}} \dot{F}_{\mathfrak{p}}^2 .$$

Similarly, $\alpha \in \Delta_{\mathfrak{q}} \dot{F}_{\mathfrak{q}}^2$ and $\alpha \in \dot{F}_{\mathfrak{q}'}^2$. So α is a non-square at $\mathfrak{q}$, it is a square at $\mathfrak{q}'$, and $\text{ord}_{\mathfrak{p}} \alpha$ is even at all $\mathfrak{p}$ in W. Since $J_F = P_F J_F^W$ there is a β in $\dot{F}$ with $2\, \text{ord}_{\mathfrak{p}} \beta = \text{ord}_{\mathfrak{p}} \alpha$ at all $\mathfrak{p}$ in W. Then α/β^2 is a W-unit which is a square at $\mathfrak{q}'$ but not at $\mathfrak{q}$. This is impossible by choice of $\mathfrak{q}$ and $\mathfrak{q}'$. So $O^+(K) = O(K)$. Hence $\text{cls}^+ K \subset \text{cls} K$.

§ 101 C. Special subgroups of $O_n(V)$

Once again we return to the questions raised in § 56. This time we must describe the groups

$$\Omega_n \cap Z_n, \quad O_n'/\Omega_n, \quad O_n^+/O_n'$$

over the global field F which is now under discussion[1]. Of course if F is a function field and $n \geqq 5$, then V is isotropic and all is already known. However, some extra effort will be needed before we obtain the general case.

101:11. **Lemma.** *Let X be a finite subset of $\dot{F}$ and suppose that at each spot $\mathfrak{p}$ on F there is a regular ternary subspace of $V_{\mathfrak{p}}$ which represents all of X. Then there is a regular ternary subspace of V which represents all of X.*

Proof. 1) For $n = 3$ the result is an immediate consequence of Theorem 66:3. Suppose $n = 4$. Let T be a finite set of spots on F which contains all archimedean and dyadic spots and is such that, at each spot $\mathfrak{p}$ outside T, the discriminant $d V_{\mathfrak{p}}$ is a unit, the Hasse symbol $S_{\mathfrak{p}} V_{\mathfrak{p}}$ is 1, and every element of X is a unit. By hypothesis there is a scalar $\gamma_{\mathfrak{p}} \in \dot{F}_{\mathfrak{p}}$ and a regular ternary space $U_{\mathfrak{p}}$ at each $\mathfrak{p}$ in T such that

$$V_{\mathfrak{p}} = \langle \gamma_{\mathfrak{p}} \rangle \perp U_{\mathfrak{p}}, \quad X \subseteq Q(U_{\mathfrak{p}}) .$$

[1] For the structure of $\Omega_n/\Omega_n \cap Z_n$ ($n \geqq 5$) over an algebraic number field see M. Kneser, *Crelle's J.*, **196** (1956), pp. 213—220.

By the Weak Approximation Theorem and fact that $\dot{F}_{\mathfrak{p}}^2$ is open in $F_{\mathfrak{p}}$ there is a γ in $\dot{F}$ such that $\gamma \in \gamma_{\mathfrak{p}} \dot{F}_{\mathfrak{p}}^2$ for all $\mathfrak{p}$ in T. Then

$$V_{\mathfrak{p}} = \langle\gamma\rangle \perp U_{\mathfrak{p}} \quad \forall\, \mathfrak{p} \in T\,.$$

Now $V_{\mathfrak{p}}$ represents γ at each discrete spots $\mathfrak{p}$ on F by Remark 63:18, hence V represents γ by Theorem 66:3, hence there is a regular ternary subspace W of V and a splitting $V = \langle\gamma\rangle \perp W$. At each $\mathfrak{p}$ in T we have $W_{\mathfrak{p}} \cong U_{\mathfrak{p}}$ by Witt's theorem, and so $X \subseteq Q(W_{\mathfrak{p}})$; at each of the remaining $\mathfrak{p}$ on F we have $X \subseteq Q(W_{\mathfrak{p}})$ by Example 63:24; hence $X \subseteq Q(W)$. This proves the case $n = 4$.

2) Now the case $n \geq 5$. Here let T be any finite set of spots on F which contains all archimedean spots and is such that every element of X is a unit at all spots outside T. Using the Weak Approximation Theorem one can easily construct a regular binary space U over F which is isotropic at each $\mathfrak{p}$ in T at which $V_{\mathfrak{p}}$ is isotropic, which is positive definite at each real $\mathfrak{p}$ at which $V_{\mathfrak{p}}$ is positive definite, and which is negative definite at each real $\mathfrak{p}$ at which $V_{\mathfrak{p}}$ is negative definite. Then there is a representation $U_{\mathfrak{p}} \rightarrowtail V_{\mathfrak{p}}$ at each $\mathfrak{p}$ on F by Theorem 63:21, hence there is a representation $U \rightarrowtail V$ by Theorem 66:3. In effect this allows us to assume that $U \subseteq V$. Let W be any regular quaternary subspace of V which contains U. Then at each $\mathfrak{p}$ in T we have $X \subseteq Q(U_{\mathfrak{p}})$; and at each remaining spot $\mathfrak{p}$ on F there will be a regular binary subspace of $W_{\mathfrak{p}}$ which represents all of $\mathfrak{u}_{\mathfrak{p}}$ (Example 63:15 and Proposition 63:17) and hence all of X. Hence there is a regular ternary (in fact binary) subspace of $W_{\mathfrak{p}}$ at each $\mathfrak{p}$ on F which represents all of X. Hence by step 1) there is a regular ternary subspace of W (and hence of V) which represents all X. **q. e. d.**

101:12. *Let V be a regular n-ary quadratic space over the global field F and let σ be an element of $O_n^+(V)$. Then σ is in $\Omega_n(V)$ if and only if $\sigma_{\mathfrak{p}}$ is in $\Omega_n(V_{\mathfrak{p}})$ at each spot $\mathfrak{p}$ on F.*

Proof. Only the sufficiency really needs proof. So let us assume that $\sigma_{\mathfrak{p}} \in \Omega_n(V_{\mathfrak{p}})$ at each $\mathfrak{p}$ on F and let us deduce from this that $\sigma \in \Omega_n(V)$. Express σ as a product of symmetries with respect to vectors $y_1, \ldots, y_r$ of V:

$$\sigma = \tau_{y_1} \cdots \tau_{y_r} \quad (r \leq n)\,.$$

Then $\sigma \in O_n'(V_{\mathfrak{p}})$ at each $\mathfrak{p}$, hence $Q(y_1) \ldots Q(y_r) \in \dot{F}_{\mathfrak{p}}^2$, hence by the Global Square Theorem we must have

$$Q(y_1) \ldots Q(y_r) \in \dot{F}^2.$$

So $\sigma \in O_n'(V)$. In particular this proves the proposition when $1 \leq n \leq 3$. Let us now assume that $n \geq 4$. Since $\sigma_{\mathfrak{p}} \in \Omega_n(V_{\mathfrak{p}})$ we can conclude from Proposition 95:2 at the discrete spots and from the archimedean theory of quadratic forms at the archimedean spots that there is a regular ternary subspace of $V_\mathfrak{p}$ at each $\mathfrak{p}$ on F which represents all the scalars $Q(y_1), \ldots, Q(y_r)$. By Lemma 101:11 there is a regular ternary subspace W of V which represents all these scalars. Pick $z_i \in W$ with $Q(z_i) = Q(y_i)$ for $1 \leqq i \leqq r$. By Proposition 55:3 it is enough to prove that $\tau_{z_1} \ldots \tau_{z_r}$ is in $\Omega_n(V)$. And this follows easily from the fact that $\tau_{z_1} \ldots \tau_{z_r}$ induces an element of $\Omega_3(W) = O_3(W)$ on W. **q. e. d.**

Now we can describe the groups $\Omega_n \cap Z_n$, etc. We have $\Omega_n \cap Z_n = \{\pm 1_V\}$ if $-1_{V_\mathfrak{p}}$ is in $\Omega_n(V_\mathfrak{p})$ at each $\mathfrak{p}$ on F, otherwise $\Omega_n \cap Z_n = 1_V$. The group O'_n/Ω_n is isomorphic to the group

$$\prod_\mathfrak{p} O'_n(V_\mathfrak{p})/\Omega_n(V_\mathfrak{p})$$

(this follows easily from Propositions 101:9 and 101:12). In particular $O'_n = \Omega_n$ when $n \neq 4$, while if $n = 4$ the group O'_4/Ω_4 is a direct product of a finite number of groups of order 2. Finally the group O^+_n/O'_n can be described by Proposition 101:8 (for $n \geqq 3$): it is the group of elements which are positive at all real spots $\mathfrak{p}$ at which $V_\mathfrak{p}$ is anisotropic, modulo the group $\dot{F}^2$.

§ 101 D. The group of split rotations J_V

We have to work with idèle groups again, particularly with the groups

$$J_F,\ P_F,\ J_F^S,\ P_F^S,\ J_F^{S,2}$$

defined in §§ 33 D, 33 I and 65 A. Here we are assuming that S is a Dedekind set consisting of almost all spots on the global field F. The idèle concept can be extended to the orthogonal group in the following way. Start with a regular quadratic space V over the global field F. Take a base $x_1, \ldots, x_n$ for V and let all the norms $\| \ \|_\mathfrak{p}$ on $V_\mathfrak{p}$ and $L_{F_\mathfrak{p}}(V_\mathfrak{p})$ be with respect to this fixed underlying base. Consider the multiplicative group

$$\prod_{\mathfrak{p} \in \Omega} O^+(V_\mathfrak{p})$$

consisting of the direct product of all the groups $O^+(V_\mathfrak{p})$. A typical element of this group is defined by its coordinates, say

$$\Sigma = (\Sigma_\mathfrak{p})_{\mathfrak{p} \in \Omega} \qquad (\Sigma_\mathfrak{p} \in O^+(V_\mathfrak{p}))\,,$$

and multiplication in the direct product is, by definition, coordinatewise. If we are just told that Σ is a typical element of the direct product, then $\Sigma_\mathfrak{p}$ will denote its $\mathfrak{p}$-coordinate. For two such elements Σ, Λ we have

$$(\Sigma\Lambda)_\mathfrak{p} = \Sigma_\mathfrak{p}\Lambda_\mathfrak{p}\,, \quad (\Sigma^{-1})_\mathfrak{p} = \Sigma_\mathfrak{p}^{-1}\,,$$

for all $\mathfrak{p}$ in Ω. We shall call Σ a split rotation of the quadratic space V if Σ is an element of the above direct product with the property

$$\|\Sigma_\mathfrak{p}\|_\mathfrak{p} = 1 \quad \text{for almost all } \mathfrak{p}.$$

This definition is independent of the underlying base since any two systems of norms agree at almost all $\mathfrak{p}$ by Example 101:1. The set of all split rotations is a subgroup of the above direct product by Example 101:4, it is called the group of split rotations, and it will be written J_V.[1] The set of all split rotations Σ with the property

$$\Sigma_{\mathfrak{p}} \in O'(V_{\mathfrak{p}}) \quad \forall\, \mathfrak{p} \in \Omega$$

is clearly a subgroup of J_V; we shall denote this subgroup with the letter J'_V. It is evident that J'_V contains the commutator subgroup of J_V; in particular, J'_V is a normal subgroup of J_V and the quotient group J_V/J'_V is abelian.

Consider a typical element σ of $O^+(V)$. Then σ has a localization $\sigma_{\mathfrak{p}}$ at each $\mathfrak{p}$ in Ω. And $\|\sigma_{\mathfrak{p}}\|_{\mathfrak{p}} = 1$ for almost all $\mathfrak{p}$, by the Product Formula. Hence σ determines a split rotation

$$(\sigma) = (\sigma_{\mathfrak{p}})_{\mathfrak{p} \in \Omega}\,.$$

The rule $\sigma \rightarrowtail (\sigma)$ therefore provides a natural multiplicative isomorphism of $O^+(V)$ into J_V. We shall call the split rotation Σ principal if there is a rotation σ such that $\Sigma = (\sigma)$. The principal split rotations form a subgroup P_V of J_V. And we have $O^+(V) \rightarrowtail P_V$.

We shall let D stand for the subgroup $\theta(O^+(V))$ of $\dot{F}$, and we let P_D be the group of principal idèles of the form $(\alpha)_{\mathfrak{p} \in \Omega}$ with α in D. In other words, P_D is the image of D under the natural isomorphism $\dot{F} \rightarrowtail P_F$.

101:13. **Example.** Suppose $n \geqq 3$. Proposition 101:8 tells us that D is then the set of all elements of $\dot{F}$ which are positive at those real spots $\mathfrak{p}$ at which $V_{\mathfrak{p}}$ is anisotropic. The Weak Approximation Theorem of § 11E shows that $(\dot{F} : D)$ is finite. Hence $(P_F : P_D) < \infty$. We know from Corollary 33:14a that there is a Dedekind set of spots S_0 which consists of almost all spots on F and is such that $J_F = P_F J_F^S$ whenever $S \subseteq S_0$. By considering a finite set of representatives of $P_F \bmod P_D$ we see that there is a set S_0 of the above type such that

$$J_F = P_D J_F^S \qquad \text{whenever } S \subseteq S_0.$$

Now consider lattices $L, K, \ldots$ on V with respect to the set of spots S under discussion. We define the subgroup J_L of J_V by the equation

$$J_L = \{\Sigma \in J_V \mid \Sigma_{\mathfrak{p}} \in O^+(L_{\mathfrak{p}}) \quad \forall\, \mathfrak{p} \in S\}\,.$$

If σ is an element of $O^+(L)$, then $\sigma_{\mathfrak{p}} \in O^+_n(L_{\mathfrak{p}})$ holds for all $\mathfrak{p}$ in S by Example 101:6, hence (σ) is an element of $P_V \cap J_L$. On the other hand, if (σ) denotes a typical element of $P_V \cap J_L$, then $(\sigma L)_{\mathfrak{p}} = \sigma_{\mathfrak{p}} L_{\mathfrak{p}} = L_{\mathfrak{p}}$ holds for all $\mathfrak{p}$ in S, hence $\sigma L = L$ by § 81E, hence $\sigma \in O^+(L)$. Hence the natural isomorphism of $O^+(V)$ onto P_V carries $O^+(L)$ onto $P_V \cap J_L$. So

[1] For a discussion and application of the topology on J_V see M. Kneser, *Math. Z.* (1961), pp. 188—194.

we have the diagram

$$\begin{array}{ccc} O^+(V) & \rightarrowtail & P_V \\ | & & | \\ O^+(L) & \rightarrowtail & P_V \cap J_L\,. \end{array}$$

We define the subgroup J_F^L of J_F by the equation

$$J_F^L = \{\mathfrak{i} \in J_F \mid \mathfrak{i}_\mathfrak{p} \in \theta(O^+(L_\mathfrak{p})) \quad \forall\, \mathfrak{p} \in S\}\,.$$

Take a typical split rotation Σ and a typical lattice L on V. We know that $\Sigma_\mathfrak{p} L_\mathfrak{p}$ is a lattice on $V_\mathfrak{p}$ at each $\mathfrak{p}$ in S, we claim that there is exactly one lattice K on V with $K_\mathfrak{p} = \Sigma_\mathfrak{p} L_\mathfrak{p}$ for all $\mathfrak{p}$ in S, and we then define ΣL to be this lattice K. In order to prove the existence of K it is enough, in view of Proposition 81:14, to show that $\Sigma_\mathfrak{p} L_\mathfrak{p} = L_\mathfrak{p}$ for almost all $\mathfrak{p}$ in S. Put $M = \mathfrak{o} x_1 + \cdots + \mathfrak{o} x_n$. Then the condition $\|\Sigma_\mathfrak{p}\|_\mathfrak{p} = 1$ is equivalent to $\Sigma_\mathfrak{p} M_\mathfrak{p} = M_\mathfrak{p}$, hence $\Sigma_\mathfrak{p} M_\mathfrak{p} = M_\mathfrak{p}$ for almost all $\mathfrak{p}$ in S. But $M_\mathfrak{p} = L_\mathfrak{p}$ for almost all $\mathfrak{p}$ in S by § 81E. Hence $\Sigma_\mathfrak{p} L_\mathfrak{p} = L_\mathfrak{p}$ for almost all $\mathfrak{p}$ in S. Hence K exists. It is unique by § 81E. So the lattice ΣL is defined; its defining equation is

$$(\Sigma L)_\mathfrak{p} = \Sigma_\mathfrak{p} L_\mathfrak{p} \quad \forall\, \mathfrak{p} \in S\,.$$

Incidentally, note that

$$\Sigma_\mathfrak{p} L_\mathfrak{p} = L_\mathfrak{p} \quad \text{for almost all } \mathfrak{p} \in S.$$

We have

$$(\Sigma\, \Lambda)\, L = \Sigma(\Lambda L) \quad \forall\, \Sigma, \Lambda \in J_V\,.$$

If σ is a rotation of V, then

$$\sigma L = (\sigma)\, L\,.$$

The group J_L can be described as

$$J_L = \{\Sigma \in J_V \mid \Sigma L = L\}\,.$$

§ 102. The genus and the spinor genus

§ 102A. Definition of gen L and spn L

We define the genus $\operatorname{gen} L$ of the lattice L on V to be the set of all lattices K on V with the following property: for each $\mathfrak{p}$ in S there exists an isometry $\Sigma_\mathfrak{p} \in O(V_\mathfrak{p})$ such that $K_\mathfrak{p} = \Sigma_\mathfrak{p} L_\mathfrak{p}$. The set of all lattices on V is thereby partitioned into genera. We immediately have

$$\operatorname{gen} K = \operatorname{gen} L \;\Leftrightarrow\; \operatorname{cls} K_\mathfrak{p} = \operatorname{cls} L_\mathfrak{p} \quad \forall\, \mathfrak{p} \in S\,.$$

The proper genus can be defined in the same way: we say that K is in the same proper genus as L if for each $\mathfrak{p}$ in S there is a rotation $\Sigma_\mathfrak{p} \in O^+(V_\mathfrak{p})$ such that $K_\mathfrak{p} = \Sigma_\mathfrak{p} L_\mathfrak{p}$. This leads to a partition of the set of all lattices on V into proper genera. The proper genus of L will be written $\operatorname{gen}^+ L$. We immediately have

$$\operatorname{gen}^+ K = \operatorname{gen}^+ L \;\Leftrightarrow\; \operatorname{cls}^+ K_\mathfrak{p} = \operatorname{cls}^+ L_\mathfrak{p} \quad \forall\, \mathfrak{p} \in S\,.$$

But we already know that the class and the proper class coincide over local fields. Hence we always have

$$\operatorname{gen} L = \operatorname{gen}^+ L .$$

The genus can be described in terms of split rotations:

$$K \in \operatorname{gen} L \quad \Leftrightarrow \quad K = \Sigma L \quad \text{for some } \Sigma \text{ in } J_V$$

(to prove this use the fact that $K_\mathfrak{p} = L_\mathfrak{p}$ for almost all $\mathfrak{p}$ in S).

We say that the lattice K on V is in the same spinor genus as L if there is an isometry σ in $O(V)$ and a rotation $\Sigma_\mathfrak{p}$ in $O'(V_\mathfrak{p})$ at each $\mathfrak{p}$ in S such that

$$K_\mathfrak{p} = \sigma_\mathfrak{p} \Sigma_\mathfrak{p} L_\mathfrak{p} \quad \forall\, \mathfrak{p} \in S .$$

This condition can be expressed in the language of split rotations: there is a σ in $O(V)$ and a Σ in J'_V such that $K = \sigma \Sigma L$. We shall use $\operatorname{spn} L$ to denote the set of lattices in the same spinor genus as L. It can be verified without difficulty that the set of all lattices on V is partitioned into spinor genera. It is an immediate consequence of the definitions that lattices in the same class are in the same spinor genus, and lattices in the same spinor genus are in the same genus. So the partition into classes is finer than the partition into spinor genera, and the partition into spinor genera is finer than the partition into genera. We have

$$\operatorname{cls} L \subseteq \operatorname{spn} L \subseteq \operatorname{gen} L .$$

We let $h(L)$ be the number of classes in $\operatorname{gen} L$, and $g(L)$ the number of spinor genera in $\operatorname{gen} L$. We shall see later that $h(L)$ and $g(L)$ are always finite.

We say that K is in the same proper spinor genus as L if there is a rotation σ in $O^+(V)$ and a rotation $\Sigma_\mathfrak{p}$ in $O'(V_\mathfrak{p})$ at each $\mathfrak{p}$ in S such that

$$K_\mathfrak{p} = \sigma_\mathfrak{p} \Sigma_\mathfrak{p} L_\mathfrak{p} \quad \forall\, \mathfrak{p} \in S .$$

This condition can be expressed in the language of split rotations by saying that there is a σ in $O^+(V)$ and a Σ in J'_V such that $K = \sigma \Sigma L$, or

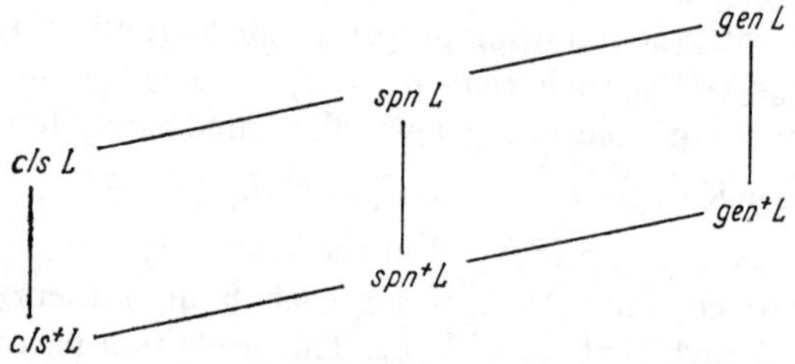

equivalently by saying that there is a Λ in P_V and a Σ in J'_V such that $K = \Lambda \Sigma L$. We shall use $\operatorname{spn}^+ L$ to denote the set of lattices in the same proper spinor genus as L. It is easily seen that the set of all lattices on V

is partitioned into proper spinor genera. Again we find that the partition into proper classes is finer than the partition into proper spinor genera, and the partition into proper spinor genera is finer than the partition into genera. We have

$$\text{cls}^+ L \subseteq \text{spn}^+ L \subseteq \text{gen}^+ L\,.$$

We let $h^+(L)$ be the number of proper classes in gen L, and $g^+(L)$ the number of proper spinor genera in gen$L$. We shall see that $h^+(L)$ and $g^+(L)$ are always finite.

All lattices in the same genus have the same volume. For consider $K \in \text{gen} L$. Then $K_\mathfrak{p} \cong L_\mathfrak{p}$ for all $\mathfrak{p}$ in S, hence

$$(\mathfrak{v}K)_\mathfrak{p} = \mathfrak{v}K_\mathfrak{p} = \mathfrak{v}L_\mathfrak{p} = (\mathfrak{v}L)_\mathfrak{p} \quad \forall\, \mathfrak{p} \in S\,,$$

hence $\mathfrak{v}K = \mathfrak{v}L$. We define the volume of a genus to be the common volume of all lattices in the genus. In the same way we can define the volume of a proper class, of a class, of a proper spinor genus, or of a spinor genus, since each of these sets is contained in a single genus. Similarly we can define the scale and norm of a genus, proper class, etc. If a genus contains an $\mathfrak{a}$-maximal (resp. $\mathfrak{a}$-modular) lattice, then every lattice in that genus is $\mathfrak{a}$-maximal (resp. $\mathfrak{a}$-modular).

102:1. **Example.** If Σ is any element of J_V, then

$$\Sigma\, \text{spn} L = \text{spn}\, \Sigma L\,, \quad \Sigma\, \text{spn}^+ L = \text{spn}^+ \Sigma L\,.$$

102:2. **Example.** Each class contains either one or two proper classes, hence

$$h(L) \leqq h^+(L) \leqq 2h(L)\,.$$

It is easily seen that each spinor genus contains either one or two proper spinor genera, so

$$g(L) \leqq g^+(L) \leqq 2g(L)\,.$$

But we can actually say more, namely that $g^+(L)$ is either $g(L)$ or $2g(L)$. For suppose that spn$L$ contains two proper spinor genera. Then $\text{spn}^+ L \subset \text{spn} L$, hence by Example 102:1 we have $\text{spn}^+ \Sigma L \subset \text{spn}\, \Sigma L$ for every Σ in J_V, hence every spinor genus in gen$L$ contains two proper spinor genera, hence $g^+(L) = 2g(L)$. In the same way we find $g^+(L) = g(L)$ when $\text{spn}^+ L = \text{spn} L$.

102:3 **Example.** Consider the genus gen$L$ of an $\mathfrak{a}$-maximal lattice L on V. We have already mentioned that every lattice in gen$L$ is also $\mathfrak{a}$-maximal. Now consider any $\mathfrak{a}$-maximal lattice K on V. Then $K_\mathfrak{p}$ and $L_\mathfrak{p}$ are $\mathfrak{a}_\mathfrak{p}$-maximal at all $\mathfrak{p}$ in S, hence $K_\mathfrak{p} \cong L_\mathfrak{p}$ by Theorem 91:2, hence $K \in \text{gen} L$. So the genus of an $\mathfrak{a}$-maximal lattice on V consists precisely of all $\mathfrak{a}$-maximal lattices on V. In particular, all $\mathfrak{a}$-maximal lattices on the same quadratic space have the same scale, norm and volume.

102:4. **Example.** Suppose $K \in \text{gen} L$. Consider a finite subset T of the underlying set of spots S. We claim that there is a lattice K' in

$\text{cls}^+ K$ such that $K'_\mathfrak{p} = L_\mathfrak{p}$ for all $\mathfrak{p}$ in T. By definition of the genus there is a rotation $\varphi_\mathfrak{p} \in O^+(V_\mathfrak{p})$ at each $\mathfrak{p}$ in T such that $\varphi_\mathfrak{p} K_\mathfrak{p} = L_\mathfrak{p}$. By the Weak Approximation Theorem for Rotations there is a rotation σ in $O^+(V)$ such that $\|\sigma - \varphi_\mathfrak{p}\|_\mathfrak{p}$ is arbitrarily small at each $\mathfrak{p}$ in T. If the approximations are good enough we will have, in virtue of Examples 101:4 and 101:6,

$$(\sigma K)_\mathfrak{p} = \sigma_\mathfrak{p} K_\mathfrak{p} = \varphi_\mathfrak{p} K_\mathfrak{p} = L_\mathfrak{p}$$

for all $\mathfrak{p}$ in T. Then σK is in $\text{cls}^+ K$, hence it is the desired lattice K'.

102:5. **Example.** Let L be a lattice on the quadratic space V, let K be a regular lattice in V. Suppose there is a representation $K_\mathfrak{p} \rightarrowtail L_\mathfrak{p}$ at each $\mathfrak{p}$ in S. We claim that there is a representation $K \rightarrowtail L'$ of K into a lattice L' in $\text{gen}\, L$. In fact we shall find a lattice L' in $\text{gen}\, L$ such that $K \subseteq L'$. To do this we take a finite subset T of S such that $K_\mathfrak{p} \subseteq L_\mathfrak{p}$ for all $\mathfrak{p}$ in $S - T$. Since there is a representation $K_\mathfrak{p} \rightarrowtail L_\mathfrak{p}$ at each $\mathfrak{p}$ in T, there is an isometry $\varphi_\mathfrak{p} \in O(V_\mathfrak{p})$ such that $\varphi_\mathfrak{p} L_\mathfrak{p} \supseteq K_\mathfrak{p}$. Define L' to be that lattice on V for which

$$L'_\mathfrak{p} = \begin{cases} L_\mathfrak{p} & \forall\, \mathfrak{p} \in S - T \\ \varphi_\mathfrak{p} L_\mathfrak{p} & \forall\, \mathfrak{p} \in T\,. \end{cases}$$

Then $L'_\mathfrak{p} \supseteq K_\mathfrak{p}$ for all $\mathfrak{p}$ in S, hence $L' \supseteq K$. Clearly $L' \in \text{gen}\, L$. Hence we have proved our assertion. We have the following special case: suppose that the scalar $\alpha \in F$ is represented by V and also by $L_\mathfrak{p}$ at each $\mathfrak{p}$ in S; then α is represented by some lattice L' in the genus of L.

102:6. **Example.** It is possible for a scalar in $Q(V)$ to be represented by $L_\mathfrak{p}$ at all $\mathfrak{p}$ in S without its being represented by L. For instance, consider the set S of all discrete spots on the field of rational numbers $\mathbf{Q}$. Let L be a lattice with respect to S on a quadratic space V over $\mathbf{Q}$ with

$$L \cong \langle 1 \rangle \perp \langle 11 \rangle\,.$$

Then the equation

$$3 = (8/5)^2 + 11\,(1/5)^2$$

shows that V represents 3, and also that L_p represents 3 for all $p \neq 5$; but L_5 also represents 3 by Corollary 92:1b. There is clearly no rational integral solution to the equation $\xi^2 + 11\eta^2 = 3$. Hence V represents 3, L_p represents 3 at all spots p, yet L does not represent 3.

§ 102B. Counting the spinor genera in a genus

102:7. $g^+(L) = (J_F : P_D J_F^L)$ *if* $n \geqq 3$.

Proof. 1) First a remark about abstract groups. Let G be any group and let H be any subgroup of G which contains the commutator subgroup of G. If x, y are typical elements of G, then the normality of H in G implies that $xH = Hx$, and the fact that H contains the commutator subgroup of G implies that $xyH = yxH$, hence the set Hxy is independent of the order of H, x, y. From this it follows for any subgroups X, Y

of G that the set HXY is independent of the order of H, X, Y, and that this set is actually the group generated by H, X, Y. In particular, this applies to the subgroups J'_V, P_V, J_L of J_V. So the group generated by these three groups is equal to $J'_V P_V J_L$, this is a normal subgroup of J_V, we can form the quotient $J_V/J'_V P_V J_L$, and we can write down the index $(J_V : J'_V P_V J_L)$. Our next claim is that this index is equal to $g^+(L)$.

2) Consider two typical proper spinor genera in $\operatorname{gen} L$. They can be written $\operatorname{spn}^+ \Sigma_1 L$ and $\operatorname{spn}^+ \Sigma_2 L$ with $\Sigma_1, \Sigma_2 \in J_V$ since ΣL runs through $\operatorname{gen} L$ as Σ runs through J_V. We then have $\operatorname{spn}^+ \Sigma_1 L = \operatorname{spn}^+ \Sigma_2 L$ if and only if $L \in \operatorname{spn}^+ \Sigma_1^{-1} \Sigma_2 L$, and this is equivalent to saying that $L = \Lambda T \Sigma_1^{-1} \Sigma_2 L$ for some $\Lambda \in P_V$, $T \in J'_V$. Hence

$$\operatorname{spn}^+ \Sigma_1 L = \operatorname{spn}^+ \Sigma_2 L \quad \Leftrightarrow \quad \Sigma_2 \in \Sigma_1 J'_V P_V J_L .$$

Therefore, if we let Σ run through a complete set of representatives of distinct cosets of J_V modulo $J'_V P_V J_L$ we obtain each proper spinor genus $\operatorname{spn}^+ \Sigma L$ in $\operatorname{gen} L$ exactly once. So we have the formula

$$g^+(L) = (J_V : J'_V P_V J_L)$$

for the number of proper spinor genera in a genus.

3) We are going to construct a group homomorphism of J_V into $J_F/P_D J_F^L$ in a certain natural way. Take a typical element Σ in J_V. Then $\Sigma_\mathfrak{p} L_\mathfrak{p} = L_\mathfrak{p}$ for almost all $\mathfrak{p}$, hence $\theta(\Sigma_\mathfrak{p}) \subseteq \mathfrak{u}_\mathfrak{p} \dot{F}_\mathfrak{p}^2$ for almost all $\mathfrak{p}$ by Proposition 92:5. We can therefore choose an idèle $\mathfrak{i}$ in J_F with

$$\mathfrak{i}_\mathfrak{p} \in \theta(\Sigma_\mathfrak{p}) \quad \forall\, \mathfrak{p} \in \Omega .$$

If $\mathfrak{j}$ is any other idèle that is associated with Σ in this way, then $\mathfrak{j} \in \mathfrak{i} J_F^2$ by definition of the spinor norm. But $J_F^2 \subseteq J_F^L \subseteq P_D J_F^L$. Hence the natural images of $\mathfrak{i}$ and $\mathfrak{j}$ in $J_F/P_D J_F^L$ are equal. We therefore have a well-defined map

$$\Phi : J_V \rightarrowtail J_F/P_D J_F^L ,$$

obtained by sending Σ to the natural image of $\mathfrak{i}$ in $J_F/P_D J_F^L$. It is immediately verified that Φ is a homomorphism.

4) Let us show that Φ is surjective. We must consider a typical idèle $\mathfrak{i}$ and we must find $\Sigma \in J_V$ such that $\mathfrak{i}_\mathfrak{p} \in \theta(\Sigma_\mathfrak{p})$ for all $\mathfrak{p}$ in Ω. By definition of J_F^L we can assume that $\mathfrak{i}_\mathfrak{p} = 1$ for all $\mathfrak{p}$ in $\Omega - S$; we define $\Sigma_\mathfrak{p}$ to be the identity map on $V_\mathfrak{p}$ for all $\mathfrak{p}$ in $\Omega - S$. What about the remaining $\mathfrak{p}$? Since $n \geq 3$ we have $\theta(O^+(V_\mathfrak{p})) = \dot{F}_\mathfrak{p}$ for all $\mathfrak{p}$ in S by Proposition 91:6 and $\theta(O^+(L_\mathfrak{p})) = \mathfrak{u}_\mathfrak{p} \dot{F}_\mathfrak{p}^2$ for almost all $\mathfrak{p}$ in S by Proposition 92:5; we can therefore choose $\Sigma_\mathfrak{p} \in O^+(V_\mathfrak{p})$ at each $\mathfrak{p}$ in S, with almost all $\Sigma_\mathfrak{p}$ in $O^+(L_\mathfrak{p})$, such that $\mathfrak{i}_\mathfrak{p} \in \theta(\Sigma_\mathfrak{p})$ for all $\mathfrak{p}$ in S. Then $\|\Sigma_\mathfrak{p}\|_\mathfrak{p} = 1$ for almost all $\mathfrak{p}$ in S. Hence $\Sigma = (\Sigma_\mathfrak{p})_{\mathfrak{p} \in \Omega}$ is a split rotation. And $\Phi \Sigma$ is the natural image of $\mathfrak{i}$ in $J_F/P_D J_F^L$. So Φ is a surjective homomorphism.

5) It is clear that J'_V, P_V, J_L are all part of the kernel of Φ, hence $J'_V P_V J_L$ is. We leave it to the reader to verify that $J'_V P_V J_L$ is the entire kernel. Hence

$$g^+(L) = (J_V : J'_V P_V J_L) = (J_F : P_D J_F^L) .$$

q. e. d.

102:7a. *$g^+(L) = (J_V : J'_V P_V J_L)$, and $g^+(L)$ divides $(J_F : P_D J_F^L)$, for any $n \geqq 1$.*

Proof. The assumption $n \geqq 3$ in Proposition 102:7 is used only in showing that the map Φ in the proof is surjective. **q. e. d.**

102:8. **Theorem.** *V is a regular quadratic space over a global field with $\dim V \geqq 3$, and L is a lattice on V. Then the number of proper spinor genera in* gen L *is of the form 2^r with $r \geqq 0$. Any value of r can be obtained by taking a suitable L on V.*

Proof. 1) Write the discriminant dV in the form $dV = \varepsilon$ with ε in $\dot{F}$. We know from Example 66:6 that the Hasse symbol $S_\mathfrak{p} V$ is 1 at almost all spots $\mathfrak{p}$ on F. We fix a non-dyadic set of spots W on F which consists of almost all spots in S, such that ε is a W-unit and $S_\mathfrak{p} V = 1$ at all $\mathfrak{p}$ in W, such that $J_F = P_D J_F^W$, and finally such that $L_\mathfrak{p}$ is unimodular at all $\mathfrak{p}$ in W. Proposition 92:5 tells us that.

$$J_F^{W,2} \subseteq J_F^L .$$

2) Let us prove two formulas that will be needed in the course of the proof. Consider a lattice K on V with respect to S, and let T be any set of spots on F with $T \subseteq S$ and $J_F = P_D J_F^T$. Then

$$\begin{aligned} g^+(K) &= (J_F : P_D J_F^K) \\ &= (J_F J_F^K : P_D J_F^K) \\ &= (P_D J_F^K J_F^T : P_D J_F^K) \\ &= (J_F^T : J_F^T \cap P_D J_F^K) . \end{aligned}$$

This is our first formula.

Now let M be some other lattice on V with respect to S, and suppose that $J_F^K \subseteq J_F^M$. Then

$$\begin{aligned} g^+(K) &= (J_F : P_D J_F^K) \\ &= (J_F : P_D J_F^M)\,(P_D J_F^M : P_D J_F^K) \\ &= g^+(M)\,(P_D J_F^K J_F^M : P_D J_F^K) . \end{aligned}$$

Hence

$$g^+(K) = g^+(M)\,(J_F^M : J_F^M \cap P_D J_F^K)$$

if $J_F^K \subseteq J_F^M$. This is our second formula.

3) The proof that $g^+(L)$ is a power of 2 is at hand. We have

$$g^+(L) = (J_F^W : J_F^W \cap P_D J_F^L)$$

by the first formula in step 2). But $J_F^{W,2} \subseteq J_F^L$. Hence

$$g^+(L) = (J_F^W : J_F^{W,2}) \div (J_F^W \cap P_D J_F^L : J_F^{W,2}) .$$

This is a power of 2 by Proposition 65:7.

4) Next we construct a lattice K on V with $g^+(K) = 1$. Start with the given lattice L and use Propositions 81:14 and 91:7 to obtain a lattice K on V with $K_\mathfrak{p} = L_\mathfrak{p}$ for all $\mathfrak{p}$ in W and

$$\theta(O^+(K_\mathfrak{p})) = \dot{F}_\mathfrak{p} \quad \forall\, \mathfrak{p} \in S - W .$$

Then $J_F^W \subseteq J_F^K$ since

$$\theta(O^+(K_\mathfrak{p})) = \mathfrak{u}_\mathfrak{p} \dot{F}_\mathfrak{p}^2 \quad \forall\, \mathfrak{p} \in W$$

by Proposition 92:5. Hence $J_F^W \cap P_D J_F^K = J_F^W$. Hence $g^+(K) = 1$ by the first formula of step 2).

5) We must digress for a moment in order to prove the following: let M and K be lattices on V, let $\mathfrak{q}$ be a non-dyadic spot in S at which

$$\theta(O^+(M_\mathfrak{q})) = \mathfrak{u}_\mathfrak{q} \dot{F}_\mathfrak{q}^2 , \quad \theta(O^+(K_\mathfrak{q})) = \dot{F}_\mathfrak{q}^2 ,$$

and suppose that $M_\mathfrak{p} = K_\mathfrak{p}$ for all $\mathfrak{p}$ in $S - \mathfrak{q}$, then $g^+(K)$ is either equal to $g^+(M)$ or to $2g^+(M)$. By definition of M and K we have $J_F^K \subseteq J_F^M$, hence by the second formula of step 2) it is enough to prove that

$$(J_F^M : J_F^M \cap P_D J_F^K) \leqq 2 ,$$

hence that $(J_F^M : J_F^K) \leqq 2$. But this last inequality follows easily from the fact that $(\mathfrak{u}_\mathfrak{q} : \mathfrak{u}_\mathfrak{q}^2) = 2$. Hence our contention is proved.

6) Now we can complete the proof. It is enough to show how to obtain a lattice K from L with $g^+(K) = 2g^+(L)$. For L is arbitrary, so we can start with an L with $g^+(L) = 1$, then obtain a new L with $g^+(L) = 2$, then an L with $g^+(L) = 4$, and so on. So consider the given lattice L. Pick $\mathfrak{q}$ and $\mathfrak{q}'$ in W in such a way that ε is a square at $\mathfrak{q}$ and $\mathfrak{q}'$, and such that a W-unit is a square at $\mathfrak{q}$ if and only if it is a square at $\mathfrak{q}'$. The existence of such a pair of spots was established in Proposition 65:20. By Proposition 81:14 and Example 92:7 there is a lattice K on V with $K_\mathfrak{p} = L_\mathfrak{p}$ for all $\mathfrak{p}$ in $S - (\mathfrak{q} \cup \mathfrak{q}')$ and with

$$\theta(O^+(K_\mathfrak{q})) = \dot{F}_\mathfrak{q}^2 , \quad \theta(O^+(K_{\mathfrak{q}'})) = \dot{F}_{\mathfrak{q}'}^2 .$$

Clearly $J_F^K \subseteq J_F^L$. We claim that

$$J_F^L \cap P_D J_F^K \subset J_F^L .$$

Suppose not. Then $J_F^L \subseteq P_D J_F^K$. Take $\mathfrak{i} \in J_F^L$ with $\mathfrak{i}_\mathfrak{p} = 1$ for all $\mathfrak{p} \in \Omega - \mathfrak{q}$ and $\mathfrak{i}_\mathfrak{q}$ a non-square unit in $F_\mathfrak{q}$. Then $\mathfrak{i} \in P_D J_F^K$, hence there is an α in D and a $\mathfrak{j}$ in J_F^K with $\mathfrak{i} = (\alpha)\mathfrak{j}$. So we have a field element α which is a square at $\mathfrak{q}'$, a non-square at $\mathfrak{q}$, and of even order at all $\mathfrak{p}$ in W. Now $J_F = P_F J_F^W$, hence there is an element β in F with $2\,\mathrm{ord}_\mathfrak{p}\, \beta = \mathrm{ord}_\mathfrak{p}\, \alpha$ for all $\mathfrak{p}$ in W.

Then α/β^2 is a W-unit, it is a square at $\mathfrak{q}'$, it is not a square at $\mathfrak{q}$. This contradicts the choice of $\mathfrak{q}$ and $\mathfrak{q}'$. Hence we do indeed have $J_F^L \cap P_D J_F^K \subset J_F^L$. So by the second formula in step 2), $g^+(K) \geqq 2g^+(L)$.

Let K' be the lattice on V with $K'_\mathfrak{p} = L_\mathfrak{p}$ for all $\mathfrak{p}$ in $S - \mathfrak{q}'$, and $K'_{\mathfrak{q}'} = K_{\mathfrak{q}'}$. Then by step 5),

$$g^+(K) = g^+(K') \quad \text{or} \quad 2g^+(K') ; \quad g^+(K') = g^+(L) \quad \text{or} \quad 2g^+(L) .$$

Hence either $g^+(K') = 2g^+(L)$, or else $g^+(K) = 2g^+(L)$. **q. e. d.**

102:8a. **Corollary.** *The number of proper spinor genera in* gen$L$ *is a power of 2 in general, i. e. when* $\dim V \geqq 1$.

Proof. The case $n = 3$ is covered by the theorem. The case $n = 1$ is trivial since then gen$L = L$. There remains $n = 2$. By Corollary 102:7a it is enough to show that $(J_F : P_D J_F^L)$ is a power of 2, and this follows as in steps 1)—3) of the proof of this theorem. **q. e. d.**

102:9. **Theorem.** *L is a lattice on the regular quadratic space V over the global field F. Suppose that the underlying set of spots S satisfies the equation $J_F = P_D J_F^S$. If*

$$\theta(O^+(L_\mathfrak{p})) \supseteq \mathfrak{u}_\mathfrak{p} \quad \forall\, \mathfrak{p} \in S ,$$

then spn^+L = gen$L$.

Proof. We must show that $g^+(L) = 1$. We use the fact that $g^+(L)$ divides $(J_F : P_D J_F^L)$. Then $J_F^S \subseteq J_F^L$ since $\theta(O^+(L_\mathfrak{p})) \supseteq \mathfrak{u}_\mathfrak{p}$ for all $\mathfrak{p}$ in S. Hence $J_F = P_D J_F^S \subseteq P_D J_F^L$. Hence $g^+(L) = 1$. **q. e. d.**

102:10. **Example.** The last theorem can be used to give sufficient conditions under which spn^+L = gen$L$. Consider the lattice L on the regular quadratic space V over the global field F with $\dim V \geqq 3$. (i) First let us examine the condition $J_F = P_D J_F^S$. This is satisfied in the function theoretic case whenever the class number $h_F(S)$ is 1 since then $J_F = P_F J_F^S$ with $\dot{F} = D$, hence $J_F = P_D J_F^S$. It is also satisfied when F is the field of rational numbers for then $J_F = P_F J_F^S$ with $\dot{F} = (\pm 1) D$, hence $P_F = (\pm 1) P_D$, but $(-1) \in J_F^S$, hence $J_F = P_D J_F^S$. (ii) The local theory has provided sufficient criteria for testing the condition $\theta(O^+(L_\mathfrak{p})) \supseteq \mathfrak{u}_\mathfrak{p}$. For instance this condition is satisfied at all $\mathfrak{p}$ in S whenever L is either modular or maximal on V (see Propositions 91:8 and 92:5, and Example 93:20). Again, it is satisfied at $\mathfrak{p}$ if there is a Jordan splitting for $L_\mathfrak{p}$ with a modular component of dimension $\geqq 2$ when $\mathfrak{p}$ is non-dyadic and of dimension $\geqq 3$ when $\mathfrak{p}$ is dyadic. Example 92:9 shows how to derive from this a simpler, but weaker, criterion involving volumes (the formula given there is for the non-dyadic case only; but a similar, though not identical, result can be obtained at the dyadic spots).

§ 103. Finiteness of class number

This paragraph is devoted to a single purpose: the proof that the number of proper classes of integral scale with given rank and volume is essentially finite[1]. This will imply that the number of proper classes in a genus is finite. Hence all the numbers $h(L)$, $h^+(L)$, $g(L)$, $g^+(L)$ of § 102A are finite.

We shall need the counting number $N\mathfrak{a}$ introduced for fractional ideals in § 33C. For any non-zero scalar α we define $N\alpha = N(\alpha\mathfrak{o})$. Thus $N(\alpha\beta) = (N\alpha)(N\beta)$. By Proposition 33:3 and the Product Formula we have

$$N\alpha = \prod_{\mathfrak{p}\in S} \frac{1}{|\alpha|_{\mathfrak{p}}} = \prod_{\mathfrak{p}\in\Omega-S} |\alpha|_{\mathfrak{p}}.$$

103:1. *Let L be a lattice on the abstract vector space V over the global field F, and let φ be a non-singular linear transformation of V into V such that $\varphi L \subseteq L$. Then*

$$(L:\varphi L) = N(\det\varphi)\,.$$

Proof. φL is a lattice on V since φ is non-singular. By applying the Invariant Factor Theorem to L and φL we can find a base $x_1, \ldots, x_n$ for V and fractional ideals $\mathfrak{a}_1, \ldots, \mathfrak{a}_n$ and $\mathfrak{b}_1, \ldots, \mathfrak{b}_n$ such that

$$\begin{cases} L = \mathfrak{b}_1 x_1 + \cdots + \mathfrak{b}_n x_n \\ \varphi L = \mathfrak{a}_1 x_1 + \cdots + \mathfrak{a}_n x_n\,. \end{cases}$$

Now write

$$y_j = \varphi x_j = \sum_i a_{ij} x_i \quad (a_{ij} \in F)\,.$$

Then

$$\varphi L = \mathfrak{b}_1 y_1 + \cdots + \mathfrak{b}_n y_n\,.$$

By comparing the above expressions for φL we obtain, from Proposition 81:8,

$$\mathfrak{a}_1 \ldots \mathfrak{a}_n = \mathfrak{b}_1 \ldots \mathfrak{b}_n \det(a_{ij})\,.$$

Then by Proposition 33:2,

$$\begin{aligned}(L:\varphi L) &= (\mathfrak{b}_1:\mathfrak{a}_1)\ldots(\mathfrak{b}_n:\mathfrak{a}_n)\\ &= (N\mathfrak{a}_1\ldots N\mathfrak{a}_n)/(N\mathfrak{b}_1\ldots N\mathfrak{b}_n)\\ &= N(\det\varphi)\,.\end{aligned}$$

q. e. d.

103:2. *There is a positive constant γ which depends only on F and S, and which has the following property: given any $n\times n$ matrix (a_{ij}) with*

[1] Using "reduction theory" it is possible to give a short and elementary proof of the finiteness of class number in the classical situation where the ring of integers in question is the ring of rational integers **Z**. See G. L. Watson, *Integral quadratic forms* (Cambridge, 1960).

entries in F *and* $\det(a_{ij}) \in \mathfrak{u}$, *there are elements* $\xi_1, \ldots, \xi_n$ *in* $\mathfrak{o}$, *not all of them* 0, *such that*

$$|a_{i1}\xi_1 + \cdots + a_{in}\xi_n|_\mathfrak{p} \leqq \gamma$$

for $1 \leqq i \leqq n$ *and for all* $\mathfrak{p}$ *in* $\Omega - S$.

Proof. 1) Fix a spot $\mathfrak{q}_0$ in $\Omega - S$. By Theorem 33:5 there is a constant $C\,(0 < C < 1)$ such that the density $M(\mathfrak{i})/\|\mathfrak{i}\|$ of the set of field elements bounded by any idèle $\mathfrak{i}$ satisfies

$$M(\mathfrak{i})/\|\mathfrak{i}\| > C\,.$$

We can assume, by taking a smaller C if necessary, that C is in the value group $|F_{\mathfrak{q}_0}|_{\mathfrak{q}_0}$. Put $\gamma = 4C^{-1}$. This will be the constant of the proposition.

2) Consider the given matrix (a_{ij}). Take a non-zero scalar A in $\mathfrak{o}$ such that all the elements $b_{ij} = A\,a_{ij}$ are in $\mathfrak{o}$. We have to find elements $\xi_1, \ldots, \xi_n$ in $\mathfrak{o}$, not all of them 0, such that

$$|b_{i1}\xi_1 + \cdots + b_{in}\xi_n|_\mathfrak{p} \leqq \gamma\,|A|_\mathfrak{p}$$

for $1 \leqq i \leqq n$ and for all $\mathfrak{p}$ in $\Omega - S$. Form the cartesian n-space $V = F \times \cdots \times F$ over F and let L be the lattice

$$L = \{(\alpha_1, \ldots, \alpha_n) \mid \alpha_i \in \mathfrak{o} \quad \text{for} \quad 1 \leqq i \leqq n\}$$

on V. Define a linear transformation $\varphi \in L_F(V)$ by the equation

$$\varphi(\alpha_1, \ldots, \alpha_n) = (\beta_1, \ldots, \beta_n)$$

where

$$\beta_i = \sum_j b_{ij}\alpha_j \quad (1 \leqq i \leqq n)\,.$$

Then

$$\det\varphi = \det(b_{ij}) = A^n \det(a_{ij})$$

and $\varphi L \subseteq L$, hence by Proposition 103:1

$$(L : \varphi L) = N(\det\varphi) = (NA)^n.$$

3) Construct an idèle $\mathfrak{i}$ with

$$|\mathfrak{i}_\mathfrak{p}|_\mathfrak{p} = \begin{cases} 1 & \text{if } \mathfrak{p} \in S \\ C^{-1}\,|A|_{\mathfrak{q}_0} & \text{if } \mathfrak{p} = \mathfrak{q}_0 \\ |A|_\mathfrak{p} & \text{if } \mathfrak{p} \in \Omega - (S \cup \mathfrak{q}_0)\,. \end{cases}$$

Thus $|\mathfrak{i}_\mathfrak{p}|_\mathfrak{p} \leqq C^{-1}\,|A|_\mathfrak{p}$ for all $\mathfrak{p}$ in $\Omega - S$ since $0 < C < 1$. The volume of $\mathfrak{i}$ is given by

$$\|\mathfrak{i}\| = C^{-1} \prod_{\mathfrak{p} \in \Omega - S} |A|_\mathfrak{p} = C^{-1}(NA)\,.$$

Then by Theorem 33:5 the idèle $\mathfrak{i}$ bounds strictly more that NA field elements. Hence there are strictly more than $(NA)^n$ vectors $(\alpha_1, \ldots, \alpha_n)$ in L which satisfy

$$|\alpha_i|_\mathfrak{p} \leqq C^{-1}\,|A|_\mathfrak{p}$$

for $1 \leqq i \leqq n$ and for all $\mathfrak{p}$ in $\Omega - S$. But $(L : \varphi L) = (NA)^n$. Hence at least two of these vectors, say $(\alpha_1, \ldots, \alpha_n)$ and $(\alpha'_1, \ldots, \alpha'_n)$, are congruent modulo φL. Put $\eta_i = \alpha_i - \alpha'_i$ for $1 \leqq i \leqq n$. Then

$$(\eta_1, \ldots, \eta_n) \in \varphi L ,$$

hence we can find $\xi_1, \ldots, \xi_n$ in $\mathfrak{o}$ such that

$$\eta_i = \sum_j b_{ij} \xi_j \quad (1 \leqq i \leqq n) .$$

On the other hand,

$$|\eta_i|_{\mathfrak{p}} = |\alpha_i - \alpha'_i|_{\mathfrak{p}} \leqq 2(|\alpha_i|_{\mathfrak{p}} + |\alpha'_i|_{\mathfrak{p}}) \leqq 4C^{-1} |A|_{\mathfrak{p}}$$

for $1 \leqq i \leqq n$ and for all $\mathfrak{p}$ in $\Omega - S$. In other words,

$$|b_{i1} \xi_1 + \cdots + b_{in} \xi_n|_{\mathfrak{p}} \leqq \gamma |A|_{\mathfrak{p}} ,$$

as required. q. e. d.

103:3. **Lemma.** *Let V be a regular quadratic space over the global field F, let $\mathfrak{c}$ be a given fractional ideal. Then there is a finite subset Φ of $\dot{F}$ such that $Q(L) \cap \Phi \neq \emptyset$ for every lattice L on V which satisfies $\mathfrak{s}L \subseteq \mathfrak{o}$, $\mathfrak{v}L \supseteq \mathfrak{c}$.*

Proof. 1) We have $\mathfrak{c} \subseteq \mathfrak{v}L \subseteq \mathfrak{o}$ since $\mathfrak{s}L \subseteq \mathfrak{o}$. Now the number of fractional ideals between $\mathfrak{c}$ and $\mathfrak{o}$ is finite. It therefore suffices to prove the lemma for all lattices L on V which satisfy the condition $\mathfrak{v}L = \mathfrak{c}$ (instead of the condition $\mathfrak{v}L \supseteq \mathfrak{c}$).

2) First suppose that V is isotropic. All $\mathfrak{o}$-maximal lattices on V have the same volume, let it be $\mathfrak{b}$. Consider any lattice L on V of the type under discussion in this lemma. Then L is contained in an $\mathfrak{o}$-maximal lattice M since $\mathfrak{n}L \subseteq \mathfrak{s}L \subseteq \mathfrak{o}$. By Proposition 82:11 there is an integral ideal $\mathfrak{a}$ such that $\mathfrak{c} = \mathfrak{a}^2\mathfrak{b}$ and $\mathfrak{a}M \subseteq L$. The ideal $\mathfrak{a}$ obtained in this way will be the same for all lattices L under discussion since $\mathfrak{a}^2 = \mathfrak{c}\mathfrak{b}^{-1}$. Take a non-zero scalar α in $\mathfrak{a}$. Then

$$M \subseteq \mathfrak{a}^{-1} L \subseteq \alpha^{-1} L .$$

It therefore suffices to prove the following: there is a non-zero field element which is represented by all $\mathfrak{o}$-maximal lattices on V.

What is this field element to be? Take a complete set of representatives $\mathfrak{a}_1, \ldots, \mathfrak{a}_k$ of the group of fractional ideals modulo the subgroup of principal ideals, i. e. of the ideal class group of F at S. Let β be a non-zero scalar which belongs to all the ideals $\mathfrak{a}_1, \ldots, \mathfrak{a}_k$ and $\mathfrak{a}_1^{-1}, \ldots, \mathfrak{a}_k^{-1}$. We claim that every $\mathfrak{o}$-maximal lattice M on V represents β^2. By Proposition 82:20 there is a splitting $M = K \perp \cdots$ in which FK is a hyperbolic plane. By Proposition 82:21 and its corollary there is an ideal $\mathfrak{a}_i$ for some i $(1 \leqq i \leqq k)$, and there is a base x, y for FK with

$$Q(x) = Q(y) = 0 , \quad B(x, y) = 1 ,$$

such that

$$K = \mathfrak{a}_i x + \frac{1}{2} \mathfrak{a}_i^{-1} y .$$

Then $\beta x + \frac{1}{2} \beta y$ is in K and hence in M. But $Q\left(\beta x + \frac{1}{2} \beta y\right) = \beta^2$. So M represents β^2 as asserted.

3) Now let V be anisotropic. By Proposition 81:5 every lattice L on V can be written in the form

$$L = \mathfrak{a}_i x_1 + \mathfrak{o} x_2 + \cdots + \mathfrak{o} x_n$$

where $x_1, \ldots, x_n$ is a base for V and $\mathfrak{a}_i$ is one of a finite number of fractional ideals $\mathfrak{a}_1, \ldots, \mathfrak{a}_k$. We may therefore restrict ourselves to the following situation: given a fractional ideal $\mathfrak{a}$ prove that there is a finite subset Φ of $\dot{F}$ such that $Q(L) \cap \Phi \neq \emptyset$ holds for every lattice L on V which satisfies $\mathfrak{s} L \subseteq \mathfrak{o}$, $\mathfrak{v} L = \mathfrak{c}$ and which has the form

$$L = \mathfrak{a} x_1 + \mathfrak{o} x_2 + \cdots + \mathfrak{o} x_n$$

in some base $x_1, \ldots, x_n$ for V. We can assume that $\mathfrak{a} \supseteq \mathfrak{o}$.

Take a lattice K of the above type, write it in the form

$$K = \mathfrak{a} z_1 + \mathfrak{o} z_2 + \cdots + \mathfrak{o} z_n ,$$

then fix it and fix the base $z_1, \ldots, z_n$ for the rest of the proof. We have

$$\mathfrak{a} \supseteq \mathfrak{o} , \quad \mathfrak{s} K \subseteq \mathfrak{o} , \quad \mathfrak{v} K = \mathfrak{c} .$$

Let $\mathfrak{i}$ be an idèle with $\mathfrak{i}_\mathfrak{p} = 1$ for all $\mathfrak{p}$ in S and

$$|\mathfrak{i}_\mathfrak{p}|_\mathfrak{p} \geq 2^{n^2} n^2 \gamma^2 \max_{i,j} |B(z_i, z_j)|_\mathfrak{p}$$

for all $\mathfrak{p}$ in $\Omega - S$, where γ denotes the constant of Proposition 103:2. The idèle $\mathfrak{i}$ bounds just a finite set of field elements by Theorem 33:4. It therefore suffices to show that every lattice L of the type under consideration represents at least one non-zero field element that is bounded by $\mathfrak{i}$.

So consider the lattice L expressed in the form

$$L = \mathfrak{a} x_1 + \mathfrak{o} x_2 + \cdots + \mathfrak{o} x_n$$

with

$$\mathfrak{a} \supseteq \mathfrak{o} , \quad \mathfrak{s} L \subseteq \mathfrak{o} , \quad \mathfrak{v} L = \mathfrak{c} .$$

Let $\varphi : V \rightarrowtail V$ be the linear transformation defined by the equations $\varphi z_j = x_j$ for $1 \leq j \leq n$. Thus $\varphi K = L$. Put

$$x_j = \varphi z_j = \sum_i a_{ij} z_i \quad (a_{ij} \in F) .$$

Now $\mathfrak{v} K = \mathfrak{c} = \mathfrak{v} L$, hence $\det(B(z_i, z_j))$ is a unit times $\det(B(x_i, x_j))$, hence $\det(a_{ij})$ is a unit. By Proposition 103:2 we can find elements $\xi_1, \ldots, \xi_n$ in $\mathfrak{o}$, not all of them 0, such that

$$|a_{i1} \xi_1 + \cdots + a_{in} \xi_n|_\mathfrak{p} \leq \gamma$$

for $1 \leqq i \leqq n$ and for all $\mathfrak{p}$ in $\Omega - S$. Put

$$z = \xi_1 z_1 + \cdots + \xi_n z_n .$$

Then $Q(\varphi z) \neq 0$ since φ is non-singular and V is anisotropic. And z is in K, hence φz is in L, hence $Q(\varphi z) \in \mathfrak{s}L \subseteq \mathfrak{o}$, hence

$$|Q(\varphi z)|_{\mathfrak{p}} \leqq 1 \quad \forall\, \mathfrak{p} \in S .$$

Now we also have

$$\varphi z = \sum_i \eta_i z_i \quad \text{where} \quad \eta_i = \sum_j a_{ij} \xi_j .$$

Here we have $|\eta_i|_{\mathfrak{p}} \leqq \gamma$ for $1 \leqq i \leqq n$ and for all $\mathfrak{p}$ in $\Omega - S$. A direct calculation then gives

$$|Q(\varphi z)|_{\mathfrak{p}} \leqq 2^{n^2} n^2 \gamma^2 \max_{i,j} |B(z_i, z_j)|_{\mathfrak{p}}$$

for all $\mathfrak{p}$ in $\Omega - S$. Therefore $Q(\varphi z)$ is a non-zero scalar which is bounded by $\mathfrak{i}$ and represented by L. **q. e. d.**

103:4. Theorem. *Let V be a regular quadratic space over a global field. Then the number of proper classes of lattices on V with integral scale and given volume is finite. In particular, the number of proper classes in a genus is finite.*

Proof. 1) Let $\mathfrak{c}$ be an integral ideal. We shall actually prove the following: the number of classes of lattices on V with integral scale and with volume containing $\mathfrak{c}$ is finite. This of course gives the theorem since each class consists of either one or two proper classes. The proof is by induction on $n = \dim V$. For $n = 1$ the result is trivial. Assume it for $n - 1$ and deduce it for the given n-ary space V.

In virtue of Lemma 103:3 it is enough to prove that the lattices L on V which represent a fixed non-zero scalar α and which satisfy $\mathfrak{s}L \subseteq \mathfrak{o}$, $\mathfrak{v}L \supseteq \mathfrak{c}$ fall into a finite number of classes.

2) Fix a vector y in V with $Q(y) = \alpha$ and take the splitting $V = Fy \perp U$. By the inductive hypothesis we can find lattices $K_1, \ldots, K_r$ on U such that every lattice K on U with

$$\mathfrak{s}K \subseteq \mathfrak{o}, \quad \mathfrak{v}K \supseteq \alpha^{2n}\mathfrak{c}$$

is isometric to one of them. Define lattices $L_1, \ldots, L_r$ on V by the equations

$$L_i = \mathfrak{o}y \perp K_i \quad (1 \leqq i \leqq r) .$$

We claim that for each of the lattices L under consideration there is a lattice L_i $(1 \leqq i \leqq r)$ and a σ in $O(V)$ such that $\sigma L \supseteq L_i$. Once this has been demonstrated we shall be through for the following reason: we will have

$$B(\sigma L, L_i) \subseteq B(\sigma L, \sigma L) = B(L, L) \subseteq \mathfrak{o} ,$$

hence $L_i \subseteq \sigma L \subseteq L_i^{\#}$; but the number of lattices between L_i and $L_i^{\#}$ is finite; hence L will be isometric to one of a finite number of lattices; hence the lattices L under consideration will fall into a finite number of classes.

3) So we must find L_i and σ. By Witt's Theorem we can assume that $y \in L$. Define the sublattice

$$K' = \{\alpha x - B(x, y)\, y \mid x \in L\}$$

of L. Clearly K' is a lattice on U. Put

$$L' = \mathfrak{a}_y y \perp K'$$

where $\mathfrak{a}_y$ is the coefficient of y in L. Thus $\mathfrak{a}_y \supseteq \mathfrak{o}$ since $y \in L$. For each x in L we have

$$\alpha x = B(x, y)\, y + (\alpha x - B(x, y)\, y) \in L'\,,$$

hence

$$\alpha L \subseteq L' \subseteq L\,.$$

Therefore $\alpha^{2n}\mathfrak{c} \subseteq \mathfrak{v}L' = \mathfrak{a}_y^2\alpha(\mathfrak{v}K')$. But $\mathfrak{a}_y^2\alpha \subseteq \mathfrak{s}L \subseteq \mathfrak{o}$. So $\mathfrak{v}K' \supseteq \alpha^{2n}\mathfrak{c}$. Now K' has integral scale since L does. There is therefore an isometry of U onto U which carries K' to K_i for some i $(1 \leqq i \leqq r)$. Hence there is a σ in $O(V)$ such that

$$\sigma L' = \mathfrak{a}_y y \perp K_i \supseteq L_i\,.$$

Then $\sigma L \supseteq \sigma L' \supseteq L_i$. We have therefore found the desired L_i and σ.

q. e. d.

103:5. **Remark.** Suppose the global field F and the underlying set of spots S are kept fixed. Let $\mathfrak{c}$ be a given integral ideal, let n be a given natural number. Then the number of quadratic spaces V of dimension n which can support a lattice L with integral scale and with volume $\mathfrak{c}$ is finite (at least up to isometry). For let us take a set of non-dyadic spots T which consists of almost all spots in S, such that $\mathfrak{c}_\mathfrak{p} = \mathfrak{o}_\mathfrak{p}$ for all $\mathfrak{p}$ in T, and such that $J_F = P_F J_F^T$. Consider an n-ary quadratic space V over F and suppose there is a lattice L on V with integral scale and with $\mathfrak{v}L = \mathfrak{c}$. Put $dV = \alpha$ with α in $\dot{F}$. Then $L_\mathfrak{p}$ is a unimodular lattice on $V_\mathfrak{p}$ at each $\mathfrak{p}$ in T. Hence the Hasse symbol $S_\mathfrak{p} V$ is 1 and the order $\operatorname{ord}_\mathfrak{p}\alpha$ is even at all $\mathfrak{p}$ in T. The information $J_F = P_F J_F^T$ gives us a β in F with $2\operatorname{ord}_\mathfrak{p}\beta = \operatorname{ord}_\mathfrak{p}\alpha$ at all $\mathfrak{p}$ in T. Hence we can write $dV = \varepsilon$ for some ε in $\mathfrak{u}(T)$, i.e. for some T-unit ε. But $\mathfrak{u}(T)$ modulo $\mathfrak{u}(T)^2$ is finite by Proposition 65:6. Therefore there are just a finite number of possibilities for the discriminant dV of a quadratic space V with the given properties. Consider those quadratic spaces V which have the given properties and have fixed discriminant $dV = \varepsilon$ with ε in $\mathfrak{u}(T)$. Then $S_\mathfrak{p} V_\mathfrak{p} = 1$ at each $\mathfrak{p}$ in T, hence $V_\mathfrak{p}$ is unique up to isometry at each $\mathfrak{p}$ in T by Theorem 63:20. Now the number of quadratic spaces of given dimension and given discriminant over a local field or over a complete archimedean field is clearly finite (up to isometry). In particular this is true over the fields $F_\mathfrak{p}$ at each $\mathfrak{p}$ in $\Omega - T$. Hence by the Hasse-Minkowski Theorem there are only a finite number of possibilities for V.

§ 104. The class and the spinor genus in the indefinite case

104:1. **Lemma.** *L is a lattice on the quadratic space V under discussion. Suppose* $\dim V \geqq 3$. *Let T be a finite subset of the underlying set of spots S. Then there is a scalar μ in $\mathfrak{o}$ which is a unit at every spot in T and has the following property: every element of $\mu\mathfrak{o} \cap Q(V)$ which is represented by $L_\mathfrak{p}$ at each $\mathfrak{p}$ in S is represented by L.*

Proof. By enlarging T if necessary we can suppose that $S - T$ contains only non-dyadic spots and that $L_\mathfrak{p}$ is unimodular at each $\mathfrak{p}$ in $S - T$. Hence $Q(L_\mathfrak{p}) = \mathfrak{o}_\mathfrak{p}$ at each $\mathfrak{p}$ in $S - T$ by Corollary 92:1b. Take lattices $L_1, \ldots, L_h$ on V, one from each of the classes contained in $\operatorname{gen} L$, and let these lattices be chosen in such a way that $L_{i\mathfrak{p}} = L_\mathfrak{p}$ for all $\mathfrak{p}$ in T and for $1 \leqq i \leqq h$ (this is possible by Example 102:4).

Using Corollary 21:2a we can find a λ in $\mathfrak{o}$ which is a unit at each spot in T and such that $\lambda L_i \subseteq L$ for $1 \leqq i \leqq h$. Put $\mu = \lambda^2$. This will be our μ. To prove this we consider an element α of $\mu\mathfrak{o}$ which is represented by V and also by $L_\mathfrak{p}$ at each $\mathfrak{p}$ in S, and we must prove that α is represented by L. Now at each $\mathfrak{p}$ in $S - T$ we have

$$\alpha/\mu \in \mathfrak{o} \subseteq \mathfrak{o}_\mathfrak{p} = Q(L_\mathfrak{p}) .$$

And at each $\mathfrak{p}$ in T we have $\alpha/\lambda^2 \in Q(L_\mathfrak{p})$ since α is represented by $L_\mathfrak{p}$ and λ^2 is a unit at $\mathfrak{p}$. So α/λ^2 is represented by V and also by $L_\mathfrak{p}$ at each $\mathfrak{p}$ in S. Hence α/λ^2 is represented by some lattice in $\operatorname{gen} L$ by Example 102:5, hence $\alpha/\lambda^2 \in Q(L_i)$ for some i $(1 \leqq i \leqq h)$. Then

$$\alpha \in Q(\lambda L_i) \subseteq Q(L) .$$

q. e. d.

104:2. **Definition.** Let S be a Dedekind set consisting of almost all spots on the global field F. Let V be a regular quadratic space over F. We say that S is indefinite for V if there is at least one spot $\mathfrak{p}$ (archimedean or discrete) in $\Omega - S$ at which $V_\mathfrak{p}$ is isotropic. If $V_\mathfrak{p}$ is anisotropic at each $\mathfrak{p}$ in $\Omega - S$, then we say that S is a definite set of spots for V.

104:3. **Theorem.** *V is a regular quadratic space over the global field F with* $\dim V \geqq 4$, *S is an indefinite set of spots for V, and T is a finite subset of S. Let a be a non-zero element of $Q(V)$ and suppose that at each $\mathfrak{p}$ in S there is a $z_\mathfrak{p}$ in $V_\mathfrak{p}$ with $Q(z_\mathfrak{p}) = a$ such that*

$$\|z_\mathfrak{p}\|_\mathfrak{p} \leqq 1 \quad \forall\, \mathfrak{p} \in S - T .$$

Then for each $\varepsilon > 0$ *there is a* z *in* V *with* $Q(z) = a$ *such that*

$$\|z\|_{\mathfrak{p}} \leqq 1 \quad \forall\, \mathfrak{p} \in S - T$$

and

$$\|z - z_{\mathfrak{p}}\|_{\mathfrak{p}} < \varepsilon \quad \forall\, \mathfrak{p} \in T\,.$$

Proof. 1) By scaling we can assume that $a = 1$. We may take $0<\varepsilon<1$. Since S is indefinite for V there is a spot $\mathfrak{q}_0$ in $\Omega - S$ at which $V_{\mathfrak{q}_0}$ is isotropic. We fix this spot $\mathfrak{q}_0$. By adjoining all discrete spots in $\Omega - (S \cup \mathfrak{q}_0)$ to T (and hence to S) we see that we can make the following assumption: S consists of all discrete spots when $\mathfrak{q}_0$ is archimedean, $S \cup \mathfrak{q}_0$ consists of all discrete spots when $\mathfrak{q}_0$ is discrete.

Let $x_1, \ldots, x_n$ be a base for V which determines all the given norms $\|\ \|_{\mathfrak{p}}$. Suppose this base is replaced by the base $\delta x_1, \ldots, \delta x_n$ where δ is any $(S - T)$-unit, and let $\|\ \|'_{\mathfrak{p}}$ denote norms with respect to this new base. Clearly $\|\ \|_{\mathfrak{p}} = \|\ \|'_{\mathfrak{p}}$ for all $\mathfrak{p}$ in $S - T$. Now Proposition 33:8 provides us with an $(S - T)$-unit δ which is arbitrarily large at all $\mathfrak{p}$ in T. In fact δ can be chosen in such a way that $\|z_{\mathfrak{p}}\|'_{\mathfrak{p}} \leqq 1$ for all $\mathfrak{p}$ in T, and hence for all $\mathfrak{p}$ in S. In effect this allows us to assume that the given norms satisfy $\|z_{\mathfrak{p}}\|_{\mathfrak{p}} \leqq 1$ for all $\mathfrak{p}$ in S.

We define the lattice L with respect to S on V by the equation

$$L = \mathfrak{o} x_1 + \cdots + \mathfrak{o} x_n\,.$$

If $\mathfrak{p}$ is in S and x is in $V_{\mathfrak{p}}$, then $x \in L_{\mathfrak{p}}$ if and only if $\|x\|_{\mathfrak{p}} \leqq 1$. So each of the given vectors $z_{\mathfrak{p}}$ is assumed to be in $L_{\mathfrak{p}}$. The required vector z will actually be found in L, i. e. it will satisfy $\|z\|_{\mathfrak{p}} \leqq 1$ for all $\mathfrak{p}$ in S.

We adjoin to T all those spots $\mathfrak{p}$ in S which are either dyadic or such that $L_{\mathfrak{p}}$ is not unimodular. The enlarged set T is still finite. It suffices to prove the theorem for the new T (since $\varepsilon < 1$).

We therefore assume for the rest of the proof that $\mathfrak{q}_0$ is either discrete or archimedean, that $S \cup \mathfrak{q}_0$ contains all discrete spots on F, that every spot $\mathfrak{p}$ in $S - T$ is non-dyadic with $L_{\mathfrak{p}}$ unimodular, and that $z_{\mathfrak{p}} \in L_{\mathfrak{p}}$ for all $\mathfrak{p}$ in S. We illustrate these facts in the spot diagram

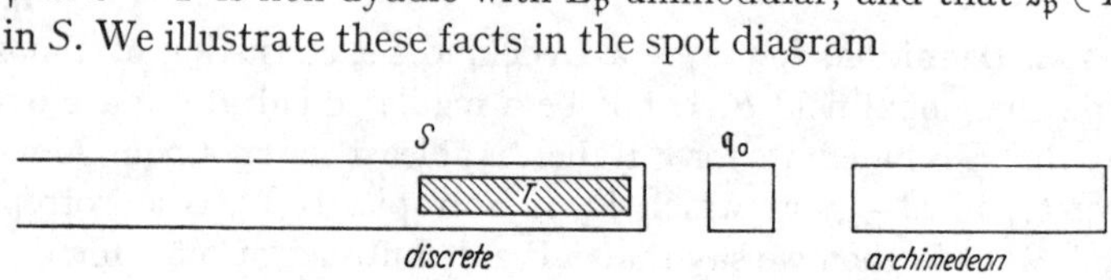

2) Pick $x \in V$ with $Q(x) = 1$. By Witt's theorem there is a rotation $\varphi_{\mathfrak{p}}$ in $O^+(V_{\mathfrak{p}})$ at each $\mathfrak{p}$ in T such that $z_{\mathfrak{p}} = \varphi_{\mathfrak{p}} x$. By the Weak Approximation Theorem for Rotations there is a rotation σ in $O^+(V)$ with $\|\sigma - \varphi_{\mathfrak{p}}\|_{\mathfrak{p}}$ arbitrarily small at each $\mathfrak{p}$ in T. Hence

$$\|\sigma x - z_{\mathfrak{p}}\|_{\mathfrak{p}} \leqq \|\sigma - \varphi_{\mathfrak{p}}\|_{\mathfrak{p}}\, \|x\|_{\mathfrak{p}}$$

is arbitrarily small at all $\mathfrak{p}$ in T. So there exists a vector y in V with $Q(y) = 1$ and

$$\|y - z_{\mathfrak{p}}\|_{\mathfrak{p}} < \varepsilon \quad \forall\, \mathfrak{p} \in T\,.$$

This implies that $\|y\|_{\mathfrak{p}} \leqq 1$ for all $\mathfrak{p}$ in T. By the Product Formula, $\|y\|_{\mathfrak{p}} = 1$ holds at almost all $\mathfrak{p}$; hence by the Strong Approximation Theorem there is a λ in $\mathfrak{o}$ such that

$$\begin{cases} |\lambda - 1|_{\mathfrak{p}} < \dfrac{\varepsilon}{\|y\|_{\mathfrak{p}} + 1} & \forall\, \mathfrak{p} \in T \\ \|\lambda y\|_{\mathfrak{p}} \leqq 1 & \forall\, \mathfrak{p} \in S - T\,. \end{cases}$$

Put $v = \lambda y$. Then $\|v - z_{\mathfrak{p}}\|_{\mathfrak{p}} < \varepsilon$ holds for all $\mathfrak{p}$ in T. The rest of the proof now consists in using this λ and this v to find a vector z in L with $Q(z) = 1$ and such that $\|z - v\|_{\mathfrak{p}} < \varepsilon$ for all $\mathfrak{p}$ in T. Once this is done we shall have our vector z and we shall be through. It is easily seen that λ and v have the following properties: we have a scalar λ and a vector v such that

$$\lambda \in \mathfrak{o}, \quad v \in L, \quad Q(v) = \lambda^2$$

with

$$|\lambda - 1|_{\mathfrak{p}} < \varepsilon \quad \forall\, \mathfrak{p} \in T\,.$$

Only these properties of λ and v will be used in the construction of z.

3) Obviously λ is a unit at all spots in T. Let T_λ denote the set of spots $\mathfrak{p}$ in S at which λ is not a unit, i. e. at which $|\lambda|_{\mathfrak{p}} < 1$. Then T_λ is a finite, possibly empty, subset of $S - T$.

Take the splitting $V = Fv \perp U$. Then $L \cap U$ is a lattice on U with respect to S, and $L_{\mathfrak{p}} \cap U_{\mathfrak{p}}$ is a lattice on $U_{\mathfrak{p}}$ with respect to $\mathfrak{p}$. (All localizations are taken in $V_{\mathfrak{p}}$.) It is easily seen, say by Theorem 81:3, that

$$(L \cap U)_{\mathfrak{p}} = L_{\mathfrak{p}} \cap U_{\mathfrak{p}} \quad \forall\, \mathfrak{p} \in S\,.$$

So by Proposition 81:14 there is a lattice K on U such that

$$K_{\mathfrak{p}} = L_{\mathfrak{p}} \cap U_{\mathfrak{p}} \quad \forall\, \mathfrak{p} \in S - T$$

and

$$K_{\mathfrak{p}} \subseteq L_{\mathfrak{p}} \cap U_{\mathfrak{p}} \quad \text{with} \quad \|K_{\mathfrak{p}}\|_{\mathfrak{p}} < \varepsilon \quad \forall\, \mathfrak{p} \in T\,.$$

Thus $K \subseteq L$. This lattice K has the following properties at the different spots in S: at each $\mathfrak{p}$ in $S - (T \cup T_\lambda)$ the lattice $K_{\mathfrak{p}}$ is unimodular since the unimodular lattice $L_{\mathfrak{p}}$ is split by the unimodular sublattice $\mathfrak{o}_{\mathfrak{p}} v$, hence by Corollary 92:1b

$$Q(K_{\mathfrak{p}}) = \mathfrak{o}_{\mathfrak{p}} \quad \forall\, \mathfrak{p} \in S - (T \cup T_\lambda)\,;$$

at each $\mathfrak{p}$ in T_λ the lattice $L_{\mathfrak{p}}$ is unimodular, hence one can use Proposition 82:17 to find a binary unimodular sublattice of $L_{\mathfrak{p}}$ that contains v, this sublattice splits $L_{\mathfrak{p}}$ by Proposition 82:15, hence $K_{\mathfrak{p}}$ contains a binary unimodular sublattice, hence by Corollary 92:1b

$$Q(K_{\mathfrak{p}}) \supseteq \mathfrak{u}_{\mathfrak{p}} \quad \forall\, \mathfrak{p} \in T_\lambda\,;$$

at each $\mathfrak{p}$ in T the lattice $K_\mathfrak{p}$ will contain a maximal lattice of some norm, hence by the last part of Example 91:3 there is an element $\alpha_\mathfrak{p}$ in $F_\mathfrak{p}$ with

$$0 < |\alpha_\mathfrak{p}|_\mathfrak{p} < |4|_\mathfrak{p}\, \varepsilon$$

such that

$$Q(K_\mathfrak{p}) \supseteq \alpha_\mathfrak{p}\, \mathfrak{u}_\mathfrak{p} \quad \forall\, \mathfrak{p} \in T\,.$$

By Lemma 104:1 there is a scalar μ in $\mathfrak{o}$ which is a unit at all spots in $T_\lambda \cup T$ and which has the following property: every element of $\mu\mathfrak{o}$ which is represented by U and also by $K_\mathfrak{p}$ at each $\mathfrak{p}$ in S is represented by K. We let T_μ denote the set of spots $\mathfrak{p}$ in S at which μ is not a unit, i. e. at which $|\mu|_\mathfrak{p} < 1$. Clearly T_μ is a finite, possibly empty, subset of $S - (T_\lambda \cup T)$. We now have the spot diagram

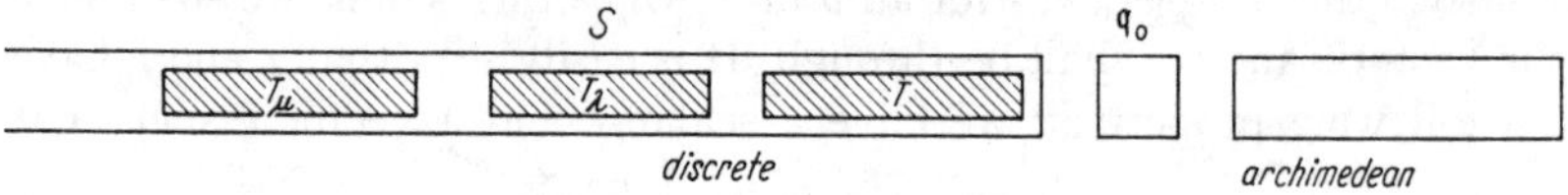

4) We claim that there exists a scalar β in F such that

$$|1 - \lambda^2\beta^2|_\mathfrak{p} \leqq |\mu|_\mathfrak{p} \leqq 1\,, \quad |\beta|_\mathfrak{p} \leqq 1\,, \quad 1 - \lambda^2\beta^2 \in Q(K_\mathfrak{p}) \quad \forall\, \mathfrak{p} \in S$$

and

$$|1 - \beta|_\mathfrak{p} < \varepsilon \quad \forall\, \mathfrak{p} \in T$$

and

$$1 - \lambda^2\beta^2 \in Q(U_\mathfrak{p}) \quad \forall\, \mathfrak{p} \in \Omega - S\,.$$

In order to prove this we use the Very Strong Approximation Theorem and its corollary (§ 33G) to find a β in F in which approximations are made in the following way at the various spots on F:

1. if $\mathfrak{p} \in S - (T_\mu \cup T_\lambda \cup T)$ make $|\beta|_\mathfrak{p} \leqq 1$,
2. if $\mathfrak{p} \in T_\mu$ make $|\beta - \lambda^{-1}|_\mathfrak{p}$ so small that $|1 - \lambda^2\beta^2|_\mathfrak{p} \leqq |\mu|_\mathfrak{p}$,
3. if $\mathfrak{p} \in T_\lambda$ make $|\beta|_\mathfrak{p} \leqq 1$,
4. if $\mathfrak{p} \in T$ make $\left|\beta - \lambda^{-1}\left(1 + \frac{1}{2}\alpha_\mathfrak{p}\right)\right|_\mathfrak{p}$ so small that $|1 - \beta|_\mathfrak{p} < \varepsilon$ and $|1 - \lambda^2\beta^2|_\mathfrak{p} = |\alpha_\mathfrak{p}|_\mathfrak{p}$,
5. if $\mathfrak{p} = \mathfrak{q}_0$ make $|\beta|_{\mathfrak{q}_0}$ so large that $-(1 - \lambda^2\beta^2) = \lambda^2\beta^2(1 - 1/\lambda^2\beta^2) \in \dot{F}^2_{\mathfrak{q}_0}$,
6. if $\mathfrak{p}$ is a real spot in $\Omega - (S \cup \mathfrak{q}_0)$ at which $U_\mathfrak{p}$ represents 1 make $|\beta|_\mathfrak{p}$ so small that $1 - \lambda^2\beta^2 \in \dot{F}^2_\mathfrak{p}$,
7. if $\mathfrak{p}$ is one of the remaining real spots in $\Omega - (S \cup \mathfrak{q}_0)$ make $|\beta|_\mathfrak{p}$ so large that $-(1 - \lambda^2\beta^2) \in \dot{F}^2_\mathfrak{p}$,
8. if $\mathfrak{p}$ is a complex spot in $\Omega - (S \cup \mathfrak{q}_0)$ nothing special is needed.

Using the results of step 3) it follows easily by direct calculation at each stage of the above approximation that the element β chosen in the above way satisfies the conditions stated at the beginning of step 4).

5) Consider the element β with the properties stated at the beginning of step 4). Then $1-\lambda^2\beta^2 \in Q(U_\mathfrak{p})$ at all $\mathfrak{p}$ in Ω, hence $1-\lambda^2\beta^2 \in Q(U)$ by Theorem 66:3. Now $1-\lambda^2\beta^2 \in \mu\,\mathfrak{o}$. And $1-\lambda^2\beta^2$ is represented by $K_\mathfrak{p}$ at all $\mathfrak{p}$ in S. Hence $1-\lambda^2\beta^2 \in Q(K)$ by the choice of μ in step 3). Take y in K with $Q(y) = 1-\lambda^2\beta^2$ and put $z = \beta v + y$. Then z is an element of L with $Q(z) = 1$. Also $\|y\|_\mathfrak{p} < \varepsilon$ for all $\mathfrak{p}$ in T by the choice of K in step 3), hence we have

$$\|z-v\|_\mathfrak{p} \leqq \max(\|v-\beta v\|_\mathfrak{p}, \|y\|_\mathfrak{p}) = \max(|1-\beta|_\mathfrak{p}\,\|v\|_\mathfrak{p}, \|y\|_\mathfrak{p}) < \varepsilon$$

for all $\mathfrak{p}$ in T. Therefore z satisfies the conditions mentioned in step 2). So the theorem is proved. **q. e. d.**

104:4. **Strong Approximation Theorem for Rotations.** *V is a regular quadratic space over the global field F with $\dim V \geqq 3$, S is an indefinite set of spots for V, and T is a finite subset of S. A rotation $\varphi_\mathfrak{p}$ is given in $O'(V_\mathfrak{p})$ at each $\mathfrak{p}$ in T. Then for each $\varepsilon > 0$ there is a rotation σ in $O'(V)$ such that*

$$\|\sigma-\varphi_\mathfrak{p}\|_\mathfrak{p} < \varepsilon \quad \forall\, \mathfrak{p} \in T$$

and

$$\|\sigma\|_\mathfrak{p} = 1 \quad \forall\, \mathfrak{p} \in S-T\,.$$

Proof. The proof consists of two major steps, the first for $n \geqq 4$ and the second for $n = 3$. We let $x_1, \ldots, x_n$ denote a base for V which determines all the norms $\|\ \|_\mathfrak{p}$. We can assume that $0 < \varepsilon < 1$.

1a) We start with $n \geqq 4$. First suppose that each $\varphi_\mathfrak{p}$ is a "short commutator", i. e. that each $\varphi_\mathfrak{p}$ has the form

$$\varphi_\mathfrak{p} = \tau_{u_\mathfrak{p}}\tau_{v_\mathfrak{p}}\tau_{u_\mathfrak{p}}\tau_{v_\mathfrak{p}}$$

where $\tau_{u_\mathfrak{p}}$ and $\tau_{v_\mathfrak{p}}$ are symmetries of $V_\mathfrak{p}$ with respect to the anisotropic vectors $u_\mathfrak{p}$ and $v_\mathfrak{p}$ of $V_\mathfrak{p}$. By using the Weak Approximation Theorem on the coordinates of the $u_\mathfrak{p}$ in the base $x_1, \ldots, x_n$ we can find a vector u in V which is arbitrarily close to $u_\mathfrak{p}$ at each $\mathfrak{p}$ in T. So by the continuity of the map $x \rightarrow \tau_x$ (see Example 101:2) we can assume that τ_u is arbitrarily close to $\tau_{u_\mathfrak{p}}$ at each $\mathfrak{p}$ in T. Now do the same thing with the $v_\mathfrak{p}$ to obtain a vector v in V such that the symmetry τ_v is arbitrarily close to $\tau_{v_\mathfrak{p}}$ at each $\mathfrak{p}$ in T. Then by the continuity of multiplication in $L_{F_\mathfrak{p}}(V_\mathfrak{p})$ we can assume that we have found vectors u and v in V such that

$$\|\tau_u\tau_v\tau_u\tau_v - \tau_{u_\mathfrak{p}}\tau_{v_\mathfrak{p}}\tau_{u_\mathfrak{p}}\tau_{v_\mathfrak{p}}\|_\mathfrak{p} < \varepsilon \quad \forall\, \mathfrak{p} \in T\,.$$

Put $w = \tau_v u$. Then $\tau_w = \tau_v\tau_u\tau_v$. Hence we have a pair of vectors u and w in V with $Q(u) = Q(w) \neq 0$ such that

$$\|\tau_u\tau_w - \varphi_\mathfrak{p}\|_\mathfrak{p} < \varepsilon \quad \forall\, \mathfrak{p} \in T\,.$$

Therefore it is enough to find $\sigma \in O'(V)$ with $\|\sigma\|_\mathfrak{p} = 1$ for all $\mathfrak{p}$ in $S-T$ such that

$$\|\sigma - \tau_u\tau_w\|_\mathfrak{p} < \varepsilon \quad \forall\, \mathfrak{p} \in T\,.$$

Let L be the lattice $L = \mathfrak{o}x_1 + \cdots + \mathfrak{o}x_n$ on V. We can assume that u and w are in L (replace them by λu and λw with a suitable λ in $\mathfrak{o}$ if necessary). Let T_1 be a finite subset of $S - T$ such that $L_\mathfrak{p}$ is unimodular with $Q(u) = Q(w)$ a unit at each $\mathfrak{p}$ in $S - (T_1 \cup T)$. By Theorem 104:3 we can find a vector z in L with $Q(z) = Q(u) = Q(w)$ such that z is arbitrarily close to u at all $\mathfrak{p}$ in T_1 and arbitrarily close to w at all $\mathfrak{p}$ in T. So by the continuity of the map $x \rightarrowtail \tau_x$ we can assume that the symmetry τ_z with respect to this vector z is arbitrarily close to τ_u at all $\mathfrak{p}$ in T_1 and arbitrarily close to τ_w at all $\mathfrak{p}$ in T. Put $\sigma = \tau_u \tau_z$. Then σ is in $O'(V)$. By the continuity of multiplication we can suppose that we have found a σ which is arbitrarily close to $1_{V_\mathfrak{p}}$ at all $\mathfrak{p}$ in T_1 and arbitrarily close to $\tau_u \tau_w$ at all $\mathfrak{p}$ in T, in other words such that

$$\begin{cases} \|\sigma\|_\mathfrak{p} = 1 & \forall\, \mathfrak{p} \in T_1 \\ \|\sigma - \tau_u \tau_w\|_\mathfrak{p} < \varepsilon & \forall\, \mathfrak{p} \in T\,. \end{cases}$$

It remains for us to prove that $\|\sigma\|_\mathfrak{p} = 1$ for all $\mathfrak{p}$ in $S - (T_1 \cup T)$. Now at each such $\mathfrak{p}$ both u and z are elements of $L_\mathfrak{p}$ with $Q(u) = Q(z)$ a unit at $\mathfrak{p}$, hence τ_u and τ_z are units of $L_\mathfrak{p}$ by § 91B, hence $\sigma_\mathfrak{p}$ is a unit of $L_\mathfrak{p}$, hence $\|\sigma\|_\mathfrak{p} = 1$.

1b) We continue with the case $n \geqq 4$, but now we let the $\varphi_\mathfrak{p}$ be arbitrary elements of the commutator subgroups $\Omega_n(V_\mathfrak{p})$ at the spots $\mathfrak{p}$ in T. We can express each $\varphi_\mathfrak{p}$ in the form

$$\varphi_\mathfrak{p} = \psi_\mathfrak{p}^1 \cdots \psi_\mathfrak{p}^r$$

where $\psi_\mathfrak{p}^i$ is a short commutator in $O_n(V_\mathfrak{p})$ (see Proposition 43:6). We can assume that the same r applies at all $\mathfrak{p}$ in T, by adjoining trivial short commutators wherever necessary. By step 1a) we can find ψ^i in $O'(V)$ with ψ^i arbitrarily close to $\psi_\mathfrak{p}^i$ at each $\mathfrak{p}$ in T and $\|\psi^i\|_\mathfrak{p} = 1$ at each $\mathfrak{p}$ in $S - T$. Do this for $i = 1, \ldots, r$. Then $\sigma = \psi^1 \ldots \psi^r$ is the required element.

1c) We conclude the case $n \geqq 4$ by considering arbitrary elements $\varphi_\mathfrak{p}$ in $O'(V_\mathfrak{p})$ at each $\mathfrak{p}$ in T. We can assume that $n = 4$, for otherwise $O'_n(V_\mathfrak{p}) = \Omega_n(V_\mathfrak{p})$ for all $\mathfrak{p}$ in T and this is covered by step 1b). We enlarge T by adjoining to it all spots $\mathfrak{p}$ in $S - T$ at which $V_\mathfrak{p}$ is anisotropic, and we define $\varphi_\mathfrak{p}$ to be $1_{V_\mathfrak{p}}$ at each of the new spots in T. In effect this allows us to assume that $V_\mathfrak{p}$ is isotropic, hence that $O'_4(V_\mathfrak{p}) = \Omega_4(V_\mathfrak{p})$, at all $\mathfrak{p}$ in $S - T$.

By Propositions 95:1 and 101:9 we can find ϱ in $O'_4(V)$ such that $\varrho_\mathfrak{p} \varphi_\mathfrak{p}$ is in $\Omega_4(V_\mathfrak{p})$ at all $\mathfrak{p}$ in T. Let T_1 be the set of spots in $S - T$ at which $\|\varrho^{-1}\|_\mathfrak{p} > 1$. By step 1b) we can find σ in $O'_4(V)$ such that $\|\sigma\|_\mathfrak{p} = 1$ at all $\mathfrak{p}$ in $S - (T_1 \cup T)$, with σ arbitrarily close to $\varrho_\mathfrak{p}$ at each $\mathfrak{p}$ in T_1, and arbitrarily close to $\varrho_\mathfrak{p} \varphi_\mathfrak{p}$ at each $\mathfrak{p}$ in T. Then at each $\mathfrak{p}$ in $S - (T_1 \cup T)$

we have $\|\varrho^{-1}\sigma\|_{\mathfrak{p}} \leqq \|\varrho^{-1}\|_{\mathfrak{p}}\, \|\sigma\|_{\mathfrak{p}} = 1$, hence $\|\varrho^{-1}\sigma\|_{\mathfrak{p}} = 1$; at each $\mathfrak{p}$ in T_1 we have

$$\|\varrho^{-1}\sigma - 1_{V_{\mathfrak{p}}}\|_{\mathfrak{p}} = \|\varrho^{-1}(\sigma - \varrho)\|_{\mathfrak{p}} \leqq \|\varrho^{-1}\|_{\mathfrak{p}}\, \|\sigma - \varrho\|_{\mathfrak{p}}$$

and this is arbitrarily small, hence $\|\varrho^{-1}\sigma\|_{\mathfrak{p}} = 1$; similarly we can obtain $\|\varrho^{-1}\sigma - \varphi_{\mathfrak{p}}\|_{\mathfrak{p}} < \varepsilon$ at each $\mathfrak{p}$ in T. Hence $\varrho^{-1}\sigma$ is the required element.

2a) Now the case $n = 3$. By scaling V we can assume that the discriminant dV is 1. All norms are determined by the base x_1, x_2, x_3 for V. Let x_1', x_2', x_3' be an orthogonal base for V, and let $\|\ \|_{\mathfrak{p}}'$ denote norms with respect to this new base. Take a finite set of spots T' with $T \subseteq T' \subset S$ such that $\|\ \|_{\mathfrak{p}} = \|\ \|_{\mathfrak{p}}'$ and $Q(x_i') \in \mathfrak{u}_{\mathfrak{p}}$ at each $\mathfrak{p}$ in $S - T'$ for $1 \leqq i \leqq 3$. Define $\varphi_{\mathfrak{p}} = 1_{V_{\mathfrak{p}}}$ for all $\mathfrak{p}$ in $T' - T$. If we can prove the theorem for the new set T' and the new system of norms, then we can find a σ in $O'(V)$ with $\|\sigma\|_{\mathfrak{p}} = \|\sigma\|_{\mathfrak{p}}' = 1$ for all $\mathfrak{p}$ in $S - T'$ and $\|\sigma - \varphi_{\mathfrak{p}}\|_{\mathfrak{p}}'$ arbitrarily small for all $\mathfrak{p}$ in T'; in particular we can make $\|\sigma - 1_{V_{\mathfrak{p}}}\|_{\mathfrak{p}} < 1$ for all $\mathfrak{p}$ in $T' - T$ and $\|\sigma - \varphi_{\mathfrak{p}}\|_{\mathfrak{p}} < \varepsilon$ for all $\mathfrak{p}$ in T; this means that $\|\sigma\|_{\mathfrak{p}} = 1$ for all $\mathfrak{p}$ in $T' - T$, and hence for all $\mathfrak{p}$ in $S - T$. In other words, it is enough to prove the theorem for the enlarged set of spots T' and the new system of norms. In effect this allows us to make the following assumption: the base x_1, x_2, x_3 used in determining the norms $\|\ \|_{\mathfrak{p}}$ is an orthogonal base for V in which

$$V \cong \langle a_1 \rangle \perp \langle a_2 \rangle \perp \langle a_1 a_2 \rangle$$

with a_1, a_2 in $\mathfrak{u}_{\mathfrak{p}}$ at each $\mathfrak{p}$ in $S - T$.

2b) Suppose the localization $V_{\mathfrak{p}}$ of V at $\mathfrak{p}$ is replaced by some other localization $V_{\mathfrak{p}}^0$ of V at $\mathfrak{p}$. Then there is a unique isometry $V_{\mathfrak{p}} \rightarrowtail V_{\mathfrak{p}}^0$ which induces the identity map on V. This isometry determines an algebra isomorphism $L_{F_{\mathfrak{p}}}(V_{\mathfrak{p}}) \rightarrowtail L_{F_{\mathfrak{p}}}(V_{\mathfrak{p}}^0)$ in a natural way. The algebra isomorphism so obtained preserves isometries, rotations, symmetries, spinor norms, norms $\|\ \|_{\mathfrak{p}}$, and localizations of elements of $L_F(V)$. It sends $O(V_{\mathfrak{p}})$ to $O(V_{\mathfrak{p}}^0)$ and $O'(V_{\mathfrak{p}})$ to $O'(V_{\mathfrak{p}}^0)$. From this it follows that the theorem holds for the given localizations $V_{\mathfrak{p}}\,(\mathfrak{p} \in \Omega)$ if and only if it holds for some other system of localizations $V_{\mathfrak{p}}^0\,(\mathfrak{p} \in \Omega)$.

Hence we can assume that each $V_{\mathfrak{p}}$ is taken in the localization $C_{\mathfrak{p}}$ of a quaternary quadratic space C which is defined in the following way: fix a 4-dimensional F-space C containing V, fix a vector 1_C in $C - V$, and make C into a quadratic space over F in such a way that

$$C = F 1_C \perp V \quad \text{with} \quad Q(1_C) = 1\,.$$

The localization $C_{\mathfrak{p}}$ of C is taken and fixed at each $\mathfrak{p}$ in Ω. The space $C_{\mathfrak{p}}$ is regarded as a quadratic space over $F_{\mathfrak{p}}$, and $V_{\mathfrak{p}}$ is the subspace of $C_{\mathfrak{p}}$ spanned by V. A norm $\|\ \|_{\mathfrak{p}}$ is put on $C_{\mathfrak{p}}$ with respect to the base $1_C, x_1, x_2, x_3$. This induces the given norm on $V_{\mathfrak{p}}$.

Recall that in the theory of quaternion algebras (§ 57) we started with a 4-dimensional vector space and a base, we fixed them, we put a multiplication on the vector space by means of a multiplication table, and we called the resulting object a quaternion algebra. The initial choice of vector space and base was quite arbitrary. Now do all this starting with

$$C = F1_C + Fx_1 + Fx_2 + Fx_3$$

and make C into the quaternion algebra $\left(\frac{-a_1, -a_2}{F}\right)$ in the defining base $1_C, x_1, x_2, x_3$. Similarly regard $C_\mathfrak{p}$ as the quaternion algebra $\left(\frac{-a_1, -a_2}{F_\mathfrak{p}}\right)$ in the defining base $1_C, x_1, x_2, x_3$. Clearly the quaternion algebra $C_\mathfrak{p}$ is the $F_\mathfrak{p}$-ification of the quaternion algebra C. Let bar stand for conjugation in C and in each $C_\mathfrak{p}$.

The space C can be regarded as a quadratic space in two different ways. First we have the quadratic structure used in defining C, namely $C = F1_C \perp V$ with $Q(1_C) = 1$. Secondly, we have the quadratic structure associated with the quaternion algebra C in the manner of § 57B. It is easily seen that these two quadratic structures are the same, namely

$$B(x, y)\, 1_C = \frac{1}{2}(x\bar{y} + y\bar{x}) \quad \forall\, x, y \in C\,.$$

The same holds for each $C_\mathfrak{p}$.

2c) So much for the logical niceties. We see from the multiplication table used in defining $C_\mathfrak{p}$ that

$$\|xy\|_\mathfrak{p} \leqq \|x\|_\mathfrak{p}\, \|y\|_\mathfrak{p}$$

holds for any x, y in $C_\mathfrak{p}$ at any $\mathfrak{p}$ in $S - T$. And there is a positive constant Γ such that

$$\|xy\|_\mathfrak{p} \leqq \Gamma\, \|x\|_\mathfrak{p}\, \|y\|_\mathfrak{p}$$

for any x, y in $C_\mathfrak{p}$ at any $\mathfrak{p}$ in T.

By Proposition 57:13 there is a vector $z_\mathfrak{p}$ in $C_\mathfrak{p}$ at each $\mathfrak{p}$ in T with $Q(z_\mathfrak{p}) = 1$ and

$$\varphi_\mathfrak{p} x = z_\mathfrak{p}\, x z_\mathfrak{p}^{-1} \quad \forall\, x \in V_\mathfrak{p}\,.$$

Apply Theorem 104:3 to the quadratic space C. We obtain a vector z in C with $Q(z) = 1$ such that $\|z\|_\mathfrak{p} \leqq 1$ at all $\mathfrak{p}$ in $S - T$ and $\|z - z_\mathfrak{p}\|_\mathfrak{p} < \varepsilon'$ at all $\mathfrak{p}$ in T, where ε' is a positive number with

$$\varepsilon' < \|z_\mathfrak{p}\|_\mathfrak{p}\,, \quad \varepsilon' < \varepsilon/\Gamma^2\, \|z_\mathfrak{p}\|_\mathfrak{p}$$

for all $\mathfrak{p}$ in T. By Proposition 57:13 there is an element σ in $O'(V)$ with

$$\sigma x = z\, x\, z^{-1} \quad \forall\, x \in V\,.$$

Then for any x in V and any $\mathfrak{p}$ in $S - T$ we have

$$\|\sigma x\|_\mathfrak{p} = \|z\, x\, z^{-1}\|_\mathfrak{p} = \|z\, x\, \bar{z}\|_\mathfrak{p} \leqq \|z\|_\mathfrak{p}\, \|x\|_\mathfrak{p}\, \|\bar{z}\|_\mathfrak{p} \leqq \|x\|_\mathfrak{p}\,,$$

hence $\|\sigma x_i\|_{\mathfrak{p}} \leqq 1$ for $1 \leqq i \leqq 3$, hence $\|\sigma\|_{\mathfrak{p}} \leqq 1$, hence $\|\sigma\|_{\mathfrak{p}} = 1$. A similar argument (essentially an argument of continuity of multiplication) gives $\|\sigma - \varphi_{\mathfrak{p}}\|_{\mathfrak{p}} < \varepsilon$ at all $\mathfrak{p}$ in T. **q. e. d.**

104:5. **Theorem.** *V is a regular quadratic space over the global field F with $\dim V \geqq 3$, S is an indefinite set of spots for V, and L is a lattice on V with respect to S. Then*

$$\text{cls}^+ L = \text{spn}^+ L \quad \textit{and} \quad \text{cls}\, L = \text{spn}\, L\,.$$

Proof. We shall prove $\text{cls}\, L = \text{spn}\, L$. The equation $\text{cls}^+ L = \text{spn}^+ L$ is done in the same way. So we consider a lattice K in $\text{spn}\, L$ and we must prove that $K \in \text{cls}\, L$. By the definition of the spinor genus we have σ in $O(V)$ and $\Sigma_{\mathfrak{p}}$ in $O'(V_{\mathfrak{p}})$ at each $\mathfrak{p}$ in S such that

$$K_{\mathfrak{p}} = \sigma_{\mathfrak{p}} \Sigma_{\mathfrak{p}} L_{\mathfrak{p}} \quad \forall\, \mathfrak{p} \in S\,.$$

Then $(\sigma^{-1} K)_{\mathfrak{p}} = \Sigma_{\mathfrak{p}} L_{\mathfrak{p}}$ for all $\mathfrak{p}$ in S. But $\sigma^{-1} K \in \text{cls}\, K$. Hence we may assume that $\sigma = 1_V$. So we now have

$$K_{\mathfrak{p}} = \Sigma_{\mathfrak{p}} L_{\mathfrak{p}} \quad \forall\, \mathfrak{p} \in S\,.$$

Fix a base $x_1, \ldots, x_n$ for V, let norms $\| \;\|_{\mathfrak{p}}$ be defined with respect to this base on $V_{\mathfrak{p}}$ at each $\mathfrak{p}$ in S, and let M be the lattice $M = \mathfrak{o} x_1 + \ldots + \mathfrak{o} x_n$. Take a finite subset T of S such that

$$K_{\mathfrak{p}} = L_{\mathfrak{p}} = M_{\mathfrak{p}} \quad \forall\, \mathfrak{p} \in S - T\,.$$

Then by the Strong Approximation Theorem for Rotations there is a rotation ϱ in $O'(V)$ such that $\|\varrho\|_{\mathfrak{p}} = 1$ for all $\mathfrak{p}$ in $S - T$, with $\|\varrho - \Sigma_{\mathfrak{p}}\|_{\mathfrak{p}}$ arbitrarily small for all $\mathfrak{p}$ in T. The first condition informs us that

$$(\varrho L)_{\mathfrak{p}} = \varrho_{\mathfrak{p}} M_{\mathfrak{p}} = K_{\mathfrak{p}} \quad \forall\, \mathfrak{p} \in S - T\,.$$

The second condition informs us, in virtue of Example 101:4, that we can obtain $\varrho_{\mathfrak{p}} L_{\mathfrak{p}} = K_{\mathfrak{p}}$ at all $\mathfrak{p}$ in T by taking good enough approximations in the selection of ϱ. Then $(\varrho L)_{\mathfrak{p}} = K_{\mathfrak{p}}$ for all $\mathfrak{p}$ in S. Hence $\varrho L = K$. Hence $K \in \text{cls}\, L$. **q. e. d.**

104:6. **Example.** The purpose of this example is to show that the condition $\dim V \geqq 3$ in Theorem 104:5 is essential. Let F be the field of rational numbers, let S be the set of all discrete spots on F, let V be a hyperbolic plane. So S is certainly indefinite for V. And $\mathfrak{o}$ is the ring of rational integers. Take a lattice $L = \mathfrak{o} x + \mathfrak{o} y$ on V and suppose that

$$L \cong \begin{pmatrix} 0 & 9 \\ 9 & 2 \end{pmatrix} \quad \text{in } x, y\,.$$

Put $K = \mathfrak{o}\left(\frac{1}{4} x\right) + \mathfrak{o}(4y)$. So

$$K \cong \begin{pmatrix} 0 & 9 \\ 9 & 32 \end{pmatrix} \quad \text{in } \frac{1}{4} x,\; 4y\,.$$

We claim that

$$K \notin \operatorname{cls}^+ L \quad \text{but} \quad K \in \operatorname{spn}^+ L\,.$$

It is easily seen that L does and that K does not represent 2, hence $K \notin \operatorname{cls}^+ L$. Let us show that $K \in \operatorname{spn}^+ L$. In fact we shall find $\varphi_{\mathfrak{p}} \in O'(V_{\mathfrak{p}})$ at each $\mathfrak{p}$ in S such that $K_{\mathfrak{p}} = \varphi_{\mathfrak{p}} L_{\mathfrak{p}}$. If $\mathfrak{p}$ is non-dyadic this is trivial since then $K_{\mathfrak{p}} = L_{\mathfrak{p}}$ by definition of K and L. So consider the 2-adic spot $\mathfrak{p}$. Then $K_{\mathfrak{p}}$ and $L_{\mathfrak{p}}$ are unimodular with $\mathfrak{n} K_{\mathfrak{p}} = \mathfrak{n} L_{\mathfrak{p}} = 2\mathfrak{o}_{\mathfrak{p}}$, hence by Corollary 82:21a there is a vector w in $V_{\mathfrak{p}}$ with $Q(w) = 0$ and

$$L_{\mathfrak{p}} = \mathfrak{o}_{\mathfrak{p}} x + \mathfrak{o}_{\mathfrak{p}} w\,, \quad K_{\mathfrak{p}} = \mathfrak{o}_{\mathfrak{p}} \left(\frac{1}{4} x\right) + \mathfrak{o}_{\mathfrak{p}} (4w)\,.$$

By Example 55:1 we have a rotation $\varphi_{\mathfrak{p}}$ of $V_{\mathfrak{p}}$ onto itself such that

$$\varphi_{\mathfrak{p}} x = \frac{1}{4} x \quad \text{and} \quad \varphi_{\mathfrak{p}} w = 4w\,,$$

and the spinor norm of this rotation is equal to $\frac{1}{4} \dot{F}_{\mathfrak{p}}^2$. Hence $\varphi_{\mathfrak{p}} \in O'(V_{\mathfrak{p}})$. Moreover $\varphi_{\mathfrak{p}} L_{\mathfrak{p}} = K_{\mathfrak{p}}$. Hence $K \in \operatorname{spn}^+ L$.

104:7. **Remark.** The reader will be able to use Theorem 106:13 to show that the assumption of indefiniteness in Theorem 104:5 is essential.

104:8. **Remark.** Theorem 104:5 tells us that we have $\operatorname{cls}^+ L = \operatorname{spn}^+ L$ in the indefinite case in 3 or more variables. On the other hand, Theorem 102:9 and Example 102:10 give sufficient conditions for $\operatorname{spn}^+ L$ and $\operatorname{gen} L$ to be equal. Hence we have sufficient conditions for determining when $\operatorname{cls}^+ L$ and $\operatorname{gen} L$ are equal. Now the genus is completely described by the local theory. Hence we have certain sufficient conditions that can be used to describe the proper class (in general these conditions are not necessary).

104:9. **Theorem.** *V is a regular quadratic space over the global field F with* $\dim V \geqq 3$, *and S is an indefinite set of spots for V which satisfies the condition $J_F = P_D J_F^S$ where $D = \theta(O^+(V))$. Suppose that L is either a modular or a maximal lattice on V with respect to S. Then*

$$\operatorname{cls}^+ L = \operatorname{gen} L\,.$$

Proof. We have $\operatorname{cls}^+ L = \operatorname{spn}^+ L$ by Theorem 104:5. But $\operatorname{spn}^+ L = \operatorname{gen} L$ by Theorem 102:9 and Example 102:10. Hence $\operatorname{cls}^+ L = \operatorname{gen} L$.

q. e. d.

104:10. **Theorem.** *V is a regular quadratic space over the field of rational numbers $\mathbb{Q}$ with* $\dim V \geqq 3$, *and S is an indefinite set of spots for V. L and K are lattices of the same norm and scale on V with respect to S. Suppose that L and K are either both modular or both maximal. Then*

$$\operatorname{cls}^+ L = \operatorname{cls}^+ K\,.$$

Proof. In the modular case we have $\operatorname{gen} L = \operatorname{gen} K$ by Theorems 92:2 and 93:29. In the maximal case we note that both L and K must be

$\mathfrak{a}$-maximal where $\mathfrak{a}$ is the common norm $\mathfrak{a} = \mathfrak{n}L = \mathfrak{n}K$, hence $\text{gen}\, L = \text{gen}\, K$ by Theorem 91:2. So in either case we have $\text{gen}\, L = \text{gen}\, K$. And $J_F = P_D J_F^S$ by Example 102:10. Hence by Theorem 104:9,

$$\text{cls}^+ L = \text{gen}\, L = \text{gen}\, K = \text{cls}^+ K\,.$$

q. e. d.

104:11. **Example.** Let L be a unimodular lattice on a binary space V over the field of rational numbers, with respect to the set of all discrete spots S on $\mathbf{Q}$. Suppose S is indefinite for V. (Thus L is a $\mathbf{Z}$-lattice and V_∞ is isotropic.) Prove from first principles that

$$L \cong \langle 1 \rangle \perp \langle -1 \rangle \quad \text{or} \quad L \cong \begin{pmatrix} 0 & 1 \\ 1 & 0 \end{pmatrix},$$

and hence $\text{cls}^+ L = \text{gen}\, L$.

104:12 **Example.** Let L be a unimodular lattice with respect to $\mathbf{Z}$ on a quadratic space V over the field of rational numbers $\mathbf{Q}$. Suppose that L has norm $\mathbf{Z}$ and that V_∞ is isotropic. Use the Hilbert Reciprocity Law, Theorem 63:20, and the Hasse-Minkowski Theorem to show that

$$V \cong \langle 1 \rangle \perp \cdots \perp \langle 1 \rangle \perp \langle -1 \rangle \perp \cdots \perp \langle -1 \rangle\,.$$

Deduce that

$$L \cong \langle 1 \rangle \perp \cdots \perp \langle 1 \rangle \perp \langle -1 \rangle \perp \cdots \perp \langle -1 \rangle\,.$$

§ 105. The indecomposable splitting of a definite lattice

Consider a lattice L in a quadratic space V. We say that L is decomposable if there exist non-zero lattices K_1 and K_2 contained in L such that

$$L = K_1 \perp K_2\,.$$

If L is not decomposable we call it indecomposable. It is clear that every lattice L is the orthogonal sum of at most n indecomposable components, where n is the rank of L. A splitting of this sort is called an indecomposable splitting of L.

105:1. **Theorem.** *L is a lattice on the regular quadratic space V over an algebraic number field F. Suppose that the underlying sets of spots S is definite for V and also that it contains all dyadic spots on F. Then the components $L_1, \ldots, L_r$ of an indecomposable splitting $L = L_1 \perp \cdots \perp L_r$ are unique (but for their order).*

Proof. 1) By scaling V we can assume that $\mathfrak{s}L \subseteq \mathfrak{o}$.

We shall again need the counting number $N\mathfrak{a}$ of § 33C. As in § 103 we put $N\alpha = N(\alpha\mathfrak{o})$ for any α in $\dot{F}$. We have

$$N\alpha = \prod_{\mathfrak{p} \in S} \frac{1}{|\alpha|_\mathfrak{p}} = \prod_{\mathfrak{p} \in \Omega - S} |\alpha|_\mathfrak{p}\,.$$

The assumption that $\mathfrak{s}L \subseteq \mathfrak{o}$ implies that $N(Qx)$ is a natural number for all non-zero x in L.

2) We shall call a vector x in L reducible if there are non-zero vectors y and z in L with $B(y, z) = 0$ such that $x = y + z$. We call x irreducible if it is not reducible. Our purpose here is to show that every vector in L is a sum of irreducible vectors of L.

First consider the sum $y + z$ of non-zero vectors y and z of V with $B(y, z) = 0$. Then $Q(y + z) = Q(y) + Q(z)$. Here $\dim V \geqq 2$, so definiteness implies that the spots in $\Omega - S$ are either real or discrete. If $\mathfrak{p}$ is any real spot on F, then $V_\mathfrak{p}$ is anisotropic and so $Q(y)$ and $Q(z)$ are either both positive or both negative at $\mathfrak{p}$, hence $|Q(y+z)|_\mathfrak{p} > |Q(y)|_\mathfrak{p}$. Now consider a discrete spot $\mathfrak{p}$ in $\Omega - S$. We say that $|Q(y + z)|_\mathfrak{p} \geqq |Q(y)|_\mathfrak{p}$. Suppose not. Then $|Q(y + z)|_\mathfrak{p} < |Q(y)|_\mathfrak{p}$ and so

$$Q(y) + Q(z) \equiv Q(y + z) \equiv 0 \bmod Q(y)\, \mathfrak{m}_\mathfrak{p}\,.$$

Hence by the Local Square Theorem $Q(y)$ is a square times $-Q(z)$. This is of course absurd since $V_\mathfrak{p}$ is not isotropic. We have therefore proved that $|Q(y + z)|_\mathfrak{p} \geqq |Q(y)|_\mathfrak{p}$ holds at all $\mathfrak{p}$ in $\Omega - S$ with strict inequality at least once. Hence

$$N(Q(y + z)) > N(Q(y))$$

Similarly $N(Q(y + z)) > N(Q(z))$.

The proof that every vector x in L is a sum of irreducible vectors of L is now done by induction on the natural number $m = N(Qx)$: if x is reducible write $x = y + z$ with y and z in L and $1 \leqq N(Qy) \leqq m - 1$, $1 \leqq N(Qz) \leqq m - 1$.

3) We put an equivalence relation on the set of non-zero irreducible vectors of L as follows: write $x \sim y$ if there is a chain of irreducible vectors

$$x = z_1, \quad z_2, \ldots, z_q = y \quad (q \geqq 1)$$

in which $B(z_i, z_{i+1}) \neq 0$ for $1 \leqq i \leqq q - 1$. Let $C_1, C_2, \ldots$ denote the equivalence classes associated with this equivalence relation. Let K_1, $K_2, \ldots$ denote the sublattice of L that is generated by the vectors in $C_1, C_2, \ldots$. Then $B(K_i, K_j) = 0$ since $B(C_i, C_j) = 0$ for $i \neq j$. Hence the number of equivalence classes is finite, say $C_1, \ldots, C_t$. And the sum of the lattices $K_1, \ldots, K_t$ is actually an orthogonal sum: $K_1 \perp \cdots \perp K_t$. Now we proved in step 2) that every vector in L is a sum of irreducible vectors of L. Hence

$$L = K_1 \perp \cdots \perp K_t\,.$$

4) Consider x in C_1. Then x is in L, i. e. x is in $L_1 \perp \cdots \perp L_r$. But x is an irreducible element of L. Hence x falls in exactly one of the above components of L, say $x \in L_1$. It follows from the definition of $\sim$ that $C_1 \subseteq L_1$. Hence $K_1 \subseteq L_1$. Hence each K_i is contained in some L_j. Since L

is also equal to $K_1 \perp \cdots \perp K_t$ we therefore see that each L_i is the orthogonal sum of all the K_j contained in it. But L_i is indecomposable. Hence each L_i is a K_j. **q. e. d.**

105:2. Example. Let V be a quadratic space over the field of rational numbers $\mathbf{Q}$ and suppose that V has a base $x_1, \ldots, x_4$ in which

$$V \cong \langle 1 \rangle \perp \langle 1 \rangle \perp \langle 1 \rangle \perp \langle 1 \rangle .$$

Consider an underlying set of spots S consisting of all non-dyadic spots on $\mathbf{Q}$ and let L be the lattice

$$L = \mathfrak{o} x_1 \perp \cdots \perp \mathfrak{o} x_4 .$$

It follows easily from the local theory that S is a definite set of spots for V. And the four vectors

$$\frac{1}{2}(x_1 \pm x_2 + x_3 \pm x_4), \quad \frac{1}{2}(x_1 \pm x_2 - x_3 \mp x_4)$$

form a base for L in which

$$L \cong \langle 1 \rangle \perp \langle 1 \rangle \perp \langle 1 \rangle \perp \langle 1 \rangle .$$

So the assumption made in Theorem 105:1 that S contain all dyadic spots cannot be relaxed. The reader may easily verify that L also has a splitting in which

$$L \cong \langle 2 \rangle \perp \langle 2 \rangle \perp \langle 1 \rangle \perp \langle 1 \rangle .$$

It is also easily verified that the assumption of definiteness in Theorem 105:1 is essential.

§ 106. Definite unimodular lattices over the rational integers

We conclude with some very special results on the class of a unimodular lattice of small dimension over the ring of rational integers $\mathbf{Z}$. If the underlying set of spots is indefinite for the quadratic space in question, then the class is equal to the genus and all is known. This is no longer true in the definite case (for instance we shall see that there is a unimodular lattice of dimension 9 whose genus contains two distinct classes). We shall confine ourselves to the definite case.

The situation then is this. F is now the field of rational numbers $\mathbf{Q}$, S is the set of all discrete spots on $\mathbf{Q}$, and $\mathbf{Z}$ is the ring of integers $\mathfrak{o}(S)$ of F at S. Lattice theory is with respect to S. As usual we use the same letter p for the prime number p and the prime spot p which it determines. V is a regular n-ary quadratic space over $\mathbf{Q}$ and it is assumed that S is a definite set of spots for V. This is equivalent to saying that the localization V_∞ is either positive definite or negative definite since S consists of all discrete spots on $\mathbf{Q}$. By scaling we can assume that V_∞ is actually positive definite. We shall assume that there is at least one unimodular lattice on V. Now the discriminant of any unimodular lattice over the ring $\mathbf{Z}$ is either $+1$ or -1, so in the situation under discussion it has to be $+1$. In particular, $dV = 1$. Furthermore there is a unimodular lattice on the localization V_p at each discrete spot p, hence $S_p V = 1$ for $p = 3, 5, 7, \ldots$; but $S_\infty V = 1$ since V_∞ is positive definite; hence $S_2 V = 1$ by the Hilbert Reciprocity Law. Hence by Theorem 63:20 and the Hasse-Minkowski Theorem we have

$$V \cong \langle 1 \rangle \perp \cdots \perp \langle 1 \rangle .$$

We therefore assume throughout this paragraph that V has the above form.

The symbol I_n will denote the $n \times n$ identity matrix. Thus we have $V \cong I_n$. We call a lattice D on V completely decomposable if it splits into an orthogonal sum of lattices of dimension 1. Thus in the situation under discussion the unimodular lattice D on V is completely decomposable if and only if it has the matrix I_n.

§ 106 A. Even and odd lattices

Consider a unimodular lattice L with respect to $\mathbf{Z}$ on the given quadratic space V over $\mathbf{Q}$. Then $\mathfrak{s}L = \mathbf{Z}$ and $2\mathbf{Z} \subseteq \mathfrak{n}L \subseteq \mathbf{Z}$ so that $\mathfrak{n}L$ is either $\mathbf{Z}$ or $2\mathbf{Z}$. We call the unimodular lattice L odd if $\mathfrak{n}L = \mathbf{Z}$, we call it even if $\mathfrak{n}L = 2\mathbf{Z}$. Thus L is even if and only if $Q(L) \subseteq 2\mathbf{Z}$. An analogous argument leading to an analogous definition can be employed for unimodular lattices over $\mathbf{Z}_2$ (but there is no distinction between odd and even over $\mathbf{Z}_p$ when $p > 2$). It is easily seen that the unimodular lattice L over $\mathbf{Z}$ is even if and only if the localization L_2 over $\mathbf{Z}_2$ is even.

106:1. *V is a regular quadratic space with matrix I_n over $\mathbf{Q}$. Then there is an even unimodular lattice with respect to $\mathbf{Z}$ on V if and only if $n \equiv 0 \bmod 8$.*

Proof. 1) In the course of the proof it will be found necessary to use the 2-adic evaluations of the Hilbert symbol, also the fact that 1, 3, 5, 7 are representatives of the four square classes of 2-adic units, and finally the fact that 5 is a 2-adic unit of quadratic defect $4\,\mathbf{Z}_2$. All these things were established in the statement and proof of Proposition 73:2. As in § 93B, we let $A(\alpha, \beta)$ stand for the 2-adic matrix $\begin{pmatrix} \alpha & 1 \\ 1 & \beta \end{pmatrix}$.

2) First suppose there is an even unimodular lattice on V. Then there is an even unimodular lattice on the localization V_2, hence by the local theory (Examples 93:11 and 93:18) we must have either

$$V_2 \cong \langle A(0,0) \rangle \perp \cdots \perp \langle A(0,0) \rangle$$

or

$$V_2 \cong \langle A(0,0) \rangle \perp \cdots \perp \langle A(2,2) \rangle .$$

But $dV_2 = +1$ and each of the numbers $-1, -3, +3$ is a non-square in $\mathbf{Q}_2$, hence we must actually have

$$V_2 \cong \langle A(0,0)\rangle \perp \cdots \perp \langle A(0,0)\rangle$$

with $n \equiv 0 \bmod 4$. A computation of Hasse symbols over $\mathbf{Q}_2$ shows that $n \equiv 4 \bmod 8$ is impossible. Hence $n \equiv 0 \bmod 8$.

3) Conversely let us assume that $n \equiv 0 \bmod 8$. Then the criterion of Theorem 63:20 informs us that

$$V_2 \cong \langle A(0,0)\rangle \perp \cdots \perp \langle A(0,0)\rangle$$

Hence by Proposition 81:14 there is a lattice L on V with $L_p \cong I_n$ when $p = 3, 5, 7, \ldots$ and

$$L_2 \cong \langle A(0,0)\rangle \perp \cdots \perp \langle A(0,0)\rangle$$

This L is clearly unimodular and even. **q. e. d.**

§ 106B. Adjacent lattices

We continue our investigation of unimodular lattices with respect to $\mathbf{Z}$ on the quadratic space V with matrix I_n over $\mathbf{Q}$. For any such lattice L we define $\mathfrak{F}(L)$ to be the set of all unimodular lattices K with respect to $\mathbf{Z}$ on V such that

$$K_p = L_p \quad \text{for} \quad p = 3, 5, 7, \ldots.$$

Note that $\sigma\mathfrak{F}(L) = \mathfrak{F}(\sigma L)$ for any σ in $O_n(V)$.

106:2. *Suppose $n \geqq 5$ and let K and L be any two unimodular lattices on the space V under discussion. Then there is a lattice J in* cls^+K *such that $J \in \mathfrak{F}(L)$.*

Proof. Let S' denote the set of all non-dyadic spots on $\mathbf{Q}$ and put $\mathbf{Z}' = \mathfrak{o}(S')$. Then $\mathbf{Z} \subseteq \mathbf{Z}' \subseteq \mathbf{Z}_p$ holds for each odd p. The $\mathbf{Z}'$-module L' generated by L in V is clearly a lattice on V with respect to $\mathbf{Z}'$, and $L \subseteq L' \subseteq L_p$ holds for each odd p. Hence L' has the property that $L'_p = L_p$ for each odd p. Now do the same with K to obtain a lattice K' with respect to $\mathbf{Z}'$ with analogous properties.

The set of spots S' is indefinite for V since the localization V_2, having dimension $\geqq 5$, is isotropic. But L' and K' are unimodular of norm $\mathbf{Z}'$. Hence $\text{cls}^+L' = \text{cls}^+K'$ by Theorem 104:10. So we can pick $\sigma \in O_n^+(V)$ such that $\sigma K' = L'$. Put $J = \sigma K$. So $J \in \text{cls}^+K$. And for each odd p we have

$$J_p = (\sigma K)_p = \sigma_p K_p = \sigma_p K'_p = (\sigma K')_p = L'_p = L_p\,.$$

Therefore $J \in \mathfrak{F}(L)$. **q. e. d.**

If K and L are unimodular lattices on V, then K is in $\mathfrak{F}(L)$ if and only if the invariant factors of K in L are of the form

$$2^{r_1}\mathbf{Z}, \ldots, 2^{r_n}\mathbf{Z}$$

(with $r_1 \leqq 0$ and $r_n \geqq 0$). We say that K is adjacent to L if the invariant factors of K in L are equal to

$$\frac{1}{2}\mathbf{Z}, \mathbf{Z}, \ldots, \mathbf{Z}, 2\mathbf{Z}$$

(assume $n \geqq 2$ for this definition). Suppose K is adjacent to L. Then it follows immediately from the definitions that L is adjacent to K, that K is in $\mathfrak{F}(L)$, that there is a base $x_1, \ldots, x_n$ for V in which

$$\begin{cases} L = \mathbf{Z}x_1 + \cdots + \mathbf{Z}x_n \\ K = \mathbf{Z}\left(\frac{1}{2}x_1\right) + \cdots + \mathbf{Z}(2x_n), \end{cases}$$

that $2L \subsetneqq K \subsetneqq \frac{1}{2}L$, and finally that σK is adjacent to σL for any σ in $O_n(V)$. The number of lattices adjacent to L is finite since the index $\left(\frac{1}{2}L : 2L\right)$ is finite.

106:3. *Suppose $n \equiv 0 \bmod 8$ and let L be any unimodular lattice on the space V under discussion. Then there are even and odd unimodular lattices on V which are adjacent to L.*

Proof. By the criterion of Theorem 63:20 we know that V_2 is the orthogonal sum of hyperbolic planes, hence by Example 93:18 we have either

$$L_2 \cong \langle A(0,0)\rangle \perp \cdots \perp \langle A(4,0)\rangle$$

or

$$L_2 \cong \langle A(0,0)\rangle \perp \cdots \perp \langle A(1,0)\rangle .$$

There is clearly a unimodular lattice K_2 on V_2 which is either odd or even as desired, and whose invariant factors in L_2 are equal to

$$\frac{1}{2}\mathbf{Z}_2, \mathbf{Z}_2, \ldots, \mathbf{Z}_2, 2\mathbf{Z}_2 .$$

Take a lattice J on V with

$$J_p = \begin{cases} K_2 & \text{if } p = 2 \\ L_p & \text{if } p = 3, 5, 7, \ldots . \end{cases}$$

This J is the required lattice on V. **q. e. d.**

106:4. *Suppose $n \geqq 2$ and let K and L be any two unimodular lattices on the space V under discussion with $K \in \mathfrak{F}(L)$ and $K \neq L$. Then there is a chain of unimodular lattices*

$$L = J_1, J_2, \ldots, J_t = K$$

on V with J_{i+1} adjacent to J_i.

Proof. It is enough to find a chain of unimodular lattices

$$J^1, J^2, \ldots, J^t \quad (\text{over } \mathbf{Z}_2)$$

on the localization V_2 with $J^1 = L_2$ and $J^t = K_2$, and such that the invariant factors of J^{i+1} in J^i are equal to

$$\frac{1}{2}\mathbf{Z}_2, \mathbf{Z}_2, \ldots, \mathbf{Z}_2, 2\mathbf{Z}_2 .$$

For then we can define lattices J_i $(1 \leqq i \leqq t)$ with respect to $\mathbf{Z}$ by the equations

$$J_{ip} = \begin{cases} J^i & \text{if } p = 2 \\ L_p & \text{if } p = 3, 5, 7, \ldots, \end{cases}$$

and $L = J_1, J_2, \ldots, J_t = K$ will provide the desired chain from L to K. If $n = 2$ we have

$$\begin{cases} L_2 = \mathbf{Z}_2 x + \mathbf{Z}_2 y \\ K_2 = \mathbf{Z}_2\left(\frac{x}{2^r}\right) + \mathbf{Z}_2(2^r y) \end{cases}$$

and the required local chain is obvious. We may therefore assume that $n \geqq 3$.

By Proposition 106:3 we can assume that both L and K are odd, hence that the localizations L_2 and K_2 are odd. The local theory (Theorem 93:29) then gives an isometry σ in $O_n(V_2)$ such that $K_2 = \sigma L_2$. By expressing σ as a product of symmetries on V_2 we see that it is enough if we assume that σ is itself a symmetry such that $\sigma L_2 \neq L_2$. Thus $\sigma = \tau_u$ with u a maximal anisotropic vector in L_2, say. It follows easily from the local theory that there is either a 1- or a 2-dimensional unimodular sublattice M of L_2 which contains u. In the first event we would have $\tau_u L_2 = L_2$. Hence M is actually binary. Take the splitting $L_2 = M \perp N$. Then $N = \tau_u N \subseteq K_2$ and we therefore have a splitting $K_2 = M' \perp N$. Write

$$\begin{cases} M \;= \mathbf{Z}_2 x + \mathbf{Z}_2 y \\ M' = \mathbf{Z}_2\left(\frac{x}{2^r}\right) + \mathbf{Z}_2(2^r y) . \end{cases}$$

The required local chain from M to M', hence from L_2 to K_2, is now obvious. **q. e. d.**

We are beginning to see how the theory of adjacent lattices can be used in determining the unimodular classes on V. We start with a fixed lattice D of the form

$$D \cong \langle 1 \rangle \perp \cdots \perp \langle 1 \rangle$$

on V. By the finiteness of class number (Theorem 103:4) and by Proposition 106:2 there are lattices $K_1, \ldots, K_t$ in $\mathfrak{F}(D)$ such that

$$\text{cls}^+ K_1, \ldots, \text{cls}^+ K_t$$

are all the distinct proper unimodular classes on V (at least for $n \geqq 5$). So we shall certainly achieve our purpose if we can find the lattices

$K_1, \ldots, K_t$. How is this to be done? By a step-by-step construction of adjacent lattices starting with the fixed lattice D. First we construct by a certain procedure all the lattices adjacent to D (we have already seen that these are finite in number); then all lattices adjacent to these; and so on. By Proposition 106:4 we can obtain $K_1, \ldots, K_t$ by steps of the above type. The essential technical features of this procedure are the following: the method of construction, deciding when all K_i have been found, eliminating duplications. The rest of the chapter is devoted to these matters and their application up to $n = 9$.

§ 106C. Rules of construction

Here we give rules for operating with adjacent lattices. Throughout this subparagraph we assume without further reference that K and L are two unimodular lattices with respect to $\mathbf{Z}$ on the quadratic space V with matrix I_n over $\mathbf{Q}$.

106:5. *The following assertions are equivalent:*

(1) *K is adjacent to L*

(2) $\mathfrak{v}(K + L) = \frac{1}{4}\mathbf{Z}$

(3) $\mathfrak{v}(K \cap L) = 4\mathbf{Z}$.

Proof. That (1) implies (2) is immediate from the definition of adjacent lattices and the definition of the volume $\mathfrak{v}$. To prove that (2) implies (3) we use the fact that $L^{\#} = L$ holds for any unimodular lattice L; then

$$\mathfrak{v}(K \cap L) = \mathfrak{v}(K^{\#} \cap L^{\#}) = \mathfrak{v}(K + L)^{\#} = 4\mathbf{Z}\,.$$

That (3) implies (1) follows easily from the Invariant Factor Theorem.

q. e. d.

106:6. *Suppose K is adjacent to L and let y be a vector in $L - K$. Then*

(1) $L + K = \mathbf{Z}y + K$

(2) $L = \mathbf{Z}y + (L \cap K)$

(3) $L \cap K = (\mathbf{Z}y + K)^{\#} = \{w \in K \mid B(w, y) \in \mathbf{Z}\}$.

Proof. (1) We have $K \subset \mathbf{Z}y + K \subseteqq L + K$. But there are no lattices properly between K and $L + K$ by Proposition 106:5. Hence $\mathbf{Z}y + K = L + K$.

(2) We have $L \cap K \subset \mathbf{Z}y + (L \cap K) \subseteqq L$. But there are no lattices properly between $L \cap K$ and L. Hence $\mathbf{Z}y + (L \cap K) = L$.

(3) Since L and K are assumed to be unimodular we have

$$L \cap K = L^{\#} \cap K^{\#} = (L + K)^{\#} = (\mathbf{Z}y + K)^{\#}.$$

The second equation follows from the definition of the dual $(\mathbf{Z}y + K)^{\#}$.

q. e. d.

106:7. *Let y be any vector in $\left(\frac{1}{2}K\right) - K$ with $Q(y)$ in $\mathbf{Z}$. Then there is exactly one unimodular lattice J on V that contains y and is adjacent* *to K. This lattice can be constructed by forming*

$$J = \mathbf{Z}y + (\mathbf{Z}y + K)^{\#}.$$

Proof. The fact that J, if it exists, will have the above form is an immediate consequence of Proposition 106:6. Incidentally, this also proves the uniqueness. We now define a lattice J on V by the above equation and we prove that the J so defined is first of all unimodular, and secondly that it is adjacent to K.

A direct computation shows that $B(J, J) \subseteq \mathbf{Z}$, hence $\mathfrak{s}J \subseteq \mathbf{Z}$. And if we write $K = \mathbf{Z}w + \mathbf{Z}x_2 + \cdots + \mathbf{Z}x_n$ with $2y = mw$ $(m \in \mathbf{Z})$, then

$$\mathbf{Z}y + K \subseteq \mathbf{Z}\left(\frac{1}{2}w\right) + \mathbf{Z}x_2 + \cdots + \mathbf{Z}x_n$$

so that

$$\mathbf{Z} \subset \mathfrak{v}(\mathbf{Z}y + K) \subseteq \frac{1}{4}\mathbf{Z},$$

hence $\mathfrak{v}(\mathbf{Z}y + K) = \frac{1}{4}\mathbf{Z}$, hence $\mathfrak{v}(\mathbf{Z}y + K)^{\#} = 4\mathbf{Z}$. Now y is not in $K^{\#} = K$, hence it is certainly not in $(\mathbf{Z}y + K)^{\#}$, hence $4\mathbf{Z} \subset \mathfrak{v}J \subseteq \mathbf{Z}$. Therefore $\mathfrak{v}J = \mathbf{Z}$. So J is unimodular.

Since $(\mathbf{Z}y + K)^{\#} \subseteq J \cap K$ we must have $4\mathbf{Z} \subseteq \mathfrak{v}(J \cap K) \subset \mathbf{Z}$, hence $\mathfrak{v}(J \cap K) = 4\mathbf{Z}$, hence J is adjacent to K. **q. e. d.**

106:8. *Suppose K is adjacent to L, let y be a vector in $L - K$, let z be any vector in V. Then $y + z$ is in $L - K$ if and only if z is in*

$$L \cap K = \{w \in K \mid B(w, y) \in \mathbf{Z}\}.$$

Proof. If $y + z$ is in $L - K$, then by the definition of adjacent lattices we have $(L + K : K) = \left(\frac{1}{2}\mathbf{Z} : \mathbf{Z}\right) = 2$, hence $z = (y + z) - y$ is in K, hence z is in $L \cap K$. The converse is obvious. **q. e. d.**

106:8a. *Let L and L' be unimodular lattices adjacent to K, let y be an element of $L - K$, let y' be an element of $L' - K$. Then $L = L'$ if and only if*

$$y - y' \in K \quad \textit{with} \quad B(y, y') \in \mathbf{Z}.$$

We now have a procedure for constructing all unimodular lattices adjacent to K. Start with a complete set of representatives of $\frac{1}{2}K$ modulo $2K$. Eliminate all those on which the quadratic form Q is not integral and also all those that fall in K. Let $y_1, \ldots, y_r$ be the remaining representatives. Form the lattices

$$L_i = \mathbf{Z}y_i + (\mathbf{Z}y_i + K)^{\#}$$

for $1 \leqq i \leqq r$. This gives every unimodular lattice adjacent to K at least once. And when do we have duplication, say $L_i = L_j$? If and only if $y_i - y_j \in K$ with $B(y_i, y_j) \in \mathbf{Z}$.

106:9. *Suppose K is adjacent to L and let y be a vector in $L - K$ which has the form $y = w + z$ with*

$$Q(w) = 1, \quad z \in K, \quad B(y, w) \in \left(\tfrac{1}{2}\,\mathbf{Z}\right) - \mathbf{Z}.$$

Then $\operatorname{cls} L = \operatorname{cls} K$.

Proof. In fact we shall prove that $K = \tau L$ where τ is the symmetry τ_w. Here

$$\tau x = x - 2B(x, w)\, w \quad \forall\, x \in V.$$

Note that $2w = 2y - 2z$ is in $L \cap K$. If we let m denote the odd integer $2B(y, w)$ we have

$$\tau y = y - mw = (y - w) - (m - 1)\, w = z - (m - 1)\, w,$$

hence τy is an element of K. And for each x in $L \cap K$ we have

$$B(x, w) = B(x, y) - B(x, z) \in \mathbf{Z}$$

so that τx is an element of $L \cap K$. Hence

$$\tau L = \tau(\mathbf{Z}y + (L \cap K)) \subseteq K.$$

q. e. d.

§ 106D. $1 \leq n \leq 7$

106:10. *Let L be a unimodular lattice on the space V under discussion, and suppose that $1 \leq n \leq 7$. Then $L \cong I_n$.*

Proof. By adjoining a lattice with matrix I_{7-n} to L we can assume, in virtue of Theorem 105:1, that we have $n = 7$. Let D be a completely decomposable unimodular lattice on V. By Proposition 106:2 we can assume that $D \in \mathfrak{F}(L)$. By Proposition 106:4 it will be enough if we prove that $\operatorname{cls} D = \operatorname{cls} L$ under the assumption that D is adjacent to L.

Take an orthogonal base for D, say $D = \mathbf{Z}x_1 \perp \cdots \perp \mathbf{Z}x_7$. Let y be any vector in $L - D$. By reordering the above base for D if necessary, we can write

$$y = \frac{1}{2}(a_1 x_1 + \cdots + a_r x_r) + a_{r+1} x_{r+1} + \cdots + a_7 x_7$$

with all a_i in $\mathbf{Z}$ and $a_1, \ldots, a_r$ odd. Here we must have $r = 4$ since $Q(y) \in \mathbf{Z}$ and since y is not in D. Put $w = \frac{1}{2}(x_1 + x_2 + x_3 \pm x_4)$ and then define z by the equation $y = w + z$. We have

$$Q(w) = 1, \quad z \in D, \quad B(y, w) \in \left(\tfrac{1}{2}\,\mathbf{Z}\right) - \mathbf{Z},$$

provided the sign $\pm$ in the definition of w is correctly chosen. Then $\operatorname{cls} L = \operatorname{cls} D$ by Proposition 106:9.

q. e. d.

§ 106E. The matrix $\Phi_n (n = 8, 12, 16, \ldots)$

Throughout this subparagraph we assume that the dimension n of the space V under discussion is one of the numbers 8, 12, 16, ... We take a completely decomposable unimodular lattice D on V and we fix it. We take a base for D in which

$$D = \mathbf{Z}x_1 \perp \cdots \perp \mathbf{Z}x_n$$

and we fix it. Our first purpose is to show that *there is always an indecomposable unimodular lattice adjacent to D*, that *all such lattices are in the same proper class*, that *all have the matrix*

$$\Phi_n = \left(\begin{array}{cc|c} \frac{n}{4} & 1 & \\ 1 & 4 & 2 \\ \hline & 2 & \Gamma_{n-2} \end{array}\right)$$

where Γ_m is the matrix

$$\Gamma_m = \begin{pmatrix} 2 & 1 & & & & \\ 1 & 2 & 1 & & & \\ & 1 & 2 & & & \\ & & & \ddots & & \\ & & & & 2 & 1 \\ & & & & 1 & 2 \end{pmatrix},$$

in particular that the matrix Φ_n just defined is unimodular for the specified values of n. Needless to say, every lattice on V with matrix Φ_n can be obtained from a suitable completely decomposable lattice in the above way.

Put $y = \frac{1}{2}(x_1 + \cdots + x_n)$. Then y is a vector in $\left(\frac{1}{2}D\right) - D$, $Q(y) \in \mathbf{Z}$ since $n \equiv 0 \bmod 4$, hence by Proposition 106:7 there is exactly one unimodular lattice E on V which contains y and is adjacent to D. (This E will be a lattice with the desired properties.) Now

$$D = \mathbf{Z}(2y) + \mathbf{Z}x_2 + \mathbf{Z}(x_3 - x_2) + \cdots + \mathbf{Z}(x_n - x_{n-1}) .$$

And

$$\mathbf{Z}y + \mathbf{Z}(2x_2) + \mathbf{Z}(x_3 - x_2) + \cdots + \mathbf{Z}(x_n - x_{n-1})$$

is a unimodular lattice adjacent to D which contains y, hence

$$E = \mathbf{Z}y + \mathbf{Z}(2x_2) + \mathbf{Z}(x_3 - x_2) + \cdots + \mathbf{Z}(x_n - x_{n-1}) .$$

By inspection we see that the matrix of E in the base

$$y,\ 2x_2,\ -(x_3 - x_2),\ (x_4 - x_3), \ldots, (x_n - x_{n-1})$$

is equal to Φ_n.

Before we prove that E is indecomposable we must give a coordinate description of the vectors in E.

106:11. $\sum_1^n A_i x_i$ *is in* E *if and only if*

$$A_i \in \frac{1}{2}\mathbf{Z}, \quad A_i - A_j \in \mathbf{Z}, \quad \sum_1^n A_i \in 2\mathbf{Z}$$

hold for $1 \leqq i \leqq n$ *and* $1 \leqq j \leqq n$.

Proof. Put $x = \Sigma A_i x_i$. First suppose that x is in E. Then x is in $\frac{1}{2} D$ and so all $A_i \in \frac{1}{2}\mathbf{Z}$. And $x_i - x_j \in (\mathbf{Z}y + D)^{\#} \subseteq E$, so $B(x, x_i - x_j) \in \mathbf{Z}$, hence $A_i - A_j \in \mathbf{Z}$. And $B(x, y) \in \mathbf{Z}$ so that $\frac{1}{2}\Sigma A_i \in \mathbf{Z}$.

Now the converse. If one A_i is in $\left(\frac{1}{2}\mathbf{Z}\right) - \mathbf{Z}$, then so are they all since $A_i - A_j \in \mathbf{Z}$ for all j. In this event replace x by $x + y$. We may therefore assume that all A_i are in $\mathbf{Z}$. We still have $\Sigma A_i \in 2\mathbf{Z}$. Now $x = \Sigma A_i x_i$ is an element of D and it has the property $B(x, y) \in \mathbf{Z}$, hence $x \in D \cap E \subseteq E$ by Proposition 106:6. q. e. d.

The last proposition can be used to show that E does not represent 1. Suppose we have $\Sigma A_i^2 = 1$ with $\Sigma A_i x_i$ in E. If one A_i is in $\mathbf{Z}$, then so are they all, and in this event

$$A_1 = \pm 1, \quad A_2 = A_3 = \cdots = A_n = 0,$$

say. But then ΣA_i is not in $2\mathbf{Z}$, and this contradicts the fact that $\Sigma A_i x_i$ is in E. So E cannot represent 1 in the above way. On the other hand, if each A_i is in $\left(\frac{1}{2}\mathbf{Z}\right) - \mathbf{Z}$ we have

$$A_1^2 + \cdots + A_n^2 \geqq \frac{n}{4} \geqq 2.$$

So E does not represent 1 in any way, as we asserted.

Why is E indecomposable? Consider the vectors

$$x_2 - x_1, x_3 - x_2, \ldots, x_n - x_{n-1}, x_1 + x_n \text{ in } E.$$

Call them $y_1, \ldots, y_n$. These vectors obviously span V. Now $Q(y_i) = 2$, and E does not represent 1, hence each y_i must fall in exactly one component of the indecomposable splitting of E. But $B(y_i, y_{i+1}) \neq 0$. Hence all y_i fall in the same component of the indecomposable splitting. Hence this component has dimension n, so it is equal to E. Hence E is indecomposable.

Finally we must prove that any other indecomposable unimodular lattice E' adjacent to D is in the same proper class as E. It is enough to prove that E' is in the same class as E since the symmetry $\tau_{x_1 - x_2}$ is a unit of E. Let

$$y' = \frac{1}{2}(a_1 x_1 + \cdots + a_n x_n) \quad (a_i \in \mathbf{Z})$$

be an element of $E' - D$. If one of the a_i were even we would have $x_i \in E' \cap D \subseteq E'$ by Proposition 106:6 and E' would then be split by $\mathbf{Z}x_i$. So all a_i are odd. By making suitable choices of sign we can write

$$a_i = \pm 1 - 4b_i \quad (b_i \in \mathbf{Z})$$

for $1 \leqq i \leqq n$. Put $z = 2(b_1 x_1 + \cdots + b_n x_n) \in 2D \subseteq E' \cap D$. Then

$$y' + z = \frac{1}{2}(\pm x_1 \pm \cdots \pm x_n)$$

is also in $E' - D$. Define an isometry $\sigma \in O_n(V)$ by the equations $\sigma x_i = \pm x_i$ for $1 \leqq i \leqq n$. Then $\sigma D = D$ and $\sigma y = y' + z$. Hence σE is a unimodular lattice adjacent to $D = \sigma D$ which contains $y' + z$. Therefore $\sigma E = E'$. So E' is in the same class as E.

106:12. *Suppose $n = 8$. If L is a unimodular lattice adjacent to E, then L is either in the class of D or in the class of E.*

Proof. Take a vector z in $L - E$. Then $2z$ is in E and so we can write

$$z = \frac{1}{2}(A_1 x_1 + \cdots + A_8 x_8)$$

with all

$$A_i \in \frac{1}{2}\mathbf{Z}, \quad A_i - A_j \in \mathbf{Z}, \quad \sum_1^8 A_i \in 2\mathbf{Z}.$$

1) First suppose that all A_i are even integers. Then $\Sigma\left(\frac{1}{2}A_i\right)$ is an odd integer since z is not in E. Put $z' = x_1$. Then $z - z'$ is in E and $B(z, z') \in \mathbf{Z}$. Now L is a unimodular lattice adjacent to E with $z \in L - E$, and D is a unimodular lattice adjacent to E with $z' \in D - E$. Hence $L = D$ by Corollary 106:8a.

2) Next suppose that all A_i are odd integers. Then $\Sigma\left(\frac{1}{2}A_i\right)$ is again an odd integer since z is not in E. Put

$$z' = \frac{1}{2}(-x_1 + x_2 + \cdots + x_8).$$

There is a unimodular lattice adjacent to E that contains z', and by Corollary 106:8a this lattice must be equal to L, hence z' is in $L - E$. Then

$$z' = -x_1 + \frac{1}{2}(x_1 + \cdots + x_8).$$

So $\operatorname{cls} L = \operatorname{cls} E$ by Proposition 106:9.

3) Let us complete the case of integral A_i. Since $Q(z) \in \mathbf{Z}$ we can suppose that exactly four A_i are odd, say $A_1, \ldots, A_4$ odd and $A_5, \ldots, A_8$ even. By successively adding and subtracting suitable vectors of the form $x_i + x_j$ to z we can find a new z in $L - E$ which has one of the four forms

$$z = \frac{1}{2}(\pm x_1 + x_2 + x_3 + x_4), \quad z = \frac{1}{2}(\pm x_1 + x_2 + x_3 + x_4) + x_5$$

(apply Proposition 106:8). In the first two cases L contains a vector z with $Q(z) = 1$, hence it is decomposable, hence $L \cong I_8$ by Proposition 106:10, hence $\operatorname{cls} L = \operatorname{cls} D$. In the second two cases write

$$z = \frac{1}{2}(\mp x_1 + x_2 + x_3 + x_4) + (\pm x_1 + x_5),$$

and apply Proposition 106:9 to find $\operatorname{cls} L = \operatorname{cls} E$.

4) Finally assume that all A_i are in $\left(\frac{1}{2}\mathbf{Z}\right) - \mathbf{Z}$. All the vectors

$$4x_1, \pm x_1 + x_2, \ldots, \pm x_1 + x_8$$

are in E; using Proposition 106:8 we can successively add and subtract certain of the above vectors (with correct signs attached) to obtain a new z in $L - E$ of the form

$$z = \frac{1}{4}(\pm a x_1 \pm x_2 \pm \cdots \pm x_8)$$

where a is one of the numbers 1, 3, 5, 7. Now a cannot be 1 or 7 since $Q(z) \in \mathbf{Z}$. If $a = 3$ we have $Q(z) = 1$, hence L splits, hence $L \cong I_8$ by Proposition 106:10, hence $\operatorname{cls} L = \operatorname{cls} D$. If $a = 5$ we write

$$z = \frac{1}{4}(\mp 3 x_1 \pm x_2 \pm \cdots \pm x_8) \pm 2x_1$$

and apply Proposition 106:9 to find $\operatorname{cls} L = \operatorname{cls} E$. **q. e. d.**

§ 106 F. Summary

106:13. **Theorem.** *V is an n-ary quadratic space over the field of rational numbers $\mathbf{Q}$ with positive definite localization V_∞, and L is a unimodular lattice with respect to $\mathbf{Z}$ on V. Suppose $1 \leq n \leq 9$. Then L has exactly one of the forms*

$$I_n, \quad \Phi_8, \quad \Phi_8 \perp I_1 .$$

Proof. The fact that L cannot have more than one the above forms is clear from Theorem 105:1 and the fact that any lattice with matrix Φ_8 is indecomposable. If $1 \leq n \leq 7$ we have $L \cong I_n$ by Proposition 106:10. Next let $n = 8$. Consider adjacent lattices J and K on V. If $J \cong I_8$, then $K \cong I_8$ if K splits and $K \cong \Phi_8$ if K does not split. If $J \cong \Phi_8$, then K has matrix I_8 or Φ_8 by Proposition 106:12. Hence by Propositions 106:2 and 106:4 we have $L \cong I_8$ or $L \cong \Phi_8$.

Finally $n = 9$. Here it is enough to prove that L represents 1, since then L will split. This will be achieved if we can prove the following: if J and K are adjacent unimodular lattices on V such that J represents 1, then K also represents 1. Since J represents 1 we have a splitting $J = \mathbf{Z}x_1 \perp J'$. Let y be any vector in $K - J$. If $J' \cong \Phi_8$ we write $2y = m x_1 + z$ with $m \in \mathbf{Z}$ and $z \in J'$; then J' is even and so the fact that $Q(y) \in \mathbf{Z}$ implies that $m \in 2\mathbf{Z}$; hence

$$x_1 \in (\mathbf{Z}y + J)^\# = J \cap K \subseteq K$$

and K therefore represents 1 as required. Otherwise $J' \cong I_8$; then we have a base for J in which

$$J = \mathbf{Z}x_1 \perp \cdots \perp \mathbf{Z}x_9\,;$$

this time write

$$2y = m_1 x_1 + \cdots + m_9 x_9 \quad (m_i \in \mathbf{Z})\,;$$

then at least one m_i must be even since $Q(y) \in \mathbf{Z}$; for this x_i we have

$$x_i \in (\mathbf{Z}y + J)^\# = J \cap K \subseteq K$$

and K therefore represents 1 as required. **q. e. d.**

With some perseverance it is possible to extend these results[1] to higher dimensions using the general principles of this paragraph. For $1 \leq n \leq 13$ the distinct proper classes on V are determined by lattices of the form

$$I_n\,, \quad \Phi_8 \perp I_{n-8}\,, \quad \Phi_{12} \perp I_{n-12}\,.$$

One obtains new indecomposable lattices in 14, 15, and 16 dimensions.

---

[1] See M. Kneser, *Arch. Math.* (1957), pp. 241—250. For an example of the classical approach using "reduction theory" we refer the reader to B. W. Jones, *The arithmetic theory of quadratic forms* (Buffalo, 1950).

Bibliography

Our original intention was to provide a complete bibliography with full documentation, but we were soon discouraged by the complexity of such a task. Instead we have decided to give the following short list of books and articles[1] in order to enable the reader to trace individual results to their source and to lead him to other fields of interest in number theory and the theory of quadratic forms.

E. Artin, *Algebraic numbers and algebraic functions* (Princeton University, 1951).

E. Artin, *Geometric algebra* (New York, 1957).

C. Chevalley, *The algebraic theory of spinors* (New York, 1954).

L. E. Dickson, *Studies in the theory of numbers* (Chicago, 1930).

L. E. Dickson, *History of the theory of numbers* vol. III (New York, 1952).

J. Dieudonné, *La géométrie des groupes classiques* (Berlin, 1955).

M. Eichler, *Quadratische Formen und orthogonale Gruppen* (Berlin, 1952).

H. Hasse, *Zahlentheorie* (Berlin, 1949).

B. W. Jones, *The arithmetic theory of quadratic forms* (Buffalo, 1950).

M. Kneser, Klassenzahlen indefiniter quadratischer Formen in drei oder mehr Veränderlichen, *Arch. Math.* (1956), pp. 323—332.

H. Minkowski, *Gesammelte Abhandlungen* (Berlin, 1911).

O. T. O'Meara, Integral equivalence of quadratic forms in ramified local fields, *Am. J. Math.* (1957), pp. 157—186.

C. L. Siegel, Über die analytische Theorie der quadratischen Formen, *Ann. Math.* **36** (1935), pp. 527—606; **37** (1936), pp. 230–263; **38** (1937), pp. 212—291.

B. L. van der Waerden, Die Reduktionstheorie der positiven quadratischen Formen, *Acta Math.* (1956), pp. 265—309.

G. L. Watson, *Integral quadratic forms* (Cambridge, 1960).

A. Weil, *Adeles and algebraic groups* (Institute for Advanced Study, 1961).

E. Witt, Theorie der quadratischen Formen in beliebigen Körpern, *Crelle's J.* **176** (1937), pp. 31–44.

[1] Additional references to specific points of interest are given in footnotes in the text.

Index

Printing: Saladruck, Berlin
Binding: Buchbinderei Lüderitz & Bauer, Berlin